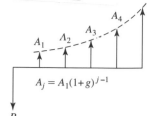
$c(g) = V_g + F$

$R(g) = P_g$

$\dfrac{(P-S)}{n}$

$\dfrac{\text{puning } \#(P-S)}{SOYD}$

$BV_t = P - \dfrac{t}{n}(P-S)$

SYD

$V_t = (P-S)\dfrac{(N-1)(N-1+1)}{N}$ $\dfrac{3}{2}$ $\dfrac{2}{10}$

$(P-S)$

$\dfrac{1}{10}$

Geometric Gradient

Geometric Series Present Worth:

To Find P $(P/A, g, i, n)$
Given A_1, g When $i = g$ $P = A_1[n(1+i)^{-1}]$

To Find P $(P/A, g, i, n)$
Given A_1, g When $i \neq g$ $P = A_1\left[\dfrac{1 - (1+g)^n(1+i)^{-n}}{i - g}\right]$

$A_j = A_1(1+g)^{j-1}$

Continuous Compounding at Nominal Rate r

Single Payment: $F = P[e^{rn}]$ $P = F[e^{-rn}]$

Uniform Series: $A = F\left[\dfrac{e^r - 1}{e^{rn} - 1}\right]$ $A = P\left[\dfrac{e^{rn}(e^r - 1)}{e^{rn} - 1}\right]$

$F = A\left[\dfrac{e^{rn} - 1}{e^r - 1}\right]$ $P = A\left[\dfrac{e^{rn} - 1}{e^{rn}(e^r - 1)}\right]$

Compound Interest

i = Interest rate per interest period*.

n = Number of interest periods.

P = A present sum of money.

F = A future sum of money. The future sum F is an amount, n interest periods from the present, that is equivalent to P with interest rate i.

A = An end-of-period cash receipt or disbursement in a uniform series continuing for n periods, the entire series equivalent to P or F at interest rate i.

G = Uniform period-by-period increase or decrease in cash receipts or disbursements; the arithmetic gradient.

g = Uniform *rate* of cash flow increase or decrease from period to period; the geometric gradient.

r = Nominal interest rate per interest period*.

m = Number of compounding subperiods per period*.

---

*Normally the interest period is one year, but it could be something else.

To Find	Excel
P	$-\text{PV}(i, n, A, F, \text{Type})$
A	$-\text{PMT}(i, n, P, F, \text{Type})$
F	$-\text{FV}(i, n, A, P, \text{Type})$
n	$\text{NPER}(i, A, P, F, \text{Type})$
i	$\text{RATE}(n, A, P, F, \text{Type}, \text{guess})$
P	$\text{NPV}(i, \text{CF}_1 : \text{CF}_n)$
i	$\text{IRR}(\text{CF}_0 : \text{CF}_n)$

ENGINEERING ECONOMIC ANALYSIS

ENGINEERING ECONOMIC ANALYSIS

TENTH EDITION

Donald G. Newnan
San Jose State University

Jerome P. Lavelle
North Carolina State University

Ted G. Eschenbach
University of Alaska Anchorage

New York   Oxford
OXFORD UNIVERSITY PRESS
2009

Oxford University Press, Inc., publishes works that further Oxford University's
objective of excellence in research, scholarship, and education.

Oxford   New York
Auckland   Cape Town   Dar es Salaam   Hong Kong   Karachi
Kuala Lumpur   Madrid   Melbourne   Mexico City   Nairobi
New Delhi   Shanghai   Taipei   Toronto

With offices in
Argentina   Austria   Brazil   Chile   Czech Republic   France   Greece
Guatemala   Hungary   Italy   Japan   Poland   Portugal   Singapore
South Korea   Switzerland   Thailand   Turkey   Ukraine   Vietnam

Published by Oxford University Press, Inc.
198 Madison Avenue, New York, New York 10016
http://www.oup.com

Oxford is a registered trademark of Oxford University Press

Library of Congress Cataloging-in-Publication Data

Newnan, Donald G.
    Engineering economic analysis/Donald G. Newnan, Jerome P. Lavelle, Ted G. Eschenbach.
    –10th ed.
        p. cm.
    ISBN 978-0-19-539463-4 (hardcover : alk. paper) 1. Engineering economy. I. Eschenbach, Ted.
II. Lavelle, Jerome P. III. Title.
    TA177.4.N48 2008
    658.15—dc22

                                                                                    2007030977

Printing number: 9 8 7 6 5 4 3 2 1

Printed in the United States of America
on acid-free paper

Contents in Brief

Contents

2   ENGINEERING COSTS AND COST ESTIMATING

3   INTEREST AND EQUIVALENCE

11 DEPRECIATION

13 REPLACEMENT ANALYSIS

14 INFLATION AND PRICE CHANGE

15 SELECTION OF A MINIMUM ATTRACTIVE RATE OF RETURN

17 Accounting and Engineering Economy

Preface

Our goal has been, and still is, to provide an easy-to-understand and up-to-date presentation of engineering economic analysis for today's students. That means the book's writing style must promote the reader's understanding. We humbly note that our approach has been well received by engineering professors—and more importantly—by engineering students through nine previous editions.

Hallmarks of this Book

Since it was first published in 1976, this text has become the market-leading book for the engineering economy course. It has always been characterized by

- A **focus on practical applications.** One way to encourage students to read the book, and to remember and apply what they have learned in this course, is to make the book interesting. And there is no better way to do that than to infuse the book with real-world examples, problems, and vignettes.
- **Accessibility.** We meet students where they are. Most don't have any expertise in accounting or finance. We take the time to explain concepts carefully while helping them apply them to engineering situations.
- **Superior instructor and student support packages.** To make this course easier to understand, learn, and teach Oxford University Press has produced the best support package available. We offer more for students and instructors than any competing text.

Changes to the 10th Edition

This edition has many significant improvements in coverage. Before going chapter-by-chapter, we'd like to point out a few global changes made to keep this text the most current and most useful in today's courses:

- We have incorporated a new focus on ethics throughout the book. Ethics is an increasingly important part of engineering decision-making, and many people are already highlighting it in their courses. We have added discussions and questions about ethics starting in chapter 1 and continuing in many of the text's vignettes.

- We have increased the book's focus on the real world, in both chapter discussions and in the problem sets. You'll see this throughout the text, beginning with the initial examples of each chapter. This will make this edition even more engaging to your students.
- We have added practice FE problems to both the text and the student's website that accompanies the text. Students have been asking for help preparing for this exam, and we are pleased to provide them plenty of opportunity. A comprehensive set in Appendix C divides problems by specific chapters as a stage in the preparation process, and the problems on the website are also linked to chapters from the text.
- We have added more spreadsheet problems and explanations. This will increase the level of problems students can tackle in the course and will more closely model how they will solve such problems once they enter the workforce.

In addition, we have made extensive changes to the chapter coverage based on feedback from fellow instructors, both users and non-users:

- Chapter 1 (*Making Economic Decisions*) now includes an extensive discussion of ethical issues that arise in the practice of engineering economic analysis. Questions to consider that are linked to ethics have been added to many of the text's vignettes.
- Chapter 7 (*Rate of Return Analysis*) now includes a section on computing interest rates when there are fees or discounts, as well as a section on computing interest rates for common situations such as leasing, buying insurance quarterly vs. annually, and comparing 1-semester parking passes with annual passes.
- Chapter 8 (*Choosing the Best Alternative*) has been completely rewritten to take advantage of the power of spreadsheets in graphically comparing multiple alternatives. The material, written by John Whittaker for this text's Canadian edition, was further revised for this volume based on class testing. An explanation of the graphical approach precedes the material on numerical incremental analysis, which now rests on a foundation that is much more intuitive and easier to understand.
- Chapter 9 (*Other Analysis Techniques*) now includes a section on using spreadsheets to do what-if sensitivity analysis; it also includes an example of the benefit–cost ratio for the private sector (present worth index).
- Chapter 10 (*Uncertainty in Future Events*) now includes a short section that conceptually introduces real options and their application to economic analysis.
- Chapter 12 (*Income Taxes*) has been updated to reflect the latest U.S. tax legislation and rates.
- Chapter 13 (*Replacement Analysis*) has been revised and shortened to deal with this challenging topic more clearly and simply.
- Chapter 15 (*Selection of a Minimum Attractive Rate of Return*) has been rewritten to include the material on rationing capital (Chapter 17 in the 9th edition) for a more concise and integrated treatment. Data on the use of risk-adjusted MARRs has been added.
- Chapter 16 (*Economic Analysis in the Public Sector*) includes a new section on the difficulties in quantifying and valuing benefits and disbenefits in the public sector.
- Six vignettes are new, and the others have all been revised and updated.

In this edition, we have also made substantial changes to increase student interest and understanding. These include:

- End-of-chapter problems have been grouped under subheadings to make them easier to find and assign. This reorganization has helped us target more than 150 new problems for a better mix of problem difficulty and coverage. Many other problems were improved and updated.
- More chapters begin with examples that are directly linked to student lives.
- A new appendix of 113 multiple choice questions intended to help students prepare for the Fundamentals of Engineering exam (Appendix C).
- "Trust Me, You'll Use This!" testimonials from former engineering economy students (who had used this text in their class) have been added between Chapters 2 and 3.

The Superior Newnan Support Package

The supplement package for this text was expanded for the 9th edition, and it has been updated and expanded even further for this edition. No competing text has a more extensive support package.

For student use there are the following:

- A Study Guide by Ed Wheeler of the University of Tennessee, Martin (ISBN 9780195336245) is provided free with every new copy of the text.
- Spreadsheet problem modules on the free student CD included with each text, written by Thomas Lacksonen of the University of Wisconsin–Stout.
- Interactive multiple-choice problems on the student CD, written by Paul Schnitzler of the University of South Florida and William Smyer of Mississippi State University.
- Additional Practice FE Exam problems on the student Companion Website, authored by Karen Thorsett, University of Phoenix. www.oup.com/us/newnan10

For instructors there is now an updated and expanded set of supplemental materials:

- An exam file edited by Meenakshi Sundaram of Tennessee Technological University.
- PowerPoint lecture notes for all chapters by Shih Ming Lee of Florida International University and David Mandeville of Oklahoma State University.
- An Instructor's Manual by Lawrence Samuelson of Tri-State University and the authors, with complete solutions to all end-of-chapter problems. This is available in print, on the instructor's CD, and to instructors on the secure website.
- The compound interest tables from the textbook are available for adopting professors who prefer to give closed-book exams. The tables are on the instructor's and student's CD as PDF files that can be printed in part or in total.

For students and instructors a companion website is available with these supplements and other updates:

- www.oup.com/us/newnan10

Spreadsheets

This edition maintains the approach to spreadsheets that was established in the 8th edition. Rather than relying on spreadsheet templates, the emphasis is on helping students learn to use the enormous capabilities of the spreadsheet software available on nearly every computer. This approach reinforces the traditional engineering economy factor approach, as the equivalent spreadsheet functions (PMT, PV, RATE, etc.) are used frequently.

For students who would benefit from a refresher or introduction on how to write good spreadsheets, there is an appendix to introduce spreadsheets. In Chapter 2, spreadsheets are used to draw cash flow diagrams. Then every chapter from Chapter 4 to Chapter 15 has a concluding section on spreadsheet use. Each section is designed to support the other material in the chapter and to add to the student's knowledge of spreadsheets. If spreadsheets are used, the student will be very well prepared to apply this tool to real-world problems after graduation.

This approach is designed to support a range of approaches to spreadsheets. For example, professors and students can rely on the traditional tools of engineering economy and, without loss of continuity, completely ignore the material on spreadsheets. Or at the other extreme, professors can introduce the concepts and require all computations to be done with spreadsheets. Or a mix of approaches may be taken, depending on the professor, the students, and the particular chapter.

Acknowledgments

Many people have directly or indirectly contributed to the content of the book. We have been influenced by our Stanford and North Carolina State University educations and our university colleagues, and by students who have provided invaluable feedback on content and form. We are particularly grateful to the following professors for their work on previous editions:

Mohamed Aboul-Seoud, Rensselaer Polytechnic Institute

V. Dean Adams, University of Nevada Reno

Dick Bernhard, North Carolina State University

Charles Burford, Texas Tech University

Ronald Terry Cutwright, Florida State University

Jeff Douthwaite, University of Washington

Utpal Dutta, University of Detroit, Mercy

Sandra Duni Eksioglu, University of Florida

John Erjavec, University of North Dakota

Lou Freund, San Jose State University

Ashok Kumar Ghosh, University of New Mexico

Scott E. Grasman, University of Missouri–Rolla

Vernon Hillsman, Purdue University

Ted Huddleston, University of South Alabama

B. J. Kim, Louisiana Tech University

C. Patrick Koelling, Virginia Polytechnic Institute and State University

Hampton Liggett, University of Tennessee

Oscar Lopez, Florida International University

Heather Nachtmann, University of Arkansas

Nic Nigro, Cogswell College North

Ben Nwokolo, Grambling State University

T. Papagiannakis, Washington State University

Cecil Peterson, Kettering University

John A. Roth, Vanderbilt University

Malgorzata Rys, Kansas State University

Robert Seaman, New England College

Meenakshi R. Sundaram, Tennessee Technological University

William N. Smyer, Mississippi State University

Arnold L. Sweet, Purdue University

Kevin Taaffe, University of Florida

Robert E. Taylor, Virginia Polytechnic Institute and State University

Roscoe Ward, Miami University

John Whittaker, University of Alberta

Jan Wolski, New Mexico Institute of Mining and Technology
    and particularly

Bruce Johnson, U.S. Naval Academy.

We also thank the following professors for their insights, reviews, contributed problems, and contributed vignettes, all of which have dramatically improved this 10th edition:

Kate D. Abel, Stevens Institute of Technology

James Blackburn, Southern Illinois University

Peter A. Cerenzio, Cerenzio & Panaro Consulting Engineers

Ronald Deep, University of Dayton

David Elizandro, Tennessee Technological University

Peter Flynn, University of Alberta

Herbert Fox, New York Institute of Technology

Oliver Hedgepeth, University of Alaska Anchorage

Morgan E. Henrie, University of Alaska Anchorage

Joseph R. Herkert, Arizona State University

Joseph Hartman, University of Florida

Neal Lewis, University of Bridgeport

Daniel P. Loucks, Cornell University

Jessica Matson, Tennessee Technological University

Bill McManus, University of Oklahoma

Roger W. Meier, University of Memphis

Brigid Mullany, University of North Carolina–Charlotte

Kim LaScola Needy, University of Pittsburgh

Niranjani H. Patel, San Jose State University

William Peterson, Arizona State University

Richard J. Puerzer, Hofstra University

Dariush Rezaei, Kent State University–Trumbull

Duane J. Rosa, West Texas A&M University

Utpal Roy, Syracuse University

Mehdi Semsar, Savannah State University

C. Thomas Sciance, University of Texas at Austin

William F. Shughart II, University of Mississippi

Nicholas Sylvester, University of South Alabama

William R. Truran, Stevens Institute of Technology

Suleyman Tufekci, University of Florida

John S. Usher, University of Louisville

John Whittaker of the University of Alberta wrote a new Chapter 8 for the Canadian edition and suggested many improvements. John Usher of Mississippi State University identified Nick Cotton and Kim Brock, subjects of the student testimonials, "Trust Me, You'll Use This!" added in this edition. Christina Carter of the University of Alaska Anchorage provided research support for the depreciation and tax chapters of the international edition. Finally, we were helped by the professors who participated in the market survey for this book, and whose collective advice helped us shape this new edition.

Finally, our largest thanks go to the professors (and their students) who have developed the products that support this text:

Thomas Lacksonen, University of Wisconsin–Stout

Shih Ming Lee, Florida International University

David Mandeville, Oklahoma State University

William Peterson, Arizona State University

Lawrence Samuelson, Tri-State University

Paul Schnitzler, University of South Florida

William Smyer, Mississippi State University

Meenakshi Sundaram, Tennessee Technological University

Karen Thorsett, University of Phoenix

Ed Wheeler, University of Tennessee, Martin

Textbooks are produced through the efforts of many people. We thank Brian Newnan for managing the 7th and 8th editions. At Oxford University Press, our editors Danielle Christensen, Dawn Stapleton, and Rachael Zimmermann have worked with us to make this a timely and improved edition. Lauren Mine authored the "Trust Me, You'll Use This!" boxes. Barbara Mathieu effectively managed the text's design and production.

This book remains the best text on the market in large part because of critical feedback from users. We would appreciate hearing comments about the book, including being informed of any errors that have snuck in despite our best attempts to eradicate them. We also look forward to adding problems and vignettes in the next edition that adopters have found effective for their students. Please write us c/o the Engineering Editor at Oxford University Press, 198 Madison Avenue, New York, NY 10016, or email us directly or at highered.us@oup.com. Thanks for using the Newnan book!

Ted Eschenbach
aftge@uaa.alaska.edu
Jerome Lavelle
jerome_lavelle@ncsu.edu

ENGINEERING ECONOMIC ANALYSIS

CHAPTER 1

MAKING ECONOMIC DECISIONS

Alternative Fuel Vehicles

Imagine a future where all cars on U.S. roads no longer run on fossil fuel, but instead have alternate power sources. The positive effect of such a scenario on the environment is the motivation behind initiatives such as the Kyoto Protocol, which requires signatory countries to reduce greenhouse gas emissions. As another example, in 2006 California passed a law to reduce automobile emissions by 25% by the year 2020. Legal and environmental concerns have increased the visibility of global warming and Americans' dependence on foreign oil, spurring the drive toward alternative-fuel vehicles.

In 2005, there were only 11 hybrid vehicle models in the U.S, which accounted for 1% of auto sales. J.D. Power and Associates predicts an increase to 52 hybrid vehicles by 2012 with 4.2% of the U.S. market. Although early hybrid fuel models were small and economical cars, newer models include pickups, minivans, and luxury cars. Some auto manufacturers, including Renault and Nissan, say the business case for hybrids is weak because they are more expensive to produce. Currently hybrids cost three to four thousand dollars more than traditional vehicles. However, Toyota, the leading manufacturer of alternative-fuel vehicles, made money in 2006 selling hybrids such as its Prius model (about $26,200 minus rebates). Consumers will pay more for hybrid because these vehicles have twice the fuel economy of conventional cars.

Engineers who design alternative-fuel vehicles must consider engine power and acceleration, fuel mileage per gallon, distance between refuels, vehicle safety, vehicle maintenance, fuel dispensing and distribution, and environmental and ethical issues. Design decisions influence manufacturing costs and therefore the final sales price. In addition, the entire supply, production, and service chain would be affected by a conversion to alternative-fuel vehicles. Depending on the degree to which alternate sources differ from the current gasoline-based technology, the conversion will be more or less complicated.

The outlook for alternative-fuel vehicles is positive for many reasons. As manufacturing experience increases, alternative-fuel vehicles can be designed to surpass the performance of conventional cars and meet customer expectations. As gasoline prices rise, consumers' interest in these vehicles should increase—which should be good news for the environment.

Contributed by Kate D. Abel, Stevens Institute of Technology

QUESTIONS TO CONSIDER

1. What marketplace dynamics drive or suppress the development of alternative-fuel vehicles? What role, if any, does government have in these dynamics? What additional charges should government have?

2. Develop a list of concerns and questions that consumers might have regarding the conversion to alternative-fuel vehicles. Which are economic and which are noneconomic factors?

3. Are there any ethical aspects to the conversion from gasoline power vehicles to alternative-fuel vehicles? List these, and determine how they could be or should be resolved and by whom.

After Completing This Chapter...

The student should be able to:

- Distinguish between simple and complex problems.
- Discuss the role and purpose of engineering economic analysis.
- Describe and give examples of the nine steps in the *economic decision-making process*.
- Select appropriate economic criteria for use with different types of problems.
- Describe common ethical issues in engineering economic decision making.
- Solve simple problems associated with engineering decision making.

This book is about making decisions. **Decision making** is a broad topic, for it is a major aspect of everyday human existence. This book develops the tools to properly analyze and solve the economic problems that are commonly faced by engineers. Even very complex situations can be broken down into components from which sensible solutions are produced. If one understands the decision-making process and has tools for obtaining realistic comparisons between alternatives, one can expect to make better decisions.

Our focus is on solving problems that confront firms in the marketplace, but many examples are problems faced in daily life. Let us start by looking at some of these problems.

A SEA OF PROBLEMS

A careful look at the world around us clearly demonstrates that we are surrounded by a sea of problems. There does not seem to be any exact way of classifying them, simply because they are so diverse in complexity and "personality." One approach arranges problems by their *difficulty*.

Simple Problems

Many problems are pretty simple, and good solutions do not require much time or effort.

- Should I pay cash or use my credit card?
- Do I buy a semester parking pass or use the parking meters?
- Shall we replace a burned-out motor?
- If we use three crates of an item a week, how many crates should we buy at a time?

Intermediate Problems

At a higher level of complexity we find problems that are primarily economic.

- Shall I buy or lease my next car?
- Which equipment should be selected for a new assembly line?
- Which materials should be used as roofing, siding, and structural support for a new building?
- Shall I buy a 1- or 2-semester parking pass?
- What size of transformer or air conditioner is most economical?

Complex Problems

Complex problems are a mixture of *economic, political,* and *humanistic* elements.

- The decision of Mercedes-Benz to build an automobile assembly plant in Tuscaloosa, Alabama, illustrates a complex problem. Beside the economic aspects, Mercedes-Benz had to consider possible reactions in the American and German auto industries. Would the German government pass legislation to prevent the overseas plant? What about German labor unions?
- The selection of a girlfriend or a boyfriend (who may later become a spouse) is obviously complex. Economic analysis can be of little or no help.

- The annual budget of a corporation is an allocation of resources and all projects are economically evaluated. The budget process is also heavily influenced by noneconomic forces such as power struggles, geographical balancing, and impact on individuals, programs, and profits. For multinational corporations there are even national interests to be considered.

THE ROLE OF ENGINEERING ECONOMIC ANALYSIS

Engineering economic analysis is most suitable for intermediate problems and the economic aspects of complex problems. They have these qualities:

1. The problem is *important enough* to justify our giving it serious thought and effort.
2. The problem can't be worked in one's head—that is, a careful analysis *requires that we organize* the problem and all the various consequences, and this is just too much to be done all at once.
3. The problem has *economic aspects* important in reaching a decision.

When problems meet these three criteria, engineering economic analysis is useful in seeking a solution. Since vast numbers of problems in the business world (and in one's personal life) meet these criteria, engineering economic analysis is often required.

Examples of Engineering Economic Analysis

Engineering economic analysis focuses on costs, revenues, and benefits that occur at different times. For example, when a civil engineer designs a road, a dam, or a building, the construction costs occur in the near future; but the benefits to users begin only when construction is finished and then continue for a long time.

In fact nearly everything that engineers design calls for spending money in the design and building stages, and only after completion do revenues or benefits occur—usually for years. Thus the economic analysis of costs, benefits, and revenues occurring over time is called *engineering* economic analysis.

Engineering economic analysis is used to answer many different questions.

- *Which engineering projects are worthwhile?* Has the mining or petroleum engineer shown that the mineral or oil deposit is worth developing?
- *Which engineering projects should have a higher priority?* Has the industrial engineer shown which factory improvement projects should be funded with the available dollars?
- *How should the engineering project be designed?* Has the mechanical or electrical engineer chosen the most economical motor size? Has the civil or mechanical engineer chosen the best thickness for insulation? Has the aeronautical engineer made the best trade-offs between (1) lighter materials that are expensive to buy but cheaper to fly and (2) heavier materials that are cheap to buy and more expensive to fly?

Engineering economic analysis can also be used to answer questions that are personally important.

- *How to achieve long-term financial goals:* How much should you save each month to buy a house, retire, or fund a trip around the world? Is going to graduate school a good investment—Will your additional earnings in later years balance your lost income while in graduate school?
- *How to compare different ways to finance purchases:* Is it better to finance your car purchase by using the dealer's low interest rate loan or by taking the rebate and borrowing money from your bank or credit union?
- *How to make short- and long-term investment decisions:* Should you buy a 1- or 2-semester parking pass? Is a higher salary better than stock options?

THE DECISION-MAKING PROCESS

Decision making may take place by default; that is, a person may not consciously recognize that an opportunity for decision making exists. This fact leads us to a first element in a definition of decision making. To have a decision-making situation, there must be at least two alternatives available. If only one course of action is available, there is nothing to decide. There is no alternative but to proceed with the single available course of action. (It is rather unusual to find that there are no alternative courses of action. More frequently, alternatives simply are not recognized.)

At this point we might conclude that the decision-making process consists of choosing from among alternative courses of action. But this is an inadequate definition. Consider the following situation.

At a race track, a bettor was uncertain about which of the five horses to bet on in the next race. He closed his eyes and pointed his finger at the list of horses printed in the racing program. Upon opening his eyes, he saw that he was pointing to horse number 4. He hurried off to place his bet on that horse.

Does the racehorse selection represent the process of decision making? Yes, it clearly was a process of choosing among alternatives (assuming the bettor had already ruled out the "do-nothing" alternative of placing no bet). But the particular method of deciding seems inadequate and irrational. We want to deal with rational decision making.

Rational Decision Making

Rational decision making is a complex process that contains nine essential elements, which are shown in Figure 1-1. Although these nine steps are shown sequentially, it is common for a decision maker to repeat steps, take them out of order, and do steps simultaneously. For example, when a new alternative is identified more data will be required. Or when the outcomes are summarized, it may become clear that the problem needs to be redefined or new goals established.

The value of this sequential diagram is to show all the steps that are usually required, and to show them in a logical order. Occasionally we will skip a step entirely. For example, a new alternative may be so clearly superior that it is immediately adopted at Step 4 without further analysis. The following sections describe the elements listed in Figure 1-1.

FIGURE 1-1 One possible flowchart of the decision process.

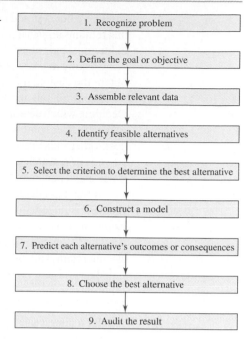

1. Recognize problem

2. Define the goal or objective

3. Assemble relevant data

4. Identify feasible alternatives

5. Select the criterion to determine the best alternative

6. Construct a model

7. Predict each alternative's outcomes or consequences

8. Choose the best alternative

9. Audit the result

1. Recognize the Problem

The starting point in rational decision making is recognizing that a problem exists.

Some years ago, for example, it was discovered that several species of ocean fish contained substantial concentrations of mercury. The decision-making process began with this recognition of a problem, and the rush was on to determine what should be done. Research revealed that fish taken from the ocean decades before and preserved in laboratories also contained similar concentrations of mercury. Thus, the problem had existed for a long time but had not been recognized.

In typical situations, recognition is obvious and immediate. An auto accident, an overdrawn check, a burned-out motor, an exhausted supply of parts all produce the recognition of a problem. Once we are aware of the problem, we can solve it as best we can. Many firms establish programs for total quality management (TQM) or continuous improvement (CI) that are designed to identify problems so that they can be solved.

2. Define the Goal or Objective

The goal or objective can be an overall goal of a person or a firm. For example, a personal goal could be to lead a pleasant and meaningful life, and a firm's goal is usually to operate profitably. The presence of multiple, conflicting goals is often the foundation of complex problems.

But an objective need not be an overall goal of a business or an individual. It may be quite narrow and specific: "I want to pay off the loan on my car by May," or "The plant must produce 300 golf carts in the next 2 weeks," are more limited objectives. Thus, defining the objective is the act of exactly describing the task or goal.

3. Assemble Relevant Data

To make a good decision, one must first assemble good information. In addition to all the published information, there is a vast quantity of information that is not written down anywhere but is stored as individuals' knowledge and experience. There is also information that remains ungathered. A question like "How many people in your town would be interested in buying a pair of left-handed scissors?" cannot be answered by examining published data or by asking any one person. Market research or other data gathering would be required to obtain the desired information.

From all this information, what is relevant in a specific decision-making process? Deciding which data are important and which are not may be a complex task. The availability of data further complicates this task. Published data are available immediately at little or no cost; other data are available from specific knowledgeable people; still other data require surveys or research to assemble the information. Some data will be of high quality—that is, precise and accurate, while other data may rely on individual judgment for an estimate.

If there is a published price or a contract, the data may be known exactly. In most cases, the data is uncertain. What will it cost to build the dam? How many vehicles will use the bridge next year and in year 20? How fast will a competing firm introduce a competing product? How will demand depend on growth in the economy? Future costs and revenues are uncertain, and the range of likely values should be part of assembling relevant data.

The problem's time horizon is part of the data that must be assembled. How long will the building or equipment last? How long will it be needed? Will it be scrapped, sold, or shifted to another use? In some cases, such as for a road or a tunnel, the life may be centuries with regular maintenance and occasional rebuilding. A shorter time period, such as 50 years, may be chosen as the problem's time horizon, so that decisions can be based on more reliable data.

In engineering decision making, an important source of data is a firm's own accounting system. These data must be examined quite carefully. Accounting data focuses on past information, and engineering judgment must often be applied to estimate current and future values. For example, accounting records can show the past cost of buying computers, but engineering judgment is required to estimate the future cost of buying computers.

Financial and cost accounting are designed to show accounting values and the flow of money—specifically **costs** and **benefits**—in a company's operations. When costs are directly related to specific operations, there is no difficulty; but there are other costs that are not related to specific operations. These indirect costs, or **overhead,** are usually allocated to a company's operations and products by some arbitrary method. The results are generally satisfactory for cost-accounting purposes but may be unreliable for use in economic analysis.

To create a meaningful economic analysis, we must determine the *true* differences between alternatives, which might require some adjustment of cost-accounting data. The following example illustrates this situation.

EXAMPLE 1-1

The cost-accounting records of a large company show the average monthly costs for the three-person printing department. The wages of the three department members and benefits, such as vacation and sick leave, make up the first category of **direct** labor. The company's indirect or overhead costs—such as heat, electricity, and employee insurance—must be distributed to its various departments in *some* manner and, like many other firms, this one uses *floor space* as the basis for its allocations. Chapter 17 will discuss allocating overhead costs in more detail.

Direct labor (including employee benefits)	$ 6,000
Materials and supplies consumed	7,000
Allocated overhead costs:	
200 m^2 of floor area at $25/m^2	5,000
	$18,000

The printing department charges the other departments for its services to recover its $18,000 monthly cost. For example, the charge to run 30,000 copies for the shipping department is:

Direct labor	$228
Materials and supplies	294
Overhead costs	291
Cost to other departments	$793

The shipping department checks with a commercial printer, which would print the same 30,000 copies for $688. The shipping department foreman wants to have the work done externally. The in-house printing department objects to this. As a result, the general manager has asked you to study the situation and recommend what should be done.

SOLUTION

Much of the printing department's output reveals the company's costs, prices, and other financial information. The company president considers the printing department necessary to prevent disclosing such information to people outside the company. The firm cannot switch to an outside printer for all needs.

A review of the cost-accounting charges reveals nothing unusual. The charges made by the printing department cover direct labor, materials and supplies, and overhead. The allocation of indirect costs is a customary procedure in cost-accounting systems, but it is potentially misleading for decision making, as the following discussion indicates.

The shipping department would reduce its cost by $105 (= $793 – $688) by using the outside printer. In that case, how much would the printing department's costs decline, and which solution is better for the firm? We will examine each of the cost components:

1. *Direct Labor.* If the printing department had been working overtime, then the overtime could be reduced or eliminated. But, assuming no overtime, how much would the saving be? It seems unlikely that a printer could be fired or even put on less than a 40-hour work week. Thus, although there might be a $228 saving, it is much more likely that there will be no reduction in direct labor.

2. *Materials and Supplies.* There would be a $294 saving in materials and supplies.

3. *Allocated Overhead Costs.* There will be no reduction in the printing department's monthly $5000 overhead, and in fact the firm will incur additional expenses in purchasing and accounting to deal with the outside printer.

The firm will save $294 in materials and supplies and may or may not save $228 in direct labor if the printing department no longer does the shipping department work. The maximum saving would be $294 + 228 = $522. Either value of $294 or $522 is less than the $688 the firm would pay the outside printer. For this reason, the shipping department should not be allowed to send its printing to the outside printer.

Gathering cost data presents other difficulties. One way to look at the financial consequences—costs and benefits—of various alternatives is as follows.

- *Market Consequences.* These consequences have an established price in the marketplace. We can quickly determine raw material prices, machinery costs, labor costs, and so forth.
- *Extra-Market Consequences.* There are other items that are not directly priced in the marketplace. But by indirect means, a price may be assigned to these items. (Economists call these prices **shadow prices.**) Examples might be the cost of an employee injury or the value to employees of going from a 5-day to a 4-day, 40-hour week.
- *Intangible Consequences.* Numerical economic analysis probably never fully describes the real differences between alternatives. The tendency to leave out consequences that do not have a significant impact on the analysis itself, or on the conversion of the final decision into actual money, is difficult to resolve or eliminate. How does one evaluate the potential loss of workers' jobs due to automation? What is the value of landscaping around a factory? These and a variety of other consequences may be left out of the numerical calculations, but they must be considered in reaching a decision.

4. Identify Feasible Alternatives

One must keep in mind that unless the best alternative is considered, the result will always be suboptimal.[1] Two types of alternatives are sometimes ignored. First, in many situations a do-nothing alternative is feasible. This may be the "Let's keep doing what we are now doing," or the "Let's not spend any money on that problem" alternative. Second, there are often feasible (but unglamorous) alternatives, such as "Patch it up and keep it running for another year before replacing it."

There is no way to ensure that the best alternative *is* among the alternatives being considered. One should try to be certain that all conventional alternatives have been listed and then make a serious effort to suggest innovative solutions. Sometimes a group of people considering alternatives in an innovative atmosphere—**brainstorming**—can be helpful. Even impractical alternatives may lead to a better possibility. The payoff from a new, innovative alternative can far exceed the value of carefully selecting between the existing alternatives.

---

[1]A group of techniques called value analysis is sometimes used to examine past decisions and the entire process that led to them.

Any good listing of alternatives will produce both practical and impractical alternatives. It would be of little use, however, to seriously consider an alternative that cannot be adopted. An alternative may be infeasible for a variety of reasons. For example, it might violate fundamental laws of science, require resources or materials that cannot be obtained, violate ethics standards, or conflict with the firm's strategy. Only the feasible alternatives are retained for further analysis.

5. Select the Criterion to Determine the Best Alternative

The central task of decision making is choosing from among alternatives. How is the choice made? Logically, to choose the best alternative, we must define what we mean by *best*. There must be a **criterion**, or set of **criteria**, to judge which alternative is best. Now, we recognize that *best* is on one end of the following relative subjective judgment:

Worst	Bad	Fair	Good	Better	Best

relative subjective judgment spectrum

Since we are dealing in *relative terms,* rather than *absolute values,* the choice will be the alternative that is relatively the most desirable. Consider a driver found guilty of speeding and given the alternatives of a $175 fine or 3 days in jail. In absolute terms, neither alternative is good. But on a relative basis, one simply makes the best of a bad situation.

There may be an unlimited number of ways that one might judge the various alternatives. Several possible criteria are:

- Create the least disturbance to the environment.
- Improve the distribution of wealth among people.
- Minimize the expenditure of money.
- Ensure that the benefits to those who gain from the decision are greater than the losses of those who are harmed by the decision.[2]
- Minimize the time to accomplish the goal or objective.
- Minimize unemployment.
- Maximize profit.

Selecting the criterion for choosing the best alternative will not be easy if different groups support different criteria and desire different alternatives. The criteria may conflict. For example, minimizing unemployment may require increasing the expenditure of money. Or minimizing environmental disturbance may conflict with minimizing time to complete the project. The disagreement between management and labor in collective bargaining (concerning wages and conditions of employment) reflects a disagreement over the objective and the criterion for selecting the best alternative.

The last criterion—maximize profit—is the one normally selected in engineering decision making. When this criterion is used, all problems fall into one of three categories: neither input nor output fixed, fixed input, or fixed output.

---

[2]This is the Kaldor criterion.

Neither Input nor Output Fixed.   The first category is the general and most common situation, in which the amount of money or other inputs is not fixed, nor is the amount of benefits or other outputs. For example:

- A consulting engineering firm has more work available than it can handle. It is considering paying the staff for working evenings to increase the amount of design work it can perform.
- One might wish to invest in the stock market, but the total cost of the investment is not fixed, and neither are the benefits.
- A car battery is needed. Batteries are available at different prices, and although each will provide the energy to start the vehicle, the useful lives of the various products are different.

What should be the criterion in this category? Obviously, to be as economically efficient as possible, we must maximize the difference between the return from the investment (benefits) and the cost of the investment. Since the difference between the benefits and the costs is simply profit, a businessperson would define this criterion as **maximizing profit.**

Fixed Input.   The amount of money or other input resources (like labor, materials, or equipment) is fixed. The objective is to effectively utilize them. For economic efficiency, the appropriate criterion is to maximize the benefits or other outputs. For example:

- A project engineer has a budget of $350,000 to overhaul a portion of a petroleum refinery.
- You have $300 to buy clothes for the start of school.

Fixed Output.   There is a fixed task (or other output objectives or results) to be accomplished. The economically efficient criterion for a situation of fixed output is to minimize the costs or other inputs. For example:

- A civil engineering firm has been given the job of surveying a tract of land and preparing a "record of survey" map.
- You must choose the most cost-effective design for a roof, an engine, a circuit, or other component.

For the three categories, the proper economic criteria are:

Category	Economic Criterion
Neither input nor output fixed	Maximize profit = value of outputs − cost of inputs.
Fixed input	Maximize the benefits or other outputs.
Fixed output	Minimize the costs or other inputs.

6. Constructing the Model

At some point in the decision-making process, the various elements must be brought together. The *objective, relevant data, feasible alternatives,* and *selection criterion* must be merged. For example, if one were considering borrowing money to pay for a car, there is a mathematical relationship between the loan's variables: amount, interest rate, duration, and monthly payment.

Constructing the interrelationships between the decision-making elements is frequently called **model building** or **constructing the model.** To an engineer, modeling may be a scaled *physical representation* of the real thing or system or a *mathematical equation,* or set of equations, describing the desired interrelationships. In economic decision making, the model is usually mathematical.

In modeling, it is helpful to represent only that part of the real system that is important to the problem at hand. Thus, the mathematical model of the student capacity of a classroom might be

$$\text{Capacity} = \frac{lw}{k}$$

where $l =$ length of classroom, in meters
$w =$ width of classroom, in meters
$k =$ classroom arrangement factor

The equation for student capacity of a classroom is a very simple model; yet it may be adequate for the problem being solved.

7. Predicting the Outcomes for Each Alternative

A model and the data are used to predict the outcomes for each feasible alternative. As was suggested earlier, each alternative might produce a variety of outcomes. Selecting a motorcycle, rather than a bicycle, for example, may make the fuel supplier happy, the neighbors unhappy, the environment more polluted, and one's savings account smaller. But, to avoid unnecessary complications, we assume that decision making is based on a single criterion for measuring the relative attractiveness of the various alternatives. If necessary, one could devise a single composite criterion that is the weighted average of several different choice criteria.

To choose the best alternative, the outcomes for each alternative must be stated in a *comparable* way. Usually the consequences of each alternative are stated in terms of money, that is, in the form of costs and benefits. This **resolution of consequences** is done with all monetary and nonmonetary consequences. The consequences can also be categorized as follows:

Market consequences—where there are established market prices available

Extra-market consequences—no direct market prices, so priced indirectly

Intangible consequences—valued by judgment, not monetary prices.

In the initial problems we will examine, the costs and benefits occur over a short time period and can be considered as occurring at the same time. In other situations the various costs and benefits take place in a longer time period. The result may be costs at one point in time followed by periodic benefits. We will resolve these in the next chapter into a *cash flow diagram* to show the timing of the various costs and benefits.

For these longer-term problems, the most common error is to assume that the current situation will be unchanged for the do-nothing alternative. In reality if we do nothing, then current profits will shrink or vanish as a result of the actions of competitors and the expectations of customers; and traffic congestion normally increases over the years as the number of vehicles increases—doing nothing does not imply that the situation will not change.

8. Choosing the Best Alternative

Earlier we said that choosing the best alternative may be simply a matter of determining which alternative best meets the selection criterion. But the solutions to most problems in economics have market consequences, extra-market consequences, and intangible consequences. Since the intangible consequences of possible alternatives are left out of the numerical calculations, they should be introduced into the decision-making process at this point. The alternative to be chosen is the one that best meets the choice criterion after considering both the numerical consequences and the consequences not included in the monetary analysis.

During the decision-making process certain feasible alternatives are eliminated because they are dominated by other, better alternatives. For example, shopping for a computer on-line may allow you to buy a custom-configured computer for less money than a stock computer in a local store. Buying at the local store is feasible, but dominated. While eliminating dominated alternatives makes the decision-making process more efficient, there are dangers.

Having examined the structure of the decision-making process, we can ask, When is a decision made, and who makes it? If one person performs *all* the steps in decision making, then he is the decision maker. *When* he makes the decision is less clear. The selection of the feasible alternatives may be the key item, with the rest of the analysis a methodical process leading to the inevitable decision. We can see that the decision may be drastically affected, or even predetermined, by the way in which the decision-making process is carried out. This is illustrated by the following example.

> Liz, a young engineer, was assigned to make an analysis of additional equipment needed for the machine shop. The single criterion for selection was that the equipment should be the most economical, considering both initial costs and future operating costs. A little investigation by Liz revealed three practical alternatives:
>
> **1.** A new specialized lathe
> **2.** A new general-purpose lathe
> **3.** A rebuilt lathe available from a used-equipment dealer
>
> A preliminary analysis indicated that the rebuilt lathe would be the most economical. Liz did not like the idea of buying a rebuilt lathe, so she decided to discard that alternative. She prepared a two-alternative analysis that showed that the general-purpose lathe was more economical than the specialized lathe. She presented this completed analysis to her manager. The manager assumed that the two alternatives presented were the best of all feasible alternatives, and he approved Liz's recommendation.

At this point we should ask: Who was the decision maker, Liz or her manager? Although the manager signed his name at the bottom of the economic analysis worksheets to authorize purchasing the general-purpose lathe, he was merely authorizing what already had been made inevitable, and thus he was not the decision maker. Rather Liz had made the key decision when she decided to discard the most economical alternative from further consideration. The result was a decision to buy the better of the two *less economically desirable* alternatives.

9. Audit the Results

An audit of the results is a comparison of what happened against the predictions. Do the results of a decision analysis reasonably agree with its projections? If a new machine tool was purchased to save labor and improve quality, did it? If so, the economic analysis seems to be accurate. If the savings are not being obtained, what was overlooked? The audit may help ensure that projected operating advantages are ultimately obtained. On the other hand, the economic analysis projections may have been unduly optimistic. We want to know this, too, so that the mistakes that led to the inaccurate projection are not repeated. Finally, an effective way to promote *realistic* economic analysis calculations is for all people involved to know that there *will* be an audit of the results!

ETHICS

You must be mindful of the ethical dimensions of engineering economic analysis and of your engineering and personal decisions. This text can only introduce the topic, and we hope that you will explore this subject in greater depth.

Ethics can be described variously; however, a common thread is the concept of distinguishing between right and wrong in decision making. Ethics includes establishing systems of beliefs and moral obligations, defining values and fairness, and determining duty and guidelines for conduct. Ethics and ethical behavior is important because when people behave in ethical ways, individuals and society benefit. Usually the ethical choice is reasonably clear, but there are ethical dilemmas with conflicting moral imperatives. Consider an overloaded and sinking lifeboat. If one or more passengers are thrown into the shark-infested waters, the entire lifeboat can be saved. How is the decision made, how is it implemented, and who if anyone goes into the water? Ethical dilemmas also exist in engineering and business contexts. Ethical decision making requires the understanding of problem context, choices, and associated outcomes.

Ethical Dimensions in Engineering Decision Making

Ethical issues can arise at every stage of the integrated process for engineering decision making described in Figure 1-1. Ethics is such an important part of professional and business decision making that ethical codes or standards of conduct exist for professional engineering societies, small and large organizations, and every individual. Written professional codes are common in the engineering profession, serving as a reference basis for new engineers and a basis for legal action against engineers who violate the code.

One such example is the Code of Ethics of the National Society of Professional Engineers (NSPE). Here is NSPE's fundamental canon of ethical behavior for engineering:

Engineers, in the fulfillment of their professional duties, shall:

- Hold paramount the safety, health and welfare of the public.
- Perform services only in areas of their competence.
- Issue public statements only in an objective and truthful manner.
- Act for each employer or client as faithful agents or trustees.
- Avoid deceptive acts.
- Conduct themselves honorably, responsibly, ethically and lawfully so as to enhance the honor, reputation, and usefulness of the profession.

In addition, NSPE has Rules of Practice and Professional Obligations for its members. Most engineering organizations have similar written standards. For all engineers difficulties arise when they act contrary to these written or internal codes, and opportunities for ethical dilemmas are found throughout the engineering decision-making process.

TABLE 1-1 Example Ethical Lapses by Decision Process Step

Decision Process Step	Example Ethical Lapses
1. Recognize the problem	• "Looking the other way," that is, not to recognize the problem—due to bribes or perhaps fear of retribution for being a "whistle-blower"
2. Define the goal or objective	• Favoring one group of stakeholders by focusing on their objective for a project
3. Assemble relevant data	• Using faulty or inaccurate data
4. Identify feasible alternatives	• Leaving legitimate alternatives out of consideration
5. Select the criterion to determine the best alternative	• Considering only monetary consequences when there are other significant consequences
6. Construct a model	• Using a short horizon that favors one alternative over another.
7. Predict each alternative's outcomes or consequences	• Using optimistic estimates for one alternative and pessimistic ones for the other alternatives
8. Choose the best alternative	• Choosing an inferior alternative, one that is unsafe, adds unnecessary cost for the end user, harms the environment, etc.
9. Audit the result	• Hiding past mistakes

Table 1-1 provides examples of ethical lapses that can occur at each step of the decision-making process.

Ethical dilemmas for engineers may arise in connection with engineering economic analysis in many situations. Following are examples of a few of these:

Gaining Knowledge and Building Trust Versus Favors for Influence
Consider these three situations:

• The salesman for a supplier of HVAC (heating, ventilating, and air conditioning) equipment invites a mechanical engineer and spouse to come along on the company jet for a users' conference at a vacation resort.
• Same salesman and same engineer, but the invitation is for a day of golfing at an exclusive club.
• Same salesman invites the same engineer to lunch.

In each case the salesman is trying to "get the order," and there is likely to be some mix of business—discussing specifications—and pleasure. The first case, which brings up the largest ethical questions, also has the largest business justification. This is the opportunity to meet other users of the products and see displays of the product line. Often, firms and government agencies have strict guidelines that dictate behavior in these situations.

Cost, Quality, and Functionality

One of the most common conflicts in the conceptual and design phase involves the trade-offs between cost, quality, and required functionality. Most modern products entail many thousands of decisions by designers that ultimately affect the cost and quality for the end user.

- A designer in an engineering consulting firm knows that a "gold-plated" solution would be very profitable for his firm (and for his bonus). This solution may also provide excellent reliability and require little maintenance cost.
- Engineers in the consumer durables division of a multinational company know that by using lower-quality connectors, fasteners, and subcomponents they can lower costs and improve the firm's market position. In addition, they know that these design elements have only a limited usable life, and the firm's most profitable business is repairs and extended warranties.

The Environment We Live In

Projects for transportation and power generation typically must consider environmental impacts in their design and in deciding whether the project should be done in any form. Who incurs the costs for the project, and who receives the benefits? Many other engineering products are designed to promote recycling, reduce energy usage, and reduce pollution. Ethical issues can be particularly difficult because there are often stakeholders with opposing viewpoints, and some of the data may be uncertain and hard to quantify.

Examples of difficult choices include:

- Protecting the habitat of an endangered species versus flood control projects that protect people, animals, and structures.
- Meeting the needs for electrical power when all choices have some negative environmental impacts:

    - Hydroelectric—reservoir covers land and habitat
    - Coal—underground mining can be dangerous, open-pit mining damages habitat, and burning the coal can cause air pollution
    - Nuclear—disposal of radioactive waste
    - Fuel oil—air pollution and economic dependence
    - Wind—visual pollution of wind farms; birds killed by whirling blades

- Determining standards for pollutants: Is 1 part per million OK, or is 1 part per billion needed?

Safety and Cost

Some of the most common and most difficult ethical dilemmas involve trade-offs between safety and cost. If a product is "too safe," then it will be too expensive, and it will not be used. Also sometimes the cost is incurred by one party and the risk by another.

- Should the oil platform be designed for the 100-year, 500-year, or 1000-year hurricane?
- Should the auto manufacturer add run-flat tires, stability control, side-cushion airbags, and rear-seat airbags to every car?
- Should a given product design go through another month of testing?

- Are stainless steel valves required, or is it economically better to use less corrosion-resistant valves and replace them more frequently?
- Is it safe to launch the Space Shuttle today?

Emerging Issues and "Solutions"

Breaches of the law by corporate leaders of Enron, Tyco, and other firms have led to attempts to prevent, limit, and expose financial wrongdoing within corporations. One part of the solution has been the Sarbanes-Oxley Act of 2002, which imposed requirements on firm executives and auditing accounting firms, as well as penalties for violations.

Globalization is another area of increasing importance for ethical considerations. One reason is that different ethical expectations prevail in the world's various countries and regions. A second reason is that jobs may be moved to another country based on differences in cost, productivity, environmental standards, and so on. What may be viewed as a sweatshop from a U.S. perspective may be viewed as a wonderful opportunity to support many families from the perspective of a less developed nation.

Importance of Ethics in Engineering and Engineering Economy

Many times engineers and firms try to act ethically, but mistakes are made—the data was wrong, the design was changed, or the operating environment was different than expected. In other cases, a choice was made between expediency (profit) and ethics. The firm and management are driven by the need to make a profit, and they expect the engineer to identify when safety will be compromised.

Ethics in engineering economic analysis focuses on how well and how honestly the decision-making process is conducted—the data, method of analysis, recommendations, and follow-up. The first step in avoiding problems is to recognize that ethical issues exist and to make them an explicit part of your decision-making process.

As a student, you had discussions about cheating on exams, plagiarism on written reports, violating university drinking and drug use policies, accepting one job while continuing to interview for others, and selling student sports tickets to nonstudents. You've made your own decisions about your behavior, and you've established patterns of behavior.

While people can and do decide to change, most of the time we continue our patterns of behavior. While engineering professors care about ethics in the classroom, a much larger concern to them and to us is the character of your work and personal behavior after graduation.

Often recent engineering graduates are asked, "What is the most important thing you want from your supervisor?" The most common response is mentoring and opportunities to learn and progress. When employees with 5, 15, 25, or more years of experience are asked the same question, the most common response at all experience levels is *integrity*. This is what your subordinates, peers, and superiors will expect and value the most from you. Integrity is the foundation for long-term career success.

ENGINEERING DECISION MAKING FOR CURRENT COSTS

Some of the easiest forms of engineering decision making deal with problems of alternative *designs, methods,* or *materials.* If results of the decision occur in a very short period of time, one can quickly add up the costs and benefits for each alternative. Then, using the suitable economic criterion, the best alternative can be identified. Three example problems illustrate these situations.

EXAMPLE 1-2

A concrete aggregate mix must contain at least 31% sand by volume for proper batching. One source of material, which has 25% sand and 75% coarse aggregate, sells for $3 per cubic meter ($m^3$). Another source, which has 40% sand and 60% coarse aggregate, sells for $4.40/$m^3$. Determine the least cost per cubic meter of blended aggregates.

SOLUTION

The least cost of blended aggregates will result from maximum use of the lower-cost material. The higher-cost material will be used to increase the proportion of sand up to the minimum level (31%) specified.

$$\text{Let } x = \text{Portion of blended aggregates from \$3.00/}m^3 \text{ source}$$

$$1 - x = \text{Portion of blended aggregates from \$4.40/}m^3 \text{ source}$$

Sand Balance

$$x(0.25) + (1 - x)(0.40) = 0.31$$

$$0.25x + 0.40 - 0.40x = 0.31$$

$$x = \frac{0.31 - 0.40}{0.25 - 0.40} = \frac{-0.09}{-0.15}$$

$$= 0.60$$

Thus the blended aggregates will contain

60% of $3.00/$m^3$ material

40% of $4.40/$m^3$ material

The least cost per cubic meter of blended aggregates is

$$0.60(\$3.00) + 0.40(\$4.40) = 1.80 + 1.76$$

$$= \$3.56/m^3$$

EXAMPLE 1-3

A machine part is manufactured at a unit cost of 40¢ for material and 15¢ for direct labor. An investment of $500,000 in tooling is required. The order calls for 3 million pieces. Halfway through the order, managers learn that a new method of manufacture can be put into effect that will reduce the unit costs to 34¢ for material and 10¢ for direct labor—but it will require $100,000 for additional tooling. This tooling will not be useful for future orders. Other costs are allocated at 2.5 times the direct labor cost. What, if anything, should be done?

SOLUTION

Since there is only one way to handle the first 1.5 million pieces, our problem concerns only the second half of the order.

Alternative *A:* **Continue with Present Method**

Material cost	1,500,000 pieces × 0.40 =	$600,000
Direct labor cost	1,500,000 pieces × 0.15 =	225,000
Other costs	2.50 × direct labor cost =	562,500
Cost for remaining 1,500,000 pieces		$1,387,500

Alternative *B:* **Change the Manufacturing Method**

Additional tooling cost		$100,000
Material cost	1,500,000 pieces × 0.34 =	510,000
Direct labor cost	1,500,000 pieces × 0.10 =	150,000
Other costs	2.50 × direct labor cost =	375,000
Cost for remaining 1,500,000 pieces		$1,135,000

Before making a final decision, one should closely examine the *Other costs* to see whether they do, in fact, vary as the *Direct labor cost* varies. Assuming they do, the decision would be to change the manufacturing method.

EXAMPLE 1-4

In the design of a cold-storage warehouse, the specifications call for a maximum heat transfer through the warehouse walls of 30,000 joules per hour per square meter of wall when there is a 30°C temperature difference between the inside surface and the outside surface of the insulation. The two insulation materials being considered are as follows:

Insulation Material	Cost per Cubic Meter	Conductivity (J-m/m^2-°C-hr)
Rock wool	$12.50	140
Foamed insulation	14.00	110

The basic equation for heat conduction through a wall is:

$$Q = \frac{K(\Delta T)}{L}$$

where

Q = heat transfer, in J/hr/m^2 of wall
K = conductivity, in J-m/m^2-°C-hr
ΔT = difference in temperature between the two surfaces, in °C
L = thickness of insulating material, in meters

Which insulation material should be selected?

SOLUTION

Two steps are needed to solve the problem. First, the required thickness of each of the available materials must be calculated. Then, since the problem is one of providing a fixed output (heat transfer through the wall limited to a fixed maximum amount), the criterion is to minimize the input (cost).

Required Insulation Thickness

Rock wool $\qquad 30{,}000 = \dfrac{140(30)}{L} \qquad L = 0.14$ m

Foamed insulation $\quad 30{,}000 = \dfrac{110(30)}{L} \qquad L = 0.11$ m

Cost of Insulation per Square Meter of Wall

Unit cost = Cost/m^3 × Insulation thickness, in meters
Rock wool $\qquad$ Unit cost = \$12.50 × 0.14 m = \$1.75/m^2
Foamed insulation $\quad$ Unit cost = \$14.00 × 0.11 m = \$1.54/m^2

The foamed insulation is the lesser-cost alternative. However, there is an intangible constraint that must be considered. How thick is the available wall space? Engineering economy and the time value of money are needed to decide what the maximum heat transfer should be. What is the cost of more insulation versus the cost of cooling the warehouse over its life?

SUMMARY

Classifying Problems

Many problems are simple and thus easy to solve. Others are of intermediate difficulty and need considerable thought and/or calculation to properly evaluate. These intermediate problems tend to have a substantial economic component and to require economic analysis. Complex problems, on the other hand, often contain people elements, along with political and economic components. Economic analysis is still very important, but the best alternative must be selected by considering all criteria—not just economics.

The Decision-Making Process

Rational decision making uses a logical method of analysis to select the best alternative from among the feasible alternatives. The following nine steps can be followed sequentially, but decision makers often repeat some steps, undertake some simultaneously, and skip others altogether.

1. Recognize the problem.
2. Define the goal or objective: What is the task?
3. Assemble relevant data: What are the facts? Are more data needed, and is it worth more than the cost to obtain it?
4. Identify feasible alternatives.
5. Select the criterion for choosing the best alternative: possible criteria include political, economic, environmental, and humanitarian. The single criterion may be a composite of several different criteria.
6. *Mathematically model* the various interrelationships.
7. Predict the outcomes for each alternative.
8. Choose the best alternative.
9. Audit the results.

Engineering decision making refers to solving substantial engineering problems in which economic aspects dominate and economic efficiency is the criterion for choosing from among possible alternatives. It is a particular case of the general decision-making process. Some of the unusual aspects of engineering decision making are as follows:

1. Cost-accounting systems, while an important source of cost data, contain allocations of indirect costs that may be inappropriate for use in economic analysis.
2. The various consequences—costs and benefits—of an alternative may be of three types:
    (a) Market consequences—there are established market prices
    (b) Extra-market consequences—there are no direct market prices, but prices can be assigned by indirect means
    (c) Intangible consequences—valued by judgment, not by monetary prices
3. The economic criteria for judging alternatives can be reduced to three cases:
    (a) When neither input nor output is fixed: maximize profit, which equals the difference between benefits and costs.
    (b) For fixed input: maximize benefits or other outputs.
    (c) For fixed output: minimize costs or other inputs.

    The first case states the general rule from which both the second and third cases may be derived.
4. To choose among the alternatives, the market consequences and extra-market consequences are organized into a cash flow diagram. We will see in Chapter 3 that engineering economic calculations can be used to compare differing cash flows. These outcomes are compared against the selection criterion. From this comparison *plus* the consequences not included in the monetary analysis, the best alternative is selected.
5. An essential part of engineering decision making is the postaudit of results. This step helps to ensure that projected benefits are obtained and to encourage realistic estimates in analyses.

Importance of Ethics in Engineering and Engineering Economy

One of the gravest responsibilities of an engineer is protecting the safety of the public, clients, and/or employees. In addition, the engineer can be responsible for the economic performance of projects and products on which bonuses and jobs depend. Not surprisingly, in this environment one of the most valued personal characteristics is integrity.

Decision Making with Current Costs

When all costs and benefits occur within a brief period of time, then the time value of money is not a consideration. We still must use the criteria of maximizing profit, minimizing cost, or maximizing benefits.

PROBLEMS

Decision Making

1-1 Think back to your first hour after awakening this morning. List 15 decision-making opportunities that existed during that hour. After you have done that, mark the decision-making opportunities that you actually recognized this morning and upon which you made a conscious decision.

1-2 Some of the following problems would be suitable for solution by engineering economic analysis. Which ones are they?

  (*a*) Would it be better to buy a car with a diesel engine or a gasoline engine?

  (*b*) Should an automatic machine be purchased to replace three workers now doing a task by hand?

  (*c*) Would it be wise to enroll for an early morning class to avoid traveling during the morning traffic rush hours?

  (*d*) Would you be better off if you changed your major?

  (*e*) One of the people you might marry has a job that pays very little money, while another has a professional job with an excellent salary. Which one should you marry?

1-3 Which one of the following problems is *most* suitable for analysis by engineering economic analysis?

  (*a*) Some 45¢ candy bars are on sale for 12 bars for $3. Sandy, who eats a couple of candy bars a week, must decide whether to buy a dozen at the lower price.

  (*b*) A woman has $150,000 in a bank checking account that pays no interest. She can either invest it immediately at a desirable interest rate or wait a week and know that she will be able to obtain an interest rate that is 0.15% higher.

  (*c*) Joe backed his car into a tree, damaging the fender. He has car insurance that will pay for the fender repair. But if he files a claim for payment, they may give him a lower "good driver" rating and charge him more for car insurance in the future.

1-4 If you have $300 and could make the right decisions, how long would it take you to become a millionaire? Explain briefly what you would do.

1-5 Many people write books explaining how to make money in the stock market. Apparently the authors plan to make *their* money selling books telling other people how to profit from the stock market. Why don't these authors forget about the books and make their money in the stock market?

1-6 The owner of a small machine shop has just lost one of his larger customers. The solution to his problem, he says, is to fire three machinists to balance his workforce with his current level of business. The owner says it is a simple problem with a simple solution. The three machinists disagree. Why?

1-7 Every college student had the problem of selecting the college or university to attend. Was this a simple, intermediate, or complex problem for you? Explain.

1-8 Toward the end of the twentieth century, the U.S. government wanted to save money by closing a small portion of its domestic military installations. While many people agreed that saving money was a desirable goal, people in areas potentially affected by a closing soon reacted negatively. Congress finally selected a panel whose task was to develop a list of installations to close, with the legislation specifying that Congress could not alter the list. Since the goal was to save money, why was this problem so hard to solve?

1-9 The college bookstore has put pads of engineering computation paper on sale at half price. What is the minimum and maximum number of pads you might buy during the sale? Explain.

1-10 Consider the seven situations described. Which one situation seems most suitable for solution by engineering economic analysis?

(a) Jane has met two college students that interest her. Bill is a music major who is lots of fun to be with. Alex is a fellow engineering student, but he does not like to dance. Jane wonders what to do.

(b) You drive periodically to the post office to pick up your mail. The parking meters require 10¢ for 6 minutes—about twice the time required to get from your car to the post office and back. If parking fines cost $8, do you put money in the meter or not?

(c) At the local market, candy bars are 45¢ each or three for $1. Should you buy them three at a time?

(d) The cost of car insurance varies widely from company to company. Should you check with several insurance companies when your policy comes up for renewal?

(e) There is a special local sales tax "sin tax" on a variety of things that the town council would like to remove from local distribution. As a result, a store has opened up just outside the town and offers an abundance of these specific items at prices about 30% less than is charged in town. Should you shop there?

(f) Your mother reminds you that she wants you to attend the annual family picnic. That same Saturday you already have a date with a person you have been trying to date for months.

(g) One of your professors mentioned that you have a poor attendance record in her class. You wonder whether to drop the course now or wait to see how you do on the first midterm exam. Unfortunately, the course is required for graduation.

1-11 A car manufacturer is considering locating an assembly plant in your region. List two simple, two intermediate, and two complex problems associated with this proposal.

1-12 Consider the following situations. Which ones appear to represent rational decision making? Explain.

(a) Joe's best friend has decided to become a civil engineer, so Joe has decided that he will also become a civil engineer.

(b) Jill needs to get to the university from her home. She bought a car and now drives to the university each day. When Jim asks her why she didn't buy a bicycle instead, she replies, "Gee, I never thought of that."

(c) Don needed a wrench to replace the spark plugs in his car. He went to the local automobile supply store and bought the cheapest one they had. It broke before he had finished replacing all the spark plugs in his car.

1-13 Identify possible objectives for NASA. For your favorite of these, how should alternative plans to achieve the objective be evaluated?

1-14 Suppose you have just 2 hours to determine how many people in your hometown would be interested in buying a pair of left-handed scissors. Give a step-by-step outline of how you would seek to find out within the time allotted.

1-15 A college student determines that he will have only $50 per month available for his housing for the coming year. He is determined to continue in the university, so he has decided to list all feasible alternatives for his housing. To help him, list five feasible alternatives.

1-16 Describe a situation in which a poor alternative was selected because there was a poor search for better alternatives.

1-17 If there are only two alternatives available and both are unpleasant and undesirable, what should you do?

1-18 The three economic criteria for choosing the best alternative are maximize the difference between output and input, minimize input, and maximize output. For each of the following situations, what is the correct economic criterion?

(a) A manufacturer of plastic drafting triangles can sell all the triangles he can produce at a fixed price. As he increases production, his unit costs increase as a result of overtime pay and so forth. The manufacturer's criterion should be _____.

(b) An architectural and engineering firm has been awarded the contract to design a wharf for a petroleum company for a fixed sum of money. The engineering firm's criterion should be _____.

(c) A book publisher is about to set the list price (retail price) on a textbook. The choice of a low list price would mean less advertising than would be used for a higher list price. The amount of advertising will affect the number of copies sold. The publisher's criterion should be _____.

(d) At an auction of antiques, a bidder for a particular porcelain statue would be trying to _____.

1-19 As in Problem 1-18, state the correct economic criterion for each of the following situations.

(a) The engineering school raffled off a car; tickets sold for 50¢ each or three for $1. When the students were selling tickets, they noted that many people had trouble deciding whether to buy one or three tickets. This indicates the buyers' criterion was _____.

(b) A student organization bought a soft-drink machine for use in a student area. There was considerable discussion over whether they should set the machine to charge 50¢, 75¢, or $1 per drink. The organization recognized that the number of soft drinks sold would depend on the price charged. Eventually the decision was made to charge 75¢. The criterion was _____.

(c) In many cities, grocery stores find that their sales are much greater on days when they advertise special bargains. However, the advertised special prices do not appear to increase the total physical volume of groceries sold by a store. This leads us to conclude that many shoppers' criterion is _____.

(d) A recently graduated engineer has decided to return to school in the evenings to obtain a master's degree. He feels it should be accomplished in a manner that will allow him the maximum amount of time for his regular day job plus time for recreation. In working for the degree, he will _____.

1-20 Seven criteria are given in the chapter for judging which is the best alternative. After reviewing the list, devise three additional criteria that might be used.

1-21 Suppose you are assigned the task of determining the route of a new highway through an older section of town. The highway will require that many older homes be either relocated or torn down. Two possible criteria that might be used in deciding exactly where to locate the highway are:

(a) Ensure that there are benefits to those who gain from the decision and that no one is harmed by the decision.

(b) Ensure that the benefits to those who gain from the decision are greater than the losses of those who are harmed by the decision.

Which criterion will you select to use in determining the route of the highway? Explain.

1-22 For the project in Problem 1-21, identify the major costs and benefits. Which are market consequences, which are extra-market consequences, and which are intangible consequences?

1-23 You must fly to another city for a Friday meeting. If you stay until Sunday morning your ticket will be $200, rather than $800. Hotel costs are $100 per night. Compare the economics with reasonable assumptions for meal expenses. What intangible consequences may dominate the decision?

1-24 In the fall, Jay Thompson decided to live in a university dormitory. He signed a dorm contract under which he was obligated to pay the room rent for the full college year. One clause stated that if he moved out during the year, he could sell his dorm contract to another student who would move into the dormitory as his replacement. The dorm cost was $2000 for the two semesters, which Jay had already paid.

A month after he moved into the dorm, he decided he would prefer to live in an apartment. That week, after some searching for a replacement to fulfill his dorm contract, Jay had two offers. One student offered to move in immediately and to pay Jay $100 per month for the eight remaining months of the school year. A second student offered to move in the second semester and pay $700 to Jay.

Jay estimates his food cost per month is $300 if he lives in the dorm and $250 if he lives in an apartment with three other students. His share of the apartment rent and utilities will be $200 per month. Assume each semester is $4\frac{1}{2}$ months long. Disregard the small differences in the timing of the disbursements or receipts.

(a) What are the three alternatives available to Jay?

(b) Evaluate the cost for each of the alternatives.

(c) What do you recommend that Jay do?

1-25 In decision making we talk about the construction of a model. What kind of model is meant?

1-26 An electric motor on a conveyor burned out. The foreman told the plant manager that the motor had to be replaced. The foreman said that there were no alternatives and asked for authorization to order the replacement. In this situation, is any decision making taking place? If so, who is making the decision(s)?

1-27 Bill Jones's parents insisted that Bill buy himself a new sport shirt. Bill's father gave specific instructions, saying the shirt must be in "good taste," that is, neither too wildly colored nor too extreme in tailoring. Bill found three types of sport shirts in the local department store:

- Rather somber shirts that Bill's father would want him to buy
- Good-looking shirts that appealed to Bill
- Weird shirts that were even too much for Bill

He wanted a good-looking shirt but wondered how to convince his father to let him keep it. The clerk suggested that Bill take home two shirts for his father to see and return the one he did not like. Bill selected a good-looking blue shirt he liked, and also a weird lavender shirt. His father took one look and insisted that Bill keep the blue shirt and return the lavender one. Bill did as his father instructed. What was the key decision in this decision process, and who made it?

1-28 A farmer must decide what combination of seed, water, fertilizer, and pest control will be most profitable for the coming year. The local agricultural college did a study of this farmer's situation and prepared the following table.

Plan	Cost/acre	Income/acre
A	$ 600	$ 800
B	1500	1900
C	1800	2250
D	2100	2500

The last page of the college's study was torn off, and hence the farmer is not sure which plan the agricultural college recommends. Which plan should the farmer adopt? Explain.

1-29 Identify the alternatives, outcomes, criteria, and process for the selection of your college major. Did you make the best choice for you?

1-30 Describe a major problem you must address in the next two years. Use the techniques of this chapter to structure the problem and recommend a decision.

1-31 Apply the steps of the decision-making process to your situation as you decide what to do after graduation.

1-32 One strategy for solving a complex problem is to break it into a group of less complex problems and then find solutions to the smaller problems. The result is the solution of the complex problem. Give an example in which this strategy will work. Then give another example in which this strategy will not work.

Ethics

1-33 When you make professional decisions involving investments in engineering projects, what criteria will you use?
Contributed by D. P. Loucks, Cornell University

1-34 What are ethics?
Contributed by D. P. Loucks, Cornell University

1-35 Suppose you are a engineer working in a private engineering firm and you are asked to sign documents verifying information that you believe is not true. You like your work and your colleagues in the firm, and your family depends on your income. What criteria can you use to guide your decision regarding this issue?
Contributed by D. P. Loucks, Cornell University

1-36 Find the ethics code for the professional society of your major.

(*a*) Summarize its key points.

(*b*) What are its similarities and differences in comparison to NSPE's ethics code?

1-37 Use a personal example or a published source to analyze what went wrong or right with respect to ethics at the assigned stage(s) of the decision-making process.

(*a*) Recognize problem.

(*b*) Define the goal or objective.

(*c*) Assemble relevant data.

(*d*) Identify feasible alternatives.

(*e*) Select the criterion for determining the best alternative.

(*f*) Construct a model.

(*g*) Predict each alternative's outcomes or consequences.

(*h*) Choose the best alternative.

(*i*) Audit the result.

1-38 One of the important responsibilities of municipal assemblies and school boards is public infrastructure, such as roads and schools. Especially for this responsibility, engineers bring skills, knowledge, and perspectives that can improve public decision making. Often the public role is a part-time one, engineers that fulfill it will also have full-time jobs as employees or owners of engineering firms.

(a) What are some of the ethical issues that can arise from conflicts between the public- and private-sector roles?

(b) What guidelines for process or outcomes are there in your hometown?

(c) Summarize and analyze an example situation from the local newspaper, your hometown newspaper, or a news magazine.

1-39 Increasing population and congestion often are addressed through road improvement projects. These may pit the interests of homeowners and business owners in the project area against the interests of people traveling through the improvement project.

(a) What are some of the ethical issues that can arise?

(b) Summarize and analyze an example situation from the local newspaper, your hometown newspaper, or a news magazine.

(c) What guidelines for process or outcomes exist to help address the ethical issues of your example?

1-40 Economic development and redevelopment often require significant acreage that is assembled by acquiring smaller parcels. Sometimes this is done through simple purchase, but the property of an "unwilling seller" can be acquired through the process of eminent domain.

(a) What are some of the ethical issues that can arise?

(b) Summarize and analyze an example situation from the local newspaper, your hometown newspaper, or a news magazine.

(c) What guidelines for process or outcomes exist to help address the ethical issues in your city or state?

1-41 State governments use a variety of advisory and regulatory bodies. Example responsibilities include oversight of professional engineering licensing and the pricing and operation of regulated utilities. Engineers bring skills, knowledge, and perspectives that are useful in this context. Often the public role is a part-time one, and engineers that fulfill it will also have full-time jobs as employees or owners of engineering firms.

(a) What are some of the ethical issues that can arise from conflicts between the public- and private-sector roles?

(b) What guidelines for process or outcomes are there in your home state?

(c) Summarize and analyze an example situation from the local newspaper, your hometown newspaper, or a news magazine.

1-42 Many engineers work in state governments, and some are in high profile roles as legislators, department commissioners, and so on. Many of these individuals move between working in the private and public sectors.

(a) What are some of the ethical issues that can arise as individuals shift between public and private sector roles?

(b) What guidelines for process or outcomes are there in your home state?

(c) Summarize and analyze an example situation from the local newspaper, your hometown newspaper, or a news magazine.

1-43 In the U.S., regulation of payment for overtime hours is done at the state and federal levels. Because most engineering work is accomplished through projects, it is common for engineers to be asked or required to work overtime as projects near deadlines. Sometimes the overtime is paid at time and a half, sometimes as straight time, and sometimes the engineer's salary is treated as a constant even when overtime occurs. In a particular firm, engineering interns, engineers, and partners may be treated the same or differently.

(a) What are some of the legal and ethical issues that can arise?

(b) What is the law or regulation in your state or home state?

(c) Summarize and analyze an example firm's policies or data on how common the problem is.

1-44 At the federal government level, the economic consequences of decisions can be very large. Firms hire lobbyists, legislators may focus on their constituents, and advocacy organizations promote their own agendas. In addition, sometimes some of the players are willing to be unethical.

(a) What are some of the ethical issues that can arise?

(b) Summarize and analyze an example situation from the local newspaper, your hometown newspaper, or a news magazine.

(c) What guidelines for process or outcomes are there for the agency or legislative body involved in your example?

1-45 At both state and federal levels, legislators can be involved in "pork barrel" funding of capital projects. These projects may even bypass the economic evaluation using engineering economy that normal projects are subject to.

(a) What are some of the ethical issues that can arise?

(b) Summarize and analyze an example situation from the local newspaper, your hometown newspaper, or a news magazine.

(c) What guidelines for process or outcomes are there for the agency or legislative body involved in your example?

1-46 At the international level, a common ethics issue important to engineering and project justification is that of environmental regulation. Often different nations have different environmental standards, and a project or product might be built in either location.

(a) What are some of the ethical issues that can arise?

(b) Summarize and analyze an example situation from the local newspaper, your hometown newspaper, or a news magazine.

(c) What guidelines for process or outcomes exist to help address the ethical issues of your example?

1-47 At the international level, a common ethics issue important to engineering and project justification is that of worker health and safety. Often different nations have different standards, and a project or product could be built in either location.

(a) What are some of the ethical issues that can arise?

(b) Summarize and analyze an example situation from the local newspaper, your hometown newspaper, or a news magazine.

(c) What guidelines for process or outcomes exist to help address the ethical issues of your example?

1-48 At the international level, engineering decisions are critical in matters of "sustainable development," a common ethics issue.

(a) What are some of the ethical issues that can arise?

(b) Summarize and analyze an example situation from the local newspaper, your hometown newspaper, or a news magazine.

(c) What guidelines for process or outcomes exist to help address the ethical issues of your example?

1-49 At the international level, questions arise about whether the U.S. ban on bribery is practical or appropriate. In some countries government workers are very poorly paid, and they can support their families only by accepting money to "grease" a process.

(a) What are some of the ethical issues that can arise?

(b) Summarize and analyze an example situation from the local newspaper, your hometown newspaper, or a news magazine.

(c) What guidelines for process or outcomes exist to help address the ethical issues of your example?

1-50 In the 1970s the Ford Motor Company sold its subcompact Pinto model with known design defects. In particular, the gas tank's design and location led to rupture, leaks, and explosion in low-speed, rear-impact collisions. Fifty-nine people burned to death in Pinto accidents. In a cost-benefit analysis weighing the cost of fixing the defects ($11 per vehicle) versus the firm's potential liability for lawsuits on behalf of accident victims, Ford had placed the value of a human life at $200,000. Ford eventually recalled 1.4 million Pintos to fix the gas tank problem for a cost of $30 million to $40 million. In addition the automaker ultimately paid out millions more in liability settlements and incurred substantial damage to its reputation.

(a) Critique Ford's actions from the perspective of the NSPE Code of Ethics.

(b) One well-known ethical theory, utilitarianism, suggests that an act is ethically justified if it results in the "greatest good for the greatest number" when all relevant stakeholders are considered. Did Ford's cost-benefit analysis validly apply this theory?

(c) What should engineers do when the product they are designing has a known safety defect with an inexpensive remedy?

Contributed by Dr. Joseph R. Herkert, Arizona State University

Current Costs

1-51 Willie Lohmann travels from city to city in the conduct of his business. Every other year he buys a used car for about $12,000. The dealer allows about $8000 as a trade-in allowance, with the result that the salesman spends $4000 every other year for a car. Willie keeps accurate records, which show that all other expenses on his car amount to 22.3¢ per mile for each mile he drives. Willie's employer has two plans by which salesmen are reimbursed for their car expenses:

A. Willie will receive all his operating expenses, and in addition will receive $2000 each year for the decline in value of the automobile.

B. Willie will receive 32¢ per mile but no operating expenses and no depreciation allowance.

If Willie travels 18,000 miles per year, which method of computation gives him the larger reimbursement? At what annual mileage do the two methods give the same reimbursement?

1-52 Maria, a college student, is getting ready for three final examinations at the end of the school year. Between now and the start of exams, she has 15 hours of study time available. She would like to get as high a grade average as possible in her math, physics, and engineering economy classes. She feels she must study at least 2 hours for each course and, if necessary, will settle for the low grade that the limited study would yield. If Maria estimates her grade in each subject as follows, how much time should she devote to each class?

Mathematics		Physics		Engineering Economy	
Study Hours	Grade	Study Hours	Grade	Study Hours	Grade
2	25	2	35	2	50
3	35	3	41	3	61
4	44	4	49	4	71
5	52	5	59	5	79
6	59	6	68	6	86
7	65	7	77	7	92
8	70	8	85	8	96

1-53 Two manufacturing companies, located in cities 90 miles apart, have discovered that they both send their trucks four times a week to the other city full of cargo and return empty. Each company pays its driver $185 a day (the round trip takes all day) and have truck operating costs (excluding the driver) of 60¢ a mile. How much could each company save each week if they shared the task, with each sending its truck twice a week and hauling the other company's cargo on the return trip?

1-54 A city needs to increase its rubbish disposal facilities. There is a choice of two rubbish disposal areas, as follows.

Area A: A gravel pit with a capacity of 16 million cubic meters. Owing to the possibility of high groundwater, however, the Regional Water Pollution Control Board has restricted the lower 2 million cubic meters of fill to inert material only (earth, concrete, asphalt, paving, brick, etc.). The inert material, principally clean earth, must be purchased and hauled to this area for the bottom fill.

Area B: Capacity is 14 million cubic meters. The entire capacity may be used for general rubbish disposal. This area will require an average increase in a round-trip haul of 5 miles for 60% of the city, a decreased haul of 2 miles for 20% of the city. For the remaining 20% of the city, the haul is the same distance as for Area A.

Assume the following conditions:

- Cost of inert material placed in Area A will be $2.35/m^3.
- Average speed of trucks from last pickup to disposal site is 15 miles per hour.
- The rubbish truck and a two-man crew will cost $35 per hour.
- Truck capacity of $4\frac{1}{2}$ tons per load or 20 m^3.
- Sufficient cover material is available at all areas; however, inert material for the bottom fill in Area A must be hauled in.

Which of the sites do you recommend?   (*Answer:* Area *B*)

1-55 An oil company is considering adding an additional grade of fuel at its service stations. To do this, an additional 3000-gallon tank must be buried at each station. Discussions with tank fabricators indicate that the least expensive tank would be cylindrical with minimum surface area. What size tank should be ordered?

(*Answer:* 8 ft diameter by 8 ft length)

1-56 The vegetable buyer for a group of grocery stores has decided to sell packages of sprouted grain in the vegetable section of the stores. The product is perishable, and any remaining unsold after one week in the store is discarded. The supplier will deliver the packages to the stores, arrange them in the display space, and remove and dispose of any old packages. The price the supplier will charge the stores depends on the size of the total weekly order for all the stores.

Weekly Order	Price per Package
Less than 1000 packages	35¢
1000–1499	28
1500–1999	25
2000 or more	20

The vegetable buyer estimates the quantity that can be sold per week, at various selling prices, as follows:

Selling Price	Packages Sold per Week
60¢	300
45	600
40	1200
33	1700
26	2300

The sprouted grain will be sold at the same price in all the grocery stores. How many packages should be purchased per week, and at which of the five prices listed above should they be sold?

1-57   Cathy Gwynn, a recently graduated engineer, decided to invest some of her money in a "Quick Shop" grocery store. The store emphasizes quick service, a limited assortment of grocery items, and rather high prices. Cathy wants to study the business to see if the store hours (currently 0600 to 0100) can be changed to make the store more profitable. Cathy assembled the following information.

Time Period	Daily Sales in the Time Period
0600–0700	$ 20
0700–0800	40
0800–0900	60
0900–1200	200
1200–1500	180
1500–1800	300
1800–2100	400
2100–2200	100
2200–2300	30
2300–2400	60
2400–0100	20

The cost of the groceries sold averages 70% of sales. The incremental cost to keep the store open, including the clerk's wage and other incremental operating costs, is $10 per hour. To maximize profit, when should the store be opened, and when should it be closed?

1-58   Jim Jones, a motel owner, noticed that just down the street the "Motel 36" advertises a $36-per-night room rental rate on its sign. As a result, this competitor has rented all 80 rooms every day by late afternoon. Jim,

on the other hand, does not advertise his rate, which is $54 per night, and he averages only a 68% occupancy of his 50 rooms.

There are a lot of other motels nearby, but only Motel 36 advertises its rate on its sign. (Rates at the other motels vary from $48 to $80 per night.) Jim estimates that his actual incremental cost per night for each room rented, rather than remaining vacant, is $12. This $12 pays for all the cleaning, laundering, maintenance, utilities, and so on. Jim believes his eight alternatives are:

Alternative		Resulting Occupancy Rate
	Advertise and Charge	
1	$35 per night	100%
2	42 per night	94
3	48 per night	80
4	54 per night	66
	Do Not Advertise and Charge	
5	$48 per night	70%
6	54 per night	68
7	62 per night	66
8	68 per night	56

What should Jim do? Show how you reached your conclusion.

1-59   A firm is planning to manufacture a new product. The sales department estimates that the quantity that can be sold depends on the selling price. As the selling price is increased, the quantity that can be sold decreases. Numerically they estimate:

$$P = \$35.00 - 0.02Q$$

where P = selling price per unit
Q = quantity sold per year

On the other hand, management estimates that the average cost of manufacturing and selling the product will decrease as the quantity sold increases. They estimate

$$C = \$4.00Q + \$8000$$

where C = cost to produce and sell Q per year

The firm's management wishes to produce and sell the product at the rate that will maximize profit, that is, income minus cost will be a maximum. What quantity should the decision makers plan to produce and sell each year?   (*Answer:* 775 units)

1-60 A manufacturing firm has received a contract to assemble 1000 units of test equipment in the next year. The firm must decide how to organize its assembly operation. Skilled workers, at $22 per hour each, could be assigned to individually assemble the test equipment. Each worker would do all the assembly steps, and it would take 2.6 hours to complete one unit. An alternate approach would be to set up teams of four less skilled workers (at $13 per hour each) and organize the assembly tasks so that each worker does part of the assembly. The four-man team would be able to assemble a unit in one hour. Which approach would result in more economical assembly?

1-61 A grower estimates that if he picks his apple crop now, he will obtain 1000 boxes of apples, which he can sell at $3 per box. However, he thinks his crop will increase by 120 boxes of apples for each week he delays picking, but that the price will drop at a rate of 15¢ per box per week; in addition, he estimates that approximately 20 boxes per week will spoil for each week he delays picking. When should he pick his crop to obtain the largest total cash return? How much will he receive for his crop at that time?

1-62 On her first engineering job, Joy Hayes was given the responsibility of determining the production rate for a new product. She has assembled the data presented in the following graphs.

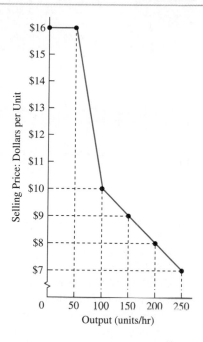

(a) Select an appropriate economic criterion and estimate the production rate based upon it.

(b) Joy's boss told Joy: "I want you to maximize output with minimum input." Joy wonders if it is possible to meet her boss's criterion. She asks for your advice. What would you tell her?

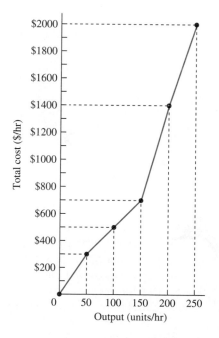

CHAPTER 2

ENGINEERING COSTS AND COST ESTIMATING

North Slope Natural Gas Pipeline

Crude oil from Alaska's North Slope has long been important to the U.S. Now a major project to bring Alaska North Slope natural gas to the lower 48 is under active consideration. The numbers associated with this private sector endeavor are: 35 trillion cubic feet (TCF) of producible gas reserves, $20 billion of estimated infrastructure costs needed to access and deliver the gas, and a 9-year engineer, design, and build process to make it all happen.

A significant driver for this massive project is consumer demand and supply capabilities. By the end of the decade the annual U.S. new sources of natural gas demand will increase to 18 TCF, and experts estimate that the consumption rate will continue to increase at a 2 to 3% annual rate. All this adds up to a case of demand soon exceeding domestic supply capabilities. That is where the North Slope project comes into play. At a design capacity of 4.5 billion cubic feet per day, the project would not completely fill the projected shortfall, but it would reduce reliance on foreign suppliers.

While the pipeline design configuration is not complete, the leading candidate is the Prudhoe Bay to Canada to Chicago route—approximately 2140 miles. An alternate route is the Prudhoe Bay to Valdez pipeline (approximately 800 miles) to a liquefaction plant for shipping on liquefied natural gas (LNG) oceangoing tankers. Under either route, the proposed pipeline design daily capacity is 4.5 billion cubic feet of gas. Delivering 4.5 billion cubic feet daily puts the estimated revenue increase for the state of Alaska between $2 billion and $3 billion annually.

As is often the case with a private-sector, large-scale project, this one faces many risks. One major risk is the development of accurate cost estimates for engineering and building the pipeline. Another risk is providing the project owners an adequate return on investment when energy prices are volatile and demands are uncertain.

The natural gas resource supply is available, and the projected consumer needs are real. Yet, this project must compete for private investors' capital with other global investment projects. Final decisions on project viability will be based on cost estimates and calculations of rate of return on investment in comparison to other investment opportunities. That is why accurate estimates are essential in the decision process, and ultimately will determine whether North Slope gas finds its way into U.S. markets.

Contributed by Morgan E. Henrie, University of Alaska Anchorage

QUESTIONS TO CONSIDER

1. When large-scale projects compete for limited corporate capital funds, what type of cost estimating should be utilized? Can you develop a philosophical statement, from the firm's perspective, regarding the detail needed and confidence justifiable in a cost estimate?

2. When a project is estimated to take 5 to 10 (or more) years from concept to completion, should cost estimates be adjusted for the greater influences and impacts of inflation, adverse government regulatory changes, and changing local economic environment? Why or why not?

3. Should large-scale infrastructure projects be required to meet the same internal rate of return requirements as smaller, shorter-completion-time projects? Why or why not?

4. Are there any ethical issues related to economics, the environment, safety, and so on that should be considered by the project sponsors, consumers, and/or the U.S. government?

After Completing This Chapter...
The student should be able to:

- Define various cost concepts.
- Provide specific examples of how and why these engineering cost concepts are important.
- Define engineering cost estimating.
- Explain the three types of engineering estimate, as well as common difficulties encountered in making engineering cost estimates.
- Use several common mathematical estimating models in cost estimating.
- Discuss the impact of the *learning curve* on cost estimates.
- State the relationship between cost estimating and estimating project benefits.
- Draw *cash flow diagrams* to show project costs and benefits.

T his chapter defines fundamental cost concepts. These include fixed and variable costs, marginal and average costs, sunk and opportunity costs, recurring and non-recurring costs, incremental cash costs, book costs, and life-cycle costs. We then describe the various types of estimates and difficulties sometimes encountered. The models that are described include unit factor, segmenting, cost indexes, power sizing, triangulation, and learning curves. The chapter discusses estimating benefits, developing cash flow diagrams, and drawing these diagrams with spreadsheets.

Understanding engineering costs is fundamental to the engineering economic analysis process, and therefore this chapter addresses an important question: Where do the numbers come from?

ENGINEERING COSTS

Evaluating a set of feasible alternatives requires that many costs be analyzed. Examples include costs for initial investment, new construction, facility modification, general labor, parts and materials, inspection and quality, contractor and subcontractor labor, training, computer hardware and software, material handling, fixtures and tooling, data management, and technical support, as well as general support costs (overhead). In this section we describe several concepts for classifying and understanding these costs.

Fixed, Variable, Marginal, and Average Costs

Fixed costs are constant or unchanging regardless of the level of output or activity. In contrast, **variable** costs depend on the level of output or activity. A **marginal** cost is the variable cost for one more unit, while the **average** cost is the total cost divided by the number of units.

In a production environment, for example, fixed costs, such as those for factory floor space and equipment, remain the same even though production quantity, number of employees, and level of work-in-process may vary. Labor costs are classified as a *variable* cost because they depend on the number of employees and the number of hours they work. Thus *fixed* costs are level or constant regardless of output or activity, and *variable* costs are changing and related to the level of output or activity.

As another example, many universities charge full-time students a fixed cost for 12 to 18 hours and a cost per credit hour for each credit hour over 18. Thus for full-time students who are taking an overload (>18 hours), there is a variable cost that depends on the level of activity, but for most full-time students tuition is a fixed cost.

This example can also be used to distinguish between *marginal* and *average* costs. A marginal cost is the cost of one more unit. This will depend on how many credit hours the student is taking. If currently enrolled for 12 to 17 hours, adding one more is free. The marginal cost of an additional credit hour is $0. However, for a student taking 18 or more hours, the marginal cost equals the variable cost of one more hour.

To illustrate average costs, the fixed and variable costs need to be specified. Suppose the cost of 12 to 18 hours is $1800 per term and overload credits are $120/hour. If a student takes 12 hours, the *average* cost is $1800/12 = $150 per credit hour. If the student were to take 18 hours, the *average* cost would decrease to $1800/18 = $100 per credit hour. If the student takes 21 hours, the *average* cost is $102.86 per credit hour [$1800 + (3 × $120)]/21. Average cost is thus calculated by dividing the total cost for all units by the total number of units. Decision makers use **average** cost to attain an overall cost picture of the investment on a per unit basis.

   Marginal cost is used to decide whether an additional unit should be made, purchased, or enrolled in. For our example full-time student, the marginal cost of another credit is $0 or $120 depending on how many credits the student has already signed up for.

EXAMPLE 2-1

The Federation of Student Societies of Engineering (FeSSE) wants to offer a one-day training course to help students in job hunting and to raise funds. The organizing committee is sure that they can find alumni, local business people, and faculty to provide the training at no charge. Thus the main costs will be for space, meals, handouts, and advertising.

   The organizers have classified the costs for room rental, room setup, and advertising as fixed costs. They also have included the meals for the speakers as a fixed cost. Their total of $225 is pegged to a room that will hold 40 people. So if demand is higher, the fixed costs will also increase.

   The variable costs for food and bound handouts will be $20 per student. The organizing committee believes that $35 is about the right price to match value to students with their budgets. Since FeSSE has not offered training courses before, they are unsure how many students will reserve seats.

   Develop equations for FeSSE's total cost and total revenue, and determine the number of registrations that would be needed for revenue to equal cost.

SOLUTION

Let x equal the number of students who sign up. Then,

$$\text{Total cost} = \$225 + \$20x$$

$$\text{Total revenue} = \$35x$$

To find the number of student registrants for revenue to equal cost, we set the equations equal to each other and solve.

$$\text{Total cost} = \text{Total revenue}$$

$$\$225 + \$20x = \$35x$$

$$\$225 = (\$35 - \$20)x$$

$$x = 225/15 = 15 \text{ students}$$

   While this example has been defined for a student engineering society, we could just as easily have described this as a training course to be sponsored by a local chapter of a professional technical society. The fixed cost and the revenue would increase by a factor of about 10, while the variable cost would probably double or triple.

If a firm were considering an in-house short course, the cost of the in-house course would be compared with the cost per employee (a variable cost) for enrolling employees in external training.

From Example 2-1 we see how it is possible to calculate total fixed and total variable costs. Furthermore, these values can be combined into a single **total** cost equation as follows:

$$\text{Total cost} = \text{Total fixed cost} + \text{Total variable cost} \qquad (2\text{-}1)$$

Example 2-1 developed *total cost* and *total revenue* equations to describe a training course proposal. These equations can be used to create what is called a *profit–loss breakeven chart* (see Figure 2-1). A plot of revenues against costs for various levels of output (activity) allows one to illustrate a *breakeven point* (in terms of costs and revenue) and regions of *profit* and *loss*. These terms can be defined as follows.

Breakeven point: The level of activity at which total costs for the product, good, or service are *equal to* the revenue (or savings) generated. This is the level at which one "just breaks even."

Profit region: Values of the variable x greater than the breakeven point, where total revenue is greater than total costs.

Loss region: Values of the variable x less than the breakeven point, where total revenue is less than total costs.

Notice in Figure 2-1 that the *breakeven* point for the number of persons in the training course is 15 people. For more than 15 people, FeSSE will make a profit. If fewer than 15 sign up, there will be a net loss. At the breakeven level, the total cost to provide the course equals the revenue received from the 15 students.

The fixed costs of our simple model are in reality *fixed* over a range of values for x. In Example 2-1, that range was 1 to 40 students. If *zero* students signed up, then the course

FIGURE 2-1 Profit–loss breakeven chart for Example 2-1.

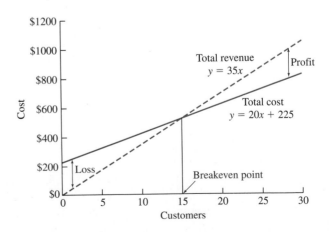

FIGURE 2-2 Nonlinear variable costs.

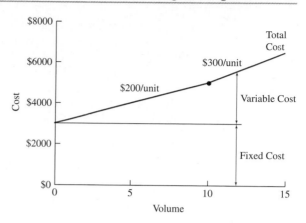

could be cancelled and many of the fixed costs would not be incurred. Some costs such as advertising might already have been spent, and there might be cancellation fees. If more than 40 students signed up, then greater costs for larger rooms or multiple sessions would be incurred. The model is valid only within the range named.

When modeling a specific situation, we often use *linear* variable costs and revenues. However, sometimes the relationship may be nonlinear. For example, employees are often paid at 150% of their hourly rate for overtime hours, so that production levels requiring overtime have higher variable costs. Total cost in Figure 2-2 is a fixed cost of $3000 plus a variable cost of $200 per unit for straight-time production of up to 10 units and $300 per unit for overtime production of up to 5 more units.

Figure 2-2 can also be used to illustrate marginal and average costs. At a volume of 5 units the marginal cost is $200 per unit, while at a volume of 12 units the marginal cost is $300 per unit. At 5 units the average cost is $800 per unit, or $(3000 + 200 \times 5)/5$. At 12 units the average cost is $467 per unit, or $(3000 + 200 \times 10 + 300 \times 2)/12$.

Sunk Costs

A **sunk cost** is money already spent as a result of a *past* decision. If only 5 students signed up for the training course in Example 2-1, the advertising costs would be a *sunk cost*.

Sunk costs must be ignored in engineering economic analysis because current decisions cannot change the past. For example, dollars spent last year to purchase new production machinery is money that is *sunk:* the money has already been spent—there is nothing that can be done now to change that action. As engineering economists we deal with present and future opportunities.

Many times it is difficult not to be influenced by sunk costs. Consider 100 shares of stock in XYZ, Inc., purchased for $15 per share last year. The share price has steadily declined over the past 12 months to a price of $10 per share today. Current decisions must focus on the $10 per share that could be attained today (as well as future price potential), not the $15 per share that was paid last year. The $15 per share paid last year is a *sunk cost* and has no influence on present opportunities.

As another example, when Regina was a sophomore, she purchased a newest-generation laptop from the college bookstore for $2000. By the time she graduated, the most anyone would pay her for the computer was $400 because the newest models were faster and cheaper and had more capabilities. For Regina, the original purchase price was a *sunk cost* that has no influence on her present opportunity to sell the laptop at its current market value of $400.

When we get to Chapters 11 and 12 on depreciation and income taxes, we will find an exception to the rule of *ignore sunk costs*. When an asset is sold or disposed of, then the sunk cost of what was paid for it is important in figuring out how much is owed in taxes. This exception only applies to the after-tax analysis of capital assets.

Opportunity Costs

An **opportunity cost** is associated with using a resource in one activity instead of another. Every time we use a business resource (equipment, dollars, manpower, etc.) in one activity, we give up the opportunity to use the same resources at that time in some other activity.

Every day businesses use resources to accomplish various tasks—forklifts are used to transport materials, engineers are used to design products and processes, assembly lines are used to make a product, and parking lots are used to provide parking for employees' vehicles. Each of these resources costs the company money to maintain for those intended purposes. However, the amount is not just made up of the dollar cost; it also includes the opportunity cost. Each resource that a firm owns can feasibly be used in several alternative ways. For instance, the assembly line could produce a different product, and the parking lot could be rented out, used as a building site, or converted into a small airstrip. Each of these alternative uses would provide some benefit to the company.

A firm that chooses to use the resource in one way is giving up the benefits that would be derived from using it in those other ways. The benefit that would be derived by using the resource in this "other activity" is the **opportunity cost** of using it in the chosen activity. Opportunity cost may also be considered a **forgone opportunity cost** because we are forgoing the benefit that could have been realized. A formal definition of opportunity cost might be:

> An *opportunity cost* is the benefit that is *forgone* by engaging a business resource in a chosen activity instead of engaging that same resource in the forgone activity.

As an example, suppose that friends invite a college student to travel through Europe over the summer break. In considering the offer, the student might calculate all the *out-of-pocket* cash costs that would be incurred. Cost estimates might be made for items such as air travel, lodging, meals, entertainment, and train passes. Suppose this amounts to $3000 for a 10-week period. After checking his bank account, the student reports that indeed he can afford the $3000 trip. However, the *true* cost to the student includes not only his *out-of-pocket* cash costs but also his *opportunity cost*. By taking the trip, the student is giving up the *opportunity* to earn $5000 as a summer intern at a local business. The student's total cost will comprise the $3000 cash cost as well as the $5000 opportunity cost (wages forgone)—the total cost to our traveler is thus $8000.

EXAMPLE 2-2

A distributor of electric pumps must decide what to do with a "lot" of old electric pumps purchased 3 years ago. Soon after the distributor purchased the lot, technology advances made the old pumps less desirable to customers. The pumps are becoming obsolescent as they sit in inventory. The pricing manager has the following information.

Distributor's purchase price 3 years ago	$ 7,000
Distributor's storage costs to date	1,000
Distributor's list price 3 years ago	9,500
Current list price of the same number of new pumps	12,000
Amount offered for the old pumps from a buyer 2 years ago	5,000
Current price the lot of old pumps would bring	3,000

Looking at the data, the pricing manager has concluded that the price should be set at $8000. This is the money that the firm has "tied up" in the lot of old pumps ($7000 purchase and $1000 storage), and it was reasoned that the company should at least recover this cost. Furthermore, the pricing manager has argued that an $8000 price would be $1500 less than the list price from 3 years ago, and it would be $4000 less than what a lot of new pumps would cost ($12,000 − $8000). What would be your advice on price?

SOLUTION

Let's look more closely at each of the data items.

Distributor's purchase price 3 years ago: This is a sunk cost that should not be considered in setting the price today.

Distributor's storage costs to date: The storage costs for keeping the pumps in inventory are sunk costs; that is, they have been paid. Hence they should not influence the pricing decision.

Distributor's list price 3 years ago: If there have been no willing buyers in the past 3 years at this price, it is unlikely that a buyer will emerge in the future. This past list price should have no influence on the current pricing decision.

Current list price of newer pumps: Newer pumps now include technology and features that have made the older pumps less valuable. Directly comparing the older pumps to those with new technology is misleading. However, the price of the new pumps and the value of the new features help determine the market value of the old pumps.

Amount offered by a buyer 2 years ago: This is a forgone opportunity. At the time of the offer, the company chose to keep the lot and thus the $5000 offered became an opportunity cost for keeping the pumps. This amount should not influence the current pricing decision.

Current price the lot could bring: The price a willing buyer in the marketplace offers is called the asset's *market value*. The lot of old pumps in question is believed to have a current market value of $3000.

From this analysis, it is easy to see the flaw in the pricing manager's reasoning. In an engineering economist analysis we deal only with *today's* and prospective *future* opportunities. It is impossible to go back in time and change decisions that have been made. Thus, the pricing manager should recommend to the distributor that the price be set at the current value that a buyer assigns to the item: $3000.

Recurring and Nonrecurring Costs

Recurring costs refer to any expense that is known and anticipated, and that occurs at regular intervals. **Nonrecurring costs** are one-of-a-kind expenses that occur at irregular intervals and thus are sometimes difficult to plan for or anticipate from a budgeting perspective.

Examples of recurring costs include those for resurfacing a highway and reshingling a roof. Annual expenses for maintenance and operation are also recurring expenses. Examples of nonrecurring costs include the cost of installing a new machine (including any facility modifications required), the cost of augmenting equipment based on older technology to restore its usefulness, emergency maintenance expenses, and the disposal or close-down costs associated with ending operations.

In engineering economic analyses *recurring costs* are modeled as cash flows that occur at regular intervals (such as every year or every 5 years). Their magnitude can be estimated, and they can be included in the overall analysis. *Nonrecurring costs* can be handled easily in our analysis if we are able to anticipate their timing and size. However, this is not always so easy to do.

Incremental Costs

One of the fundamental principles in engineering economic analysis is that in choosing between competing alternatives, the focus is on the *differences* between those alternatives. This is the concept of **incremental costs.** For instance, one may be interested in comparing two options to lease a vehicle for personal use. The two lease options may have several specifics for which costs are the same. However, there may be incremental costs associated with one option but not with the other. In comparing the two leases, the focus should be on the differences between the alternatives, not on the costs that are the same.

EXAMPLE 2-3

Philip is choosing between model *A* (a budget model) and model *B* (with more features and a higher purchase price). What *incremental costs* would Philip incur if he chose model *B* instead of the less expensive model *A*?

Cost Items	Model *A*	Model *B*
Purchase price	$10,000	$17,500
Installation costs	3,500	5,000
Annual maintenance costs	2,500	750
Annual utility expenses	1,200	2,000
Disposal costs after useful life	700	500

SOLUTION

We are interested in the incremental or *extra* costs that are associated with choosing model *B* instead of model *A*. To obtain these we subtract model *A* costs from model *B* costs for each category (cost item) with the following results.

Cost Items	(Model *B* Cost − *A* Cost)	Incremental Cost of *B*
Purchase price	17,500 − 10,000	$7500
Installation costs	5,000 − 3,500	1500
Annual maintenance costs	750 − 2,500	−1750/yr
Annual utility expenses	2,000 − 1,200	800/yr
Disposal costs after useful life	500 − 700	−200

Notice that for the cost categories given, the incremental costs of model *B* are both positive and negative. Positive incremental costs mean that model *B* costs more than model *A*, and negative incremental costs mean that there would be a *savings* (reduction in cost) if model *B* were chosen instead.

Because model *B* has more features, a decision would also have to consider the incremental benefits offered by those features.

Cash Costs Versus Book Costs

A *cash cost* requires the cash transaction of dollars "out of one person's pocket" into "the pocket of someone else." When you buy dinner for your friends or make your monthly car payment you are incurring a **cash cost** or **cash flow.** Cash costs and cash flows are the basis for engineering economic analysis.

Book costs do not require the transaction of dollars "from one pocket to another." Rather, **book costs** are cost effects from past decisions that are recorded "in the books" (accounting books) of a firm. In one common book cost, asset depreciation (which we discuss in Chapter 11), the expense paid for a particular business asset is "written off" on a company's accounting system over a number of periods. Book costs do not ordinarily represent cash flows and thus are not included in engineering economic analysis. One exception to this is the impact of asset depreciation on tax payments—which are cash flows and are included in after-tax analyses.

Life-Cycle Costs

The products, goods, and services designed by engineers all progress through a **life cycle** very much like the human life cycle. People are conceived, go through a growth phase, reach their peak during maturity, and then gradually decline and expire. The same general pattern holds for products, goods, and services. As with humans, the duration of the different phases, the height of the peak at maturity, and the time of the onset of decline and termination all vary depending on the individual product, good, or service. Figure 2-3 illustrates the typical phases that a product, good, or service progresses through over its life cycle.

Life-cycle costing refers to the concept of designing products, goods, and services with a full and explicit recognition of the associated costs over the various phases of their life cycles. Two key concepts in life-cycle costing are that the later design changes are made, the higher the costs, and that decisions made early in the life cycle tend to "lock in" costs that will be incurred later. Figure 2-4 illustrates how costs are committed early in the product life cycle—nearly 70–90% of all costs are set during the design phases. At the same time, as the figure shows, only 10–30% of cumulative life-cycle costs have been spent.

Figure 2-5 reinforces these concepts by illustrating that later product changes are more costly and that earlier changes are easier (and less costly) to make. When planners try to

Beginning ——————————————— Time ——————————————→ End					
Needs Assessment and Justification Phase	Conceptual or Preliminary Design Phase	Detailed Design Phase	Production or Construction Phase	Operational Use Phase	Decline and Retirement Phase
Requirements	Impact Analysis	Allocation of Resources	Product, Goods, & Services Built	Operational Use	Declining Use
Overall Feasibility	Proof of Concept	Detailed Specifications	All Supporting Facilities Built	Use by Ultimate Customer	Phase Out
Conceptual Design Planning	Prototype/ Breadboard	Component and Supplier Selection	Operational Use Planning	Maintenance and Support	Retirement
	Development and Testing	Production or Construction Phase		Processes, Materials and Methods Use	Responsible Disposal
	Detailed Design Planning			Decline and Retirement Planning	

FIGURE 2-3 Typical life cycle for products, goods, and services.

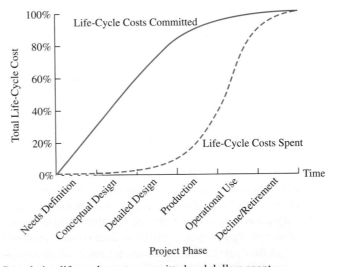

FIGURE 2-4 Cumulative life-cycle costs committed and dollars spent.

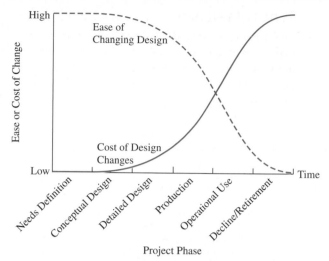

FIGURE 2-5 Life-cycle design change costs and ease of change.

save money at an early design stage, the result is often a poor design that results in change orders during construction and prototype development. These changes, in turn, are more costly than working out a better design would have been.

From Figures 2-4 and 2-5 we see that the time to consider all life-cycle effects and make design changes is during the needs and conceptual/preliminary design phases—before a lot of dollars are committed. Some of the life-cycle effects that engineers should consider at design time include product costs for liability, production, material, testing and quality assurance, and maintenance and warranty. Other life-cycle effects include product features based on customer input and product disposal effects on the environment. The key point is that engineers who design products and the systems that produce them should consider all life-cycle costs.

COST ESTIMATING

Engineering economic analysis focuses on the future consequences of current decisions. Because these consequences are in the future, usually they must be estimated and cannot be known with certainty. Estimates that may be needed in engineering economic analysis include purchase costs, annual revenue, yearly maintenance, interest rates for investments, annual labor and insurance costs, equipment salvage values, and tax rates.

Estimating is the foundation of economic analysis. As in any analysis procedure, the outcome is only as good as the numbers used to reach the decision. For example, a person who wants to estimate her federal income taxes for a given year could do a very detailed analysis, including social security deductions, retirement savings deductions, itemized personal deductions, exemption calculations, and estimates of likely changes to the tax code. However, this very technical and detailed analysis will be grossly inaccurate if poor data are used to predict the next year's income. Thus, to ensure that an analysis is a reasonable evaluation of future events, it is very important to make careful estimates.

Types of Estimate

The American poet and novelist Gertrude Stein wrote in *The Making of Americans* in 1925 that "a rose is a rose is a rose is a rose." However, what holds for roses does not necessarily hold for estimates because "an estimate is not an estimate." Rather, we can define three general types of estimate whose purposes, accuracies, and underlying methods are quite different.

Rough estimates: order-of-magnitude estimates used for high-level planning, for determining project feasibility, and in a project's initial planning and evaluation phases. Rough estimates tend to involve back-of-the-envelope numbers with little detail or accuracy. The intent is to quantify and consider the order of magnitude of the numbers involved. These estimates require minimum resources to develop, and their accuracy is generally −30 to +60%.

Notice the nonsymmetry in the estimating error. This is because decision makers tend to underestimate the magnitude of costs (negative economic effects). Also, as Murphy's law predicts, there seem to be more ways for results to be worse than expected than there are for the results to be better than expected.

Semidetailed estimates: used for budgeting purposes at a project's conceptual or preliminary design stages. These estimates are more detailed, and they require additional time and resources. Greater sophistication is used in developing semidetailed estimates than the rough-order type, and their accuracy is generally −15 to +20%.

Detailed estimates: used during a project's detailed design and contract bidding phases. These estimates are made from detailed quantitative models, blueprints, product specification sheets, and vendor quotes. Detailed estimates involve the most time and resources to develop and thus are much more accurate than rough or semidetailed estimates. The accuracy of these estimates is generally −3 to +5%.

The upper limits of +60% for rough order, +20% for semidetailed, and +5% for detailed estimates are based on construction data for plants and infrastructure. Final costs for software, research and development, and new military weapons often have much higher corresponding percentages.

In considering the three types of estimate it is important to recognize that each has its unique purpose, place, and function in a project's life. Rough estimates are used for general feasibility activities; semidetailed estimates support budgeting and preliminary design decisions and detailed estimates are used for establishing design details and contracts. As one moves from rough to detailed design, one moves from less to much more accurate estimates.

However, this increased accuracy requires added time and resources. Figure 2-6 illustrates the trade-off between accuracy and cost. In engineering economic analysis, the resources spent must be justified by the need for detail in the estimate. As an illustration, during the project feasibility stages we would not want to use our people, time, and money to develop detailed estimates for infeasible alternatives that will be quickly eliminated from further consideration. However, regardless of how accurate an estimate is assumed to be, it is only an estimate of what the future will be. There will be some error even if ample resources and sophisticated methods are used.

Difficulties in Estimation

Estimating is difficult because the future is unknown. With few exceptions (such as with legal contracts) it is difficult to foresee future economic consequences exactly. In this section we discuss several aspects of estimating that make it a difficult task.

One-of-a-Kind Estimates

Estimated parameters can be for one-of-a-kind or first-run projects. The first time something is done, it is difficult to estimate costs required to design, produce, and maintain a product over its life cycle. Consider the projected cost estimates that were developed for the first NASA missions. The U.S. space program initially had no experience with human flight in outer space; thus the development of the cost estimates for design, production, launch, and recovery of the astronauts, flight hardware, and payloads was a "first-time experience." The same is true for any endeavor lacking local or global historical cost data. New products or processes that are unique and fundamentally different make estimating costs difficult.

The good news is that there are very few one-of-a-kind estimates to be made in engineering design and analysis. Nearly all new technologies, products, and processes have "close cousins" that have led to their development. The concept of **estimation by analogy** allows one to use knowledge about well-understood activities to anticipate costs for new activities. In the 1950s, at the start of the military missile program, aircraft companies drew on their in-depth knowledge of designing and producing aircraft when they bid on missile contracts. As another example, consider the problem of estimating the production labor requirements for a brand new product, X. A company may use its labor knowledge about Product Y, a similar type of product, to build up the estimate for X. Thus, although "first-run" estimates are difficult to make, estimation by analogy can be an effective tool.

Time and Effort Available

Our ability to develop engineering estimates is constrained by time and person-power availability. In an ideal world, it would *cost nothing* to use *unlimited resources* over an *extended period of time*. However, reality requires the use of limited resources in fixed intervals of time. Thus for a rough estimate only limited effort is used.

Constraints on time and person-power can make the overall estimating task more difficult. If the estimate does not require as much detail (such as when a rough estimate is

FIGURE 2-6 Accuracy versus cost trade-off in estimating.

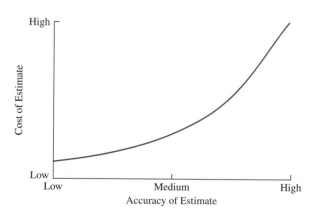

the goal), then time and personnel constraints may not be a factor. When detail is necessary and critical (such as in legal contracts), however, requirements must be anticipated and resource use planned.

Estimator Expertise

Consider two common expressions: *The past is our greatest teacher* and *knowledge is power*. These simple axioms hold true for much of what we encounter during life, and they are true in engineering estimating as well. The more experienced and knowledgeable the engineering estimator is, the less difficult the estimating process will be, the more accurate the estimate will be, the less likely it is that a major error will occur, and the more likely it is that the estimate will be of high quality.

How is experience acquired in industry? One approach is to assign inexperienced engineers relatively small jobs, to create expertise and build familiarity with products and processes. Another strategy used is to pair inexperienced engineers with mentors who have vast technical experience. Technical boards and review meetings conducted to "justify the numbers" also are used to build knowledge and experience. Finally, many firms maintain databases of their past estimates and the costs that were actually incurred.

ESTIMATING MODELS

This section develops several estimating models that can be used at the rough, semidetailed, or detailed design levels. For rough estimates the models are used with rough data, likewise for detailed design estimates they are used with detailed data.

Per-Unit Model

The **per-unit model** uses a "per unit" factor, such as cost per square foot, to develop the estimate desired. This is a very simplistic yet useful technique, especially for developing estimates of the rough or order-of-magnitude type. The per-unit model is commonly used in the construction industry. As an example, you may be interested in a new home that is constructed with a certain type of material and has a specific construction style. Based on this information a contractor may quote a cost of $65 per square foot for your home. If you are interested in a 2000 square foot floor plan, your cost would thus be: $2000 \times 65 = \$130,000$. Other examples where per unit factors are used include

- Service cost per customer
- Safety cost per employee
- Gasoline cost per mile
- Cost of defects per batch
- Maintenance cost per window
- Mileage cost per vehicle
- Utility cost per square foot of floor space
- Housing cost per student

It is important to note that the per-unit model does not make allowances for economies of scale (the fact that higher quantities usually cost less on a per-unit basis). In most cases, however, the model can be effective at getting the decision maker "in the ballpark" of likely costs, and it can be very accurate if accurate data are used.

EXAMPLE 2-4

Use the per-unit model to estimate the cost per student that you will incur for hosting 24 foreign exchange students at a local island campground for 10 days. During camp you are planning the following activities:

- 2 days of canoeing
- 3 campsite-sponsored day hikes
- 3 days at the lake beach (swimming, volleyball, etc.)
- Nightly entertainment

After calling the campground and collecting other information, you have accumulated the following data:

- Van rental from your city to the camp (one way) is $50 per 15-person van plus gas.
- Camp is 50 miles away, the van gets 10 miles per gallon, and gas is $1 per gallon.
- Each cabin at the camp holds 4 campers, and rent is $10 per day per cabin.
- Meals are $10 per day per camper; no outside food is allowed.
- Boat transportation to the island is $2 per camper (one way).
- Insurance/grounds fee/overhead is $1 per day per camper.
- Canoe rentals are $5 per day per canoe; a canoe holds 3 campers.
- Day hikes are $2.50 per camper (plus the cost for meals).
- Beach rental is $25 per group per half-day.
- Nightly entertainment is free.

SOLUTION

You are asked to use the per-unit factor to estimate the cost per student on this trip. For planning purposes we assume that there will be 100% participation in all activities. We will break the total cost down into categories of transportation, living, and entertainment.

Transportation Costs

Van travel to and from camp: 2 vans/trip × 2 trips × ($50/van + 50 miles/van × 1 gal/10 miles × $1/gal) = $220

Boat travel to and from island: 2 trips × $2/camper × 24 campers = $96

$$\text{Transportation costs} = 220 + 96 = \$316$$

Living Costs

Meals for the 10-day period: 24 campers × $10/camper-day × 10 days = $2400

Cabin rental for the 10-day period: 24 campers × 1 cabin/4 campers × $10/day/cabin × 10 days = $600

Insurance/Overhead expense for the 10-day period: 24 campers × $1/camper-day × 10 days = $240

$$\text{Living costs} = 2400 + 600 + 240 = \$3240$$

Entertainment Costs

Canoe rental costs: 2 canoe days × 24 campers × 1 canoe/3 campers × $5/day/canoe = $80

Beach rental costs: 3 days × 2 half-days/day × $25/half-day = $150

Day hike costs: 24 campers × 3 day hikes × $2.50/camper-day hike = $180

Nightly entertainment: This is free! Can you believe it?

$$\text{Entertainment costs} = 80 + 150 + 180 + 0 = \$410$$

Total cost

$$\text{Total cost for 10-day period} = \text{Transportation costs} + \text{Living costs} + \text{Entertainment costs}$$
$$= 316 + 3240 + 410 = \$3966$$

Thus, the cost per student would be $3966/24 = $165.25.

Thus, it would cost you $165.25 per student to host the students at the island campground for the 10-day period. In this case the per-unit model gives you a very detailed cost estimate (although its accuracy depends on the accuracy of your data and assumptions you've made).

Segmenting Model

The **segmenting model** can be described as "divide and conquer." An estimate is decomposed into its individual components, estimates are made at those lower levels, and then the estimates are aggregated (added) back together. It is much easier to estimate at the lower levels because they are more readily understood. This approach is common in engineering estimating in many applications and for any level of accuracy needed. In planning the camp trip of Example 2-4, the overall estimate was **segmented** into the costs for travel, living, and entertainment. The example illustrated the segmenting model (division of the overall estimate into the various categories) together with the unit factor model to make the subestimates for each category. Example 2-5 provides another example of the segmenting approach.

EXAMPLE 2-5

Clean Lawn Corp., a manufacturer of yard equipment, is planning to introduce a new high-end industrial-use lawn mower called the Grass Grabber. The Grass Grabber is designed as a walk-behind, self-propelled mower. Clean Lawn engineers have been asked by the accounting department to estimate the material costs for the new mower. The material cost estimate will be used, along with estimates for labor and overhead to evaluate the potential of this new model.

SOLUTION

The engineers decide to decompose the design specifications for the Grass Grabber into its components, estimate the material costs for each of the components, and then sum these costs up to obtain the overall estimate. The engineers are using a segmenting approach to build up their estimate. After careful consideration, the engineers have divided the mower into the following major subsystems: chassis, drive train, controls, and cutting/collection system. Each of these was further divided as appropriate, and unit material costs were estimated at this lowest of levels as follows:

Cost Item	Unit Material Cost Estimate	Cost Item	Unit Material Cost Estimate
A. Chassis		**C. Controls**	
A.1 Deck	$ 7.40	C.1 Handle assembly	$ 3.85
A.2 Wheels	10.20	C.2 Engine linkage	8.55
A.3 Axles	4.85	C.3 Blade linkage	4.70
	$22.45	C.4 Speed control linkage	21.50
B. Drive train		C.5 Drive control assembly	6.70
B.1 Engine	$38.50	C.6 Cutting height adjuster	7.40
B.2 Starter assembly	5.90		$52.70
B.3 Transmission	5.45		
B.4 Drive disc assembly	10.00	**D. Cutting/Collection system**	
B.5 Clutch linkage	5.15	D.1 Blade assembly	$10.80
B.6 Belt assemblies	7.70	D.2 Side chute	7.05
	$72.70	D.3 Grass bag and adapter	7.75
			$25.60

The total material cost estimate of $173.45 was calculated by summing up the estimates for each of the four major subsystem levels (chassis, drive train, controls, and cutting/collection system). It should be noted that this cost represents only the material portion of the overall cost to produce the mowers. Other costs would include labor and overhead items.

In Example 2-5 the engineers at Clean Lawn Corp. decomposed the cost estimation problem into logical elements. The scheme they used of decomposing cost items and numbering the material components (A.1, A.1, A.2, etc.) is known as a **work breakdown structure.** This technique is commonly used in engineering cost estimating and project management of large products, processes, or projects. A work breakdown structure decomposes a large "work package" into its constituent parts, which can then be estimated or managed individually. In Example 2-5 the work breakdown structure of the Grass Grabber has three levels. At the top level is the product itself, at the second level are the four major subsystems, and at the third level are the individual cost items. Imagine what the product work breakdown structure for a Boeing 787 looks like. Then imagine trying to manage the 787's design, engineering, construction, and costing without a tool like the work breakdown structure.

Cost Indexes

Cost indexes are numerical values that reflect historical change in engineering (and other) costs. The cost index numbers are dimensionless, and they reflect relative price change in either individual cost items (labor, material, utilities) or groups of costs (consumer prices, producer prices). Indexes can be used to update historical costs with the basic ratio relationship given in Equation 2-2.

$$\frac{\text{Cost at time } A}{\text{Cost at time } B} = \frac{\text{Index value at time } A}{\text{Index value at time } B} \qquad (2\text{-}2)$$

Equation 2-2 states that the ratio of the cost index numbers at two points in time (A and B) is equivalent to the dollar cost ratio of the item at the same times (see Example 2-6).

EXAMPLE 2-6

Miriam is interested in estimating the annual labor and material costs for a new production facility. She was able to obtain the following cost data:

Labor Costs

- Labor cost index value was at 124 ten years ago and is 188 today.
- Annual labor costs for a similar facility were $575,500 ten years ago.

Material Costs

- Material cost index value was at 544 three years ago and is 715 today.
- Annual material costs for a similar facility were $2,455,000 three years ago.

SOLUTION

Miriam will use Equation 2-2 to develop her cost estimates for annual labor and material costs.

Labor

$$\frac{\text{Annual cost today}}{\text{Annual cost 10 years ago}} = \frac{\text{Index value today}}{\text{Index value 10 years ago}}$$

$$\text{Annual cost today} = \frac{188}{124} \times \$575{,}500 = \$871{,}800$$

Materials

$$\frac{\text{Annual cost today}}{\text{Annual cost 3 years ago}} = \frac{\text{Index value today}}{\text{Index value 3 years ago}}$$

$$\text{Annual cost today} = \frac{715}{544} \times \$2{,}455{,}000 = \$3{,}227{,}000$$

Cost index data are collected and published by several private and public sources in the United States (and world). The U.S. government publishes data through the Bureau of Labor Statistics of the Department of Commerce. The *Statistical Abstract of the United States* publishes cost indexes for labor, construction, and materials. Another useful source for engineering cost index data is the *Engineering News Record*.

Power-Sizing Model

The **power-sizing model** is used to estimate the costs of industrial plants and equipment. The model "scales up" or "scales down" known costs, thereby accounting for economies of scale that are common in industrial plant and equipment costs. Consider the cost to build a refinery. Would it cost twice as much to build the same facility with double the capacity? It is unlikely. The *power-sizing model* uses the exponent (*x*), called the *power-sizing exponent,* to reflect economies of scale in the size or capacity:

$$\frac{\text{Cost of equipment } A}{\text{Cost of equipment } B} = \left(\frac{\text{Size (capacity) of equipment } A}{\text{Size (capacity) of } B} \right)^x \qquad (2\text{-}3)$$

where *x* is the power-sizing exponent, costs of *A* and *B* are at the same point in time (same dollar basis), and size or capacity is in the same physical units for both *A* and *B*.

The power-sizing exponent (*x*) can be 1.0 (indicating a linear cost-versus-size/capacity relationship) or greater than 1.0 (indicating *dis*economies of scale), but it is usually less than 1.0 (indicating economies of scale). Generally the size ratio should be less than 2, and it should never exceed 5. This model works best in a "middle" range—not very small or very large.

Exponent values for plants and equipment of many types may be found in several sources, including industry reference books, research reports, and technical journals. Such exponent values may be found in *Perry's Chemical Engineers' Handbook, Plant Design and Economics for Chemical Engineers,* and *Preliminary Plant Design in Chemical Engineering.* Table 2-1 gives power-sizing exponent values for several types of industrial facilities and equipment. The exponent given applies only to equipment within the size range specified.

In Equation 2-3 equipment costs for both *A* and *B* occur at the same point in time. This equation is useful for scaling equipment costs but *not* for updating those costs. When the time of the desired cost estimate is different from the time in which the scaling occurs (per Equation 2-3) cost indexes accomplish the time updating. Thus, in cases like Example 2-7 involving both scaling and updating, we use the power-sizing model together with cost indexes.

TABLE 2-1 Example Power-Sizing Exponent Values

Equipment or Facility	Size Range	Power-Sizing Exponent
Blower, centrifugal	10,000–100,000 ft³/min	0.59
Compressor	200–2100 hp	0.32
Crystallizer, vacuum batch	500–7000 ft²	0.37
Dryer, drum, single atmospheric	10–100 ft²	0.40
Fan, centrifugal	20,000–70,000 ft²/min	1.17
Filter, vacuum rotary drum	10–1500 ft²	0.48
Lagoon, aerated	0.05–20 million gal/day	1.13
Motor	5–20 hp	0.69
Reactor, 300 psi	100–1000 gal	0.56
Tank, atmospheric, horizontal	100–40,000 gal	0.57

EXAMPLE 2-7

Based on her work in Example 2-6, Miriam has been asked to estimate the cost today of a 2500 ft^2 heat exchange system for the new plant being analyzed. She has the following data.

- Her company paid $50,000 for a 1000 ft^2 heat exchanger 5 years ago.
- Heat exchangers within this range of capacity have a power-sizing exponent (x) of 0.55.
- Five years ago the Heat Exchanger Cost Index (HECI) was 1306; it is 1487 today.

SOLUTION

Miriam will first use Equation 2-3 to scale up the cost of the 1000 ft^2 exchanger to one that is 2500 ft^2 using the 0.55 power-sizing exponent.

$$\frac{\text{Cost of 2500 ft}^2 \text{ equipment}}{\text{Cost of 1000 ft}^2 \text{ equipment}} = \left(\frac{2500 \text{ ft}^2 \text{ equipment}}{1000 \text{ ft}^2 \text{ equipment}}\right)^{0.55}$$

$$\text{Cost of 2500 ft}^2 \text{ equipment} = \left(\frac{2500}{1000}\right)^{0.55} \times 50,000 = \$82,800$$

Miriam knows that the $82,800 reflects only the scaling up of the cost of the 1000 ft^2 model to a 2500 ft^2 model. Now she will use Equation 2-2 and the HECI data to estimate the cost of a 2500 ft^2 exchanger today. Miriam's cost estimate would be:

$$\frac{\text{Equipment cost today}}{\text{Equipment cost 5 years ago}} = \frac{\text{Index value today}}{\text{Index value 5 years ago}}$$

$$\text{Equipment cost today} = \frac{1487}{1306} \times \$82,800 = \$94,300$$

Triangulation

Triangulation is used in engineering surveying. A geographical area is divided into triangles from which the surveyor is able to map points within that region by using three fixed points and horizontal angular distances to locate fixed points of interest (e.g., property line reference points). Since any point can be located with two lines, the third line represents an extra perspective and check. We will not use trigonometry to arrive at our cost estimates, but we can use the concept of triangulation. We should approach our economic estimate from different perspectives because such varied perspectives add richness, confidence, and quality to the estimate. **Triangulation** in cost estimating might involve using different sources of data or using different quantitative models. As decision makers, we should always seek out varied perspectives.

Improvement and the Learning Curve

One common phenomenon observed, regardless of the task being performed, is that as the number of repetitions increases, performance becomes faster and more accurate. This is the concept of learning and improvement in the activities that people perform. From our own experience we all know that our fiftieth repetition is done in much less time than we needed to do the task the first time.

The **learning curve** captures the relationship between task performance and task repetition. In general, as output *doubles*, the unit production time will be reduced to some fixed percentage, the **learning curve percentage** or **learning curve rate.** For example, it may take 300 minutes to produce the third unit in a production run involving a task with a 95% learning time curve. In this case the sixth unit doubles the output, so it will take $300(0.95) = 285$ minutes to produce. Sometimes the learning curve is also known as the progress curve, improvement curve, experience curve, or manufacturing progress function.

Equation 2-4 gives an expression that can be used for time estimating for repetitive tasks.

$$T_N = T_{\text{initial}} \times N^b \tag{2-4}$$

where T_N = time required for the Nth unit of production
  T_{initial} = time required for the first (initial) unit of production
  N = number of completed units (cumulative production)
  b = learning curve exponent (slope of the learning curve on a log–log plot)

As just given, a learning curve is often referred to by its percentage learning slope. Thus, a curve with $b = -0.074$ is a 95% learning curve because $2^{-0.074} = 0.95$. This equation uses 2 because the learning curve percentage applies for doubling cumulative production. The learning curve exponent is calculated by using Equation 2-5.

$$b = \frac{\log (\text{learning curve expressed as a decimal})}{\log 2.0} \tag{2-5}$$

EXAMPLE 2-8

Calculate the time required to produce the hundredth unit of a production run if the first unit took 32.0 minutes to produce and the learning curve rate for production is 80%.

SOLUTION

Using Equation 2-4, we write

$$T_{100} = T_1 \times 100^b$$

$$T_{100} = T_1 \times 100^{\log 0.80/\log 2.0}$$

$$T_{100} = 32.0 \times 100^{-0.3219}$$

$$T_{100} = 7.27 \text{ minutes}$$

It is particularly important to account for the learning-curve effect if the production run involves a small number of units instead of a large number. When thousands or even millions of units are being produced, early inefficiencies tend to be "averaged out" because of the larger batch sizes. However, in the short run, inefficiencies of the same magnitude can lead to rather poor estimates of production time requirements, and thus production cost estimates may be understated. Consider Example 2-9 and the results that might be observed if the learning-curve effect is ignored. Notice in this example that a "steady state" time is given. Steady state is the time at which the physical constraints of performing the task prevent the achievement of any more learning or improvement.

EXAMPLE 2-9

Estimate the overall labor cost portion due to a task that has a learning-curve rate of 85% and reaches a steady state value after 16 units of 5.0 minutes per unit. Labor and benefits are $22 per hour, and the task requires two skilled workers. The overall production run is 20 units.

SOLUTION

Because we know the time required for the 16th unit, we can use Equation 2-4 to calculate the time required to produce the first unit.

$$T_{16} = T_1 \times 16^{\log 0.85/\log 2.0}$$

$$5.0 = T_1 \times 16^{-0.2345}$$

$$T_1 = 5.0 \times 16^{0.2345}$$

$$T_1 = 9.6 \text{ minutes}$$

Now we use Equation 2-4 to calculate the time requirements for the first 16 units in the production run, and we sum for the total production time required.

$$T_N = 9.6 \times N^{-0.2345}$$

Unit Number, N	Time (min) to Produce N^{th} Unit	Cumulative Time from 1 to N	Unit Number, N	Time (min) to Produce N^{th} Unit	Cumulative Time from 1 to N
1	9.6	9.6	11	5.5	74.0
2	8.2	17.8	12	5.4	79.2
3	7.4	24.2	13	5.3	84.5
4	6.9	32.1	14	5.2	89.7
5	6.6	38.7	15	5.1	94.8
6	6.3	45.0	16	5.0	99.8
7	6.1	51.1	17	5.0	104.8
8	5.9	57.0	18	5.0	109.8
9	5.7	62.7	19	5.0	114.8
10	5.6	68.3	20	5.0	119.8

The total cumulative time of the production run is 119.8 minutes (2.0 hours). Thus the total labor cost estimate would be:

$$2.0 \text{ hours} \times \$22/\text{hour per worker} \times 2 \text{ workers} = \$88$$

If we ignore the learning-curve effect and calculate the labor cost portion based only on the steady state labor rate, the estimate would be

$$0.083 \text{ hours/unit} \times 20 \text{ units} \times \$22/\text{hour per worker} \times 2 \text{ workers} = \$73.04$$

This estimate is understated by about 20% from what the true cost would be.

Figures 2-7 and 2-8 illustrate learning curves plotted on linear scales and as a *log–log* plot. On the log–log plot the relationship is a straight line, since doubling the number of units on the x axis reduces the time required by a constant percentage.

FIGURE 2-7 Learning curve of time vs. number of units.

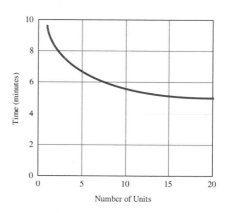

FIGURE 2-8 Learning curve on log–log scale.

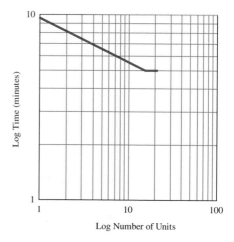

ESTIMATING BENEFITS

This chapter has focused on cost terms and cost estimating. However, engineering economists must often also estimate benefits. Benefits include sales of products, revenues from bridge tolls and electric power sales, cost reductions from reduced material or labor costs, less time spent in traffic jams, and reduced risk of flooding. Many engineering projects are undertaken precisely to secure these benefits.

The cost concepts and cost estimating models can also be applied to economic benefits. Fixed and variable benefits, recurring and nonrecurring benefits, incremental benefits, and life-cycle benefits all have meaning. Also, issues regarding the type of estimate (rough, semidetailed, and detailed), as well as difficulties in estimation (one of a kind, time and effort, and estimator expertise), all apply directly to estimating benefits. Last, per-unit, segmented, and indexed models are used to estimate benefits. The concept of triangulation is particularly important for estimating benefits.

The uncertainty in benefit estimates is also typically asymmetric, with a broader limit for negative outcomes. Benefits are more likely to be overestimated than underestimated, so an example set of limits might be (–50%, +20%). One difference between cost and benefit estimation is that many costs of engineering projects occur in the near future (for design and construction), but the benefits are further in the future. Because benefits are often further in the future, they are more difficult to estimate accurately, and more uncertainty is typical.

The estimation of economic benefits for inclusion in our analysis is an important step that should not be overlooked. Many of the models, concepts, and issues that apply in the estimation of costs also apply in the estimation of economic benefits.

CASH FLOW DIAGRAMS

The costs and benefits of engineering projects occur over time and are summarized on a cash flow diagram (CFD). Specifically, a CFD illustrates the size, sign, and timing of individual cash flows. In this way the CFD is the basis for engineering economic analysis.

A **cash flow diagram** is created by first drawing a segmented time-based horizontal line, divided into time units. The time units on the CFD can be years, months, quarters, or any other consistent time unit. Then at each time at which a cash flow will occur, a vertical arrow is added—pointing down for costs and up for revenues or benefits. These cash flows are drawn to scale.

The cash flows are **assumed** to occur at time 0 or at the **end** of each period. Consider Figure 2-9, the CFD for a specific investment opportunity whose cash flows are described as follows:

Timing of Cash Flow	Size of Cash Flow
At time zero (now or today)	A positive cash flow of $100
1 time period from today	A negative cash flow of $100
2 time periods from today	A positive cash flow of $100
3 time periods from today	A negative cash flow of $150
4 time periods from today	A negative cash flow of $150
5 time periods from today	A positive cash flow of $50

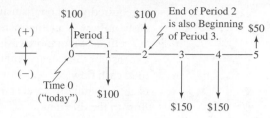

FIGURE 2-9 An example cash flow diagram (CFD).

Categories of Cash Flows

The expenses and receipts due to engineering projects usually fall into one of the following categories:

First cost $\equiv$ expense to build or to buy and install

Operations and maintenance (O&M) $\equiv$ annual expense, such as electricity, labor, and minor repairs

Salvage value $\equiv$ receipt at project termination for sale or transfer of the equipment (can be a salvage cost)

Revenues $\equiv$ annual receipts due to sale of products or services

Overhaul $\equiv$ major capital expenditure that occurs during the asset's life

Individual projects will often have specific costs, revenues, or user benefits. For example, annual operations and maintenance (O&M) expenses on an assembly line might be divided into direct labor, power, and other. Similarly, a public sector dam project might have its annual benefits divided into flood control, agricultural irrigation, and recreation.

Drawing a Cash Flow Diagram

The cash flow diagram shows when all cash flows occur. Look at Figure 2-9 and the $100 positive cash flow at the end of period 2. From the time line one can see that this cash flow can also be described as occurring at the *beginning* of period 3. Thus, in a CFD the end of *period t* is the same time as the beginning of *period t + 1*. Beginning-of-period cash flows (such as rent, lease, and insurance payments) are thus easy to handle: just draw your CFD and put them in where they occur. Thus O&M, salvages, revenues, and overhauls are assumed to be end-of-period cash flows.

The choice of time 0 is arbitrary. For example, it can be when a project is analyzed, when funding is approved, or when construction begins. When construction periods are assumed to be short, first costs are assumed to occur at time 0, and the first annual revenues and costs start at the end of the first period. When construction periods are long, time 0 is usually the date of commissioning—when the facility comes on stream.

Perspective is also important when one is drawing a CFD. Consider the simple transaction of paying $5000 for some equipment. To the firm buying the equipment, the cash flow is a cost and hence negative in sign. To the firm selling the equipment, the cash flow is a revenue and positive in sign. This simple example shows that a consistent perspective is required when one is using a CFD to model the cash flows of a problem. One person's cash outflow is another person's inflow.

Often two or more cash flows occur in the same year, such as an overhaul and an O&M expense or the salvage value and the last year's O&M expense. Combining these into one total cash flow per year would simplify the cash flow diagram. However, it is better to show each individually, to ensure a clear connection from the problem statement to each cash flow in the diagram.

Drawing Cash Flow Diagrams with a Spreadsheet

One simple way to draw cash flow diagrams with "arrows" proportional to the size of the cash flows is to use a spreadsheet to draw a stacked bar chart. The data for the cash flows are entered, as shown in the table part of Figure 2-10. To make a quick graph, select cells B1 to D8, which are the three columns of the cash flow. Then select the graph menu and choose column chart and select the stack option. Except for labeling axes (using the cells for year 0 to year 6), choosing the scale for the y axis, and adding titles, the cash flow diagram is done. Refer to the appendix for a review of basic spreadsheet use. (*Note:* A bar chart labels periods rather than using an *x* axis with arrows at times 0, 1, 2, . . .)

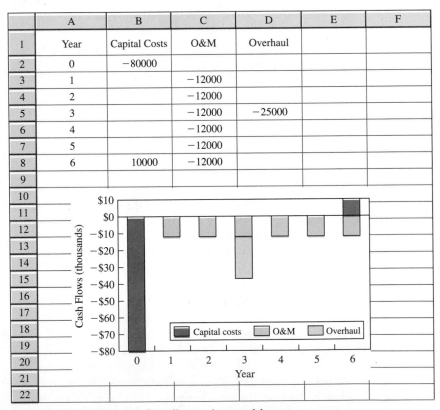

	A	B	C	D	E	F
1	Year	Capital Costs	O&M	Overhaul		
2	0	−80000				
3	1		−12000			
4	2		−12000			
5	3		−12000	−25000		
6	4		−12000			
7	5		−12000			
8	6	10000	−12000			
9						

FIGURE 2-10 Example of cash flow diagram in spreadsheets.

SUMMARY

This chapter has introduced the following cost concepts: fixed and variable, marginal and average, sunk, opportunity, recurring and nonrecurring, incremental, cash and book, and life-cycle. **Fixed costs** are constant and unchanging as volumes change, while **variable costs** change as output changes. Fixed and variable costs are used to find a breakeven value between costs and revenues, as well as the regions of net profit and loss. A **marginal cost** is for one more unit, while the **average cost** is the total cost divided by the number of units.

Sunk costs result from past decisions; they should not influence our attitude toward current and future opportunities. Remember, "sunk costs are sunk." **Opportunity costs** involve the benefit that is forgone when we choose to use a resource in one activity instead of another. **Recurring costs** can be planned and anticipated expenses; **nonrecurring costs** are one-of-a-kind costs that are often more difficult to anticipate.

Incremental costs are economic consequences associated with the differences between two choices of action. **Cash costs** are also known as **out-of-pocket costs** that represent actual cash flows. **Book costs** do not result in the exchange of money, but rather are costs listed in a firm's accounting books. **Life-cycle costs** are all costs that are incurred over the life of a product, process, or service. Thus engineering designers must consider life-cycle costs when choosing materials and components, tolerances, processes, testing, safety, service and warranty, and disposal.

Cost estimating is the process of "developing the numbers" for engineering economic analysis. Unlike a textbook, the real world does not present its challenges with neat problem statements that provide all the data. **Rough estimates** give us order-of-magnitude numbers and are useful for high-level and initial planning as well as judging the feasibility of alternatives. **Semidetailed estimates** are more accurate than rough-order estimates, thus requiring more resources (people, time, and money) to develop. These estimates are used in preliminary design and budgeting. **Detailed estimates** generally have an accuracy of -3% to 5%. They are used during the detailed design and contract bidding phases of a project.

Difficulties are common in developing estimates. **One-of-a-kind estimates** will have no basis in earlier work, but this disadvantage can be addressed through **estimation by analogy.** Lack of time is best addressed by planning and by matching the estimate's detail to the purpose—one should not spend money developing a detailed estimate when only a rough estimate is needed. **Estimator expertise** must be developed through work experiences and mentors.

Several general models and techniques for developing cost estimates were discussed. The **per-unit** and **segmenting models** use different levels of detail and costs per square foot or other unit. **Cost index data** are useful for updating historical costs to formulate current estimates. The **power-sizing model** is useful for scaling up or down a known cost quantity to account for economies of scale, with different power-sizing exponents for industrial plants and equipment of different types. **Triangulation** suggests that one should seek varying perspectives when developing cost estimates. Different information sources, databases, and analytical models can all be used to create unique perspectives. As the number of task repetitions increases, efficiency improves because of learning or improvement. This is summarized in the **learning curve percentage,** where doubling the cumulative production reduces the time to complete the task, which equals the learning curve percentage times the current production time.

Cash flow estimation must include project benefits. These include labor cost savings, avoided quality costs, direct revenue from sales, reduced catastrophic risks, improved traffic flow, and cheaper power supplies. **Cash flow diagrams** are used to model the positive and negative cash flows of potential investment opportunities. These diagrams provide a consistent view of the problem (and the alternatives) to support economic analysis.

PROBLEMS

Fixed, Variable, Average, and Marginal Costs

2-1 One of your firm's suppliers discounts prices for larger quantities. The first 1000 parts are $13 each. The next 2000 are $12 each. All parts in excess of 3000 cost $11 each. What are the average cost and marginal cost per part for the following quantities?

(a) 500

(b) 1500

(c) 2500

(d) 3500

2-2 A new machine comes with 100 free service hours over the first year. Additional time costs $75 per hour. What are the average and marginal costs per hour for the following quantities?

(a) 75

(b) 125

(c) 250

2-3 Venus Computer can produce 23,000 personal computers a year on its daytime shift. The fixed manufacturing costs per year are $2 million and the total labor cost is $9,109,000. To increase its production to 46,000 computers per year, Venus is considering adding a second shift. The unit labor cost for the second shift would be 25% higher than the day shift, but the total fixed manufacturing costs would increase only to $2.4 million from $2 million.

(a) Compute the unit manufacturing cost for the daytime shift.

(b) Would adding a second shift increase or decrease the unit manufacturing cost at the plant?

2-4 A small machine shop, with 30 hp of connected load, purchases electricity at the following monthly rates (assume any demand charge is included in this schedule):

First 50 kWh per hp of connected load at 8.6¢ per kWh

Next 50 kWh per hp of connected load at 6.6¢ per kWh

Next 150 kWh per hp of connected load at 4.0¢ per kWh

All electricity over 250 kWh per hp of connected load at 3.7¢ per kWh

The shop uses 2800 kWh per month.

(a) Calculate the monthly bill for this shop. What are the marginal and average costs per kilowatt-hour?

(b) Suppose Jennifer, the proprietor of the shop, has the chance to secure additional business that will require her to operate her existing equipment more hours per day. This will use an extra 1200 kWh per month. What is the lowest figure that she might reasonably consider to be the "cost" of this additional energy? What is this per kilowatt-hour?

(c) She contemplates installing certain new machines that will reduce the labor time required on certain operations. These will increase the connected load by 10 hp, but since they will operate only on certain special jobs, will add only 100 kWh per month. In a study to determine the economy of installing these new machines, what should be considered as the "cost" of this energy? What is this per kilowatt-hour?

2-5 Two automatic systems for dispensing maps are being compared by the state highway department. The accompanying breakeven chart of the comparison of these systems (System I vs. System II) shows total yearly costs for the number of maps dispensed per year for both alternatives. Answer the following questions.

(a) What is the fixed cost for System I?

(b) What is the fixed cost for System II?

(c) What is the variable cost per map dispensed for System I?

(d) What is the variable cost per map dispensed for System II?

(e) What is the breakeven point in terms of maps dispensed at which the two systems have equal annual costs?

(f) For what range of annual number of maps dispensed is System I recommended?

(g) For what range of annual number of maps dispensed is System II recommended?

(h) At 3000 maps per year, what are the marginal and average map costs for each system?

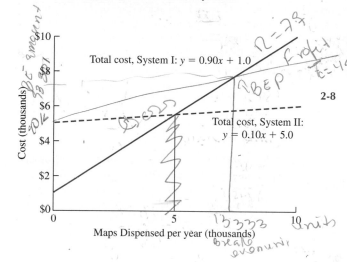

Total cost, System I: $y = 0.90x + 1.0$

Total cost, System II: $y = 0.10x + 5.0$

Cost (thousands)

Maps Dispensed per year (thousands)

2-6 A privately owned summer camp for youngsters has the following data for a 12-week session:

Charge per camper	$120 per week
Fixed costs	$48,000 per session
Variable cost per camper	$80 per week
Capacity	200 campers

(a) Develop the mathematical relationships for total cost and total revenue.

(b) What is the total number of campers that will allow the camp to *just break even*?

(c) What is the profit or loss for the 12-week session if the camp operates at 80% capacity?

(d) What are marginal and average costs per camper at 80% capacity?

2-7 Two new rides are being compared by a local amusement park in terms of their annual operating costs. The two rides are assumed to be able to generate the same level of revenue (thus the focus on costs). The Tummy Tugger has fixed costs of $10,000 per year and variable costs of $2.50 per visitor. The Head Buzzer has fixed costs of $4000 per year and variable costs of $4 per visitor. Provide answers to the following questions so the amusement park can make the needed comparison.

(a) Mathematically determine the breakeven number of visitors per year for the two rides to have equal annual costs.

(b) Develop a breakeven graph that illustrates the following:

- Accurate total cost lines for the two alternatives (show line, slopes, and equations).
- The breakeven point for the two rides in terms of number of visitors.
- The ranges of visitors per year where each alternative is preferred.

2-8 Consider the accompanying breakeven graph for an investment, and answer the following questions.

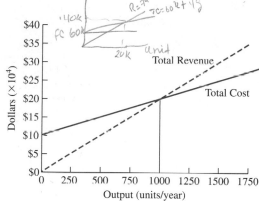

Total Revenue

Total Cost

Dollars ($\times 10^4$)

Output (units/year)

(a) Give the equation for total revenue for x units per year.

(b) Give the equation for total costs for x units per year.

(c) What is the "breakeven" level of x?

(d) If you sell 1500 units this year, will you have a profit or loss? How much?

(e) At 1500 units, what are your marginal and average costs?

2-9 Quatro Hermanas, Inc. is investigating implementing some new production machinery as part of its operations. Three alternatives have been identified, and they have the following fixed and variable costs:

Alternative	Annual Fixed Costs	Annual Variable Costs per Unit
A	$100,000	$20.00
B	200,000	5.00
C	150,000	7.50

Determine the ranges of production (units produced per year) over which each alternative would be recommended. Be exact. (*Note:* Consider the range of production to be 0–30,000 units per year.)

2-10 Three alternative designs have been created by Snakisco engineers for a new machine that spreads cheese between the crackers in a Snakisco snack. The costs for the three designs (where x is the annual production rate of boxes of cheese crackers) follow:

Design	Fixed Cost	Variable Cost ($/x$)
A	$100,000	20.5x
B	350,000	10.5x
C	600,000	8.0x

(a) Management is interested in the production interval of 0–150,000 boxes of crackers per year. Mathematically determine the production volume over which each design (A or B or C) would be chosen.

(b) Depict your solution from part (a) graphically, clearly labeling your axes and including a *title* for the graph, so that management can easily see the following:

    i. Accurate total cost lines for each alternative (show line, slopes, and line equations).

    ii. Any relevant breakeven, or crossover points.

    iii. Ranges of annual production where each alternative is preferred.

2-11 A painting operation is performed by a production worker at a labor cost of $1.40 per unit. A robot spray-painting machine, costing $15,000, would reduce the labor cost to $0.20 per unit. If the device would be valueless at the end of 3 years, what is the minimum number of units that would have to be painted each year to justify the purchase of the robot machine?

2-12 Company A has fixed expenses of $15,000 per year and each unit of product has a $0.002 variable cost. Company B has fixed expenses of $5000 per year and can produce the same product at a $0.05 variable cost. At what number of units of annual production will Company A have the same overall cost as Company B?

2-13 Mr. Sam Spade, the president of Ajax, recently read in a report that a competitor named Bendix has the following relationship between cost and production quantity:

$$C = \$3,000,000 - \$18,000Q + \$75Q^2$$

where C = total manufacturing cost per year and Q = number of units produced per year.

    A newly hired employee, who previously worked for Bendix, tells Mr. Spade that Bendix is now producing 110 units per year. If the selling price remains unchanged, Sam wonders if Bendix is likely to increase the number of units produced per year, in the near future. He asks you to look at the information and tell him what you are able to deduce from it.

2-14 A small company manufactures a certain product. Variable costs are $20 per unit and fixed costs are $10,875. The price–demand relationship for this product is $P = -0.25D + 250$, where P is the unit sales price of the product and D is the annual demand. Use the data (and helpful hints) that follow to work out answers to parts (a)–(e).

- Total cost = Fixed cost + Variable cost
- Revenue = Demand × Price
- Profit = Revenue − Total cost

Set up your graph with dollars on the y axis (between 0 and $70,000) and, on the x axis, demand D: (units produced or sold), between 0 and 1000 units.

(a) Develop the equations for total cost and total revenue.

(b) Find the breakeven quantity (in terms of profit and loss) for the product.

(c) What profit would the company obtain by maximizing its total revenue?

(d) What is the company's maximum possible profit?

(e) Neatly graph the solutions from parts (a), (b), (c), and (d).

2-15 A firm believes the sales volume (S) of its product depends on its unit selling price (P) and can be determined from the equation $P = \$100 - S$. The cost ($C$) of producing the product is $\$1000 + 10S$.

(a) Draw a graph with the sales volume (S) from 0 to 100 on the x axis, and total cost and total income from $0 to $2500 on the y axis. On the graph draw the line $C = \$1000 + 10S$. Then plot the curve of total income [which is sales volume (S) × unit selling price ($\$100 - S$)]. Mark the breakeven points on the graph.

(b) Determine the breakeven point (lowest sales volume at which total sales income just equals total production cost). (*Hint:* This may be done by trial and error or by using the quadratic equation to locate the point at which profit is zero.)

(c) Determine the sales volume (S) at which the firm's profit is a maximum. (*Hint:* Write an equation for profit and solve it by trial and error, or as a minima–maxima calculus problem.)

Sunk and Other Costs

2-16   A pump has failed in a facility that will be completely replaced in 3 years. A brass pump costing $6000 installed will last 3 years. However, a used stainless steel pump that should last 3 more years has been sitting in the maintenance shop for a year. The pump cost $13,000 new. The accountants say the pump is worth $7000 now. The maintenance supervisor says that it will cost an extra $500 to reconfigure the pump for the new use and that he could sell it used (as is) for $4000.

(a) What is the book cost of the stainless steel pump?

(b) What is the opportunity cost of the stainless steel pump?

(c) How much cheaper or more expensive would it be to use the stainless steel pump rather than a new brass pump?

2-17   Last year to help with your New Year's resolutions you purchased a $500 piece of fitness equipment. However, you use it only once a week on average. It is December, and you can sell the equipment for $200 (to someone with a New Year's resolution) and rely on the university gym until you graduate in May. If you don't sell until May, you will get only $100. If you keep the heavy piece, you'll have to pay $25 to move it to the city of your new job (where you interned last summer). There is no convenient gym at the new location. What costs and intangible consequences are relevant to your decision? What should you do?

2-18   Bob Johnson decided to buy a new home. After looking at tracts of new homes, he decided that a custom-built home was preferable. He hired an architect to prepare the drawings. In due time, the architect completed the drawings and submitted them. Bob liked the plans; he was less pleased that he had to pay the architect a fee of $4000 to design the house. Bob asked a building contractor to provide a bid to construct the home on a lot Bob already owned. While the contractor was working to assemble the bid, Bob came across a book of standard house plans. In the book was a home that he and his wife liked better than the one designed for them by the architect. Bob paid $75 and obtained a complete set of plans for this other house. Bob then asked the contractor to provide a bid to construct this "stock plan" home. In this way Bob felt he could compare the costs and make a decision. The building contractor submitted the following bids:

Custom-designed home	$128,000
Stock-plan home	128,500

Bob was willing to pay the extra $500 for it. Bob's wife, however, felt they should go ahead with the custom-designed home, for, as she put it, "We can't afford to throw away a set of plans that cost $4000." Bob agreed, but he disliked the thought of building a home that is less desirable than the stock plan home. Then he asked your advice. Which house would you advise him to build? Explain.

2-19   Consider the situation of owning rental properties that local university students rent from you for the academic year. Develop a set of costs that you could classify as recurring and others that could be classified as nonrecurring.

2-20   List and classify your costs in this academic year as recurring or nonrecurring.

2-21   Define the difference between a "cash cost" and a "book cost." Is engineering economic analysis concerned with both types of cost? Give an example of each, and provide the context in which it is important.

2-22   In your own words, develop a statement of what the authors mean by "life-cycle costs." Is it important for a firm to be aware of life-cycle costs? Explain why.

2-23   Most engineering students own a computer. What costs have you incurred at each stage of your computer's life cycle? Estimate the total cost of ownership. Estimate the benefits of ownership. Has it been worth it?

2-24 Estimating benefits is often more difficult than cost estimation. Use the example of car ownership to describe the complicating factors in estimating the costs and benefits.

2-25 In looking at Figures 2-4 and 2-5, restate in your own words what the authors are trying to get across with these graphs. Do you agree that this is an important effect for companies? Explain.

2-26 In the text we describe three effects that complicate the process of making estimates to be used in engineering economy analyses. List these three effects and comment on which might be the most influential.

2-27 Develop an estimate for each of the following situations.

(*a*) The cost of a 500-mile automobile trip, if gasoline is $3 per gallon, vehicle wear and tear is $0.50 per mile, and our vehicle gets 20 miles per gallon.

(*b*) The total number of hours in the average human life, if the average life is 75 years.

(*c*) The number of days it takes to travel around the equator using a hot air balloon, if the balloon averages 100 miles per day, the diameter of the earth is ~4000 miles.

(*d*) The total area in square miles of the United States of America, if Kansas is an average-sized state. Kansas has an area of 390 miles × 200 miles.

Segmenting Models

2-28 Northern Tundra Telephone (NTT) has received a contract to install emergency phones along a new 100-mile section of the Snow-Moose Turnpike. Fifty emergency phone systems will be installed about 2 miles apart. The material cost of a unit is $125. NTT will need to run underground communication lines that cost NTT $7500 per mile (including labor) to install. There will also be a one-time cost of $10,000 to network these phones into NTT's current communication system.

(*a*) Develop a cost estimate of the project from NTT's perspective.

(*b*) If NTT adds a profit margin of 35% to its costs, how much will it cost the state to fund the project?

2-29 You and your spouse are planning a second honeymoon to the Cayman Islands this summer and would like to have your house painted while you are away. Estimate the total cost of the paint job from the information given below, where:

$$\text{Cost}_{\text{total}} = \text{Cost}_{\text{paint}} + \text{Cost}_{\text{labor}} + \text{Cost}_{\text{fixed}}$$

Paint information: Your house has a surface area of 6000 ft^2. One can of paint can cover 300 ft^2. You are estimating the cost to put on *two coats* of paint for the entire house.

Number of Cans Purchased	Cost per Can
First 10 cans purchased	$15.00
Second 15 cans purchased	$10.00
Up to next 50 cans purchased	$7.50

Labor information: You plan to hire five painters who will paint for 10 hours per day each. You estimate that the job will require 4.5 days of their painting time. The painter's rate is $8.75 per hour.

Fixed cost information: The painting company charges $200 per job to cover travel expenses, clothing, cloths, thinner, administration, and so on.

2-30 You want a mountain cabin built for weekend trips, vacations, to host family, and perhaps eventually to retire in. After discussing the project with a local contractor, you receive an estimate that the total construction cost of your 2000 ft^2 lodge will be $150,000. The percentage of costs is broken down as follows:

Cost Items	Percentage of Total Costs
Construction permits, legal and title fees	8
Roadway, site clearing, preparation	15
Foundation, concrete, masonry	13
Wallboard, flooring, carpentry	12
Heating, ventilation, air conditioning	13
Electric, plumbing, communications	10
Roofing, flooring	12
Painting, finishing	17
	100

(a) What is the cost per square foot of the 2000 ft^2 lodge?

(b) If you are also considering a 4000 ft^2 layout option, estimate your construction costs if:

i. All cost items (in the table) change proportionately to the size increase.

ii. The first two cost items do not change at all; all others are proportionate.

2-31 SungSam, Inc. is designing a new digital camcorder that is projected to have the following per-unit costs to manufacture:

Cost Categories	Unit Costs
Materials costs	$112
Labor costs	85
Overhead costs	213
Total unit cost	$410

SungSam adds 30% to its manufacturing cost for corporate profit.

(a) What unit profit would SungSam realize on each camcorder?

(b) What is the overall cost to produce a batch of 10,000 camcorders?

(c) What would SungSam's profit be on the batch of 10,000 if historical data show that 1% of product will be scrapped in manufacturing, 3% of finished product will go unsold, and 2% of *sold* product will be returned for refund?

(d) How much can SungSam afford to pay for a contract that would lock in a 50% reduction in the unit material cost previously given? If SungSam does sign the contract, the sales price will not change.

Indexes and Sizing Models

2-32 Fifty years ago, Grandma Bell purchased a set of gold-plated dinnerware for $55, and last year you inherited it. Unfortunately a fire at your home destroyed the set. Your insurance company is at a loss to define the replacement cost and has asked your help. You do some research and find that the Aurum Flatware Cost Index (AFCI) for gold-plated dinnerware, which was 112 when Grandma Bell bought her set, is at 2050 today. Use the AFCI to update the cost of Bell's set to today's cost to show to the insurance company.

2-33 Your boss is the director of reporting for the Athens County Construction Agency (ACCA). It has been his job to track the cost of construction in Athens County. Twenty-five years ago he created the ACCA Cost Index to track these costs. Costs during the first year of the index were $12 per square foot of constructed space (the index value was set at 100 for that first year). This past year a survey of contractors revealed that costs were $72 per square foot. What index number will your boss publish in his report for this year? If the index value was 525 last year, what was the cost per square foot last year?

2-34 A new aerated sewage lagoon is required in a small town. Earlier this year one was built on a similar site in an adjacent city for $2.3 million. The new lagoon will be 60% larger. Use the data in Table 2-1 to estimate the cost of the new lagoon.

2-35 A refinisher of antiques named Constance has been so successful with her small business that she is planning to expand her shop. She is going to start enlarging her shop by purchasing the following equipment.

Equipment	Original Size	Cost of Original Equipment	Power-Sizing Exponent	New Equipment Size
Varnish bath	50 gal	$3500	0.80	75 gal
Power scraper	3/4 hp	$250	0.22	1.5 hp
Paint booth	3 ft^3	$3000	0.6	12 ft^3

What would be the *net* cost to Constance to obtain this equipment—assume that she can trade the old equipment in for 15% of its original cost. Assume there has been no inflation in equipment prices.

2-36 Refer to Problem 2-35 and now assume the prices for the equipment that Constance wants to replace have not been constant. Use the cost index data for each piece of equipment to update the costs to the price that would be paid today. Develop the overall cost for Constance, again assuming the 15% trade-in allowance for the old equipment. Trade-in value is based on original cost.

Original Equipment	Cost Index When Originally Purchased	Cost Index Today
Varnish bath	154	171
Power scraper	780	900
Paint booth	49	76

2-37 Five years ago, when the relevant cost index was 120, a nuclear centrifuge cost $40,000. The centrifuge had a capacity of separating 1500 gallons of ionized solution per hour. Today, it is desired to build a centrifuge with capacity of 4500 gallons per hour, but the cost index now is 300. Assuming a power-sizing exponent to reflect economies of scale, x, of 0.75, use the power-sizing model to determine the approximate cost (expressed in today's dollars) of the new reactor.

2-38 Padre works for a trade magazine that publishes lists of *Power-Sizing Exponents (PSE)* that reflect economies of scale for developing engineering estimates of various types of equipment. Padre has been unable to find any published data on the VMIC machine and wants to list its *PSE* value in his next issue. Given the following data (your staff was able to find data regarding costs and sizes of the VMIC machine) calculate the *PSE* value that Padre should publish. (*Note:* The VMIC-100 can handle twice the volume of a VMIC-50.)

> Cost of VMIC-100 today $100,000
> Cost of VMIC-50 5 years ago $45,000
> VMIC equipment index today $= 214$
> VMIC equipment index 5 years ago $= 151$

Learning Curve Models

2-39 If 200 labor hours was required to produce the 1st unit in a production run and 60 labor hours was required to produce the 7th unit, what was the *learning curve rate* during production?

2-40 Rose is a project manager at the civil engineering consulting firm of Sands, Gravel, Concrete, and Waters, Inc. She has been collecting data on a project in which concrete pillars were being constructed; however not all the data are available. She has been able to find out that the 10th pillar required 260 person-hours to construct, and that a 75% learning curve applied. She is interested in calculating the time required to construct the 1st and 20th pillars. Compute the values for her.

2-41 Sally Statistics is implementing a system of statistical process control (SPC) charts in her factory in an effort to reduce the overall cost of scrapped product. The current cost of scrap is X per month. If an 80% learning curve is expected in the use of the SPC charts to reduce the cost of scrap, what would the *percentage reduction* in monthly scrap cost be after the charts have been used for 12 months? (*Hint:* Model each month as a unit of production.)

2-42 Randy Duckout has been asked to develop an estimate of the *per-unit selling price* (the price that each unit will be sold for) of a new line of handcrafted booklets that offer excuses for missed appointments. His assistant Doc Duckout has collected information that Randy will need in developing his estimate:

Cost of direct labor	$20 per hour
Cost of materials	$43.75 per batch of 25 booklets
Cost of overhead items	50% of direct labor cost
Desired profit	20% of total manufacturing cost

Doc also finds out that (1) they should use a 75% learning curve for estimating the cost of direct labor, (2) the time to complete the 1st booklet is estimated at 0.60 hour, and (3) the estimated time to complete the 25th booklet should be used as their standard time for the purpose of determining the *unit selling price*. What would Randy and Doc's estimate be for the *unit selling price*?

Benefit Estimation

2-43 Develop a statement that expresses the extent to which cost estimating topics also apply to estimating benefits. Provide examples to illustrate.

Cash Flow Diagrams

2-44 On December 1, Al Smith purchased a car for $18,500. He paid $5000 immediately and agreed to pay three additional payments of $6000 each (which includes principal and interest) at the end of 1, 2, and 3 years. Maintenance for the car is projected at $1000 at the end of the first year and $2000 at the end of each subsequent year. Al expects to sell the car at the end of the fourth year (after paying for the maintenance work) for $7000. Using these facts, prepare a table of cash flows.

2-45 Bonka Toys is considering a robot that will cost $20,000 to buy. After 7 years its salvage value will be $2000. An overhaul costing $5000 will be needed in Year 4. O&M costs will be $2500 per year. Draw the cash flow diagram.

2-46 Pine Village needs some additional recreation fields. Construction will cost $225,000, and annual O&M expenses are $85,000. The city council estimates that the value of added youth leagues is about $190,000 annually. In Year 6 another $75,000 will be needed to refurbish the fields. The salvage value is estimated to be $100,000 after 10 years. Draw the cash flow diagram.

2-47 Identify your major cash flows for the current school term as first costs, O&M expenses, salvage values, revenues, overhauls, and so on. Using a week as the time period, draw the cash flow diagram.

Trust Me, You'll Use This!

As an industrial engineer for a high-profile corporation, Nick Cotton makes a good living, but he's figured out a way to make his skills work double-time.

Cotton's engineering economics class taught him how to turn his passion for math into profit. At 19, he began to invest in stocks and mutual funds, making use of principles in his undergraduate text. "I was able to calculate the future value of stocks and so could predict how well each would do," he says. He now owns two houses—one of which is still being built and has already appreciated by 15%.

Being the enterprising type, Cotton realized that he could market his solid investment skills. He started a real estate business that has turned out to be as lucrative as his day job. When he breaks down potential and interest for properties up for sale, he demonstrates to his clients how a seemingly small percentage increase in worth each year can become a significant amount of money because of compounding.

Cotton is in prime position to become a tycoon, and he has engineering economics to thank. "No other class has contributed so much to my life," he says. "Who doesn't like making money?"

Trust Me, You'll Use This!

When Kim Brock enrolled in Engineering Economy, she had no idea she would ever use the skills taught in the course, let alone apply them to a career in aeronautics.

"I was just flipping through a magazine and saw an ad for Boeing—thought I might as well send them a résumé," she says.

As it turned out, Brock's understanding of big-picture problem solving made her an ideal job candidate. She now works as an industrial engineer for the Boeing 767 line, designing, improving, and controlling systems of people, materials, and information.

She says that Engineering Economy was a particularly valuable course because it taught her how to project the favorable and unfavorable consequences of a situation. When she conducts design meetings and process improvement workshops, she helps determine the safest and most cost-effective options for Boeing mechanics, right down to the way they hold a drill. "Recently, I had to offer advice on building a fixture to check that a part was working," she says. "After running some analyses, I realized that the fixture would cost more than the part itself."

It's this kind of commonsense know-how that Engineering Economy imparts. Brock still owns her copy of an earlier edition of *Engineering Economic Analysis* and says that it's been helpful as she considers buying a condo: "It's a book about the reality of making decisions."

CHAPTER 3

INTEREST AND EQUIVALENCE

Going Up in Smoke

State governments throughout the United States agreed to a large financial windfall in 1998. Tobacco companies agreed to pay in perpetuity to settle claims arising from the health effects of smoking. Payments over the first 25 years are estimated to be nearly $250 billion.

State officials announced they would earmark these funds for goals such as health care, education and, of course, antismoking campaigns. But several states involved in the settlement were chronically short of money and were desperate to plug budget deficits. They wanted money now, instead of waiting for the payments to dribble in year by year.

That's when someone hit on the idea of raising instant money by issuing "tobacco bonds." The states would sell bonds to investors and pay them interest out of the tobacco settlement payments the states were receiving. State governments could get the whole sale price of the bond up front; investors would get ongoing income from the bond.

A growing number of states are now selling these bonds, and pocketing quick billions. To attract buyers, however, the states have to pay a high rate of interest on their tobacco bonds, since investors view them as riskier than other government-backed securities. Investors reason that tobacco companies could go broke, after all—especially if the no-smoking laws being passed all over the country cause enough people to snuff out their cigarette habit for good.

QUESTIONS TO CONSIDER

1. Tobacco bonds often pay interest of as much as 7%, while conventional state government bonds may pay only 5%. On a $1 million bond with a term of 25 years, how much does this interest differential amount to, in total?

2. Assuming a 2% interest differential, over a 25-year period, how much are states losing by opting to take the settlement money "up front" instead of collecting the long-term payout?

3. What are the potential ethical issues linked with determining who receives the benefits of the tobacco settlements?

4. When the money they get from selling bonds runs out, state governments will have to find a new way of closing their budget gaps. Some commentators predicted that states would soon look to other business sectors as "deep pockets." The tobacco industry was particularly vulnerable because it sold a product that came to be seen as dangerous and socially undesirable. What other industries might face the same fate in years to come? How about fast-food restaurant chains, or sellers of alcohol products? Should a company like Anheuser-Busch be worried? Should companies in potentially "problematic" industries consider setting aside money for future settlement costs?

After Completing This Chapter...

The student should be able to:

- Define and provide examples of the *time value of money*.
- Distinguish between *simple* and *compound interest,* and use compound interest in engineering economic analysis.
- Explain *equivalence* of cash flows.
- Solve problems using the single payment compound interest formulas.

I n the first chapter, we discussed situations where the economic consequences of an alternative were immediate or took place in a very short period of time, as in Example 1-2 (design of a concrete aggregate mix) or Example 1-3 (change of manufacturing method). We totaled the various positive and negative aspects, compared our results, and could quickly reach a decision. But can we do the same if the economic consequences occur over a considerable period of time?

No we cannot, because *money has value over time.* Would you rather (1) receive $1000 today or (2) receive $1000 ten years from today? Obviously, the $1000 today has more value. Money's value over time is expressed by an interest rate. This chapter describes two introductory concepts involving the *time value of money:* interest and cash flow equivalence.

COMPUTING CASH FLOWS

Installing expensive machinery in a plant obviously has economic consequences that occur over an extended period of time. If the machinery were bought on credit, the simple process of paying for it may take several years. What about the usefulness of the machinery? Certainly it was purchased because it would be a beneficial addition to the plant. These favorable consequences may last as long as the equipment performs its useful function. In these circumstances, we describe each alternative as cash **receipts** or **disbursements** at different points in **time.** Since earlier cash flows are more valuable than later cash flows, we cannot just add them together. Instead, each alternative is resolved into a set of **cash flows,** as in Examples 3-1 and 3-2.

EXAMPLE 3-1

The manager has decided to purchase a new $30,000 mixing machine. The machine may be paid for in one of two ways:

1. Pay the full price now *minus* a 3% discount.
2. Pay $5000 now; at the end of one year, pay $8000; at the end of each of the next four years, pay $6000.

List the alternatives in the form of a table of cash flows.

SOLUTION

The first plan represents a lump sum of $29,100 now, the second one calls for payments continuing until the end of the fifth year. Those payments total $37,000, but the $37,000 value has no clear meaning, since it has added together cash flows occurring at different times. The second alternative can only be described as a set of *cash flows.* Both alternatives are shown in the cash flow table, with disbursements given negative signs.

End of Year	Pay in Full Now	Pay Over 5 Years
0 (now)	−$29,100	−$5000
1	0	−8000
2	0	−6000
3	0	−6000
4	0	−6000
5	0	−6000

As a cash flow diagram, that is:

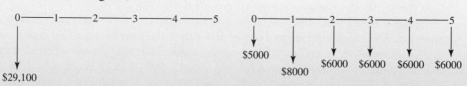

EXAMPLE 3-2

A man borrowed $1000 from a bank at 8% interest. He agreed to repay the loan in two end-of-year payments. At the end of each year, he will repay half of the $1000 principal amount plus the interest that is due. Compute the borrower's cash flow.

SOLUTION

In engineering economic analysis, we normally refer to the beginning of the first year as "time 0." At this point the man receives $1000 from the bank. (A positive sign represents a receipt of money and a negative sign, a disbursement.) The time 0 cash flow is +$1000.

At the end of the first year, the man pays 8% interest for the use of $1000 for that year. The interest is $0.08 \times \$1000 = \$80$. In addition, he repays half the $1000 loan, or $500. Therefore, the end-of-year-1 cash flow is −$580.

At the end of the second year, the payment is 8% for the use of the balance of the principal ($500) for the one-year period, or $0.08 \times 500 = \$40$. The $500 principal is also repaid for a total end-of-year-2 cash flow of −$540. The cash flow is:

End of Year	Cash Flow
0 (now)	+$1000
1	−580
2	−540

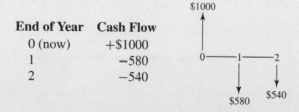

In this chapter, we will demonstrate techniques for comparing the value of money at different dates, which is the foundation of engineering economic analysis. We must be able to compare, for example, a low-cost pump with a higher-cost pump. If there were no other consequences, we would obviously prefer the low-cost one. But what if the higher-cost pump is more efficient and/or more durable? Then we must consider whether to spend more money now to reduce future power costs and/or to postpone the future cost of a replacement pump. This chapter will provide the methods for comparing the alternatives to determine which pump is preferred.

TIME VALUE OF MONEY

We often find that the monetary consequences of any alternative occur over a substantial period of time—say, a year or more. When monetary consequences occur in a short period of time, we simply add up the various sums of money and obtain a net result. But can we treat money this way when the time span is greater?

Which would you prefer, $100 cash today or the assurance of receiving $100 a year from now? You might decide you would prefer the $100 now because that is one way to be certain of receiving it. But suppose you were convinced that you would receive the $100 one year hence. Now what would be your answer? A little thought should convince you that it *still* would be more desirable to receive the $100 now. If you had the money now, rather than a year hence, you would have the use of it for an extra year. And if you had no current use for $100, you could let someone else use it.

Money is quite a valuable asset—so valuable that people are willing to pay to have money available for their use. Money can be rented in roughly the same way one rents an apartment; only with money, the charge is called **interest** instead of rent. The importance of interest is demonstrated by banks and savings institutions continuously offering to pay for the use of people's money, to pay interest.

If the current interest rate is 9% per year and you put $100 into the bank for one year, how much will you receive back at the end of the year? You will receive your original $100 together with $9 interest, for a total of $109. This example demonstrates the time preference for money: we would rather have $100 today than the assured promise of $100 one year hence; but we might well consider leaving the $100 in a bank if we knew it would be worth $109 one year hence. This is because there is a **time value of money** in the form of the willingness of banks, businesses, and people to pay interest for the use of various sums.

Simple Interest

Simple interest is interest that is computed only on the original sum, not on accrued interest. Thus if you were to loan a present sum of money P to someone at a simple annual interest rate i (stated as a decimal) for a period of n years, the amount of interest you would receive from the loan would be:

$$\text{Total interest earned} = P \times i \times n = Pin \qquad (3\text{-}1)$$

At the end of n years the amount of money due you, F, would equal the amount of the loan P plus the total interest earned. That is, the amount of money due at the end of the loan would be

$$F = P + Pin \qquad (3\text{-}2)$$

or $F = P(1 + in)$.

EXAMPLE 3-3

You have agreed to loan a friend $5000 for 5 years at a simple interest rate of 8% per year. How much interest will you receive from the loan? How much will your friend pay you at the end of 5 years?

$$\text{Total interest earned} = Pin = (\$5000)(0.08)(5 \text{ yr}) = \$2000$$

$$\text{Amount due at end of loan} = P + Pin = 5000 + 2000 = \$7000$$

In Example 3-3 the interest earned at the end of the first year is $(5000)(0.08)(1) = \$400$, but this money is not paid to the lender until the end of the fifth year. As a result, the borrower has the use of the $400 for 4 years without paying any interest on it. This is how simple interest works, and it is easy to see why lenders seldom agree to make simple interest loans.

Compound Interest

With simple interest, the amount earned (for invested money) or due (for borrowed money) in one period does not affect the principal for interest calculations in later periods. However, this is not how interest is normally calculated. In practice, interest is computed by the **compound interest** method. For a loan, any interest owed but not paid at the end of the year is added to the balance due. Then the next year's interest is calculated on the unpaid balance due, which includes the unpaid interest from the preceding period. In this way, compound interest can be thought of as *interest on top of interest*. This distinguishes compound interest from simple interest. In this section, the remainder of the book, and in practice you should **assume that the rate is a compound interest rate**. The very few exceptions will clearly state "simple interest."

EXAMPLE 3-4

To highlight the difference between simple and compound interest, rework Example 3-3 using an interest rate of 8% per year compound interest. How will this change affect the amount that your friend pays you at the end of 5 years?

Original loan amount (original principal) = $5000

Loan term = 5 years

Interest rate charged = 8% per year compound interest

In the following table we calculate on a year-to-year basis the total dollar amount due at the end of each year. Notice that this amount becomes the principal upon which interest is calculated in the next year (this is the compounding effect).

Year	Total Principal (P) on Which Interest Is Calculated in Year n	Interest (I) Owed at the End of Year n from Year n's Unpaid Total Principal	Total Amount Due at the End of Year n, New Total Principal for Year n+1
1	$5000	$5000 × 0.08 = 400	$5000 + 400 = 5400
2	5400	5400 × 0.08 = 432	5400 + 432 = 5832
3	5832	5832 × 0.08 = 467	5832 + 467 = 6299
4	6299	6299 × 0.08 = 504	6299 + 504 = 6803
5	6803	6803 × 0.08 = 544	6803 + 544 = 7347

The total amount due at the end of the fifth year is $7347. This is $347 more than you received for loaning the same amount, for the same period, at simple interest. This is because of the effect of interest being earned on top of interest.

Repaying a Debt

To better understand the mechanics of interest, let us say that $5000 is owed and is to be repaid in 5 years, together with 8% annual interest. There are a great many ways in which debts are repaid; for simplicity, we have selected four specific ways for our example. Table 3-1 tabulates the four plans.

Plan 1, like Example 3-2, repays $1/n^{\text{th}}$ of the principal each year. So in Plan 1, $1000 will be paid at the end of each year plus the interest due at the end of the year for the use of money to that point. Thus, at the end of Year 1, we will have had the use of $5000. The interest owed is 8% × $5000 = $400. The end-of-year payment is $1000 principal *plus* $400 interest, for a total payment of $1400. At the end of Year 2, another $1000 principal plus interest will be repaid on the money owed during the year. This time the amount owed has declined from $5000 to $4000 because of the $1000 principal payment at the end of Year 1. The interest payment is 8% × $4000 = $320, making the end-of-year payment a total of $1320. As indicated in Table 3-1, the series of payments continues each year until the loan is fully repaid at the end of the Year 5.

In *Plan 2* only the interest due is paid each year, with no principal payment. Instead, the $5000 owed is repaid in a lump sum at the end of the fifth year. The end-of-year payment in each of the first 4 years of Plan 2 is 8% × $5000 = $400. The fifth year, the payment is $400 interest *plus* the $5000 principal, for a total of $5400.

Plan 3 calls for five equal end-of-year payments of $1252 each. At this point, we have not shown how the figure of $1252 was computed (see later: Example 4-3). However, it is clear that there is some equal end-of-year amount that would repay the loan. By following the computations in Table 3-1, we see that a series of five payments of $1252 repays a $5000 debt in 5 years with interest at 8%.

Plan 4 repays the $5000 debt like Example 3-4. In this plan, no payment is made until the end of Year 5, when the loan is completely repaid. Note what happens at the end of

TABLE 3-1 Four Plans for Repayment of $5000 in 5 Years with Interest at 8%

(a) Year	(b) Amount Owed at Beginning of Year	(c) Interest Owed for That Year, 8% × (b)	(d) Total Owed at End of Year, (b) + (c)	(e) Principal Payment	(f) Total End-of-Year Payment
Plan 1: At end of each year pay $1000 principal *plus* interest due.					
1	$5000	$ 400	$5400	$1000	$1400
2	4000	320	4320	1000	1320
3	3000	240	3240	1000	1240
4	2000	160	2160	1000	1160
5	1000	80	1080	1000	1080
		$1200		$5000	$6200
Plan 2: Pay interest due at end of each year and principal at end of 5 years.					
1	$5000	$ 400	$5400	$ 0	$ 400
2	5000	400	5400	0	400
3	5000	400	5400	0	400
4	5000	400	5400	0	400
5	5000	400	5400	5000	5400
		$2000		$5000	$7000
Plan 3: Pay in five equal end-of-year payments.					
1	$5000	$ 400	$5400	$ 852	$1252*
2	4148	331	4479	921	1252
3	3227	258	3485	994	1252
4	2233	178	2411	1074	1252
5	1159	93	1252	1159	1252
		$1260		$5000	$6260
Plan 4: Pay principal and interest in one payment at end of 5 years.					
1	$5000	$ 400	$5400	$ 0	$ 0
2	5400	432	5832	0	0
3	5832	467	6299	0	0
4	6299	504	6803	0	0
5	6803	544	7347	5000	7347
		$2347		$5000	$7347

*The exact value is $1252.28, which has been rounded to an even dollar amount.

Year 1: the interest due for the first year—8% × $5000 = $400—is not paid; instead, it is added to the debt. At the second year then, the debt has increased to $5400. The Year 2 interest is thus 8% × $5400 = $432. This amount, again unpaid, is added to the debt, increasing it further to $5832. At the end of Year 5, the total sum due has grown to $7347 and is paid at that time.

Note that when the $400 interest was not paid at the end of Year 1, it was added to the debt and, in Year 2 there was interest charged on this unpaid interest. That is, the $400 of unpaid interest resulted in 8% × $400 = $32 of additional interest charge in Year 2. That $32, together with 8% × $5000 = $400 interest on the $5000 original debt, brought the total interest charge at the end of the Year 2 to $432. Charging interest on unpaid interest is called **compound interest.**

With Table 3-1 we have illustrated four different ways of accomplishing the same task, that is, to repay a debt of $5000 in 5 years with interest at 8%. Having described the alternatives, we will now use them to present the important concept of *equivalence*.

EQUIVALENCE

When we are indifferent as to whether we have a quantity of money now or the assurance of some other sum of money in the future, or series of future sums of money, we say that the present sum of money is **equivalent** to the future sum or series of future sums.

If an industrial firm believed 8% was a reasonable interest rate, it would have no particular preference about whether it received $5000 now or was repaid by Plan 1 of Table 3-1. Thus $5000 today is equivalent to the series of five end-of-year payments. In fact, *all four repayment plans must be equivalent to each other and to $5000 now at 8% interest.*

Equivalence is an essential factor in engineering economic analysis. In Chapter 2, we saw how an alternative could be represented by a cash flow table. How might two alternatives with different cash flows be compared? For example, consider the cash flows for Plans 1 and 2 from Table 3-1:

Year	Plan 1	Plan 2
1	−$1400	−$400
2	−1320	−400
3	−1240	−400
4	−1160	−400
5	−1080	−5400
	−$6200	−$7000

If you were given your choice between the two alternatives, which one would you choose? Obviously the two plans have cash flows that are different, and you cannot compare the −$6200 and −$7000. Plan 1 requires that there be larger payments in the first 4 years, but the total payments are smaller. To make a decision, we must use the **technique of equivalence**.

We can determine an **equivalent value** at some point in time for Plan 1 and a **comparable equivalent value** for Plan 2, based on a selected interest rate. Then we can judge the relative attractiveness of the two alternatives, not from their cash flows, but from comparable equivalent values. Since Plan 1, like Plan 2, repays a *present* sum of $5000 with interest at 8%, the plans are equivalent to $5000 *now;* therefore, the alternatives are equally attractive. This cannot be deduced from the given cash flows alone. It is necessary to learn this by determining the equivalent values for each alternative at some point in time, which in this case is "the present."

Difference in Repayment Plans

The four plans computed in Table 3-1 are equivalent in nature but different in structure. Table 3-2 repeats the end-of-year payment schedules from Table 3-1 and also graphs each

TABLE 3-2 End-of-Year Payment Schedules and Their Graphs

Plan 1: At end of each year pay $1000 principal *plus* interest due.

Year	End-of-Year Payment
1	$1400
2	1320
3	1240
4	1160
5	1080
	$6200

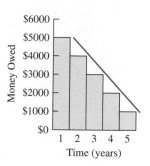

Plan 2: Pay interest due at end of each year and principal at end of 5 years.

Year	End-of-Year Payment
1	$ 400
2	400
3	400
4	400
5	5400
	$7000

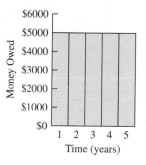

Plan 3: Pay in five equal end-of-year payments.

Year	End-of-Year Payment
1	$1252
2	1252
3	1252
4	1252
5	1252
	$6260

Plan 4: Pay principal and interest in one payment at end of 5 years.

Year	End-of-Year Payment
1	$ 0
2	0
3	0
4	0
5	7347
	$7347

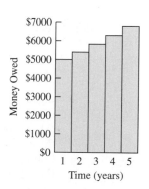

plan to show the debt still owed at any point in time. Since $5000 was borrowed at the beginning of the first year, all the graphs begin at that point. We see, however, that the four plans result in quite different amounts of money owed at any other point in time. In Plans 1 and 3, the money owed declines as time passes. With Plan 2 the debt remains constant, while Plan 4 increases the debt until the end of the fifth year. These graphs show an important difference among the repayment plans—the areas under the curves differ greatly. Since the axes are *Money Owed* and *Time,* the area is their product: Money owed × Time, in years.

In the discussion of the time value of money, we saw that the use of money over a time period was valuable, that people are willing to pay interest to have the use of money for periods of time. When people borrow money, they are acquiring the use of money as represented by the area under the curve for Money owed versus Time. It follows that, at a given interest rate, the amount of interest to be paid will be proportional to the area under the curve. Since in each case the $5000 loan is repaid, the interest for each plan is the total *minus* the $5000 principal:

Plan	Total Interest Paid
1	$1200
2	2000
3	1260
4	2347

We can use Table 3-2 and the data from Table 3-1, to compute the area under each of the four curves, that is, the area bounded by the abscissa, the ordinate, and the curve itself. We multiply the ordinate (Money owed) by the abscissa (1 year) for each of the five years, then *add:*

$$\text{Shaded area} = (\text{Money owed in Year 1})(1 \text{ year})$$

$$+ (\text{Money owed in Year 2})(1 \text{ year})$$

$$+ \cdots$$

$$+ (\text{Money owed in Year 5})(1 \text{ year})$$

or

$$\text{Shaded area } [(\text{Money owed})(\text{Time})] = \textbf{dollar-years}$$

The dollar-years for the four plans would be as follows:

	Dollar-Years			
	Plan 1	Plan 2	Plan 3	Plan 4
(Money owed in Year 1)(1 year)	$ 5,000	$ 5,000	$ 5,000	$ 5,000
(Money owed in Year 2)(1 year)	4,000	5,000	4,148	5,400
(Money owed in Year 3)(1 year)	3,000	5,000	3,227	5,832
(Money owed in Year 4)(1 year)	2,000	5,000	2,233	6,299
(Money owed in Year 5)(1 year)	1,000	5,000	1,159	6,803
Total dollar-years	$15,000	$25,000	$15,767	$29,334

With the area under each curve computed in dollar-years, the ratio of total interest paid to area under the curve may be obtained:

Plan	Total Interest Paid	Area Under Curve (dollar-years)	Ratio of Total Interest Paid to Area Under Curve
1	$1200	15,000	0.08
2	2000	25,000	0.08
3	1260	15,767	0.08
4	2347	29,334	0.08

We see that the ratio of total interest paid to the area under the curve is constant and equal to 8%. Stated another way, the total interest paid equals the interest rate *times* the area under the curve.

From our calculations, we more easily see why the repayment plans require the payment of different total sums of money, yet are actually equivalent. The key factor is that the four repayment plans provide the borrower with different quantities of dollar-years. Since dollar-years times interest rate equals the interest charge, the four plans result in different total interest charges.

Equivalence Is Dependent on Interest Rate

In the example of Plans 1 to 4, all calculations were made at an 8% interest rate. At this interest rate, it has been shown that all four plans are equivalent to a present sum of $5000. But what would happen if we were to change the problem by changing the interest rate?

If the interest rate were increased to 9%, we know that the interest payment for each plan would increase, and the calculated repayment schedules (Table 3-1, column **f**) could no longer repay the $5000 debt with the higher interest. Instead, each plan would repay a sum *less* than the principal of $5000, because more money would have to be used to repay the higher interest rate. By some calculations (to be explained later in this chapter and in Chapter 4), the equivalent present sum that each plan will repay at 9% interest is:

Plan	Repay a Present Sum of
1	$4877
2	4806
3	4870
4	4775

As predicted, at the higher 9% interest each of the repayment plans of Table 3-1 repays a present sum less than $5000. But they do not repay the *same* present sum. Plan 1 would repay $4877 with 9% interest, while Plan 2 would repay $4806. Thus, with interest at 9%, Plans 1 and 2 are no longer equivalent, for they will not repay the same present sum. The two series of payments (Plan 1 and Plan 2) were equivalent at 8%, but not at 9%. This leads to the conclusion **that equivalence is dependent on the interest rate.** Changing the interest rate destroys the equivalence between two series of payments.

Could we create revised repayment schemes that would be equivalent to $5000 now with interest at 9%? Yes, of course we could: to revise Plan 1 of Table 3-1, we need to increase the total end-of-year payment in order to pay 9% interest on the outstanding debt.

Year	Amount Owed at Beginning of Year	9% Interest for Year	Total End-of-Year Payment ($1000 plus interest)
1	$5000	$450	$1450
2	4000	360	1360
3	3000	270	1270
4	2000	180	1180
5	1000	90	1090

Plan 2 of Table 3-1 is revised for 9% interest by increasing the first four payments to $9\% \times \$5000 = \$450$ and the final payment to $5450. Two plans that repay $5000 in 5 years with interest at 9% are:

Revised Year	End-of-Year Plan 1	Payments Plan 2
1	$1450	$ 450
2	1360	450
3	1270	450
4	1180	450
5	1090	5450

We have determined that Revised Plan 1 is equivalent to a present sum of $5000 and Revised Plan 2 is equivalent to $5000 now; it follows that at 9% interest, Revised Plan 1 is equivalent to Revised Plan 2.

Application of Equivalence Calculations

To understand the usefulness of equivalence calculations, consider the following:

Year	Alternative A: Lower Initial Cost, Higher Operating Cost	Alternative B: Higher Initial Cost, Lower Operating Cost
0 (now)	−$600	−$850
1	−115	−80
2	−115	−80
3	−115	−80
.	.	.
.	.	.
.	.	.
10	−115	−80
	−$1750	−$1650

Is the least-cost alternative the one that has the lower initial cost and higher operating costs or the one with higher initial cost and lower continuing costs? Because of the time value of money, one cannot add up sums of money at different points in time directly. This means that alternatives cannot be compared in actual dollars at different points in time; instead comparisons must be made in some equivalent comparable sums of money.

It is not sufficient to compare the initial $600 against $850, nor the $1750 against the $1650. Instead, we must compute equivalent values that may be validly compared. The methods for accomplishing this will be presented later in this chapter and in Chapter 4.

Thus far we have discussed computing equivalent present sums for a cash flow. But the technique of equivalence is not limited to a present computation. Instead, we could compute the equivalent sum for a cash flow at any point in time. We could compare alternatives in "Equivalent Year 10" dollars rather than "now" (Year 0) dollars. Further, the equivalence need not be a single sum; it could be a series of payments or receipts. In Plan 3 of Table 3-1, the series of equal payments was equivalent to $5000 now. But the equivalency works both ways. Suppose we ask, What is the equivalent equal annual payment continuing for 5 years, given a present sum of $5000 and interest at 8%? The answer is $1252.

SINGLE PAYMENT COMPOUND INTEREST FORMULAS

To facilitate equivalence computations, a series of **interest formulas** will be derived. We use the following notation:

i = *interest rate per interest period;* in the equations the interest rate is stated as a decimal (that is, 9% interest is 0.09)

n = *number of interest periods*

P = *a present sum of money*

F = *a future sum of money;* the future sum F is an amount, n interest periods from the present, that is equivalent to P with interest rate i

Suppose a present sum of money P is invested for one year[1] at interest rate i. At the end of the year, we should receive back our initial investment P, together with interest equal to iP, or a total amount $P + iP$. Factoring P, the sum at the end of one year is $P(1 + i)$.

Let us assume that, instead of removing our investment at the end of one year, we agree to let it remain for another year. How much would our investment be worth at the end of Year 2? The end-of-first-year sum $P(1+i)$ will draw interest in the second year of $iP(1+i)$. This means that, at the end of Year 2, the total investment will be

$$P(1 + i) + i[P(1 + i)]$$

---

[1]A more general statement is to specify "one interest period" rather than "one year." Since it is easier to visualize one year, the derivation uses one year as the interest period.

This may be rearranged by factoring $P(1 + i)$, which gives

$$P(1 + i)(1 + i) \qquad \text{or} \qquad P(1 + i)^2$$

If the process is continued for Year 3, the end-of-the-third-year total amount will be $P(1+i)^3$; at the end of n years, we will have $P(1 + i)^n$. The progression looks like:

Year	Amount at Beginning of Interest Period	+ Interest for Period	= Amount at End of Interest Period
1	P	$+iP$	$= P(1 + i)$
2	$P(1 + i)$	$+iP(1 + i)$	$= P(1 + i)^2$
3	$P(1 + i)^2$	$+iP(1 + i)^2$	$= P(1 + i)^3$
n	$P(1 + i)^{n-1}$	$+iP(1 + i)^{n-1}$	$= P(1 + i)^n$

In other words, a present sum P increases in n periods to $P(1 + i)^n$. We therefore have a relationship between a present sum P and its equivalent future sum, F.

$$\text{Future sum} = (\text{Present sum}) (1 + i)^n$$

$$F = P(1 + i)^n \tag{3-3}$$

This is the **single payment compound amount formula** and is written in functional notation as

$$F = P(F/P, i, n) \tag{3-4}$$

The notation in parentheses $(F/P, i, n)$ can be read as follows:

To find a future sum F, given a present sum, P, at an interest rate i per interest period, and n interest periods hence.

or, Find F, given P, at i, over n.

Functional notation is designed so that the compound interest factors may be written in an equation in an algebraically correct form. In Equation 3-4, for example, the functional notation is interpreted as

$$F = P\left(\frac{F}{P}\right)$$

which is dimensionally correct. Without proceeding further, we can see that when we derive a compound interest factor to find a present sum P, given a future sum F, the factor will be $(P/F, i, n)$; so, the resulting equation would be

$$P = F(P/F, i, n)$$

which is dimensionally correct.

EXAMPLE 3-5

If \$500 were deposited in a bank savings account, how much would be in the account 3 years hence if the bank paid 6% interest compounded annually?

SOLUTION

From the viewpoint of the person depositing the \$500, the cash flows are:

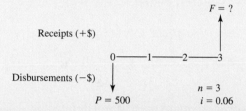

The present sum P is \$500. The interest rate per interest period is 6%, and in 3 years there are three interest periods. The future sum F is found by using Equation 3-3, where $P = \$500, i = 0.06, n = 3$, and F is unknown:

$$F = P(1 + i)^n = 500(1 + 0.06)^3 = \$595.50$$

Thus if we deposit \$500 now at 6% interest, there will be \$595.50 in the account in three years.

ALTERNATE SOLUTION

The equation $F = P(1 + i)^n$ need not be solved. Instead, *the single payment compound amount factor,* $(F/P, i, n)$, is readily found in the tables given in Appendix B.[2] In this case the factor is

$$(F/P, 6\%, 3)$$

Knowing $n = 3$, locate the proper row in the 6% table of Appendix B. To find F given P, look in the first column, which is headed "Single Payment, Compound Amount Factor": or F/P for $n = 3$, we find 1.191.
  Thus,

$$F = 500(F/P, 6\%, 3) = 500(1.191) = \$595.50$$

Before leaving this problem, let's draw another diagram of it, this time from the bank's point of view.

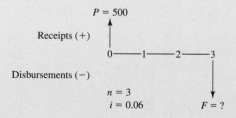

---

[2]Appendix B contains compound interest tables for rates between 1/4 and 60%.

This indicates the bank receives $500 now and must make a disbursement of F at the end of 3 years. The computation, from the bank's point of view, is

$$F = 500(F/P, 6\%, 3) = 500(1.191) = \$595.50$$

This is exactly the same as what was computed from the depositor's viewpoint, since this is just the other side of the same transaction. The bank's future disbursement equals the depositor's future receipt.

If we take $F = P(1+i)^n$ and solve for P, then

$$P = F\frac{1}{(1+i)^n} = F(1+i)^{-n}$$

This is the **single payment present worth formula.** The equation

$$P = F(1+i)^{-n} \tag{3-5}$$

in our notation becomes

$$P = F(P/F, i, n) \tag{3-6}$$

EXAMPLE 3-6

If you wish to have $800 in a savings account at the end of 4 years, and 5% interest will be paid annually, how much should you put into the savings account now?

SOLUTION

$$F = \$800 \qquad i = 0.05 \qquad n = 4 \qquad P = \text{unknown}$$

$$P = F(1+i)^{-n} = 800(1+0.05)^{-4} = 800(0.8227) = \$658.16$$

Thus to have $800 in the savings account at the end of 4 years, we must deposit $658.16 now.

ALTERNATE SOLUTION

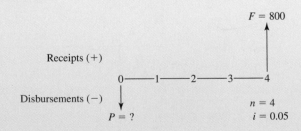

$$P = F(P/F, i, n) = \$800(P/F, 5\%, 4)$$

From the compound interest tables,

$$(P/F, 5\%, 4) = 0.8227$$

$$P = \$800(0.8227) = \$658.16$$

Here the problem has an exact answer. In many situations, however, the answer is rounded off, since it can be only as accurate as the input information on which it is based.

EXAMPLE 3-7

Suppose the bank changed the interest policy in Example 3-5 to "6% interest, compounded quarterly." For this situation, how much money would be in the account at the end of 3 years, assuming a $500 deposit now?

SOLUTION

First, we must be certain to understand the meaning of *6% interest, compounded quarterly*. There are two elements:

6% interest: Unless otherwise described, it is customary to assume that the stated interest is for a one-year period. *If the stated interest is for any other period, the time frame must be clearly stated.*

Compounded quarterly: This indicates there are four interest periods per year; that is, an interest period is 3 months long.

We know that the 6% interest is an annual rate because if it were for a different period, it would have been stated. Since we are dealing with four interest periods per year, it follows that the interest rate per interest period is $1\frac{1}{2}\%$ (=6%/4). For the total 3-year duration, there are 12 interest periods. Thus

$$P = \$500 \qquad i = 0.015 \qquad n = (4 \times 3) = 12 \qquad F = \text{unknown}$$

$$F = P(1 + i)^n = P(F/P, i, n)$$

$$= \$500(1 + 0.015)^{12} = \$500(F/P, 1\frac{1}{2}\%, 12)$$

$$= \$500(1.196) = \$598$$

A $500 deposit now would yield $598 in 3 years.

EXAMPLE 3-8

Consider the following situation, where the borrowing of P is repaid in two payments:

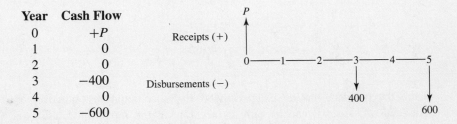

Year	Cash Flow
0	+P
1	0
2	0
3	−400
4	0
5	−600

Solve for P assuming a 12% interest rate and using the compound interest tables.

SOLUTION

$$P = 400(P/F, 12\%, 3) + 600(P/F, 12\%, 5)$$
$$= 400(0.7118) + 600(0.5674)$$
$$= \$625.16$$

Let's examine the computations further.

If $625.16 is borrowed for one year at 12% interest, the balance due will increase to $[625.16 + 0.12(625.16)] = \$700.18$. If for the second year the $700.18 is borrowed at 12%, the balance due will increase to $[700.18 + 0.12(700.18)] = \$784.20$. And if this is repeated for another year, the balance due is $[784.20 + 0.12(784.20)] = \$878.30$.

We are now at the end of Year 3. The original $625.16 has increased through the addition of interest to $878.30. It is at this point that $400 is repaid. Deducting $400 from $878.30 leaves $478.30 as the balance due.

The $478.30 can be borrowed at 12% for the fourth year and will increase to $[478.30 + 0.12(478.30)] = \$535.70$. And if left at interest for another year, it will increase to $[535.70 + 0.12(535.70)] = \$600$. We are now at the end of Year 5; with a $600 payout; there is no balance due.

In other words, the $625.16 was just enough money, at a 12% interest rate, to exactly provide for a $400 disbursement at the end of Year 3 and also a $600 disbursement at the end of Year 5. We end up neither short of money nor with money left over: this is an illustration of equivalence. The initial $625.16 is *equivalent* to the combination of a $400 disbursement at the end of Year 3 and a $600 disbursement at the end of Year 5.

ALTERNATE FORMATION OF EXAMPLE 3-8

There is another way to see what the $625.16 value of P represents.

Suppose at Year 0 you were offered a piece of paper that guaranteed you would be paid $400 at the end of 3 years and $600 at the end of 5 years. How much would you be willing to pay for this piece of paper if you wanted your money to produce a 12% interest rate?

This alternate statement of the problem changes the signs in the cash flow and the diagram:

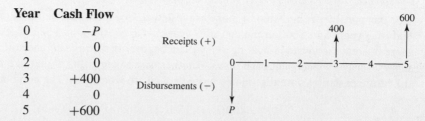

Year	Cash Flow
0	−P
1	0
2	0
3	+400
4	0
5	+600

Since the goal is to recover our initial investment P together with 12% interest per year, we can see that P must be *less* than the total amount to be received in the future (that is, $400 + 600 = \$1000$). We must calculate the present sum P that is *equivalent,* at 12% interest, to an aggregate of $400 in 3 years and $600 in 5 years.

Using Equation 3-5

$$P = (1 + i)^{-n}F$$

we write

$$P = 400(1 + 0.12)^{-3} + 600(1 + 0.12)^{-5}$$
$$= \$625.17$$

This is virtually the same amount computed from the first statement of this example. [The slight difference is due to the rounding in the compound interest tables. For example, $(1 + 0.12)^{-5} = 0.567427$, but the compound interest table for 12% shows 0.5674.]

Both problems in Example 3-8 have been solved by computing the value of P that is equivalent to $400 at the end of Year 3 and $600 at the end of Year 5. In fact, the two problems could represent the buyer and seller of the same piece of paper. The seller would receive $+P$ at Year 0 while the buyer would pay $-P$. Thus, while the problems looked different, they could have been one situation examined first from the viewpoint of the seller and then from that of the buyer. Either way, the solution is based on an equivalence computation.

EXAMPLE 3-9

The second set of cash flows in Example 3-8 was:

Year	Cash Flow
0	−P
1	0
2	0
3	+400
4	0
5	+600

At a 12% interest rate, P was computed to be $625.17. Suppose the interest rate is increased to 15%. Will the value of P be larger or smaller?

One can consider P as a sum of money invested at 15% from which one is to obtain $400 at the end of 3 years and $600 at the end of 5 years. At 12%, the P is $625.17. At 15%, P will earn more interest each year, indicating that we can begin with a *smaller P* and still accumulate enough money for the subsequent cash flows. Another way to view this is that at a higher interest rate, the future cash flows are discounted more so they are worth less. The computation is:

$$P = 400(P/F, 15\%, 3) + 600(P/F, 15\%, 5)$$
$$= 400(0.6575) + 600(0.4972)$$
$$= \$561.32$$

The value of P is smaller at 15% than at 12% interest.

SUMMARY

This chapter describes cash flow tables, the time value of money, and equivalence. The single payment compound interest formulas were derived. It is essential that these concepts and the use of the interest formulas be understood, since the remainder of this book and the practice of engineering economy are based on them.

Time value of money: The continuing offer of banks to pay interest for the temporary use of other people's money is ample proof that there is a time value of money. Thus, we would always choose to receive $100 today rather than the promise of $100 to be paid at a future date.

Equivalence: What sum would a person be willing to accept a year hence instead of $100 today? If a 9% interest rate is considered to be appropriate, he would require $109 a year hence. If $100 today and $109 a year hence are considered equally desirable, we say the two sums of money are equivalent. But, if on further consideration, we decided that a 12% interest rate is applicable, then $109 a year hence would no longer be equivalent to $100 today. This illustrates that equivalence is dependent on the interest rate.

Single Payment Formulas

These formulas are for compound interest, which is used in engineering economy.

Compound amount $F = P(1 + i)^n = P(F/P, i, n)$

Present worth $P = F(1 + i)^{-n} = F(P/F, i, n)$

where i = interest rate per interest period (stated as a decimal)
n = number of interest periods
P = a present sum of money
F = a future sum of money; the future sum F is an amount, n interest periods from the present, that is equivalent to P with interest rate i

This chapter also defined simple interest, where interest does not carry over and become part of the principal in subsequent periods. Unless otherwise specified, all interest rates in this text are compound rates.

PROBLEMS

Equivalence

3-1   In your own words explain the *time value of money*. From your own life (either now or in a situation that might occur in your future), provide an example in which the time value of money would be important.

3-2   Which is more valuable, $20,000 received now or $5000 per year for 4 years? Why?

3-3   Explain the difference between simple and compound interest. Which is more common?

3-4   Magdalen, Miriam, and Mary June were asked to consider two different cash flows: $500 that they could receive today and $1000 that would be received 3 years from today. Magdalen wanted the $500 dollars today, Miriam chose to collect $1000 in 3 years, and Mary June was indifferent between these two options. Can you offer an explanation of the choice made by each woman?

3-5   A woman borrowed $2000 and agreed to repay it at the end of 3 years, together with 10% simple interest per year. How much will she pay 3 years hence?

3-6   A $5000 loan was to be repaid with 8% simple annual interest. A total of $5350 was paid. How long had the loan been outstanding?

3-7   Solve the diagram for the unknown Q assuming a 10% interest rate.

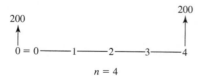

$n = 4$

(*Answer:* $Q = \$136.60$)

Single Payment Factors

3-8   A man borrowed $750 from a bank. He agreed to repay the sum at the end of 3 years, together with the interest at 8% per year. How much will he owe the bank at the end of 3 years?   (*Answer:* $945)

3-9   We know that a certain piece of equipment will cost $150,000 in 5 years. How much must be deposited today using 10% interest to pay for it?

3-10   Suppose that $2000 is deposited in an account that earns 6% interest. How much is in the account?

(*a*) After 5 years

(*b*) After 10 years

(*c*) After 20 years

(*d*) After 50 years

(*e*) After 100 years

3-11   An inheritance will be $20,000. The interest rate for the time value of money is 7%. How much is the inheritance worth now, if it will be received

(*a*) in 5 years?

(*b*) in 10 years?

(*c*) in 20 years?

(*d*) in 50 years?

3-12   Rita borrows $5000 from her parents. She repays them $6000. What is the interest rate if she pays the $6000 at the end of

(*a*) Year 2?

(*b*) Year 3?

(*c*) Year 5?

(*d*) Year 10?

3-13   How long would it take to double your money if you invest it at 4% simple interest? At 4% compound interest?

3-14   A savings account earns 8% interest. If $1000 is invested, how many years is it until each of the following amounts is on deposit?

(*a*) $1360

(*b*) $2720

(*c*) $4316

(*d*) $6848

3-15   How much must you invest now at 7.9% interest to accumulate $175,000 in 63 years?

3-16   (*a*) If $100 at Time "0" will be worth $110 a year later and was $90 a year ago, compute the interest rate for the past year and the interest rate next year.

(*b*) Assume that $90 invested a year ago will return $110 a year from now. What is the annual interest rate in this situation?

3-17 In 1995 an anonymous private collector purchased a painting by Picasso entitled *Angel Fernandez de Soto* for $29,152,000. The picture depicts Picasso's friend de Soto seated in a Barcelona cafe drinking absinthe. The painting was done in 1903 and was valued then at $600. If the painting was owned by the same family until its sale in 1995, what rate of return did they receive on the $600 investment?

3-18 The following series of payments will repay a present sum of $5000 at an 8% interest rate. Use single payment factors to find the present sum that is equivalent to this series of payments at a 10% interest rate.

End-of-Year

Year	Payment
1	$1400
2	1320
3	1240
4	1160
5	1080

3-19 What sum of money now is equivalent to $8250 two years later, if interest is 4% per 6-month period? (*Answer:* $7052)

3-20 A sum of money invested at 2% per 6-month period (semiannually), will double in amount in approximately how many years?   (*Answer:* $17\frac{1}{2}$ years)

3-21 One thousand dollars is borrowed for one year at an interest rate of 1% per month. If the same sum of money could be borrowed for the same period at an interest rate of 12% per year, how much could be saved in interest charges?

3-22 Sally Stanford is buying a car that costs $12,000. She will pay $2000 immediately and the remaining $10,000 in four annual end-of-year principal payments of $2500 each. In addition to the $2500, she must pay 15% interest on the unpaid balance of the loan each year. Prepare a cash flow table to represent this situation.

3-23 The local bank offers to pay 5% interest on savings deposits. In a nearby town, the bank pays 1.25% per 3-month period (quarterly). A man who has $3000 to put in a savings account wonders whether the higher interest paid in the nearby town justifies driving there to make the deposit. Assuming he will leave all money in the account for 2 years, how much additional interest would he obtain from the out-of-town bank over the local bank?

3-24 The following cash flows are equivalent in value if the interest rate is *i*. Which one is more valuable if the interest rate is 2*i*?

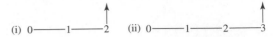

3-25 The tabulated factors stop at *n* = 100. How can they be used to calculate (*P/F, i,* 150)? (*P/F, i,* 200)?

3-26 In 1990 Mrs. John Hay Whitney sold her painting by Renoir, *Au Moulin de la Galette,* depicting an open-air Parisian dance hall, for $71 million. The buyer also had to pay the auction house commission of 10%, or a total of $78.1 million. The Whitney family had purchased the painting in 1929 for $165,000.

(*a*) What rate of return did Mrs, Whitney receive on the investment?

(*b*) Was the rate of return really as high as you computed in (*a*)? Explain.

3-27 A sum of money *Q* will be received 6 years from now. At 5% annual interest, the present worth of *Q* is $60. At the same interest rate, what would be the value of *Q* in 10 years?

3-28 The United States recently purchased $1 billion of 30-year zero-coupon bonds from a struggling foreign nation. The bonds yield $4\frac{1}{2}$% per year interest. The zero-coupon bonds pay no interest during their 30-year life. Instead, at the end of 30 years, the U.S. government is to receive back its $1 billion together with interest at $4\frac{1}{2}$% per year. A U.S. senator objected to the purchase, claiming that the correct interest rate for bonds like this is $5\frac{1}{4}$%. The result, he said, was a multimillion dollar gift to the foreign country without the approval of Congress. Assuming the senator's math is correct, how much will the foreign country have saved in interest when it repays the bonds at $4\frac{1}{2}$% instead of $5\frac{1}{4}$% at the end of 30 years?

3-29 The Apex Company sold a water softener to Marty Smith. The price of the unit was $350. Marty asked for a deferred payment plan, and a contract was written. Under the contract, the buyer could delay paying for the water softener if he purchased the coarse salt for recharging the softener from Apex. At the end of 2 years, the buyer was to pay for the unit in a lump sum, with interest at a rate of 1.5% per quarter-year. According to the contract, if the customer ceased buying salt from Apex at any time prior to 2 years, the full payment due at the end of 2 years would automatically become due.

Six months later, Marty decided to buy salt elsewhere and stopped buying from Apex, whereupon Apex asked for the full payment that was to have been due 18 months hence. Marty was unhappy about this, so Apex offered as an alternative to accept the $350 with interest at 10% per semiannual period for the 6 months that Marty had been buying salt from Apex. Which of these alternatives should Marty accept? Explain.

3-30 The local garbage company charges $6 a month for garbage collection. It had been their practice to send out bills to their 100,000 customers at the end of each 2-month period. Thus, at the end of February it would send a bill to each customer for $12 for garbage collection during January and February.

Recently the firm changed its billing date: it now sends out the 2-month bills after one month's service has been performed. Bills for January and February, for example, are sent out at the end of January. The local newspaper points out that the firm is receiving half its money before the garbage collection. This unearned money, the newspaper says, could be temporarily invested for one month at 1% per month interest by the garbage company to earn extra income.

Compute how much extra income the garbage company could earn each year if it invests the money as described by the newspaper. (*Answer:* $36,000)

CHAPTER 4

MORE INTEREST FORMULAS

Anne Scheiber's Bonanza

When Anne Scheiber died in 1995, aged 101, she left an estate worth more than $20 million. It all went to Yeshiva University in New York. The university's officials were grateful. But who, they asked, was Anne Scheiber?

She wasn't a mysterious heiress or a business tycoon, it turned out. She was a retired IRS auditor who had started investing in 1944, putting $5000—her life savings up to that point—into the stock market.

Scheiber's portfolio was not based on get-rich-quick companies. In fact, it contained mostly "garden variety" stocks like Coca-Cola and Exxon. Scheiber typically researched stock purchases carefully, and then held onto her shares for years, rather than trading them. She also continually reinvested her dividends.

Despite her increasing wealth, Scheiber never indulged in a lavish lifestyle. Acquaintances described her as extremely frugal and reported that she lived as a near recluse in her small apartment.

What motivated Scheiber to accumulate so much money? In large part, it seems to have been a reaction to her life experiences. Scheiber worked for over 20 years as a tax auditor but failed to receive promotions. At her request, the endowment she left to Yeshiva University was used to fund scholarships and interest-free loans for women students.

QUESTIONS TO CONSIDER

1. Set aside your calculator and test your intuition for a minute. What annual rate of return do you suppose it would take to turn $5000 into $20 million in 51 years? 20%, 100%?

2. Now get out the calculator and refer to an equation from Chapter 3. What was Anne Scheiber's actual average annual rate of return during the years she was investing? (Assume no added investment from her salary.)

3. How does that rate compare with the overall performance of the stock market from 1944 to 1995?

4. Anne Scheiber's endowment was given to support scholarships and loans for women. Should universities allow donors to place constraints on money they give? What ethical boundaries are appropriate?

After Completing This Chapter...
The student should be able to:

- Solve problems modeled by the uniform series compound interest formulas.
- Use arithmetic and geometric gradients to solve appropriately modeled problems.
- Apply *nominal* and *effective interest rates.*
- Use *discrete* and *continuous* compounding in appropriate contexts.
- Use spreadsheets and financial functions to model and solve engineering economic analysis problems.

hapter 3 presented the fundamental components of engineering economic analysis, including formulas to compute equivalent single sums of money at different points in time. Most problems we will encounter are much more complex. This chapter develops formulas for cash flows that are a uniform series or are increasing on an arithmetic or geometric gradient. Later in the chapter, nominal and effective interest are discussed. Finally, equations are derived for situations where interest is compounded continuously.

UNIFORM SERIES COMPOUND INTEREST FORMULAS

Many times we will find uniform series of receipts or disbursements. Automobile loans, house payments, and many other loans are based on a **uniform payment series.** Future costs and benefits are often estimated to be the *same* or *uniform* every year. This is the simplest assumption; it is often sufficiently accurate; and often there is no data to support what might be a more accurate model.

Since most engineering economy problems define "a period" as one year, the uniform cash flow is an *annual* cash flow denoted by A for all period lengths. More formally, A is defined as:

$A =$ an end-of-period cash receipt or disbursement in a uniform series, continuing for n periods

Engineering economy practice and textbooks (including this one) generally assume that cash flows after Time 0 are end-of-period cash flows. Thus the equations and tabulated values of A and F assume end-of-period timing. Tables for beginning- or middle-of-period assumptions have been built, but they are rarely used.

The horizontal line in Figure 4-1 is a representation of time with four interest periods illustrated. Uniform payments A have been placed at the end of each interest period, and there are as many A's as there are interest periods n. (Both these conditions are specified in the definition of A.) Figure 4-1 uses January 1 and December 31, but other 1-year or other length periods could be used.

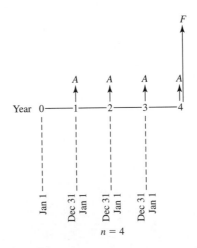

FIGURE 4-1 The general relationship between A and F.

In Chapter 3's section on single payment formulas, we saw that a sum P at one point in time would increase to a sum F in n periods, according to the equation

$$F = P(1 + i)^n$$

We will use this relationship in our uniform series derivation.

Looking at Figure 4-1, we see that if an amount A is invested at the end of each year for 4 years, the total amount F at the end of 4 years will be the sum of the compound amounts of the individual investments.

In the general case for n years,

$$F = A(1 + i)^{n-1} + \cdots + A(1 + i)^3 + A(1 + i)^2 + A(1 + i) + A \qquad (4\text{-}1)$$

Multiplying Equation 4-1 by $(1 + i)$, we have

$$(1 + i)F = A(1 + i)^n + \cdots + A(1 + i)^4$$
$$+ A(1 + i)^3 + A(1 + i)^2 + A(1 + i) \qquad (4\text{-}2)$$

Factoring out A and subtracting Equation 4-1 gives

$$(1 + i)F = A[(1 + i)^n + \cdots + (1 + i)^4 + (1 + i)^3 + (1 + i)^2 + (1 + i)] \qquad (4\text{-}3)$$
$$- \quad F = A[(1 + i)^{n-1} + \cdots + (1 + i)^3 + (1 + i)^2 + (1 + i) + 1]$$

$$iF = A[(1 + i)^n - 1] \qquad (4\text{-}4)$$

Solving Equation 4-4 for F gives

$$F = A\left[\frac{(1 + i)^n - 1}{i}\right] = A(F/A, i\%, n) \qquad (4\text{-}5)$$

Thus we have an equation for F when A is known. The term inside the brackets

$$\left[\frac{(1 + i)^n - 1}{i}\right]$$

is called the **uniform series compound amount factor** and has the notation $(F/A, i, n)$.

EXAMPLE 4-1

You deposit $500 in a credit union at the end of each year for 5 years. The credit union pays 5% interest, compounded annually. Immediately after the fifth deposit, how much can you withdraw from your account?

SOLUTION

The diagram on the left shows the situation from your point of view; the one on the right, from the credit union's point of view. Either way, the diagram of the five deposits and the desired computation of the future sum F duplicates the situation for the uniform series compound amount formula

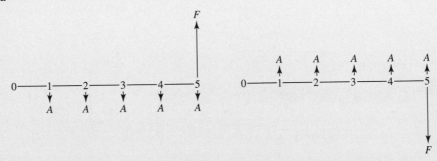

$$F = A\left[\frac{(1+i)^n - 1}{i}\right] = A(F/A, i\%, n)$$

where $A = \$500, n = 5, i = 0.05, F =$ unknown. Filling in the known variables gives

$$F = \$500(F/A, 5\%, 5) = \$500(5.526) = \$2763$$

There will be $2763 in the account following the fifth deposit.

If Equation 4-5 is solved for A, we have

$$A = F\left[\frac{i}{(1+i)^n - 1}\right]$$

$$= F(A/F, i\%, n) \tag{4-6}$$

where

$$\left[\frac{i}{(1+i)^n - 1}\right]$$

is called the **uniform series sinking fund**[1] factor and is written as $(A/F, i, n)$.

---

[1] A *sinking fund* is a separate fund into which one makes a uniform series of money deposits (A) to accumulate a desired future sum (F) by the end of period n.

EXAMPLE 4-2

Jim Hayes wants to buy some electronic equipment for $1000. Jim has decided to save a uniform amount at the end of each month so that he will have the required $1000 at the end of one year. The local credit union pays 6% interest, compounded monthly. How much does Jim have to deposit each month?

SOLUTION

In this example,

$$F = \$1000 \qquad n = 12 \qquad i = \tfrac{1}{2}\% \qquad A = \text{unknown}$$

$$A = 1000(A/F, \tfrac{1}{2}\%, 12) = 1000(0.0811) = \$81.10$$

Jim would have to deposit $81.10 each month.

If we use the sinking fund formula (Equation 4-6) and substitute for F the single payment compound amount formula (Equation 3-3), we obtain

$$A = F\left[\frac{i}{(1+i)^n - 1}\right] = P(1+i)^n\left[\frac{i}{(1+i)^n - 1}\right]$$

$$A = P\left[\frac{i(1+i)^n}{(1+i)^n - 1}\right] = P(A/P, i\%, n) \tag{4-7}$$

We now have an equation for determining the value of a series of end-of-period payments— or disbursements—A when the present sum P is known.

The portion inside the brackets

$$\left[\frac{i(1+i)^n}{(1+i)^n - 1}\right]$$

is called the **uniform series capital recovery factor** and has the notation $(A/P, i, n)$.

The name *capital recovery factor* comes from asking the question, How large does the annual return, A, have to be to "recover" the capital, P, that is invested at Time 0? In other words, find A given P. This is illustrated in Example 4-3.

EXAMPLE 4-3

An energy-efficient machine costs $5000 and has a life of 5 years. If the interest rate is 8%, how much must be saved every year to recover the cost of the capital invested in it?

SOLUTION

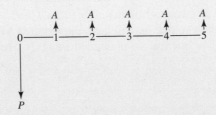

$$P = \$5000 \quad n = 5 \quad i = 8\% \quad A = \text{unknown}$$

$$A = P(A/P, 8\%, 5) = 5000(0.2505) = \$1252$$

The required annual savings to recover the capital investment is $1252.

In Example 4-3, with interest at 8%, a present sum of $5000 is equivalent to five equal end-of-period disbursements of $1252. This is another way of stating Plan 3 of Table 3-1. The method for determining the annual payment that would repay $5000 in 5 years with 8% interest has now been explained. The calculation is simply

$$A = 5000(A/P, 8\%, 5) = 5000(0.2505) = \$1252$$

If the capital recovery formula (Equation 4-7) is solved for the present sum P, we obtain the uniform series present worth formula

$$P = A\left[\frac{(1+i)^n - 1}{i(1+i)^n}\right] = A(P/A, i\%, n) \tag{4-8}$$

and

$$(P/A, i\%, n) = \left[\frac{(1+i)^n - 1}{i(1+i)^n}\right]$$

which is the **uniform series present worth factor.**

EXAMPLE 4-4

An investor holds a time payment purchase contract on some machine tools. The investor will receive $140 at the end of each month for a 5-year period. The first payment is due in one month. He offers to sell you the contract for $6800 cash today. If you otherwise can make 1% per month on your money, would you accept or reject the investor's offer?

SOLUTION

Summarizing the data in a cash flow diagram, we have:

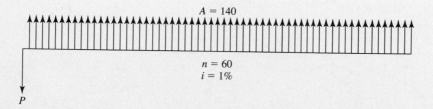

$A = 140$

$n = 60$
$i = 1\%$

P

Use the uniform series present worth formula to compute the contract's present worth.

$$P = A(P/A, i, n) = 140(P/A, 1\%, 60)$$

$$= 140(44.955)$$

$$= \$6293.70$$

Since you would be paying more than $6293.70, you would receive less than the required 1% per month interest. Reject the investor's offer.

EXAMPLE 4-5

If $6800 is paid for the time purchase contract in Example 4-4, what is the monthly rate of return?

SOLUTION

We know P, A, and n, but we do not know i. The problem may be solved by using either the uniform series present worth formula

$$P = A(P/A, i, n)$$

or the uniform series capital recovery formula

$$A = P(A/P, i, n)$$

Either way, we have one equation with one unknown.

$$P = \$6800 \qquad A = \$140 \qquad n = 60 \qquad i = \text{unknown}$$

$$P = A(P/A, i, n)$$

$$\$6800 = \$140(P/A, i, 60)$$

$$(P/A, i, 60) = \frac{6800}{140} = 48.571$$

We look through the compound interest tables to find the values of $(P/A, i, 60)$ that are closest to 48.571. Then we compute the rate of return i by interpolation. Entering values from the tables in Appendix B, we find

Interest Rate	$(P/A, i, 60)$
½%	51.726
i	48.571
¾%	48.174

The rate of return, which is between ½% and ¾%, may indeed be computed by a linear interpolation. The interest formulas are not linear, so a linear interpolation will not give an exact solution. To minimize the error, the interpolation should be computed with interest rates as close to the correct answer as possible. [Since $a/b = c/d, a = b(c/d)$], we write

$$
\begin{aligned}
\text{Rate of return} \quad i &= 0.5\% + a \\
&= 0.5\% + b(c/d) \\
&= 0.50\% + 0.25\% \left(\frac{51.726 - 48.571}{51.726 - 48.174} \right) \\
&= 0.50\% + 0.25\% \left(\frac{3.155}{3.552} \right) = 0.50\% + 0.22\% \\
&= 0.72\% \text{ per month}
\end{aligned}
$$

The monthly rate of return on our investment would be 0.72% per month.

EXAMPLE 4-6

A student is borrowing \$100 per year for 3 years. The loan will be repaid 2 years later at a 15% interest rate. Compute how much will be repaid. This is F in the following cash flow and diagram.

Year	Cash Flow
1	+100
2	+100
3	+100
4	0
5	$-F$

We see that the cash flow diagram does not match the sinking fund factor diagram: F occurs two periods later, rather than at the same time as the last A. Since the diagrams do not match, the problem is more difficult than those we've discussed so far. The approach to use in this situation is to convert the cash flow from its present form into standard forms, for which we have compound interest factors and compound interest tables.

One way to solve this problem is to consider the cash flow as a series of single payments P and then to compute their sum F. In other words, the cash flow is broken into three parts, each one of which we can solve.

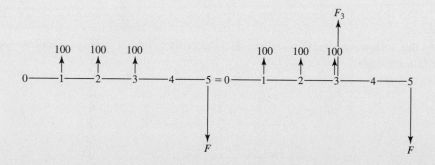

$$F = F_1 + F_2 + F_3 = 100(F/P, 15\%, 4) + 100(F/P, 15\%, 3) + 100(F/P, 15\%, 2)$$
$$= 100(1.749) + 100(1.521) + 100(1.322)$$
$$= \$459.20$$

The value of F in the illustrated cash flow is \$459.20.

A second approach is to calculate an equivalent F_3 at the end of Period 3.

Looked at this way, we first solve for F_3.

$$F_3 = 100(F/A, 15\%, 3) = 100(3.472) = \$347.20$$

Now F_3 can be considered a present sum P in the diagram

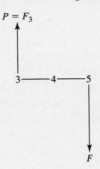

and so

$$F = F_3(F/P, 15\%, 2)$$
$$= 347.20(1.322)$$
$$= \$459.00$$

The slightly different value from the preceding computation is due to rounding in the compound interest tables.

This two-step solution can be combined into a single equation:

$$F = 100(F/A, 15\%, 3)(F/P, 15\%, 2)$$
$$= 100(3.472)(1.322)$$
$$= \$459.00$$

EXAMPLE 4-7

Consider the following situation, where P is deposited into a savings account and three withdrawals are made.

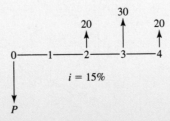

The diagram is not in a standard form, indicating that there will be a multiple-step solution. There are at least three different ways of computing the answer. (It is important that you understand how the three computations are made, so please study all three solutions.)

SOLUTION 1

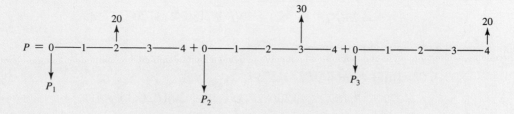

$$P = P_1 + P_2 + P_3$$
$$= 20(P/F, 15\%, 2) + 30(P/F, 15\%, 3) + 20(P/F, 15\%, 4)$$
$$= 20(0.7561) + 30(0.6575) + 20(0.5718)$$
$$= \$46.28$$

SOLUTION 2

The second approach converts the three withdrawals into an equivalent value at the end of Period 4.

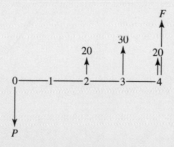

The relationship between P and F in the diagram is

$$P = F(P/F, 15\%, 4)$$

Next we compute the future sums of the three payments, as follows:

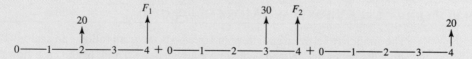

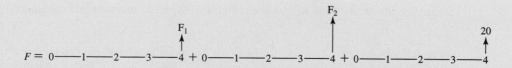

$$F = F_1 + F_2 + 20$$
$$= 20(F/P, 15\%, 2) + 30(F/P, 15\%, 1) + 20$$

Combining the two equations, we have

$$P = [F_1 + F_2 + 20](P/F, 15\%, 4)$$
$$= [20(F/P, 15\%, 2) + 30(F/P, 15\%, 1) + 20](P/F, 15\%, 4)$$
$$= [20(1.322) + 30(1.150) + 20](0.5718)$$
$$= \$46.28$$

SOLUTION 3

The third approach finds how much would have to be deposited (P_1) at $t = 1$ and then converts that into an equivalent value (P) at $t = 0$.

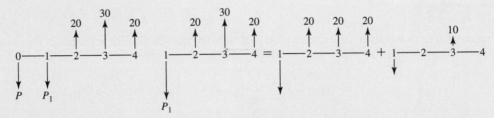

$$P = P_1(P/F, 15\%, 1) \qquad\qquad P_1 = 20(P/A, 15\%, 3) + 10(P/F, 15\%, 2)$$

Combining, we have

$$P = [20(P/A, 15\%, 3) + 10(P/F, 15\%, 2)](P/F, 15\%, 1)$$
$$= [20(2.283) + 10(0.7561)](0.8696)$$
$$= \$46.28$$

ECONOMIC EQUIVALENCE VIEWED AS A MOMENT DIAGRAM

The similarity between cash flow and free body diagrams allows an analogy that helps some students better understand economic equivalence.[2] Think of the cash flows as forces that are always perpendicular to the axis. Then the time periods become the distances along the axis.

When we are solving for unknown forces in a free body diagram, we know that a moment equation about any point will be in equilibrium. Typically moments are calculated by using

---

[2]We thank David Elizandro and Jessica Matson of Tennessee Tech for developing, testing, and describing this approach (*The Engineering Economist*, 2007, Vol. 52, No. 2, "Taking a Moment to Teach Engineering Economy," pp. 97–116).

the right-hand rule so that counterclockwise moments are positive, but it is also possible to define moments so that a clockwise rotation is positive. We are assuming that **clockwise rotations are positive**. This allows us to make the normal assumptions that positive forces point up and positive distances from the force to the pivot point are measured from left to right. Thus, negative forces point down, and negative distances are measured from the force on the right to the pivot point on the left. These assumptions are summarized in the following diagram.

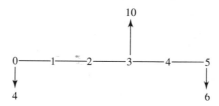

With these assumptions we can write force moment equations at equilibrium for the following diagram.

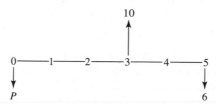

For example, in the force moment equation about Point 1, the force at 0 is –4, and the distance from the force to the pivot point is 1. Similarly the force at 5 is –6, and the distance from the force on the right to the pivot point on the left is –4. The equilibrium equation for force moments about Point 1 is:

$$0 = -4 \times 1 + 10 \times (-2) + -6 \times (-4)$$

To write the cash flow moment equation for the cash flow diagram we need:

1. A sign convention for cash flows, such that positive values point up.
2. A way to measure the moment arm for each cash flow. This moment arm must be measured as $(1 + i)^T$, where T is the number of periods measured *from* the cash flow *to* the pivot point or axis of rotation.

   a. Thus the sign of the distance is moved to the exponent.
   b. For cash flows at the pivot point, $T = 0$, and $(1 + i)^0 = 1$.

We can redraw the simple example with an unknown present cash flow, P.

Since P is drawn as a negative cash flow, we put a minus sign in front of it when we write the cash flow moment equation. To rewrite the cash flow moment equation at Year 1, we use the distances from the diagram as the exponents for $(1 + i)^t$:

$$0 = -P \times (1 + i)^1 + 10 \times (1 + i)^{-2} + -6 \times (1 + i)^{-4}$$

If $i = 5\%$, the value of P can be calculated.

$$0 = -P \times 1.05^1 + 10 \times 1.05^{-2} + -6 \times 1.05^{-4}$$
$$P = 10 \times 1.05^{-3} + -6 \times 1.05^{-5} = 8.64 - 4.70 = 3.94$$

EXAMPLE 4-6   REVISITED

For the cash flow diagram in Example 4-6 (repeated here), write the cash flow moment equations at Years 0, 3, and 5. Solve for F when $i = 15\%$.

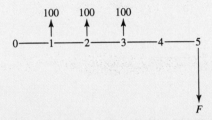

SOLUTION

With Year 0 as the pivot point, the cash flow moment equation is:

$$0 = 100 \times (1 + i)^{-1} + 100 \times (1 + i)^{-2} + 100 \times (1 + i)^{-3} - F \times (1 + i)^{-5}$$

With Year 3 as the pivot point, the cash flow moment equation is:

$$0 = 100 \times (1 + i)^2 + 100 \times (1 + i)^1 + 100 \times (1 + i)^0 - F \times (1 + i)^2$$

With Year 5 as the pivot point, the cash flow moment equation is:

$$0 = 100 \times (1 + i)^4 + 100 \times (1 + i)^3 + 100 \times (1 + i)^2 - F \times (1 + i)^0$$

In each case, the cash flow moment equation simplifies to:

$$F = 100 \times (1 + i)^4 + 100 \times (1 + i)^3 + 100 \times (1 + i)^2$$

If $i = 15\%$, then

$$F = 100 \times (1.749 + 1.521 + 1.323) = \$459.23$$

RELATIONSHIPS BETWEEN COMPOUND INTEREST FACTORS

From the derivations, we see there are several simple relationships between the compound interest factors. They are summarized here.

Single Payment

$$\text{Compound amount factor} = \frac{1}{\text{Present worth factor}}$$

$$(F/P, i, n) = \frac{1}{(P/F, i, n)} \tag{4-9}$$

Uniform Series

$$\text{Capital recovery factor} = \frac{1}{\text{Present worth factor}}$$

$$(A/P, i, n) = \frac{1}{(P/A, i, n)} \tag{4-10}$$

$$\text{Compound amount factor} = \frac{1}{\text{Sinking fund factor}}$$

$$(F/A, i, n) = \frac{1}{(A/F, i, n)} \tag{4-11}$$

The uniform series present worth factor is simply the sum of the n terms of the single payment present worth factor

$$(P/A, i, n) = \sum_{t=1}^{n} (P/F, i, t) \tag{4-12}$$

For example:

$$(P/A, 5\%, 4) = (P/F, 5\%, 1) + (P/F, 5\%, 2) + (P/F, 5\%, 3) + (P/F, 5\%, 4)$$
$$3.546 = 0.9524 + 0.9070 + 0.8638 + 0.8227$$

The uniform series compound amount factor equals 1 *plus* the sum of $(n - 1)$ terms of the single payment compound amount factor

$$(F/A, i, n) = 1 + \sum_{t=1}^{n-1} (F/P, i, t) \tag{4-13}$$

For example,

$$(F/A, 5\%, 4) = 1 + (F/P, 5\%, 1) + (F/P, 5\%, 2) + (F/P, 5\%, 3)$$
$$4.310 = 1 + 1.050 + 1.102 + 1.158$$

The uniform series capital recovery factor equals the uniform series sinking fund factor *plus i*:

$$(A/P, i, n) = (A/F, i, n) + i \qquad (4\text{-}14)$$

For example,

$$(A/P, 5\%, 4) = (A/F, 5\%, 4) + 0.05$$

$$0.2820 = 0.2320 + 0.05$$

This may be proved as follows:

$$(A/P, i, n) = (A/F, i, n) + i$$

$$\left[\frac{i(1+i)^n}{(1+i)^n - 1} \right] = \left[\frac{1}{(1+i)^n - 1} \right] + i$$

Multiply by $(1+i)^n - 1$ to get

$$i(1+i)^n = i + i(1+i)^n - i = i(1+i)^n.$$

ARITHMETIC GRADIENT

It frequently happens that the cash flow series is not of constant amount A. Instead, there is a uniformly increasing series as shown:

Cash flows of this form may be resolved into two components:

Note that by resolving the problem in this manner, the first cash flow in the arithmetic gradient series becomes zero. This is done so that G is the change from period to period, and because the gradient (G) series normally is used along with a uniform series (A). To find an equivalent present worth, we already have an equation for $(P/A, i, n)$, and we need to derive an equation for $(P/G, i, n)$. In this way, we will be able to write

$$P = A(P/A, i, n) + G(P/G, i, n)$$

Pause here and look at the tables. What is the value of $(P/G, i, 1)$ for any i?

Derivation of Arithmetic Gradient Factors

The arithmetic gradient is a series of increasing cash flows as follows:

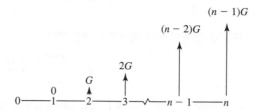

The arithmetic gradient series may be thought of as a series of individual cash flows that can individually be converted to equivalent final cash flows at the end of period n.

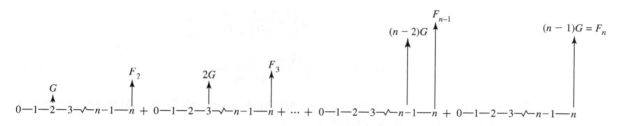

The value of F for the sum of the cash flows $= F_2 + F_3 + \cdots + F_{n-1} + F_n$, or

$$F = G(1+i)^{n-2} + 2G(1+i)^{n-3} + \cdots + (n-2)(G)(1+i)^1 + (n-1)G \qquad (4\text{-}15)$$

Multiply Equation 4-15 by $(1+i)$ and factor out G, or

$$(1+i)F = G[(1+i)^{n-1} + 2(1+i)^{n-2} + \cdots + (n-2)(1+i)^2 + (n-1)(1+i)^1] \qquad (4\text{-}16)$$

Rewrite Equation 4-15 to show other terms in the series,

$$F = G[(1+i)^{n-2} + \cdots + (n-3)(1+i)^2 + (n-2)(1+i)^1 + n - 1] \qquad (4\text{-}17)$$

Subtracting Equation 4-17 from Equation 4-16, we obtain

$$F + iF - F = G[(1+i)^{n-1} + (1+i)^{n-2} + \cdots + (1+i)^2 + (1+i)^1 + 1] - nG \qquad (4\text{-}18)$$

In the derivation of Equation 4-5, the terms inside the brackets of Equation 4-18 were shown to equal the series compound amount factor:

$$[(1+i)^{n-1} + (1+i)^{n-2} + \cdots + (1+i)^2 + (1+i)^1 + 1] = \frac{(1+i)^n - 1}{i}$$

Thus, Equation 4-18 becomes

$$iF = G\left[\frac{(1+i)^n - 1}{i}\right] - nG$$

Rearranging and solving for F, we write

$$F = \frac{G}{i}\left[\frac{(1+i)^n - 1}{i} - n\right] \tag{4-19}$$

Multiplying Equation 4-19 by the single payment present worth factor gives

$$P = \frac{G}{i}\left[\frac{(1+i)^n - 1}{i} - n\right]\left[\frac{1}{(1+i)^n}\right]$$

$$= G\left[\frac{(1+i)^n - in - 1}{i^2(1+i)^n}\right]$$

$$(P/G, i, n) = \left[\frac{(1+i)^n - in - 1}{i^2(1+i)^n}\right] \tag{4-20}$$

Equation 4-20 is the **arithmetic gradient present worth factor.** Multiplying Equation 4-19 by the sinking fund factor, we have

$$A = \frac{G}{i}\left[\frac{(1+i)^n - 1}{i} - n\right]\left[\frac{i}{(1+i)^n - 1}\right] = G\left[\frac{(1+i)^n - in - 1}{i(1+i)^n - i}\right]$$

$$(A/G, i, n) = \left[\frac{(1+i)^n - in - 1}{i(1+i)^n - i}\right] = \left[\frac{1}{i} - \frac{n}{(1+i)^n - 1}\right] \tag{4-21}$$

Equation 4-21 is the **arithmetic gradient uniform series factor.**

EXAMPLE 4-8

Andrew has purchased a new car. He wishes to set aside enough money in a bank account to pay the maintenance for the first 5 years. It has been estimated that the maintenance cost of a car is as follows:

Year	Maintenance Cost
1	$120
2	150
3	180
4	210
5	240

Assume the maintenance costs occur at the end of each year and that the bank pays 5% interest. How much should Andrew deposit in the bank now?

SOLUTION

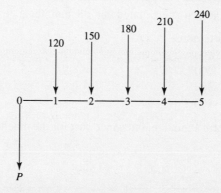

The cash flow may be broken into its two components:

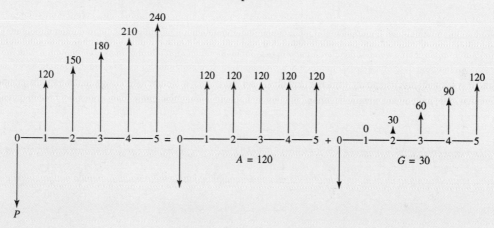

Both components represent cash flows for which compound interest factors have been derived. The first is a uniform series present worth, and the second is an arithmetic gradient series present worth:

$$P = A(P/A, 5\%, 5) + G(P/G, 5\%, 5)$$

Note that the value of n in the gradient factor is 5, not 4. In deriving the gradient factor, the cash flow in the first period is zero followed by $(n - 1)$ terms containing G. Here there are four terms containing G, and it is a 5-period gradient.

$$P = 120(P/A, 5\%, 5) + 30(P/G, 5\%, 5)$$
$$= 120(4.329) + 30(8.237)$$
$$= 519 + 247$$
$$= \$766$$

Andrew should deposit $766 in the bank now.

EXAMPLE 4-9

On a certain piece of machinery, it is estimated that the maintenance expense will be as follows:

Year	Maintenance
1	$100
2	200
3	300
4	400

What is the equivalent uniform annual maintenance cost for the machinery if 6% interest is used?

SOLUTION

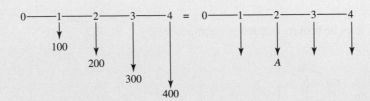

The first cash flow in the arithmetic gradient series is zero, hence the diagram is *not* in proper form for the arithmetic gradient equation. As in Example 4-8, the cash flow must be resolved into two components:

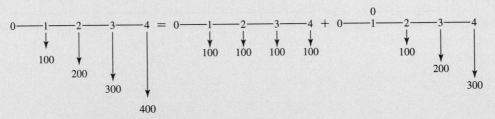

$$A = 100 + 100(A/G, 6\%, 4) = 100 + 100(1.427) = \$242.70$$

The equivalent uniform annual maintenance cost is $242.70.

EXAMPLE 4-10

Demand for a new product will decline as competitors enter the market. If interest is 10%, what is an equivalent uniform value?

Year	Revenue
1	$24,000
2	18,000
3	12,000
4	6,000

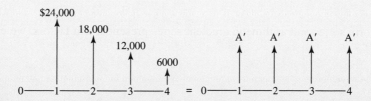

SOLUTION

The projected cash flow is still a cash flow in Period 1 ($24,000) that defines the uniform series. However, now the gradient or change each year is −$6000.

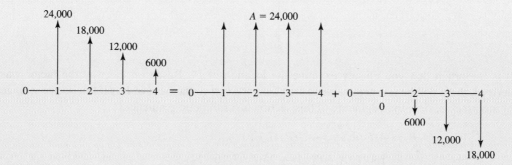

$$A' = 24,000 - 6000(A/G, 10\%, 4)$$
$$= 24,000 - 6000(1.381)$$
$$= \$15,714$$

The projected equivalent uniform value is $15,714 per year.

EXAMPLE 4-11

A car's warranty is 3 years. Upon expiration, annual maintenance starts at $150 and then climbs $25 per year until the car is sold at the end of Year 7. Use a 10% interest rate and find the present worth of these expenses.

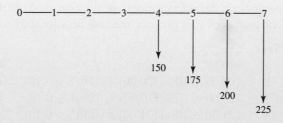

SOLUTION

With the uniform series and arithmetic gradient series present worth factors, we can compute a present sum P'.

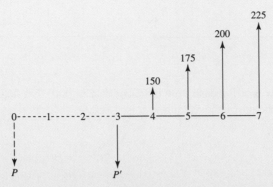

It is important that you closely examine the location of P'. Based on the way the factor was derived, there will be one zero value in the gradient series to the right of P'. (If this seems strange or incorrect, review the beginning of this section on arithmetic gradients.)

$$P' = A(P/A, i, n) + G(P/G, i, n)$$

$$= 150(P/A, 10\%, 4) + 25(P/G, 10\%, 4)$$

$$= 150(3.170) + 25(4.378) = 475.50 + 109.45 = 584.95$$

Then

$$P = P'(P/F, 10\%, 3) = 584.95(0.7513) = \$439.47$$

GEOMETRIC GRADIENT

In the preceding section, we saw that the arithmetic gradient is applicable where the period-by-period change in a cash receipt or payment is a *constant* amount. There are other situations where the period-by-period change is a **uniform rate,** g. Often geometric gradients can be traced to population levels or other levels of activity where changes over time are best modeled as a percentage of the previous year. The percentage or *rate* is constant over time, rather than the *amount* of the change as in the arithmetic gradient.

For example, if the maintenance costs for a car are $100 the first year and they increase at a uniform rate, g, of 10% per year, the cash flow for the first 5 years would be as follows:

Year			Cash Flow
1	100.00		$= \$100.00$
2	$100.00 + 10\%(100.00) = 100(1 + 0.10)^1$	$=$	110.00
3	$110.00 + 10\%(110.00) = 100(1 + 0.10)^2$	$=$	121.00
4	$121.00 + 10\%(121.00) = 100(1 + 0.10)^3$	$=$	133.10
5	$133.10 + 10\%(133.10) = 100(1 + 0.10)^4$	$=$	146.41

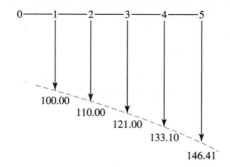

From the table, we can see that the maintenance cost in any year is

$$\$100(1 + g)^{n-1}$$

Stated in a more general form,

$$A_n = A_1(1 + g)^{n-1} \tag{4-22}$$

where g = uniform *rate* of cash flow increase/decrease from period
to period, that is, the geometric gradient
A_1 = value of cash flow at Year 1 ($100 in the example)
A_n = value of cash flow at any year n

Since the present worth P_n of any cash flow A_n at interest rate i is

$$P_n = A_n(1 + i)^{-n} \tag{4-23}$$

we can substitute Equation 4-22 into Equation 4-23 to get

$$P_n = A_1(1 + g)^{n-1}(1 + i)^{-n}$$

This may be rewritten as

$$P = A_1(1 + i)^{-1} \sum_{x=1}^{n} \left(\frac{1 + g}{1 + i} \right)^{x-1} \tag{4-24}$$

The present worth of the entire gradient series of cash flows may be obtained by expanding
Equation 4-24:

$$P = A_1(1 + i)^{-1} \sum_{x=1}^{n} \left(\frac{1 + g}{1 + i} \right)^{x-1} \tag{4-25}$$

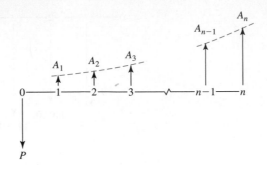

In the general case, where $i \neq g$, Equation 4-24 may be written out as follows:

$$P = A_1(1+i)^{n-1} + A_1(1+i)^{-1}\left(\frac{1+g}{1+i}\right) + A_1(1+i)^{-1}\left(\frac{1+g}{1+i}\right)^2$$

$$+ \cdots + A_1(1+i)^{-1}\left(\frac{1+g}{1+i}\right)^{n-1} \tag{4-26}$$

Let $a = A_1(1+i)^{-1}$ and $b = (1+g)/(1+i)$. Equation 4-26 becomes

$$P = a + ab + ab^2 + \cdots + ab^{n-1} \tag{4-27}$$

Multiply Equation 4-27 by b:

$$bP = ab + ab^2 + ab^3 + \cdots + ab^{n-1} + ab^n \tag{4-28}$$

Subtract Equation 4-28 from Equation 4-27:

$$P - bP = a - ab^n$$
$$P(1-b) = a(1-b^n)$$
$$P = \frac{a(1-b^n)}{1-b}$$

Replacing the original values for a and b, we obtain

$$P = A_1(1+i)^{-1}\left[\frac{1-\left(\dfrac{1+g}{1+i}\right)^n}{1-\left(\dfrac{1+g}{1+i}\right)}\right] = A_1\left[\frac{1-\left(\dfrac{1+g}{1+i}\right)^n}{(1+i)-\left(\dfrac{1+g}{1+i}\right)(1+i)}\right]$$

$$= A_1 \left[\frac{1 - (1 + g)^n (1 + i)^{-n}}{1 + i - 1 - g} \right]$$

$$P = A_1 \left[\frac{1 - (1 + g)^n (1 + i)^{-n}}{i - g} \right] \tag{4-29}$$

where $i \neq g$.

The expression in the brackets of Equation 4-29 is the **geometric series present worth factor**

$$(P/A, g, i, n) = \left[\frac{1 - (1 + g)^n (1 + i)^{-n}}{i - g} \right] \qquad \text{where } i \neq g \tag{4-30}$$

In the special case of $i = g$, Equation 4-29 becomes

$$P = A_1 n (1 + i)^{-1}$$

$$(P/A, g, i, n) = [n(1 + i)^{-1}] \qquad \text{where } i = g \tag{4-31}$$

EXAMPLE 4-12

The first-year maintenance cost for a new car is estimated to be $100, and it increases at a uniform rate of 10% per year. Using an 8% interest rate, calculate the present worth (PW) of the cost of the first 5 years of maintenance.

STEP-BY-STEP SOLUTION

Year n		Maintenance Cost		$(P/F, 8\%, n)$	PW of Maintenance
1	100.00	= 100.00	×	0.9259 =	$ 92.59
2	100.00 + 10%(100.00)	= 110.00	×	0.8573 =	94.30
3	110.00 + 10%(110.00)	= 121.00	×	0.7938 =	96.05
4	121.00 + 10%(121.00)	= 133.10	×	0.7350 =	97.83
5	133.10 + 10%(133.10)	= 146.41	×	0.6806 =	99.65
					$480.42

SOLUTION USING GEOMETRIC SERIES PRESENT WORTH FACTOR

$$P = A_1 \left[\frac{1 - (1 + g)^n (1 + i)^{-n}}{i - g} \right] \qquad \text{where } i \neq g$$

$$= 100.00 \left[\frac{1 - (1.10)^5 (1.08)^{-5}}{-0.02} \right] = \$480.42$$

The present worth of cost of maintenance for the first 5 years is $480.42.

NOMINAL AND EFFECTIVE INTEREST

EXAMPLE 4-13

Consider the situation where a person deposits $100 into a bank that pays 5% interest, compounded semiannually. How much would be in the savings account at the end of one year?

SOLUTION

Five percent interest, compounded semiannually, means that the bank pays $2^1/_2$% every 6 months. Thus, the initial amount $P = \$100$ would be credited with $0.025(100) = \$2.50$ interest at the end of 6 months, or

$$P \rightarrow P + Pi = 100 + 100(0.025) = 100 + 2.50 = \$102.50$$

The $102.50 is left in the savings account; at the end of the second 6-month period, the interest earned is $0.025(102.50) = \$2.56$, for a total in the account at the end of one year of $102.50 + 2.56 = \$105.06$, or

$$(P + Pi) \rightarrow (P + Pi) + i(P + Pi) = P(1 + i)^2 = 100(1 + 0.025)^2$$
$$= \$105.06$$

Nominal interest rate per year, r, is the annual interest rate without considering the effect of any compounding.

In Example 4-13, the bank pays $2^1/_2$% interest every 6 months. The nominal interest rate per year, r, therefore, is $2 \times 2^1/_2\% = 5\%$. The federal government mandates that lenders provide the **annual percentage rate (APR)** for any loan. This is the nominal interest rate.

Effective interest rate per year, i_a, is the *annual* interest rate taking into account the effect of any compounding during the year.

In Example 4-13, we saw that $100 left in the savings account for one year increased to $105.06, so the interest paid was $5.06. The effective interest rate per year, i_a, is $\$5.06/\$100.00 = 0.0506 = 5.06\%$. The effective annual interest rate earned on savings accounts, certificates of deposit (CD), bonds, and so on is often called the annual percentage yield (**APY**). To calculate the effective annual interest rate i_a, we use the following variables:

$r = $ Nominal interest rate per interest period (usually one year)

$i = $ Effective interest rate per interest period

$m = $ Number of compounding subperiods per time period

Using the method presented in Example 4-13, we can derive the equation for the effective interest rate. If a $1 deposit were made to an account that compounded interest m times per year and paid a nominal interest rate per year, r, the *interest rate per compounding subperiod* would be r/m, and the total in the account at the end of one year would be

$$\$1\left(1 + \frac{r}{m}\right)^m \quad \text{or simply} \quad \left(1 + \frac{r}{m}\right)^m$$

If we deduct the $1 principal sum, the expression would be

$$\left(1 + \frac{r}{m}\right)^m - 1$$

Therefore,

Effective annual interest rate $i_a = \left(1 + \dfrac{r}{m}\right)^m - 1$ (4-32)

where r = nominal interest rate per year
$\qquad m$ = number of compounding subperiods per year

Or, substituting the effective interest rate per compounding subperiod, $i = (r/m)$,

Effective annual interest rate $i_a = (1 + i)^m - 1$ (4-33)

where i = effective interest rate per compounding subperiod
$\qquad m$ = number of compounding subperiods per year

Either Equation 4-32 or 4-33 may be used to compute an effective interest rate per year. This and many other texts use i_a for the effective annual interest rate. The Fundamentals of Engineering (FE) exam uses i_e for the effective annual interest rate in its supplied material.

One should note that i was described in Chapter 3 simply as the interest rate per interest period. We were describing the effective interest rate without making any fuss about it. A more precise definition, we now know, is that i is the *effective* interest rate per interest period. Although it seems more complicated, we are describing the same exact situation, but with more care.

The nominal interest rate r is often given for a one-year period (but it could be given for either a shorter or a longer time period). In the special case of a nominal interest rate that is given per compounding subperiod, the effective interest rate per compounding subperiod, i, equals the nominal interest rate per subperiod, r.

In the typical effective interest computation, there are multiple compounding subperiods ($m > 1$). The resulting effective interest rate is either the solution to the problem or an intermediate solution, which allows us to use standard compound interest factors to proceed to solve the problem.

For **continuous compounding** (which is described in the next section),

Effective annual interest rate $i_a = e^r - 1$ (4-34)

EXAMPLE 4-14

If a savings bank pays $1^1/_2\%$ interest every 3 months, what are the nominal and effective interest rates per year?

SOLUTION

$$\text{Nominal interest rate per year} \quad r = 4 \times 1^1/_2\% = 6\%$$

$$\text{Effective annual interest rate} \quad i_a = \left(1 + \frac{r}{m}\right)^m - 1$$

$$= \left(1 + \frac{0.06}{4}\right)^4 - 1 = 0.061$$

$$= 6.1\%$$

Alternately,

$$\text{Effective annual interest rate} \quad i_a = (1 + i)^m - 1$$

$$= (1 + 0.015)^4 - 1 = 0.061$$

$$= 6.1\%$$

Table 4-1 tabulates the effective interest rate for a range of compounding frequencies and nominal interest rates. It should be noted that when a nominal interest rate is compounded annually, the nominal interest rate equals the effective interest rate. Also, it will be noted that increasing the frequency of compounding (for example, from monthly to continuously) has only a small impact on the effective interest rate. But if the amount of money is large, even small differences in the effective interest rate can be significant.

TABLE 4-1 Nominal and Effective Interest

Nominal Interest Rate per Year	Effective Annual Interest Rate, i_a When Nominal Rate Is Compounded				
r (%)	Yearly	Semiannually	Monthly	Daily	Continuously
1	1%	1.0025%	1.0046%	1.0050%	1.0050%
2	2	2.0100	2.0184	2.0201	2.0201
3	3	3.0225	3.0416	3.0453	3.0455
4	4	4.0400	4.0742	4.0809	4.0811
5	5	5.0625	5.1162	5.1268	5.1271
6	6	6.0900	6.1678	6.1831	6.1837
8	8	8.1600	8.3000	8.3278	8.3287
10	10	10.2500	10.4713	10.5156	10.5171
15	15	15.5625	16.0755	16.1798	16.1834
25	25	26.5625	28.0732	28.3916	28.4025

EXAMPLE 4-15

A loan shark lends money on the following terms: "If I give you $50 on Monday, you owe me $60 on the following Monday."

(a) What nominal interest rate per year (r) is the loan shark charging?

(b) What effective interest rate per year (i_a) is he charging?

(c) If the loan shark started with $50 and was able to keep it, as well as all the money he received, loaned out at all times, how much money did he have at the end of one year?

SOLUTION TO PART a

$$F = P(F/P, i, n)$$

$$60 = 50(F/P, i, 1)$$

$$(F/P, i, 1) = 1.2$$

Therefore, $i = 20\%$ per week.

$$\text{Nominal interest rate per year} = 52 \text{ weeks} \times 0.20 = 10.40 = 1040\%$$

SOLUTION TO PART b

$$\text{Effective annual interest rate} \quad i_a = \left(1 + \frac{r}{m}\right)^m - 1$$

$$= \left(1 + \frac{10.40}{52}\right)^{52} - 1 = 13,105 - 1$$

$$= 13,104 = 1,310,400\%$$

Or

$$\text{Effective annual interest rate} \quad i_a = (1 + i)^m - 1$$

$$= (1 + 0.20)^{52} - 1 = 13,104$$

$$= 1,310,400\%$$

SOLUTION TO PART c

$$F = P(1 + i)^n = 50(1 + 0.20)^{52}$$

$$= \$655,200$$

With a nominal interest rate of 1040% per year and effective interest rate of 1,310,400% per year, the loan shark who started with $50 had $655,200 at the end of one year.

When the various time periods in a problem match, we generally can solve the problem by simple calculations. Thus in Example 4-3, where we had $5000 in an account paying 8% interest, compounded annually, the five equal end-of-year withdrawals are simply computed as follows:

$$A = P(A/P, 8\%, 5) = 5000(0.2505) = \$1252$$

Consider how this simple problem becomes more difficult if the compounding period is changed so that it no longer matches the annual withdrawals.

EXAMPLE 4-16

On January 1, a woman deposits $5000 in a credit union that pays 8% nominal annual interest, compounded quarterly. She wishes to withdraw all the money in five equal yearly sums, beginning December 31 of the first year. How much should she withdraw each year?

SOLUTION

Since the 8% nominal annual interest rate r is compounded quarterly, we know that the effective interest rate per interest period, i, is 2%; and there are a total of $4 \times 5 = 20$ interest periods in 5 years. For the equation $A = P(A/P, i, n)$ to be used, there must be as many periodic withdrawals as there are interest periods, n. In this example we have 5 withdrawals and 20 interest periods.

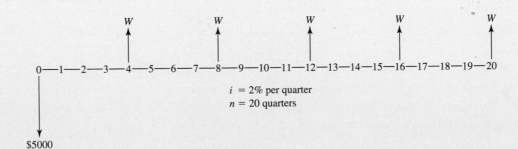

i = 2% per quarter
n = 20 quarters

To solve the problem, we must adjust it so that it is in one of the standard forms for which we have compound interest factors. This means we must first compute either an equivalent A for each 3-month interest period or an effective i for each time period between withdrawals. Let's solve the problem both ways.

SOLUTION 1

Compute an equivalent A for each 3-month time period.
If we had been required to compute the amount that could be withdrawn quarterly, the diagram would be as follows:

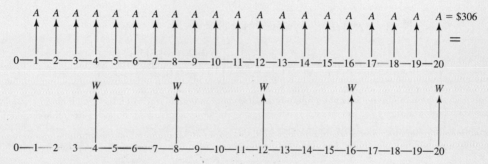

$$A = P(A/P, i, n) = 5000(A/P, 2\%, 20) = 5000(0.0612) = \$306$$

Now, since we know A, we can construct the diagram that relates it to our desired equivalent annual withdrawal, W.

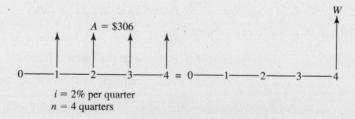

Looking at a one-year period,

$$i = 2\% \text{ per quarter}$$
$$n = 4 \text{ quarters}$$

$$W = A(F/A, i, n) = 306(F/A, 2\%, 4) = 306(4.122)$$
$$= \$1260$$

SOLUTION 2

Compute an effective i for the time period between withdrawals.

Between withdrawals, W, there are four interest periods, hence $m = 4$ compounding subperiods per year. Since the nominal interest rate per year, r, is 8%, we can proceed to compute the effective annual interest rate.

$$\text{Effective annual interest rate} \quad i_a = \left(1 + \frac{r}{m}\right)^m - 1 = \left(1 + \frac{0.08}{4}\right)^4 - 1$$

$$= 0.0824 = 8.24\% \text{ per year}$$

Now the problem may be redrawn as follows:

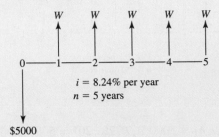

$i = 8.24\%$ per year
$n = 5$ years

This diagram may be directly solved to determine the annual withdrawal W with the capital recovery factor:

$$W = P(A/P, i, n) = 5000(A/P, 8.24\%, 5)$$

$$= P\left[\frac{i(1 + i)^n}{(1 + i)^n - 1}\right] = 5000\left[\frac{0.0824(1 + 0.0824)^5}{(1 + 0.0824)^5 - 1}\right]$$

$$= 5000(0.2520) = \$1260$$

The depositor should withdraw $1260 per year.

CONTINUOUS COMPOUNDING

Two variables we have introduced are:

$$r = \text{Nominal interest rate per interest period}$$
$$m = \text{Number of compounding subperiods per time period}$$

Since the interest period is normally one year, the definitions become:

$$r = \text{Nominal interest rate per year}$$
$$m = \text{Number of compounding subperiods per year}$$
$$\frac{r}{m} = \text{Interest rate per interest period}$$
$$mn = \text{Number of compounding subperiods in } n \text{ years}$$

Single Payment Interest Factors: Continuous Compounding

The single payment compound amount formula (Equation 3-3)

$$F = P(1 + i)^n$$

may be rewritten as

$$F = P\left(1 + \frac{r}{m}\right)^{mn}$$

If we increase m, the number of compounding subperiods per year, without limit, m becomes very large and approaches infinity, and r/m becomes very small and approaches zero.

This is the condition of **continuous compounding**: that is, the duration of the interest period decreases from some finite duration Δt to an infinitely small duration dt, and the number of interest periods per year becomes infinite. In this situation of continuous compounding:

$$F = P \lim_{m \to \infty} \left(1 + \frac{r}{m}\right)^{mn} \tag{4-35}$$

An important limit in calculus is:

$$\lim_{x \to 0} (1+x)^{1/x} = 2.71828 = e \tag{4-36}$$

If we set $x = r/m$, then mn may be written as $(1/x)(rn)$. As m becomes infinite, x becomes 0. Equation 4-35 becomes

$$F = P[\lim_{x \to 0} (1+x)^{1/x}]^{rn}$$

Equation 4-36 tells us the quantity inside the brackets equals e. So returning to Equation 3-3, we find that

$$F = P(1+i)^n \quad \text{becomes} \quad F = Pe^{rn} \tag{4-37}$$

and

$$P = F(1+i)^{-n} \quad \text{becomes} \quad P = Fe^{-rn} \tag{4-38}$$

We see that for continuous compounding,

$$(1+i) = e^r$$

or, as shown earlier,

$$\textbf{Effective interest rate per year} \quad i_a = e^r - 1 \tag{4-34}$$

To find compound amount and present worth for continuous compounding and a single payment, we write:

$$\textbf{Compound amount} \quad F = P(e^{rn}) = P[F/P, r, n] \tag{4-39}$$

$$\textbf{Present worth} \quad P = F(e^{-rn}) = F[P/F, r, n] \tag{4-40}$$

Square brackets around the factors distinguish continuous compounding. If your hand calculator does not have e^x, use the table of e^{rn} and e^{-rn}, provided at the end of Appendix B.

EXAMPLE 4-17

If you were to deposit $2000 in a bank that pays 5% nominal interest, compounded continuously, how much would be in the account at the end of 2 years?

SOLUTION

The single payment compound amount equation for continuous compounding is

$$F = Pe^{rn}$$

where $r =$ nominal interest rate $= 0.05$
       $n =$ number of years $= 2$

$$F = 2000e^{(0.05)(2)} = 2000(1.1052) = \$2210.40$$

There would be $2210.40 in the account at the end of 2 years.

EXAMPLE 4-18

A bank offers to sell savings certificates that will pay the purchaser $5000 at the end of 10 years but will pay nothing to the purchaser in the meantime. If interest is computed at 6%, compounded continuously, at what price is the bank selling the certificates?

SOLUTION

$$P = Fe^{-rn} \qquad F = \$5000 \qquad r = 0.06 \qquad n = 10 \text{ years}$$

$$P = 5000e^{-0.06 \times 10} = 5000(0.5488) = \$2744$$

Therefore, the bank is selling the $5000 certificates for $2744.

EXAMPLE 4-19

How long will it take for money to double at 10% nominal interest, compounded continuously?

$$F = Pe^{rn}$$

$$2 = 1e^{0.10n}$$

$$e^{0.10n} = 2$$

or

$$0.10n = \ln 2 = 0.693$$

$$n = 6.93 \text{ years}$$

It will take 6.93 years for money to double at 10% nominal interest, compounded continuously.

EXAMPLE 4-20

If the savings bank in Example 4-14 changes its interest policy to 6% interest, compounded continuously, what are the nominal and the effective interest rates?

SOLUTION

The nominal interest rate remains at 6% per year.

$$\text{Effective interest rate} = e^r - 1$$

$$= e^{0.06} - 1 = 0.0618$$

$$= 6.18\%$$

Uniform Payment Series: Continuous Compounding at Nominal Rate r per Period

Let us now substitute the equation $i = e^r - 1$ into the equations for end-of-period compounding.

Continuous Compounding Sinking Fund

$$[A/F, r, n] = \frac{e^r - 1}{e^{rn} - 1} \tag{4-41}$$

Continuous Compounding Capital Recovery

$$[A/P, r, n] = \frac{e^{rn}(e^r - 1)}{e^{rn} - 1} \tag{4-42}$$

Continuous Compounding Series Compound Amount

$$[F/A, r, n] = \frac{e^{rn} - 1}{e^r - 1} \tag{4-43}$$

Continuous Compounding Series Present Worth

$$[P/A, r, n] = \frac{e^{rn} - 1}{e^{rn}(e^r - 1)}$$ (4-44)

EXAMPLE 4-21

In Example 4-1, $500 per year was deposited into a credit union that paid 5% interest, compounded annually. At the end of 5 years, $2763 was in the credit union account. How much would there have been if the institution paid 5% nominal interest, compounded continuously?

SOLUTION

$$A = \$500 \qquad r = 0.05 \qquad n = 5 \text{ years}$$

$$F = A[F/A, r, n] = A\left(\frac{e^{rn} - 1}{e^r - 1}\right) = 500\left(\frac{e^{0.05(5)} - 1}{e^{0.05} - 1}\right)$$

$$= \$2769.84$$

EXAMPLE 4-22

In Example 4-2, Jim Hayes wished to save a uniform amount each month so he would have $1000 at the end of a year. Based on 6% nominal interest, compounded monthly, he had to deposit $81.10 per month. How much would he have to deposit if his credit union paid 6% nominal interest, compounded continuously?

SOLUTION

The deposits are made monthly; hence, there are 12 compounding subperiods in the one-year time period.

$$F = \$1000 \qquad r = \text{nominal interest rate/interest period} = \frac{0.06}{12} = 0.005$$

$n = 12$ compounding subperiods in the one-year period of the problem

$$A = F[A/F, r, n] = F\left(\frac{e^r - 1}{e^{rn} - 1}\right) = 1000\left(\frac{e^{0.005} - 1}{e^{0.005(12)} - 1}\right)$$

$$= 1000\left(\frac{0.005013}{0.061837}\right) = \$81.07$$

He would have to deposit $81.07 per month. Note that the difference between monthly and continuous compounding is just 3 cents per month.

SPREADSHEETS FOR ECONOMIC ANALYSIS

Spreadsheets are used in most real-world applications of engineering economy. Common tasks include the following:

1. Constructing tables of cash flows.
2. Using annuity functions to calculate a P, F, A, n, or i.
3. Using a block function to find the present worth or internal rate of return for a table of cash flows.
4. Making graphs for analysis and convincing presentations.
5. Calculating "what-if" for different assumed values of problem variables.

Constructing tables of cash flows relies mainly on spreadsheet basics that are covered in Appendix A. These basics include using and naming spreadsheet variables, understanding the difference between absolute and relative addresses when copying a formula, and formatting a cell. Appendix A uses the example of the amortization schedule shown in Table 3-1, Plan 3. This amortization schedule divides each scheduled loan payment into principal and interest and includes the outstanding balance for each period.

Because spreadsheet functions can be found through pointing and clicking on menus, those steps are not detailed. In Excel the starting point is the f_x button. Excel functions are used here; syntax differences for most other spreadsheet programs are minor.

Spreadsheet Annuity Functions

In tables of engineering economy factors, i is the table header, n is the row, and two of P, F, A, and G define a column. The spreadsheet annuity functions list four arguments chosen from n, A, P, F, and i, and solve for the fifth argument. The *Type* argument is optional. If it is omitted or 0, then the A value is assumed to be the end-of-period cash flow. If the A value represents the beginning-of-period cash flow, then a value of 1 can be entered for the Type variable.

To find the equivalent P	$-PV(i,n,A,F,\text{Type})$
To find the equivalent A	$-PMT(i,n,P,\text{Type})$
To find the equivalent F	$-FV(i,n,A,P,\text{Type})$
To find n	$NPER(i,A,P,F,\text{Type})$
To find i	$RATE(n,A,P,\text{Type,guess})$

The sign convention for the first three functions seems odd to some students. The PV of $200 per period for 10 periods is negative, and the PV of $-\$200$ per period is positive. So a minus sign is inserted to find the equivalent P, A, or F. Without this minus sign, the calculated value is not equivalent to the four given values.

EXAMPLE 4-23

A new engineer wants to save money for down payment on a house. The initial deposit is $685, and $375 is deposited at the end of each month. The savings account earns interest at an annual nominal rate of 6% with monthly compounding. How much is on deposit after 48 months?

SOLUTION

Because deposits are made monthly, the nominal annual interest rate of 6% must be converted to $\frac{1}{2}\%$ per month for the 48 months. Thus we must find F if $i = 0.5\% = 0.005$, $n = 48$, $A = 375$, and $P = 685$. Note both the initial and periodic deposits are positive cash flows for the savings account. The Excel function is multiplied by -1 or $-FV(0.005,48,375,685,0)$, and the result is $21,156.97.

EXAMPLE 4-24

A new engineer buys a car with 0% down financing from the dealer. The cost with all taxes, registration, and license fees is $15,732. If each of the 48 monthly payments is $398, what is the monthly interest rate? What is the effective annual interest rate?

SOLUTION

The RATE function can be used to find the monthly interest rate, given that $n = 48$, $A = -398$, $P = 15,732$, and $F = 0$. The Excel function is RATE(48,−398,15732,0) and the result is 0.822%. The effective annual interest rate is $1.00822^{12} - 1 = 10.33\%$.

Spreadsheet Block Functions

Cash flows can be specified period by period as a block of values. These cash flows are analyzed by **block functions** that identify the row or column entries for which a present worth or an internal rate of return should be calculated. In Excel the two functions are NPV(i,values) and IRR(values,guess).

Economic Criterion	Excel Function	Values for Periods
Net present value	NPV(i,values)	1 to n
Internal rate of return	IRR(values,guess); guess argument is optional	0 to n

Excel's IRR function can be used to find the interest rate for a loan with irregular payments (other applications are covered in Chapter 7).

These block functions make different assumptions about the range of years included. For example, NPV(i,values) assumes that Year 0 is **not** included, while IRR(values,guess) assumes that Year 0 is included. These functions require that a cash flow be identified for each period. You cannot leave cells blank even if the cash flow is $0. The cash flows for 1 to n are assumed to be end-of-period flows. All periods are assumed to be the same length of time.

Also the NPV functions returns the present worth equivalent to the cash flows, unlike the PV annuity function, which returns the negative of the equivalent value.

For cash flows involving only constant values of P, F, and A this block approach seems to be inferior to the annuity functions. However, this is a conceptually easy approach for more complicated cash flows, such as arithmetic gradients. Suppose the years (row 1) and the cash flows (row 2) are specified in columns B through E.

	A	B	C	D	E	F
1	Year	0	1	2	3	4
2	Cash flow	−25,000	6000	8000	10,000	12,000

If an interest rate of 8% is assumed, then the present worth of the cash flows can be calculated as $=$B2+NPV(.08,C2:F2), which equals $4172.95. This is the present worth equivalent to the five cash flows, rather than the negative of the present worth equivalent returned by the PV annuity function. The internal rate of return calculated using IRR(B2:F2) is 14.5%. Notice how the NPV function does not include the Year 0 cash flow in B2 but the IRR function does.

- PW $=$ B2+NPV(.08,C2:F2)   NPV range without Year 0
- IRR $=$ IRR(B2:F2)   IRR range with Year 0

Using Spreadsheets for Basic Graphing

Often we are interested in the relationship between two variables. Examples include the number and size of payments to repay a loan, the present worth of an MS degree and how long until we retire, and the interest rate and present worth for a new machine. This kind of two-variable relationship is best shown with a graph.

As we will show in Example 4-25, the goal is to place one variable on each axis of the graph and then plot the relationship. Spreadsheets automate most steps of drawing a graph, so that it is quite easy. However, there are two very similar chart types, and we must be careful to choose the *xy* chart, not the *line* chart. Both charts measure the *y* variable, but they treat the *x* variable differently. The *xy* chart measures the *x* variable; thus its *x* value is measured along the *x* axis. For the *line* chart, each *x* value is placed an equal distance along the *x* axis. Thus *x* values of 1, 2, 4, 8 would be spaced evenly, rather than doubling each distance. The *line* chart is really designed to plot *y* values for different categories, such as prices for models of cars or enrollments for different universities.

Drawing an *xy* plot with Excel is easiest if the table of data lists the *x* values before the *y* values. This convention makes it easy for Excel to specify one set of *x* values and several sets of *y* values. The block of *xy* values is selected, and then the chart tool is selected. Then the spreadsheet guides the user through the rest of the steps.

EXAMPLE 4-25

Graph the loan payment as a function of the number of payments for a possible auto loan. Let the number of monthly payments vary between 36 and 60. The nominal annual interest rate is 12%, and the amount borrowed is $18,000.

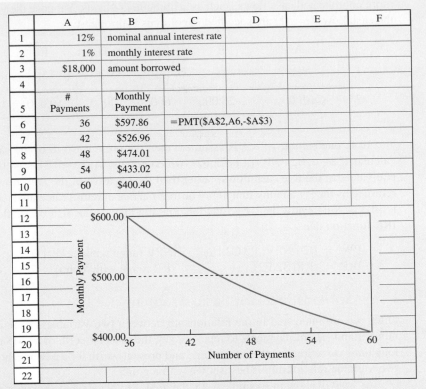

	A	B	C	D	E	F
1	12%	nominal annual interest rate				
2	1%	monthly interest rate				
3	$18,000	amount borrowed				
4						
5	# Payments	Monthly Payment				
6	36	$597.86	=PMT($A$2,A6,-$A$3)			
7	42	$526.96				
8	48	$474.01				
9	54	$433.02				
10	60	$400.40				
11						
12						
13						
14						
15						
16						
17						
18						
19						
20						
21						
22						

FIGURE 4-2 Example spreadsheet graph.

SOLUTION

The spreadsheet table shown in Figure 4-2 is constructed first. Cells A5:B10 are selected, and then the Chartwizard icon is selected. The first step is to select an *xy (scatter)* plot with smoothed lines and without markers as the chart type. (The other choices are no lines, straight lines, and adding markers for the data points.) The second step shows us the graph and allows the option of changing the data cells selected. The third step is for chart options. Here we add titles for the two axes and turn off showing the legend. (Since we have only one line in our graph, the legend is not needed. Deleting it leaves more room for the graph.) In the fourth step we choose where the chart is placed.

Because *xy* plots are normally graphed with the origin set to (0, 0), an attractive plot is best obtained if the minimum and maximum values are changed for each axis. This is done by placing the mouse cursor over the axis and left-clicking. Handles or small black boxes should appear on the axis to show that it has been selected. Right-clicking brings up a menu to select format axis. The scale tab allows us to change the minimum and maximum values.

SUMMARY

The compound interest formulas described in this chapter, along with those in Chapter 3, will be referred to throughout the rest of the book. It is very important that the reader understand the concepts presented and how these formulas are used. The following notation is used consistently:

i = effective interest rate per interest period[3] (stated as a decimal)

n = number of interest periods

P = a present sum of money

F = a future sum of money: the future sum F is an amount, n interest periods from the present, that is equivalent to P with interest rate i

A = an end-of-period cash receipt or disbursement in a uniform series continuing for n periods; the entire series equivalent to P or F at interest rate i

G = uniform period-by-period increase or decrease in cash receipts or disbursements; the arithmetic gradient

g = uniform rate of cash flow increase or decrease from period to period; the geometric gradient

r = nominal interest rate per interest period (see footnote 3)

i_a = effective annual interest rate

m = number of compounding subperiods per period (see footnote 3)

Single Payment Formulas (derived in Chapter 3)

Compound amount:

$$F = P(1+i)^n = P(F/P, i, n)$$

Present worth:

$$P = F(1+i)^{-n} = F(P/F, i, n)$$

Uniform Series Formulas

Compound amount:

$$F = A\left[\frac{(1+i)^n - 1}{i}\right] = A(F/A, i, n)$$

Sinking fund:

$$A = F\left[\frac{i}{(1+i)^n - 1}\right] = F(A/F, i, n)$$

Capital recovery:

$$A = P\left[\frac{i(1+i)^n}{(1+i)^n - 1}\right] = P(A/P, i, n)$$

Present worth:

$$P = A\left[\frac{(1+i)^n - 1}{i(1+i)^n}\right] = A(P/A, i, n)$$

---

[3]Normally the interest period is one year, but it could be some other period (e.g., quarter, month, half-year).

Arithmetic Gradient Formulas

Arithmetic gradient present worth:

$$P = G\left[\frac{(1 + i)^n - in - 1}{i^2(1 + i)^n}\right] = G(P/G, i, n)$$

Arithmetic gradient uniform series:

$$A = G\left[\frac{(1 + i)^n - in - 1}{i(1 + i)^n - i}\right] = G\left[\frac{1}{i} - \frac{n}{(1 + i)^n - 1}\right] = G(A/G, i, n)$$

Geometric Gradient Formulas

Geometric series present worth, where $i \neq g$:

$$P = A_1\left[\frac{1 - (1 + g)^n(1 + i)^{-n}}{i - g}\right] = A_1(P/A, g, i, n)$$

Geometric series present worth, where $i = g$:

$$P = A_1[n(1 + i)^{-1}] = A_1(P/A, g, i, n) = A_1(P/A, i, i, n)$$

Single Payment Formulas: Continuous Compounding at Nominal Rate r per Period

Compound amount:

$$F = P(e^{rn}) = P[F/P, r, n]$$

Present worth:

$$P = F(e^{-rn}) = F[P/F, r, n]$$

Note that square brackets around the factors are used to distinguish continuous compounding.

Uniform Payment Series: Continuous Compounding at Nominal Rate r per Period

Continuous compounding sinking fund:

$$A = F\left[\frac{e^r - 1}{e^{rn} - 1}\right] = F[A/F, r, n]$$

Continuous compounding capital recovery:

$$A = P\left[\frac{e^{rn}(e^r - 1)}{e^{rn} - 1}\right] = P[A/P, r, n]$$

Continuous compounding series compound amount:

$$F = A\left[\frac{e^{rn} - 1}{e^r - 1}\right] = A[F/A, r, n]$$

Continuous compounding series present worth:

$$P = A\left[\frac{e^{rn} - 1}{e^{rn}(e^r - 1)}\right] = A[P/A, r, n]$$

Nominal Annual Interest Rate, r

The annual interest rate without considering the effect of any compounding.

Effective Annual Interest Rate, i_a

The annual interest rate taking into account the effect of any compounding during the year.
Effective annual interest rate (periodic compounding):

$$i_a = \left(1 + \frac{r}{m}\right)^m - 1$$

or

$$i_a = (1 + i)^m - 1$$

Effective annual interest rate (continuous compounding):

$$i_a = e^r - 1$$

PROBLEMS

Most problems could be solved with a spreadsheet, but calculators and tabulated factors are often easier. The icon indicates that a spreadsheet is recommended.

Uniform Annual Cash Flows

4-1 For diagrams (a) to (c), compute the unknown values B, C, V, using the minimum number of compound interest factors.

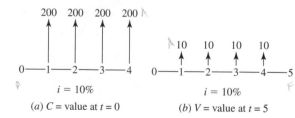

(a) C = value at $t = 0$

(b) V = value at $t = 5$

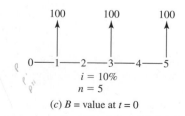

(c) B = value at $t = 0$

(*Answers: B* = \$228.13; *C* = \$634; *V* = \$51.05)

4-2 The cash flows have a present value of 0. Compute the value of J, assuming a 10% interest rate.

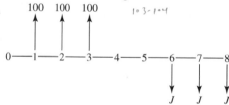

4-3 For what value of n, based on a $3\frac{1}{2}$% interest rate, do these cash flows have a present value of 0?

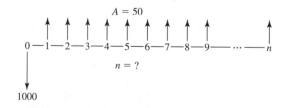

4-4 The cash flows have a present value of 0. Compute the value of n, assuming a 10% interest rate.

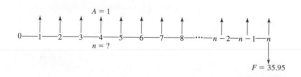

4-5   How much must be deposited now at $5\frac{1}{4}\%$ interest to produce $300 at the end of every year for 10 years?

4-6   A man buys a car for $3000 with no money down. He pays for the car in 30 equal monthly payments with interest at 12% per annum, compounded monthly. What is his monthly loan payment?
(*Answer:* $116.10)

4-7   A car may be purchased with a $3000 down payment now and 60 monthly payments of $280. If the interest rate is 12% compounded monthly, what is the price of the car?

4-8   A company deposits $2000 in a bank at the end of every year for 10 years. The company makes no deposits during the subsequent 5 years. If the bank pays 8% interest, how much would be in the account at end of 15 years?

4-9   A city engineer knows that she will need $25 million in 3 years to replace toll booths on a toll road in the city. Traffic on the road is estimated to be 20 million vehicles per year. How much per vehicle should the toll be to cover the cost of the toll booth replacement project? Interest is 10%. (Simplify your analysis by assuming that the toll receipts are received at the end of each year in a lump sum.)

4-10   A student wants to have $30,000 at graduation 4 years from now to buy a new car. His grandfather gave him $10,000 as a high school graduation present. How much must the student save each year if he deposits the $10,000 today and can earn 12% on both the $10,000 and his earnings in a mutual fund his grandfather recommends?

4-11   Using linear interpolation, determine the value of $(P/A, 6\frac{1}{2}\%, 10)$ from the compound interest tables. Compute this same value using the equation. Why do the values differ?

4-12   How many months will it take to pay off a $525 debt, with monthly payments of $15 at the end of each month, if the interest rate is 18%, compounded monthly?   (*Answer:* 50 months)

4-13   A young engineer wishes to become a millionaire by the time he is 60 years old. He believes that by careful investment he can obtain a 15% rate of return. He plans to add a uniform sum of money to his investment program each year, beginning on his 20th birthday and continuing through his 59th birthday. How much money must the engineer set aside in this project each year?

4-14   A contractor wishes to set up a special fund by making uniform semiannual end-of-period deposits for 20 years. The fund is to provide $10,000 at the end of each of the last 5 years of the 20-year period. If interest is 8%, compounded semiannually, what is the required semiannual deposit?

4-15   What amount will be required to purchase, on an engineers's 40th birthday, an annuity to provide him with 30 equal semiannual payments of $1000 each, the first to be received on his 50th birthday, if nominal interest is 4% compounded semiannually?

4-16   The first of a series of equal semiannual cash flows occurs on July 1, 2007, and the last occurs on January 1, 2020. Each cash flow is equal to $128,000. The nominal interest rate is 12% compounded semiannually. What single amount on July 1, 2011 is equivalent to this cash flow system?

4-17   On January 1, Frank Jenson bought a used car for $4200 and agreed to pay for it as follows: $\frac{1}{3}$ down payment; the balance to be paid in 36 equal monthly payments; the first payment due February 1; an annual interest rate of 9%, compounded monthly.

(*a*) What is the amount of Frank's monthly payment?

(*b*) During the summer, Frank made enough money to pay off the entire balance due on the car as of October 1. How much did Frank owe on October 1?

4-18   Mary Lavor plans to save money at her bank for use in December. She will deposit $30 a month, beginning on March 1 and continuing through November 1. She will withdraw all the money on December 1. If the bank pays $\frac{1}{2}\%$ interest each month, how much money will she receive on December 1?

4-19   If $i = 12\%$, for what value of B is the present value = 0.

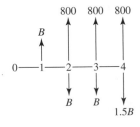

4-20   Compute E so that the cash flows have a present value of 0.

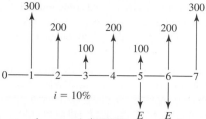

$i = 10\%$

4-21   A man borrowed $500 from a bank on October 15th. He must repay the loan in 16 equal monthly payments, due on the 15th of each month, beginning November 15th. If interest is computed at 1% per month, how much must he pay each month? (*Answer:* $33.95)

4-22   Jerry bought a house for $500,000 and made a $100,000 down payment. He obtained a 30-year loan for the remaining amount. Payments were made monthly. The nominal annual interest rate was 9%. After 10 years (120 payments) he decided to pay the remaining balance on the loan.

(*a*)   What was his monthly loan payment?

(*b*)   What must he have paid (in addition to his regular 120th monthly payment) to pay the remaining balance of his loan?

4-23   A man wants to help provide a college education for his young daughter. He can afford to invest $600/yr for the next 4 years, beginning on the girl's 4th birthday. He wishes to give his daughter $4000 on her 18th, 19th, 20th, and 21st birthdays, for a total of $16,000. Assuming 5% interest, what uniform annual investment will he have to make on the girl's 8th through 17th birthdays?   (*Answer:* $792.73)

4-24   Table 3-1 presented four plans for the repayment of $5000 in 5 years with interest at 8%. Still another way to repay the $5000 would be to make four annual end-of-year payments of $1000 each, followed by a final payment at the end of the fifth year. How much would the final payment be?

4-25   A $150 bicycle was purchased on December 1 with a $15 down payment. The balance is to be paid at the rate of $10 at the end of each month, with the first payment due on December 31. The last payment may be some amount less than $10. If interest on the unpaid balance is computed at 1$^1/_2$% per month, how many payments will there be, and what is the amount of the final payment?   (*Answers:* 16 payments; final payment: $1.99)

4-26   A company buys a machine for $12,000, which it agrees to pay for in five equal annual payments, beginning one year after the date of purchase, at an annual interest rate of 4%. Immediately after the second payment, the terms of the agreement are changed to allow the balance due to be paid off in a single payment the next year. What is the final single payment? (*Answer:* $7778)

4-27   An engineering student bought a car at a local used car lot. Including tax and insurance, the total price was $3000. He is to pay for the car in 12 equal monthly payments, beginning with the first payment immediately (in other words, the first payment was the down payment). Nominal interest on the loan is 12%, compounded monthly. After six payments (the down payment plus five additional payments), he decides to sell the car. A buyer agrees to pay a cash amount to pay off the loan in full at the time the next payment is due and also to pay the engineering student $1000. If there are no penalty charges for this early payment of the loan, how much will the car cost the new buyer?

4-28   A realtor sold a house on August 31, 2007, for $150,000 to a buyer in which a 20% down payment was made. The buyer took a 15-year mortgage on the property with an effective interest rate of 8% per annum. The buyer intends to pay off the mortgage owed in yearly payments starting on August 31, 2008.

(*a*)   How much of the mortgage will still be owed after the payment due on August 31, 2014, has been made?

(*b*)   Solve the same problem by separating the interest and the principal amounts.

4-29   To provide for a college education for his daughter, a man opened an escrow account in which equal deposits were made. The first deposit was made on January 1, 1991, and the last deposit was made on January 1, 2008. The yearly college expenses including tuition were estimated to be $8000, for each of the 4 years. Assuming the interest rate to be 5.75%, how much did the father have to deposit each year in the escrow account for the daughter to draw $8000 per year for 4 years beginning January 1, 2008?

4-30   A bank recently announced an "instant cash" plan for holders of its bank credit cards. A cardholder may receive cash from the bank up to a preset limit (about $500). There is a special charge of 4% made at the time the "instant cash" is sent to the cardholders. The debt may be repaid in monthly installments. Each month the bank charges 1$^1/_2$% on the unpaid balance. The monthly payment, including interest, may be as little as $10. Thus, for $150 of "instant cash," an initial charge of $6 is made and added to the balance due. Assume the cardholder makes a monthly payment of $10 (this includes both principal and interest). How many months are required to repay the debt? If your answer includes a fraction of a month, round up to the next month.

4-31 An engineer borrowed $3000 from the bank, payable in six equal end-of-year payments at 8%. The bank agreed to reduce the interest on the loan if interest rates declined in the United States before the loan was fully repaid. At the end of 3 years, at the time of the third payment, the bank agreed to reduce the interest rate on the remaining debt from 8% to 7%. What was the amount of the equal annual end-of-year payments for each of the first 3 years? What was the amount of the equal annual end-of-year payments for each of the last 3 years?

4-32 A bank is offering a loan of $25,000 with a nominal interest rate of 18% compounded monthly, payable in 60 months. (*Hint:* The loan origination fee of 2% will be taken out from the loan amount.)

 (*a*) What is the monthly payment?

 (*b*) If a loan origination fee of 2% is charged at the time of the loan, what is the effective interest rate?

4-33 A local finance company will loan $10,000 to a homeowner. It is to be repaid in 24 monthly payments of $499 each. The first payment is due 30 days after the $10,000 is received. What interest rate per month are they charging? (*Answer:* $1^1/2\%$)

4-34 A woman made 10 annual end-of-year purchases of $1000 worth of common stock. The stock paid no dividends. Then for 4 years she held the stock. At the end of the 4 years she sold all the stock for $28,000. What interest rate did she obtain on her investment?

4-35 One of the largest car dealers in the city advertises a 3-year-old car for sale as follows:

> Cash price $3575, or a down payment of $375 with 45 monthly payments of $93.41.

Susan DeVaux bought the car and made a down payment of $800. The dealer charged her the same interest rate used in his advertised offer. How much will Susan pay each month for 45 months? What effective interest rate is being charged? (*Answers:* $81.03; 16.1%)

Relationships Between Factors

4-36 For some interest rate i and some number of interest periods n, the uniform series capital recovery factor is 0.1408 and the sinking fund factor is 0.0408. What is the interest rate? What is n?

4-37 For some interest rate i and some number of interest periods n, the uniform series capital recovery factor is 0.1728 and the sinking fund factor is 0.0378. What is the interest rate?

4-38 Derive an equation to find the end-of-year future sum F that is equivalent to a series of n beginning-of-year payments B at interest rate i. Then use the equation to determine the future sum F equivalent to six B payments of $100 at 8% interest. (*Answer:* $F = \$792.28$)

4-39 If $200 is deposited in a savings account at the beginning of each of 15 years, and the account draws interest at 7% per year, how much will be in the account at the end of 15 years?

Arithmetic Gradients

4-40 Assume a 10% interest rate and find S, T, and x.

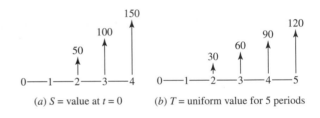

(a) S = value at $t = 0$ (b) T = uniform value for 5 periods

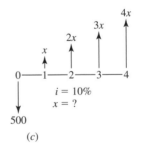

(*Answers:* $S = \$218.90$; $T = \$54.30$; $x = \$66.24$)

4-41 Compute the unknown values.

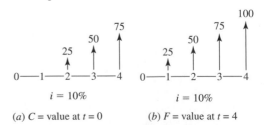

(a) C = value at $t = 0$ (b) F = value at $t = 4$

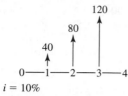

i = 10%

(c) A = uniform value from t = 1 to 4

(*Answers:* C = $109.45; F = $276.37; A = $60.78)

4-42 For diagrams (*a*) to (*d*), compute the present values of the cash flows.

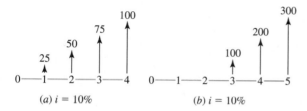

(a) i = 10%   (b) i = 10%

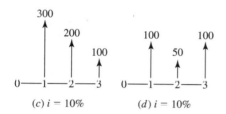

(c) i = 10%   (d) i = 10%

4-43 Compute the present value of the cash flows.

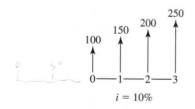

i = 10%

(*Answer:* $589.50)

4-44 Use a 15% interest rate to compute the present value of the cash flows.

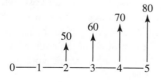

4-45 The cash flows have a present value of 0. Compute the value of C, assuming a 10% interest rate.

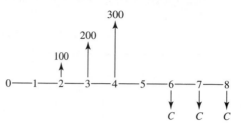

4-46 If i = 12%, for what value of G does the present value equal 0?

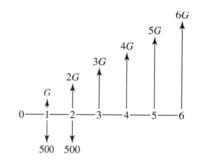

4-47 The cash flows have a present value equal 0. Compute the value of D in the diagram.

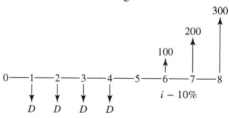

4-48 Using a 10% interest rate, for what value of B does the present value equal 0?

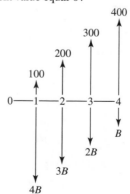

4-49 These cash flow transactions are said to be equivalent in terms of economic desirability at an interest rate of 12% compounded annually. Determine the unknown value *A*.

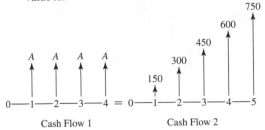

Cash Flow 1   Cash Flow 2

4-50 For what value of *P* in the cash flow diagram does the present value equal 0?

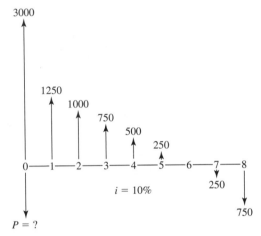

4-51 Use a 10% interest rate to compute the present value of the cash flows.

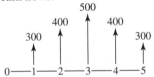

4-52 Consider the cash flow:

Year	Cash Flow
0	−$100
1	+50
2	+60
3	+70
4	+80
5	+140

Which one of the following is correct for this cash flow?

A. $100 = 50 + 10(A/G, i, 5) + 50(P/F, i, 5)$

B. $1 = \dfrac{50(P/A, i, 5) + 10(P/G, i, 5) + 50(P/F, i, 5)}{100}$

C. $100(A/P, i, 5) = 50 + 10(A/G, i, 5)$

D. None of the equations are correct.

4-53 Consider the following cash flow:

Year	Cash Flow
0	−$P
1	+1000
2	+850
3	+700
4	+550
5	+400
6	+400
7	+400
8	+400

Alice was asked to compute the value of *P* for the cash flow at 8% interest. She wrote three equations:

A. $P = 1000(P/A, 8\%, 8) - 150(P/G, 8\%, 8) + 150(P/G, 8\%, 4)(P/F, 8\%, 4)$

B. $P = 400(P/A, 8\%, 8) + 600(P/A, 8\%, 5) - 150(P/G, 8\%, 4)$

C. $P = 150(P/G, 8\%, 4) + 850(P/A, 8\%, 4) + 400(P/A, 8\%, 4)(P/F, 8\%, 4)$

Which of the equations is correct?

4-54 It is estimated that the maintenance cost on a new car will be $40 the first year. Each subsequent year, this cost is expected to increase by $10. How much would you need to set aside when you bought a new car to pay all future maintenance costs if you planned to keep the vehicle for 7 years? Assume interest is 5% per annum. (*Answer:* $393.76)

4-55 The council members of a small town have decided that the earth levee that protects the town flooding should be rebuilt and strengthened. The town engineer estimates that the cost of the work at the end of the first year will be $85,000. He estimates that in subsequent years the annual repair costs will decline by $10,000, making the second-year cost $75,000; the third-year $65,000, and so forth. The council members want to know what the equivalent present cost is for the first 5 years of repair work if interest is 4%. (*Answer:* $292,870)

4-56 A firm expects to install smog control equipment on the exhaust of a gasoline engine. The local smog control district has agreed to pay to the firm a lump sum of money to provide for the first cost of the equipment and maintenance during its 10-year useful life. At the end of 10 years the equipment, which initially cost $10,000, is valueless. The firm and the smog control district have agreed that the following are reasonable estimates of the end-of-year maintenance costs:

Year 1	$500	Year 6	$200
2	100	7	225
3	125	8	250
4	150	9	275
5	175	10	300

Assuming interest at 6% per year, how much should the smog control district pay to the firm now to provide for the first cost of the equipment and its maintenance for 10 years?   (*Answer:* $11,693)

4-57 A debt of $5000 can be repaid, with interest at 8%, by the following payments.

Year	Payment
1	$ 500
2	1000
3	1500
4	2000
5	X

The payment at the end of the fifth year is shown as X. How much is X?

4-58 A man is buying a small garden tractor. There will be no maintenance cost during the first 2 years because the tractor is sold with 2 years free maintenance. For the third year, the maintenance is estimated at $20. In subsequent years the maintenance cost will increase by $20 per year (i.e., fourth-year maintenance will be $40, fifth-year $60, etc.). How much would need to be set aside now at 8% interest to pay the maintenance costs on the tractor for the first 6 years of ownership?

4-59 A man decides to deposit $50 in the bank today and to make 10 additional deposits every 6 months beginning 6 months from now, the first of which will be $50 and increasing $10 per deposit after that. A few minutes after making the last deposit, he decides to withdraw all the money deposited. If the bank pays 6% nominal interest compounded semiannually, how much money will he receive?

4-60 A man makes an investment every 3 months at a nominal annual interest rate of 28%, compounded quarterly. His first investment was $100, followed by investments *increasing* $20 each 3 months. Thus, the second investment was $120, the third investment $140, and so on. If he continues to make this series of investments for a total of 20 years, what will be the value of the investments at the end of that time?

4-61 The following cash flows are equivalent in value if the interest rate is i. Which one is more valuable if the interest rate is $2i$?

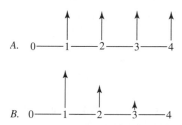

4-62 The following cash flows are equivalent in value if the interest rate is i. Which one is more valuable if the interest rate is $2i$?

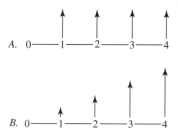

Geometric Gradients

4-63 A set of cash flows begins at $20,000 the first year, with an increase each year until $n = 10$ years. If the interest rate is 8%, what is the present value when

(*a*) the annual increase is $2000?

(*b*) the annual increase is 10%?

(*Answer:* (*b*) $201,405)

4-64 A set of cash flows begins at $50,000 the first year, with an increase each year until $n = 15$ years. If the interest rate is 7%, what is the present value when

(*a*) the annual increase is $5000?

(*b*) the annual increase is 10%?

4-65 A set of cash flows begins at $20,000 the first year, with an increase each year until $n = 10$ years. If the interest rate is 10%, what is the present value when

(*a*) the annual increase is $2000?

(*b*) the annual increase is 10%?

4-66 A set of cash flows begins at $20,000 the first year, with a decrease each year until $n = 10$ years. If the interest rate is 8%, what is the present value when

(a) the annual decrease is $2000?

(b) the annual decrease is 10%?

4-67 The market for a product is expected to increase at an annual rate of 8%. First-year sales are estimated at $60,000, the horizon is 15 years, and the interest rate is 10%. What is the present value?

4-68 Fred is evaluating whether a more efficient motor with a life of 5 years should be installed on an assembly line. Energy savings are estimated at $400 for the first year, then increasing by 6% annually. If the interest rate is 10%, what is the present value of the energy savings?

4-69 In Problem 4-68, what if the energy savings are increasing by 15% annually?

4-70 Suzanne is a recent chemical engineering graduate who has been offered a 5-year contract at a remote location. She has been offered two choices. The first is a fixed salary of $75,000 per year. The second has a starting salary of $65,000 with annual raises of 5% starting in Year 2. (For simplicity, assume that her salary is paid at the end of the year, just before her annual vacation.) If her interest rate is 9%, which should she take?

4-71 Mark Johnson saves a fixed percentage of his salary at the end of each year. This year he saved $1500. For the next 5 years, he expects his salary to increase at an 8% annual rate, and he plans to increase his savings at the same 8% annual rate. He invests his money in the stock market. Thus there will be six end-of-year investments (the initial $1500 plus five more). Solve the problem using the geometric gradient factor.

(a) How much will the investments be worth at the end of 6 years if they increase in the stock market at a 10% annual rate?

(b) How much will Mark have at the end of 6 years if his stock market investments increase only at 8% annually?

4-72 The Macintosh Company has an employee savings plan that allows every employee to invest up to 5% of his or her annual salary. The money is invested in company common stock with the company guaranteeing that the annual return will never be less than 8%. Jill was hired at an annual salary of $52,000. She immediately joined the savings plan investing the full 5% of her salary each year. If Jill's salary increases at an 8% uniform rate, and she continues to invest 5% of it each year, what amount of money is she guaranteed to have at the end of 20 years?

4-73 The football coach at a midwestern university was given a 5-year employment contract that paid $225,000 the first year, and increased at an 8% uniform rate in each subsequent year. At the end of the first year's football season, the alumni demanded that the coach be fired. The alumni agreed to buy his remaining years on the contract by paying him the equivalent present sum, computed using a 12% interest rate. How much will the coach receive?

4-74 A 25-year-old engineer is opening an individual retirement account (IRA) at a bank. Her goal is to accumulate $1 million in the account by the time she retires from work in 40 years. The bank manager estimates she may expect to receive 8% nominal annual interest, compounded quarterly, throughout the 40 years. The engineer believes her income will increase at a 7% annual rate during her career. She wishes to start her IRA with as low a deposit as possible and increase it at a 7% rate each year. Assuming end-of-year deposits, how much should she deposit the first year?

Nominal and Effective Interest Rates

4-75 A thousand dollars is invested for 7 months at an interest rate of 1% per month. What is the nominal interest rate? What is the effective interest rate? (*Answers:* 12%; 12.68%)

4-76 A firm charges its credit customers $1 3/4$% interest per month. What is the effective interest rate?

4-77 If the nominal annual interest rate is 12% compounded quarterly, what is the effective annual interest rate?

4-78 A local store charges $1 1/2$% each month on the unpaid balance for its charge account. What nominal annual interest rate is being charged? What is the effective interest rate?

4-79 What interest rate, compounded quarterly, is equivalent to a 9.31% effective interest rate?

4-80 A bank advertises it pays 7% annual interest, compounded daily, on savings accounts, provided the money is left in the account for 4 years. What is the effective annual interest rate?

4-81 At the Central Furniture Company, customers who buy on credit pay an effective annual interest rate of 16.1%, based on monthly compounding. What is the nominal annual interest rate that they pay?

4-82 Pete borrows $10,000 to purchase a car. He must repay the loan in 48 equal end-of-period monthly payments. Interest is calculated at $1^1/4\%$ per month. Determine the following:

(*a*) The nominal annual interest rate

(*b*) The effective annual interest rate

(*c*) The amount of the monthly payment

4-83 Picabo borrows $1000. To repay the amount she makes 12 equal monthly payments of $90.30. Determine the following:

(*a*) The effective monthly interest rate

(*b*) The nominal annual interest rate

(*c*) The effective annual interest rate

4-84 A student bought a $75 used guitar and agreed to pay for it with a single $85 payment at the end of 6 months. Assuming semiannual (every 6 months) compounding, what is the nominal annual interest rate? What is the effective interest rate?

4-85 The *Bawl Street Journal* costs $206, payable now, for a 2-year subscription. The newspaper is published 252 days per year (5 days per week, except holidays). If a 10% nominal annual interest rate, compounded quarterly, is used:

(*a*) What is the effective annual interest rate in this problem?

(*b*) Compute the equivalent interest rate per $1/252$ of a year.

(*c*) What is a subscriber's cost per copy of the newspaper, taking interest into account?

4-86 A bank is offering to sell 6-month certificates of deposit for $9500. At the end of 6 months, the bank will pay $10,000 to the certificate owner. Based on a 6-month interest period, compute the nominal annual interest rate and the effective annual interest rate.

4-87 You are taking a $2000 loan. You will pay it back in four equal amounts, paid every 6 months starting 3 years from now. The interest rate is 6% compounded semiannually. Calculate:

(*a*) The effective interest rate, based on both semiannual and continuous compounding

(*b*) The amount of each semiannual payment

(*c*) The total interest paid

4-88 Mr. Sansome withdrew $1000 from a savings account and invested it in common stock. At the end of 5 years, he sold the stock and received a check for $1307. If Mr. Sansome had left his $1000 in the savings account, he would have received an interest rate of 5%, compounded quarterly. Mr. Sansome would like to compute a comparable interest rate on his common stock investment. Based on quarterly compounding, what nominal annual interest rate did Mr. Sansome receive on his investment in stock? What effective annual interest rate did he receive?

4-89 The treasurer of a firm noted that many invoices were received with the following terms of payment: "2%–10 days, net 30 days." Thus, if he were to pay the bill within 10 days of its date, he could deduct 2%. On the other hand, if he did not promptly pay the bill, the full amount would be due 30 days from the date of the invoice. Assuming a 20-day compounding period, the 2% deduction for prompt payment is equivalent to what effective annual interest rate?

4-90 In 1555 King Henry borrowed money from his bankers on the condition that he pay 5% of the loan at each fair (there were four fairs per year) until he had made 40 payments. At that time the loan would be considered repaid. What effective annual interest did King Henry pay?

4-91 Jim Duggan made an investment of $10,000 in a savings account 10 years ago. This account paid interest of $5^1/2\%$ for the first 4 years and $6^1/2\%$ interest for the remaining 6 years. The interest charges were compounded quarterly. How much is this investment worth now?

4-92 A local bank offers a customer a 2-year car loan of $1000 as follows:

Money to pay for car	$1000
Two years' interest at 7%: $2 \times 0.07 \times 1000$	140
	$1140

$$24 \text{ monthly payments} = \frac{1140}{24} = \$47.50$$

The first payment must be made in 30 days. What is the nominal annual interest rate the bank is receiving?

4-93 A local lending institution advertises the "51–50 Club." A person may borrow $2000 and repay $51 for the next 50 months, beginning 30 days after receiving the money. Compute the nominal annual interest rate for this loan. What is the effective interest rate?

4-94 Ann deposits $100 at the end of each month into her bank savings account. The bank paid 6% nominal interest, compounded and paid quarterly. No interest was paid on money not in the account for the full 3-month period. How much was in Ann's account at the end of 3 years? (*Answer:* $3912.30)

4-95 The **Rule of 78's** is a commonly used method of computing the amount of interest when the balance of a loan is repaid in advance.

Adding the numbers representing 12 months gives

$$1 + 2 + 3 + 4 + 5 + \cdots + 11 + 12 = 78$$

If a 12-month loan is repaid at the end of one month, for example, the interest the borrower would be charged is 12/78 of the year's interest. If the loan is repaid at the end of 2 months, the total interest charged would be $(12 + 11)/78$, or 23/78 of the year's interest. After 11 months the interest charge would therefore be 77/78 of the total year's interest.

Helen borrowed $10,000 on January 1 at 9% annual interest, compounded monthly. The loan was to be repaid in 12 equal end-of-period payments. Helen made the first two payments and then decided to repay the balance of the loan with the third payment. Thus she will pay the third payment plus an additional sum.

You are to calculate the amount of this additional sum

(a) Based on the rule of 78s.

(b) Based on exact economic analysis methods.

Continuous Compounding

4-96 Select the best of the following five alternatives. Assume the investment is for a period of 4 years and $P = \$10,000$.

A. 11.98% interest rate compounded continuously

B. 12.00% interest rate compounded daily

C. 12.01% interest rate compounded monthly

D. 12.02% interest rate compounded quarterly

E. 12.03% interest rate compounded yearly

4-97 Traffic at a certain intersection is 2000 cars per day. A consultant has told the city that traffic is expected to grow at a continuous rate of 5% per year for the next 4 years. How much traffic will be expected at the end of 2 years?

4-98 A bank pays 10% nominal annual interest on special three-year certificates. What is the effective annual interest rate if interest is compounded:

(a) Every three months?

(b) Daily?

(c) Continuously?

4-99 A department store charges $1\frac{3}{4}\%$ interest per month, compounded continuously, on its customer's charge accounts. What is the nominal annual interest rate? What is the effective interest rate?   (*Answers:* 21%; 23.4%)

4-100 If you want a 12% rate of return, continuously compounded, on a project that will yield $6000 at the end of $2\frac{1}{2}$ years, how much must you be willing to invest now?   (*Answer:* $4444.80)

4-101 Bank North advertises, "We pay 6.50%, compounded daily." Bank South says, "We pay 6.50%, compounded continuously." If you deposit $10,000 with Bank South for one year, how much additional interest will you receive?

4-102 Bart wishes to tour the country with his friends. To do this, he is saving money for a bus.

(a) How much money must Bart deposit in a savings account paying 8% nominal annual interest, compounded continuously, in order to have $8000 in $4\frac{1}{2}$ years?

(b) A friend offers to repay Bart $8000 in $4\frac{1}{2}$ years if Bart gives him $5000 now. Assuming continuous compounding, what is the nominal annual interest rate of this offer?

4-103 The I've Been Moved Corporation receives a constant flow of funds from its worldwide operations. This money (in the form of checks) is continuously deposited in many banks with the goal of earning as much interest as possible for "IBM." One billion dollars is deposited each month, and the money earns an average of $1/2\%$ interest per month, compounded continuously. Assume all the money remains in the accounts until the end of the month.

(a) How much interest does IBM earn each month?

(b) How much interest would IBM earn each month if it held the checks and made deposits to its bank accounts just four times a month?

4-104 A forklift truck costs $29,000. A company agrees to purchase such a truck with the understanding that it will make a single payment for the balance due in 3 years. The vendor agrees to the deal and offers two different interest schedules. The first schedule uses an annual effective interest rate of 13%. The second schedule uses 12.75% compounded continuously.

(a) Which schedule should the company accept?

(b) What would be the size of the single payment?

4-105 PARC Company has money to invest in an employee benefit plan, and you have been chosen as the plan's trustee. As an employee yourself, you want to maximize the interest earned on this investment and have found an account that pays 14% compounded continuously. PARC is providing you $1200 per month to put into your account for 7 years. What will be the balance in this account at the end of the 7-year period?

4-106 Barry, a recent engineering graduate, never took engineering economics. When he graduated, he was hired by a prominent architectural firm. The earnings from this job allowed him to deposit $750 each quarter into a savings account. There were two banks that offered savings accounts in his town (a small town!). The first bank was offering 4.5% interest compounded continuously. The second bank offered 4.6% compounded monthly. Barry decided to deposit in the first bank because it offered continuous compounding. Did he make the right decision?

4-107 How long will it take for $10,000, invested at 5% per year, compounded continuously, to triple in value?

4-108 A friend was left $50,000 by his uncle. He has decided to put it into a savings account for the next year or so. He finds there are varying interest rates at savings institutions: $4\,3/8\%$ compounded annually, $4^1/_4\%$ compounded quarterly, and $4^1/_8\%$ compounded continuously. He wishes to select the savings institution that will give him the highest return on his money. What interest rate should he select?

4-109 Jack deposited $500,000 into a bank for 6 months. At the end of that time, he withdrew the money and received $520,000. If the bank paid interest based on continuous compounding:

(*a*) What was the effective annual interest rate?

(*b*) What was the nominal annual interest rate?

4-110 What single amount on October 1, 2007, is equal to a series of $1000 quarterly deposits made into an account? The first deposit occurs on October 1, 2007, and the last deposit occurs on January 1, 2021. The account earns 13% compounded continuously.

Compounding and Payment Periods Differ

4-111 Upon the birth of his first child, Dick Jones decided to establish a savings account to partly pay for his son's education. He plans to deposit $20 per month in the account, beginning when the boy is 13 months old. The savings and loan association has a current interest policy of 6% per annum, compounded monthly, paid quarterly. Assuming no change in the interest rate, how much will be in the savings account when Dick's son becomes 16 years old?

4-112 What is the present worth of a series of equal quarterly payments of $3000 that extends over a period of 8 years if the interest rate is 10% compounded monthly?

4-113 What single amount on April 1, 2008, is equivalent to a series of equal, semiannual cash flows of $1000 that starts with a cash flow on January 1, 2006, and ends with a cash flow on January 1, 2015? The interest rate is 14% and compounding is quarterly.

4-114 Paco's saving account earns 13% compounded weekly and receives quarterly deposits of $38,000. His first deposit occurred on October 1, 2006, and the last deposit is scheduled for April 1, 2022. Tisha's account earns 13% compounded weekly. Semiannual deposits of $18,000 are made into her account, with the first one occurring on July 1, 2016, and the last one occurring on January 1, 2025. What single amount on January 1, 2017, is equivalent to the sum of both cash flow series?

4-115 The first of a series of equal monthly cash flows of $2000 occurred on April 1, 2008, and the last of the monthly cash flows occurred on February 1, 2010. This series of monthly cash flows is equivalent to a series of semiannual cash flows. The first semiannual cash flow occurred on July 1, 2011, and the last semiannual cash flow will occur on January 1, 2020. What is the amount of each semiannual cash flow? Use a nominal interest rate of 12% with monthly compounding on all accounts.

4-116 A series of monthly cash flows is deposited into an account that earns 12% nominal interest compounded monthly. Each monthly deposit is equal to $2100. The first monthly deposit occurred on June 1, 2008 and the last monthly deposit will be on January 1, 2015. The account (the series of monthly deposits, 12% nominal interest, and monthly compounding) also has equivalent quarterly withdrawals from it. The first quarterly withdrawal is equal to $5000 and occurred on October 1, 2008. The last $5000 withdrawal will occur on January 1, 2015. How much remains in the account after the last withdrawal?

Spreadsheets for Economic Analysis

4-117 Develop a complete amortization table for a loan of $4500, to be paid back in 24 uniform monthly installments, based on an interest rate of 6%.

The amortization table must include the following column headings:

> Payment Number, Principal Owed (beginning of period), Interest Owed in Each Period, Total Owed (end of each period), Principal Paid in Each Payment, Uniform Monthly Payment Amount

You must also show the equations used to calculate each column of the table. You are encouraged to use spreadsheets. *The entire table must be shown.*

4-118 Using the loan and payment plan developed in Problem 4-117, determine the month that the final payment is due, and the amount of the final payment, if $500 is paid for Payment 8 and $280 is paid for Payment 10. This problem requires a separate amortization table giving the balance due, principal payment, and interest payment for each period of the loan.

4-119 The following beginning-of-month (BOM) and end-of-month (EOM) amounts are to be deposited in a savings account that pays interest at 9%, compounded monthly:

Today (BOM 1)	$400
EOM 2	270
EOM 6	100
EOM 7	180
BOM 10	200

Set up a spreadsheet to calculate the account balance at the end of the first year (EOM12). The spreadsheet must include the following column headings: Month Number, Deposit BOM, Account Balance at BOM, Interest Earned in Each Month, Deposit EOM, Account Balance at EOM. Also, use the compound interest tables to draw a cash flow diagram of this problem and solve for the account balance at the EOM 12.

4-120 What is the present worth of cash flows that begin at $20,000 and increase at 7% per year for 10 years? The interest rate is 9%. (*Answer:* $169,054)

4-121 What is the present worth of cash flows that begin at $50,000 and decrease at 12% per year for 10 years? The interest rate is 8%. (*Answer:* $217,750)

4-122 Net revenues at an older manufacturing plant will be $2 million for this year. The net revenue will decrease 15% per year for 5 years, when the assembly plant will be closed (at the end of Year 6). If the firm's interest rate is 10%, calculate the PW of the revenue stream.

4-123 What is the present worth of cash flows that begin at $10,000 and increase at 8% per year for 4 years? The interest rate is 6%.

4-124 What is the present worth of cash flows that begin at $30,000 and decrease at 15% per year for 6 years? The interest rate is 10%.

4-125 Five annual payments at an interest rate of 9% are made to repay a loan of $6000. Build the table that shows the balance due, principal payment, and interest payment for each payment. What is the annual payment? (Use a spreadsheet function, not the tables.) What interest is paid in the last year? (*Answer:* payment = $1542.55, interest = $127.37)

4-126 A newly graduated engineer bought furniture for $900 from a local store. Monthly payments for 1 year will be made. Interest is computed at a nominal rate of 6%. Build the table that shows the balance due, principal payment, and interest payment for each payment. What is the monthly payment? (Use a spreadsheet function, not the tables.) What interest is paid in the last month? (*Answer:* payment = $77.46, interest = $0.39)

4-127 Calculate and print out the balance due, principal payment, and interest payment for each period of a used-car loan. The nominal interest is 12% per year, compounded monthly. Payments are made monthly for 3 years. The original loan is for $11,000.

4-128 Calculate and print out the balance due, principal payment, and interest payment for each period of a new car loan. The nominal interest is 9% per year, compounded monthly. Payments are made monthly for 5 years. The original loan is for $17,000.

4-129 For the used car loan of Problem 4-127, graph the monthly payment.

(*a*) As a function of the interest rate (5–15%).

(*b*) As a function of the number of payments (24–48).

4-130 For the new car loan of Problem 4-128, graph the monthly payment.

(*a*) As a function of the interest rate (4–14%).

(*b*) As a function of the number of payments (36–84).

4-131 Your beginning salary is $50,000. You deposit 10% at the end of each year in a savings account that earns 6% interest. Your salary increases by 5% per year. What value does your savings book show after 40 years?

4-132 The market volume for widgets is increasing by 15% per year from current profits of $200,000. Investing in a design change will allow the profit per widget to stay steady; otherwise profits will drop 3% per year. What is the present worth of the savings over the next 5 years? Ten years? The interest rate is 10%.

4-133 A 30-year mortgage for $120,000 has been issued. The interest rate is 10% and payments are made monthly. Print out the balance due, principal payment, and interest payment for each period.

4-134 A homeowner may upgrade a furnace that runs on fuel oil to a natural gas unit. The investment will be $2500 installed. The cost of the natural gas will average $60 per month over the year, instead of the $145 per month that the fuel oil costs. If the interest rate is 9% per year, how long will it take to recover the initial investment?

4-135 Develop a general-purpose spreadsheet to calculate the balance due, principal payment, and interest payment for each period of a loan. The user inputs to the spreadsheet will be the loan amount, the number of payments per year, the number of years payments are made, and the nominal interest rate. Submit printouts of your analysis of a loan in the amount of $15,000 at 8.9% nominal rate for 36 months and for 60 months of payments.

4-136 Use the spreadsheet developed for Problem 4-135 to analyze 180-month and 360-month house loan payments. Analyze a $100,000 mortgage loan at a nominal interest rate of 7.5% and submit a graph of the interest and principal paid over time. You need not submit the printout of the 360 payments because it will not fit on one page.

CHAPTER 5

PRESENT WORTH ANALYSIS

Boeing Versus Airbus

Until December 2002, Boeing had been working on development of the Sonic Cruiser, an innovative lightweight plane that can fly near the speed of sound. This is about 15% faster than a conventional jet, so flying the Sonic Cruiser would allow airlines to cut long-haul flight times. Then Boeing shifted its focus to providing 15 to 20% better fuel efficiency at today's top commercial jet speeds. The 787 Dreamliner is now selling very well.

Meanwhile, Airbus, Boeing's rival in the passenger aircraft business, had also been working on a more efficient plane. But Airbus was betting on size rather than speed. Its new model, the A380, is known as the super-jumbo and is the largest passenger plane in the world.

By using several recently developed composite materials, Airbus hoped to reduce by 20% both the weight and the manufacturing cost. Airbus estimates that flying the composite plane will allow airlines to cut their operating costs by about 8%.

However manufacturing problems have caused significant delays in A380 deliveries, and some airlines have cancelled orders.

QUESTIONS TO CONSIDER

1. So far, Boeing has not been able to sell the Sonic Cruiser idea to airlines. Fast is good, they say, but cheap is better. Why isn't increased speed more attractive to airlines, especially over long-haul flights? Can't airlines charge more for faster flights?
2. In many cases, air travelers actually spend as much time on the ground as in the air. Going through security, changing planes, clearing customs, and picking up baggage can add hours to the travel time. How might this affect a traveler's decision about whether to pay more to fly on a faster plane like the Sonic Cruiser?
3. Initially, airlines seemed to be more interested in the advantages claimed for the new Airbus model. Can you state some reasons for this apparent preference?
4. Cutting operating costs potentially saves Airbus millions of dollars. When costs go down, profits go up. Do companies have an ethical obligation to pass savings on to customers? What are the issues involved?

After Completing This Chapter...
The student should be able to:

- Define and apply the *present worth criteria*.
- Compare two competing alternative choices using present worth (PW).
- Apply the PW model in cases with equal, unequal, and infinite project lives.
- Compare multiple alternatives using the PW criteria.
- Develop and use spreadsheets to make *present worth* calculations.

I n Chapters 3 and 4 we accomplished two important tasks. First, we presented the concept of equivalence. We can compare cash flows only if we can resolve them into equivalent values. Second, we derived a series of compound interest factors to find those equivalent values.

ASSUMPTIONS IN SOLVING ECONOMIC ANALYSIS PROBLEMS

One of the difficulties of problem solving is that most problems tend to be very complicated. It becomes apparent that *some* simplifying assumptions are needed to make complex problems manageable. The trick, of course, is to solve the simplified problem and still be satisfied that the solution is applicable to the *real* problem! In the subsections that follow, we will consider six different items and explain the customary assumptions that are made. These assumptions apply to all problems and examples, unless other assumptions are given.

End-of-Year Convention

As we indicated in Chapter 4, economic analysis textbooks and practice follow the end-of-period convention. This makes "A" a series of end-of-period receipts or disbursements.

A cash flow diagram of P, A, and F for the end-of-period convention is as follows:

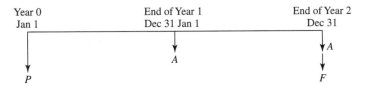

If one were to adopt a middle-of-period convention, the diagram would be:

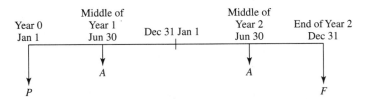

As the diagrams illustrate, only A shifts; P remains at the beginning of the period and F at the end of period, regardless of the convention. The compound interest tables in Appendix B are based on the end-of-period convention.

Viewpoint of Economic Analysis Studies

When we make economic analysis calculations, we must proceed from a point of reference. Generally, we will want to take the point of view of a total firm when doing industrial economic analyses. Example 1-1 vividly illustrated the problem: a firm's shipping department decided it could save money by outsourcing its printing work rather than by using the in-house printing department. An analysis from the viewpoint of the shipping department supported this, as it could get for $688.50 the same printing it was paying $793.50 for in-house. Further analysis showed, however, that its printing department costs would decline *less* than using the commercial printer would save. From the viewpoint of the firm, the net result would be an increase in total cost.

From Example 1-1 we see it *is* important that the **viewpoint of the study** be carefully considered. Selecting a narrow viewpoint, like that of the shipping department, may result in a suboptimal decision from the viewpoint of the firm. For this reason, the viewpoint of the total firm is used in industrial economic analyses. For public sector problems, the combined viewpoint of the government and the citizenry is chosen, since for many public projects the benefits of faster commuting, newer schools, and so on are received by individuals and the costs are paid by the government.

Sunk Costs

We know that it is the **differences between alternatives** that are relevant in economic analysis. Events that have occurred in the past really have no bearing on what we should do in the future. When the judge says, "$200 fine or 3 days in jail," the events that led to these unhappy alternatives really are unimportant. It is the current and future differences between the alternatives that *are* important. Past costs, like past events, have no bearing on deciding between alternatives unless the past costs somehow affect the present or future costs. In general, past costs do not affect the present or the future, so we refer to them as *sunk costs* and disregard them.

Borrowed Money Viewpoint

In most economic analyses, the proposed alternatives inevitably require money to be spent, and so it is natural to ask the source of that money. Thus, each problem has two monetary aspects: one is the **financing**—the obtaining of money; the other is the **investment**—the spending of money. Experience has shown that these two concerns should be distinguished. When separated, the problems of obtaining money and of spending it are both logical and straightforward. Failure to separate financing and investment sometimes produces confusing results and poor decision making.

The conventional assumption in economic analysis is that the money required to finance alternatives/solutions in problem solving is considered to be **obtained at interest rate i**.

Effect of Inflation and Deflation

For the present we will assume that prices are stable. This means that a machine that costs $5000 today can be expected to cost the same amount several years hence. Inflation and deflation can be serious problems for after-tax analysis and for cost and revenues whose inflation rates differ from the economy's inflation rates, but we assume stable prices for now.

Income Taxes

Income taxes, like inflation and deflation, must be considered to find the real payoff of a project. However, taxes will often affect alternatives similarly, allowing us to compare our choices without considering income taxes. We will introduce income taxes into economic analyses in a later chapter.

ECONOMIC CRITERIA

We have shown how to manipulate cash flows in a variety of ways, and we can now solve many kinds of compound interest problems. But engineering economic analysis is more than simply solving interest problems. The decision process (see Figure 1-1) requires that the outcomes of feasible alternatives be arranged so that they may be judged for **economic efficiency** in terms of the selection criterion. The economic criterion will be one of the following, depending on the situation:

Situation	**Criterion**
Neither input nor output fixed	Maximize (output−input)
For fixed input	Maximize output
For fixed output	Minimize input

We will now examine ways to resolve engineering problems, so that criteria for economic efficiency can be applied.

Equivalence provides the logic by which we may adjust the cash flow for a given alternative into some equivalent sum or series. We must still choose which comparable units to use. In this chapter we will learn how analysis can resolve alternatives into *equivalent present consequences*, referred to simply as **present worth analysis.** Chapter 6 will show how given alternatives are converted into an *equivalent uniform annual cash flow*, and Chapter 7 solves for the interest rate at which favorable consequences—that is, *benefits*—are equivalent to unfavorable consequences—or *costs*.

As a general rule, any economic analysis problem may be solved by any of these three methods. This is true because *present worth*, *annual cash flow*, and *rate of return* are exact methods that will always yield the same solution in selecting the best alternative from among a set of mutually exclusive alternatives. Remember that "mutually exclusive" means that selecting one alternative precludes selecting any other alternative. For example, constructing a gas station and constructing a drive-in restaurant on a particular piece of vacant land are mutually exclusive alternatives.

Some problems, however, may be more easily solved by one method. We now focus on problems that are most readily solved by present worth analysis.

APPLYING PRESENT WORTH TECHNIQUES

One of the easiest ways to compare mutually exclusive alternatives is to resolve their consequences to the present time. The three criteria for economic efficiency are restated in terms of present worth analysis in Table 5-1.

Present worth analysis is most frequently used to determine the present value of future money receipts and disbursements. It would help us, for example, to determine the present worth of an income-producing property, like an oil well or an apartment house. If the future income and costs are known, then we can use a suitable interest rate to calculate the present worth of the property. This should provide a good estimate of the price at which the property could be bought or sold. Another application is valuing stocks or bonds based on the anticipated future benefits of ownership.

In present worth analysis, careful consideration must be given to the time period covered by the analysis. Usually the task to be accomplished has a time period associated with it.

TABLE 5-1 Present Worth Analysis

	Situation	Criterion
Neither input nor output is fixed	Typical, general case	Maximize (present worth of benefits *minus* present worth of costs), that is, maximize net present worth
Fixed input	Amount of money or other input resources are fixed	Maximize present worth of benefits or other outputs
Fixed output	There is a fixed task, benefit, or other output to be accomplished	Minimize present worth of costs or other inputs

The consequences of each alternative must be considered for this period of time, which is usually called the **analysis period, planning horizon,** or **project life.**

The analysis period for an economy study should be determined from the situation. In some industries with rapidly changing technologies, a rather short analysis period or planning horizon might be in order. Industries with more stable technologies (like steelmaking) might use a longer period (say, 10–20 years), while government agencies frequently use analysis periods extending to 50 years or more.

Three different analysis-period situations are encountered in economic analysis problems with multiple alternatives:

1. The useful life of each alternative equals the analysis period.
2. The alternatives have useful lives different from the analysis period.
3. There is an infinite analysis period, $n = \infty$.

Useful Lives Equal the Analysis Period

Since different lives and an infinite analysis period present some complications, we will begin with four examples in which the useful life of each alternative equals the analysis period.

EXAMPLE 5-1

A firm is considering which of two mechanical devices to install to reduce costs. Both devices have useful lives of 5 years and no salvage value. Device *A* costs $1000 and can be expected to result in $300 savings annually. Device *B* costs $1350 and will provide cost savings of $300 the first year but will increase $50 annually, making the second-year savings $350, the third-year savings $400, and so forth. With interest at 7%, which device should the firm purchase?

SOLUTION

The analysis period can conveniently be selected as the useful life of the devices, or 5 years. The appropriate decision criterion is to choose the alternative that maximizes the net present worth of benefits minus costs.

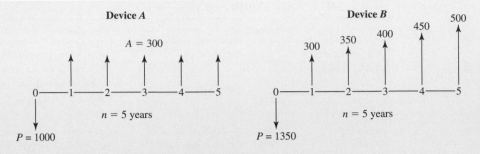

$$PW_A = 300(P/A, 7\%, 5) = -1000 + 300(4.100) = \$230$$

$$PW_B = -1350 + 300(P/A, 7\%, 5) + 50(P/G, 7\%, 5)$$

$$= -1350 + 300(4.100) + 50(7.647) = \$262.4$$

Device *B* has the larger present worth and is the preferred alternative.

EXAMPLE 5-2

Wayne County will build an aqueduct to bring water in from the upper part of the state. It can be built at a reduced size now for $300 million and be enlarged 25 years hence for an additional $350 million. An alternative is to construct the full-sized aqueduct now for $400 million.

Both alternatives would provide the needed capacity for the 50-year analysis period. Maintenance costs are small and may be ignored. At 6% interest, which alternative should be selected?

SOLUTION

This problem illustrates staged construction. The aqueduct may be built in a single stage, or in a smaller first stage followed many years later by a second stage to provide the additional capacity when needed.

For the Two-Stage Construction

$$PW \text{ of cost} = \$300 \text{ million} + 350 \text{ million}(P/F, 6\%, 25)$$

$$= \$300 \text{ million} + 81.6 \text{ million}$$

$$= \$381.6 \text{ million}$$

For the Single-Stage Construction

$$PW \text{ of cost} = \$400 \text{ million}$$

The two-stage construction has a smaller present worth of cost and is the preferred construction plan.

EXAMPLE 5-3

A purchasing agent plans to buy some new equipment for the mailroom. Two manufacturers have provided bids. An analysis shows the following:

Manufacturer	Cost	Useful Life (years)	End-of-Useful-Life Salvage Value
Speedy	$1500	5	$200
Allied	1600	5	325

The equipment of both manufacturers is expected to perform at the desired level of (fixed) output. For a 5-year analysis period, which manufacturer's equipment should be selected? Assume 7% interest and equal maintenance costs.

SOLUTION

For fixed output, the criterion is to minimize the present worth of cost.

Speedy

$$\text{PW of cost} = 1500 - 200(P/F, 7\%, 5)$$
$$= 1500 - 200(0.7130)$$
$$= 1500 - 143 = \$1357$$

Allied

$$\text{PW of cost} = 1600 - 325(P/F, 7\%, 5) = 1600 - 325(0.7130)$$
$$= 1600 - 232 = \$1368$$

Since it is only the *differences between alternatives* that are relevant, maintenance costs may be left out of the economic analysis. Although the PWs of cost for the two alternatives are nearly identical, we would still choose the one with minimum present worth of cost unless there were other tangible or intangible differences that would change the decision. Buy the Speedy equipment.

EXAMPLE 5-4

A firm is trying to decide which of two weighing scales it should install to check a package-filling operation in the plant. The ideal scale would allow better control of the filling operation, hence less overfilling. If both scales have lives equal to the 6-year analysis period, which one should be selected? Assume an 8% interest rate.

Alternatives	Cost	Uniform Annual Benefit	End-of-Useful-Life Salvage Value
Atlas scale	$2000	$450	$100
Tom Thumb scale	3000	600	700

SOLUTION

Atlas Scale

$$\text{PW of benefits} - \text{PW of cost} = 450(P/A, 8\%, 6) + 100(P/F, 8\%, 6) - 2000$$
$$= 450(4.623) + 100(0.6302) - 2000$$
$$= 2080 + 63 - 2000 = \$143$$

Tom Thumb Scale

$$\text{PW of benefits} - \text{PW of cost} = 600(P/A, 8\%, 6) + 700(P/F, 8\%, 6) - 3000$$
$$= 600(4.623) + 700(0.6302) - 3000$$
$$= 2774 + 441 - 3000 = \$215$$

The salvage value of each scale, it should be noted, is simply treated as another positive cash flow. Since the criterion is to maximize the present worth of benefits minus the present worth of cost, the preferred alternative is the Tom Thumb scale.

In Examples 5-1 and 5-4, we compared two alternatives and selected the one in which present worth of benefits *minus* present worth of cost was a maximum. The criterion is called the **net present worth criterion** and written simply as **NPW:**

$$\text{Net present worth} = \text{Present worth of benefits} - \text{Present worth of cost}$$
$$\text{NPW} = \text{PW of benefits} - \text{PW of cost} \tag{5-1}$$

The field of engineering economy and this text often use present worth (PW), present value (PV), net present worth (NPW), and net present value (NPV) as synonyms. Sometimes, as in the foregoing definition, *net* is included to emphasize that both costs and benefits have been considered.

Useful Lives Different from the Analysis Period

In present worth analysis, there always must be an identified analysis period. It follows, then, that each alternative must be considered for the entire period. In Examples 5-1 to 5-4, the useful life of each alternative was equal to the analysis period. While often this is true, in many situations at least one alternative will have a useful life different from the analysis period. This section describes one way to evaluate alternatives with lives different from the study period.

In Example 5-3, suppose that the Allied equipment was expected to have a 10-year useful life, or twice that of the Speedy equipment. Assuming the Allied salvage value would still be $325 in 10 years, which equipment should now be purchased? We can recompute the present worth of cost of the Allied equipment, starting as follows:

$$PW \text{ of cost} = 1600 - 325(P/F, 7\%, 10)$$

$$= 1600 - 325(0.5083)$$

$$= 1600 - 165 = \$1435$$

The present worth of cost has increased. This, of course, is because of the more distant recovery of the salvage value. More importantly, we now find ourselves attempting to compare Speedy equipment, with its 5-year life, against the Allied equipment with a 10-year life. Because of the variation in the useful life of the equipment, we no longer have a situation of *fixed output*. Speedy equipment in the mailroom for 5 years is certainly not the same as 10 years of service with Allied equipment.

For present worth calculations, it is important that we select an analysis period and judge the consequences of each of the alternatives during that period. As such, it is not a fair comparison to compare the NPW of the Allied equipment over its 10-year life against the NPW of the Speedy equipment over its 5-year life.

The firm, its economic environment, and the specific situation are important in selecting an analysis period. If the Allied equipment (Example 5-3) has a useful life of 10 years, and the Speedy equipment will last 5 years, one method is to select an analysis period that is the **least common multiple** of their useful lives. Thus we would compare the 10-year life of Allied equipment against an initial purchase of Speedy equipment *plus* its replacement with new Speedy equipment in 5 years. The result is to judge the alternatives on the basis of a 10-year requirement in the mailroom. On this basis the economic analysis is as follows.

Assuming the replacement Speedy equipment 5 years hence will also cost $1500,

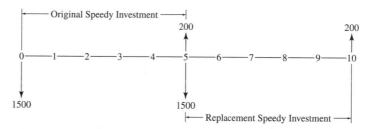

$$PW \text{ of cost} = 1500 + (1500 - 200)(P/F, 7\%, 5) - 200(P/F, 7\%, 10)$$

$$= 1500 + 1300(0.7130) - 200(0.5083)$$

$$= 1500 + 927 - 102 = \$2325$$

For the Allied equipment, on the other hand, we have the following results:

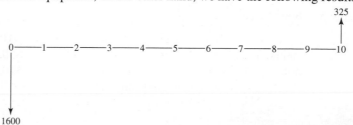

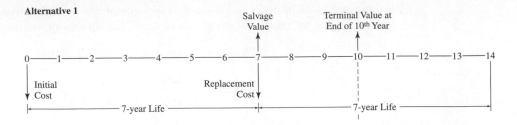

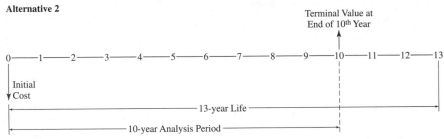

FIGURE 5-1 Superimposing a 10-year analysis period on 7- and 13-year alternatives.

$$PW \text{ of cost} = 1600 - 325(P/F, 7\%, 10) = 1600 - 325(0.5083) = \$1435$$

For the fixed output of 10 years of service in the mailroom, the Allied equipment, with its smaller present worth of cost, is preferred.

    We have seen that setting the analysis period equal to the least common multiple of the lives of the two alternatives seems reasonable in the revised Example 5-3. However, what if the alternatives had useful lives of 7 and 13 years? Here the least common multiple of lives is 91 years. An analysis period of 91 years hardly seems realistic. Instead, a suitable analysis period should be based on how long the equipment is likely to be needed. This may require that terminal values be estimated for the alternatives at some point prior to the end of their useful lives.

    As Figure 5-1 shows, it is not necessary for the analysis period to equal the useful life of an alternative or some multiple of the useful life. To properly reflect the situation at the end of the analysis period, an estimate is required of the market value of the equipment at that time. The calculations might be easier if everything came out even, but this is not essential.

EXAMPLE 5-5

A diesel manufacturer is considering the two alternative production machines graphically depicted in Figure 5-1. Specific data are as follows:

	Alt. 1	Alt. 2
Initial cost	$50,000	$75,000
Estimated salvage value at end of useful life	$10,000	$12,000
Useful life of equipment, in years	7	13

The manufacturer uses an interest rate of 8% and wants to use the PW method to compare these alternatives over an analysis period of 10 years.

	Alt. 1	**Alt. 2**
Estimated market value, end of 10-year analysis period	$20,000	$15,000

SOLUTION

In this case, the decision maker is setting the analysis period at 10 years rather than accepting a common multiple of the lives of the alternatives, or assuming that the period of needed service is infinite (to be discussed in the next section). This is a legitimate approach—perhaps the diesel manufacturer will be phasing out this model at the end of the 10-year period. In any event, we need to compare the alternatives over 10 years.

As illustrated in Figure 5-1, we may assume that Alternative 1 will be replaced by an identical machine after its 7-year useful life. Alternative 2 has a 13-year useful life. The diesel manufacturer has provided an estimated market value of the equipment at the time of the analysis period. We can compare the two choices over 10 years as follows:

$$PW \,(Alt. \,1) = -50{,}000 + (10{,}000 - 50{,}000)(P/F, 8\%, 7) + 20{,}000(P/F, 8\%, 10)$$

$$= -50{,}000 - 40{,}000(0.5835) + 20{,}000(0.4632)$$

$$= -\$64{,}076$$

$$PW \,(Alt. \,2) = -75{,}000 + 15{,}000(P/F, 8\%, 10)$$

$$= -75{,}000 + 15{,}000(0.4632)$$

$$= -\$69{,}442$$

To minimize PW of costs the diesel manufacturer should select Alt. 1.

Infinite Analysis Period: Capitalized Cost

Another difficulty in present worth analysis arises when we encounter an infinite analysis period ($n = \infty$). In governmental analyses, a service or condition sometimes must be maintained for an infinite period. The need for roads, dams, pipelines, and so on, is sometimes considered to be permanent. In these situations a present worth of cost analysis would have an infinite analysis period. We call this particular analysis **capitalized cost.**

Infinite lives are rare in the private sector, but a similar assumption of "indefinitely long" horizons is sometimes made. This assumes that the facility will need electric motors, mechanical HVAC equipment, and forklifts as long as it operates, and that the facility will last far longer than any individual unit of equipment. So the equipment can be analyzed as though the problem horizon is *infinite* or indefinitely long.

Capitalized cost is the present sum of money that would need to be set aside now, at some interest rate, to yield the funds required to provide the service (or whatever) indefinitely. To accomplish this, the money set aside for future expenditures must not decline. The interest received on the money set aside can be spent, but not the principal. When one stops to think about an infinite analysis period (as opposed to something relatively short, like a hundred years), we see that an undiminished principal sum is essential; otherwise one will of necessity run out of money prior to infinity.

In Chapter 4 we saw that

$$\text{Principal sum} + \text{Interest for the period} = \text{Amount at end of period, or}$$

$$P \quad + \quad iP \quad = \quad P + iP$$

If we spend iP, then in the next interest period the principal sum P will again increase to $P + iP$. Thus, we can again spend iP.

This concept may be illustrated by a numerical example. Suppose you deposited $200 in a bank that paid 4% interest annually. How much money could be withdrawn each year without reducing the balance in the account below the initial $200? At the end of the first year, the $200 would have earned 4%($200) = $8 interest. If this interest were withdrawn, the $200 would remain in the account. At the end of the second year, the $200 balance would again earn 4%($200) = $8. This $8 could also be withdrawn and the account would still have $200. This procedure could be continued indefinitely and the bank account would always contain $200. If more or less than $8 is withdrawn, the account will either increase to ∞ or decrease to 0.

The year-by-year situation would be depicted like this:

Year 1: $200 initial $P \rightarrow 200 + 8 = 208$
Withdrawal $iP = -\ 8$

Year 2: $200 \rightarrow 200 + 8 = 208$
Withdrawal $iP = -\ 8$
$200

and so on

Thus, for any initial present sum P, there can be an end-of-period withdrawal of A equal to iP each period, and these withdrawals can continue forever without diminishing the initial sum P. This gives us the basic relationship:

$$\text{For} \quad n = \infty, \quad A = Pi$$

This relationship is the key to capitalized cost calculations. Earlier we defined capitalized cost as the present sum of money that would need to be set aside at some interest rate to yield the funds to provide the desired task or service forever. Capitalized cost is therefore the P in the equation $A = iP$. It follows that:

$$\textbf{Capitalized cost} \quad P = \frac{A}{i} \tag{5-2}$$

If we can resolve the desired task or service into an equivalent A, the capitalized cost can be computed. The following examples illustrate such computations.

EXAMPLE 5-6

How much should one set aside to pay $50 per year for maintenance on a gravesite if interest is assumed to be 4%? For perpetual maintenance, the principal sum must remain undiminished after the annual disbursement is made.

SOLUTION

$$\text{Capitalized cost } P = \frac{\text{Annual disbursement } A}{\text{Interest rate } i}$$

$$P = \frac{50}{0.04} = \$1250$$

One should set aside $1250.

EXAMPLE 5-7

A city plans a pipeline to transport water from a distant watershed area to the city. The pipeline will cost $8 million and will have an expected life of 70 years. The city expects to keep the water line in service indefinitely. Compute the capitalized cost, assuming 7% interest.

SOLUTION

The capitalized cost equation

$$P = \frac{A}{i}$$

is simple to apply when there are end-of-period disbursements A. Here we have renewals of the pipeline every 70 years. To compute the capitalized cost, it is necessary to first compute an end-of-period disbursement A that is equivalent to $8 million every 70 years.

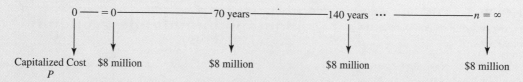

The $8 million disbursement at the end of each 70-year period may be resolved into an equivalent A.

$$A = F(A/F, i, n) = \$8 \text{ million}(A/F, 7\%, 70)$$

$$= \$8 \text{ million}(0.00062)$$

$$= \$4960$$

Each 70-year period is identical to this one, and the infinite series is shown in Figure 5-2.

$$\text{Capitalized cost } P = 8 \text{ million} + \frac{A}{i} = 8 \text{ million} + \frac{4960}{0.07}$$

$$= \$8,071,000$$

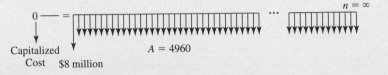

FIGURE 5-2
Using the sinking fund factor to compute an infinite series.

Capitalized Cost $8 million

$A = 4960$

$n = \infty$

ALTERNATE SOLUTION 1

Instead of solving for an equivalent end-of-period payment A based on a *future* $8 million disbursement, we could find A, given a *present* $8 million disbursement.

$$A = P(A/P, i, n) = 8 \text{ million}(A/P, 7\%, 70)$$

$$= 8 \text{ million}(0.0706) = \$565,000$$

On this basis, the infinite series is shown in Figure 5-3. Carefully note the difference between this and Figure 5-2. Now:

$$\text{Capitalized cost } P = \frac{A}{i} = \frac{565,000}{0.07} = \$8,071,000$$

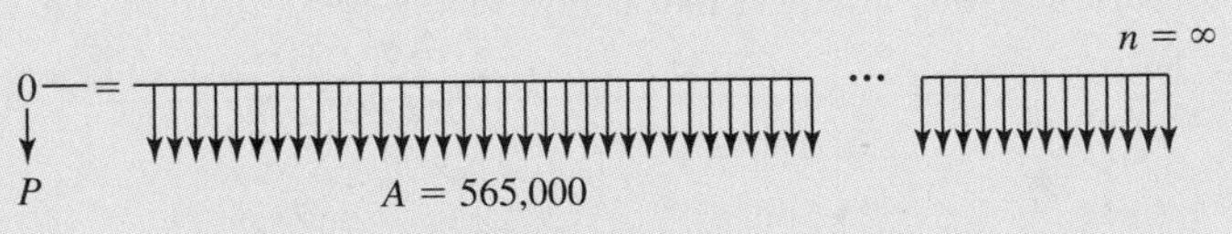

FIGURE 5-3 Using the capital recovery factor to compute an infinite series.

P

$A = 565,000$

$n = \infty$

ALTERNATE SOLUTION 2

Another way of solving the problem is to assume the interest period is 70 years long and compute an equivalent interest rate for the 70-year period. Then the capitalized cost may be computed by using Equation 4-33 for $m = 70$

$$i_{70\,\text{yr}} = (1 + i_{1\,\text{yr}})^{70} - 1 = (1 + 0.07)^{70} - 1 = 112.989$$

$$\text{Capitalized cost} = 8 \text{ million} + \frac{8 \text{ million}}{112.989} = \$8,070,803$$

Multiple Alternatives

So far the discussion has been based on examples with only two alternatives. But multiple-alternative problems may be solved by exactly the same methods. (The only reason for avoiding multiple alternatives was to simplify the examples.) Examples 5-8 and 5-9 have multiple alternatives.

EXAMPLE 5-8

A contractor has been awarded the contract to construct a 6-mile-long tunnel in the mountains. During the 5-year construction period, the contractor will need water from a nearby stream. He will construct a pipeline to carry the water to the main construction yard. An analysis of costs for various pipe sizes is as follows:

	Pipe Sizes (in.)			
	2	**3**	**4**	**6**
Installed cost of pipeline and pump	$22,000	$23,000	$25,000	$30,000
Cost per hour for pumping	$1.20	$0.65	$0.50	$0.40

At the end of 5 years, the pipe and pump will have a salvage value equal to the cost of removing them. The pump will operate 2000 hours per year. The lowest interest rate at which the contractor is willing to invest money is 7%. (The minimum required interest rate for invested money is called the **minimum attractive rate of return,** or MARR.) Select the alternative with the least present worth of cost.

SOLUTION

We can compute the present worth of cost for each alternative. For each pipe size, the present worth of cost is equal to the installed cost of the pipeline and pump plus the present worth of 5 years of pumping costs.

	Pipe Size (in.)			
	2	**3**	**4**	**6**
Installed cost of pipeline and pump	$22,000	$23,000	$25,000	$30,000
1.20×2000 hr $\times$ $(P/A, 7\%, 5)$	9,840			
0.65×2000 hr $\times$ 4.100		5,330		
0.50×2000 hr $\times$ 4.100			4,100	
0.40×2000 hr $\times$ 4.100				3,280
Present worth of cost	$31,840	$28,330	$29,100	$33,280

Select the 3 in. pipe.

EXAMPLE 5-9

An investor paid $8000 to a consulting firm to analyze possible uses for a small parcel of land on the edge of town that can be bought for $30,000. In their report, the consultants suggested four alternatives:

Alternatives		Total Investment Including Land*	Uniform Net Annual Benefit	Terminal Value at End of 20 yr
A	Do nothing	$ 0	$ 0	$ 0
B	Vegetable market	50,000	5,100	30,000
C	Gas station	95,000	10,500	30,000
D	Small motel	350,000	36,000	150,000

*Includes the land and structures but does not include the $8000 fee to the consulting firm.

Assuming 10% is the minimum attractive rate of return, what should the investor do?

SOLUTION

Alternative A is the "do-nothing" alternative. Generally, one feasible alternative in any situation is to remain in the present status and do nothing. In this problem, the investor could decide that the most attractive alternative is not to purchase the property and develop it. This is clearly a decision to do nothing.

We note, however, that if he does nothing, the total venture would not be a very satisfactory one. This is because the investor spent $8000 for professional advice on the possible uses of the property. But because the $8000 is a past cost, it is a **sunk cost.** The only relevant costs in an economic analysis are *present* and *future* costs. Past events and past, or sunk, costs are gone and cannot be allowed to affect future planning. (Past costs may be relevant in computing depreciation charges and income taxes, but nowhere else.) The past should not deter the investor from making the best decision now, regardless of the costs that brought him to this situation and point of time.

This problem is one of neither fixed input nor fixed output, so our criterion will be to maximize the present worth of benefits *minus* the present worth of cost; that is, to maximize net present worth.

Alternative A, Do Nothing

$$NPW = 0$$

Alternative B, Vegetable Market

$$NPW = -50,000 + 5100(P/A, 10\%, 20) + 30,000(P/F, 10\%, 20)$$

$$= -50,000 + 5100(8.514) + 30,000(0.1486)$$

$$= -50,000 + 43,420 + 4460$$

$$= -\$2120$$

Alternative C, Gas Station

$$NPW = -95{,}000 + 10{,}500(P/A, 10\%, 20) + 30{,}000(P/F, 10\%, 20)$$

$$= -95{,}000 + 89{,}400 + 4460$$

$$= -\$1140$$

Alternative D, Small Motel

$$NPW = -350{,}000 + 36{,}000(P/A, 10\%, 20) + 150{,}000(P/F, 10\%, 20)$$

$$= -350{,}000 + 306{,}500 + 22{,}290$$

$$= -\$21{,}210$$

The criterion is to maximize net present worth. In this situation, one alternative has NPW equal to zero, and three alternatives have negative values for NPW. We will select the best of the four alternatives, namely, the do-nothing Alt. A, with NPW equal to zero.

EXAMPLE 5-10

A piece of land may be purchased for $610,000 to be strip-mined for the underlying coal. Annual net income will be $200,000 for 10 years. At the end of the 10 years, the surface of the land will be restored as required by a federal law on strip mining. The reclamation will cost $1.5 million more than the resale value of the land after it is restored. Using a 10% interest rate, determine whether the project is desirable.

SOLUTION

The investment opportunity may be described by the following cash flow:

Year	Cash Flow (thousands)
0	−$610
1–10	+200 (per year)
10	−1500

$$NPW = -610 + 200(P/A, 10\%, 10) - 1500(P/F, 10\%, 10)$$

$$= -610 + 200(6.145) - 1500(0.3855)$$

$$= -610 + 1229 - 578$$

$$= +\$41$$

Since NPW is positive, the project is desirable. (See Appendix 7A for a more complete analysis of this type of problem. At interest rates of 4.07% and 18.29%, NPW = 0.)

EXAMPLE 5-11

Two pieces of construction equipment are being analyzed:

Year	Alt. A	Alt. B
0	−$2000	−$1500
1	+1000	+700
2	+850	+300
3	+700	+300
4	+550	+300
5	+400	+300
6	+400	+400
7	+400	+500
8	+400	+600

At an 8% interest rate, which alternative should be selected?

SOLUTION

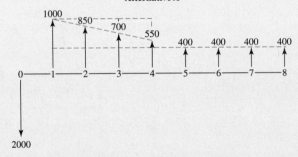

Alternative A

$$\text{PW of benefits} = 400(P/A, 8\%, 8) + 600(P/A, 8\%, 4) - 150(P/G, 8\%, 4)$$

$$= 400(5.747) + 600(3.312) - 150(4.650) = 3588.50$$

$$\text{PW of cost} = 2000$$

$$\text{Net present worth} = 3588.50 - 2000 = +\$1588.50$$

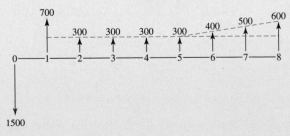

Alternative B

$$\text{PW of benefits} = 300(P/A, 8\%, 8) + (700 - 300)(P/F, 8\%, 1)$$
$$+ 100(P/G, 8\%, 4)(P/F, 8\%, 4)$$
$$= 300(5.747) + 400(0.9259) + 100(4.650)(0.7350)$$
$$= 2436.24$$
$$\text{PW of cost} = 1500$$
$$\text{Net present worth} = 2436.24 - 1500$$
$$= +\$936.24$$

To maximize NPW, choose Alt. A.

SPREADSHEETS AND PRESENT WORTH

Spreadsheets make it easy to build more accurate models with shorter time periods. When one is using factors, it is common to assume that costs and revenues are uniform for n years. With spreadsheets it is easy to use 120 months instead of 10 years, and the cash flows can be estimated for each month. For example, energy costs for air conditioning peak in the summer, and in many areas there is little construction during the winter. Cash flows that depend on population often increase at $x\%$ per year, such as for electric power and transportation costs.

In spreadsheets any interest rate is entered exactly—so no interpolation is needed. This makes it easy to calculate the monthly repayment schedule for a car loan or a house mortgage. Examples 5-12 and 5-13 illustrate using spreadsheets to calculate PWs.

EXAMPLE 5-12

NLE Construction is bidding on a project whose costs are divided into $30,000 for start-up and $240,000 for the first year. If the interest rate is 1% per month or 12.68% per year, what is the present worth with monthly compounding?

SOLUTION

Figure 5-4 illustrates the spreadsheet solution with the assumption that costs are distributed evenly throughout the year ($-20,000 = -240,000/12$).

FIGURE 5-4 Spreadsheet with monthly cash flows.

	A	B	C	D
1	1%	*i*		
2	−30,000	initial cash flow		
3	−240,000	annual amount		
4				
5	Month	Cash Flow		
6	0	−30000		
7	1	−20000		
8	2	−20000		
9	3	−20000		
10	4	−20000		
11	5	−20000		
12	6	−20000		
13	7	−20000		
14	8	−20000		
15	9	−20000		
16	10	−20000		
17	11	−20000		
18	12	−20000		
19	NPV	−$255,102	=NPV(A1,B7:B18)+B6	

Since the costs are uniform, the factor solution is:

$$PW_{mon} = -30,000 - 20,000(P/A, 1\%, 12) = -\$255,102$$

The value of monthly periods can be illustrated by computing the PW assuming an annual period. The results differ by more than $12,000, because $20,000 at the end of Months 1 through 12 is not the same as $240,000 at the end of Month 12. The timing of the cash flows makes the difference, even though the effective interest rates are the same.

$$PW_{annual} = -30,000 - 240,000(P/F, 12.68\%, 1) = -30,000 - 240,000/1.1268 = -\$242,993$$

EXAMPLE 5-13

Regina Industries has a new product whose sales are expected to be 1.2, 3.5, 7, 5, and 3 million units per year over the next 5 years. Production, distribution, and overhead costs are stable at $120 per unit. The price will be $200 per unit for the first 2 years, and then $180, $160, and $140 for the next 3 years. The remaining R&D and production costs are $300 million. If *i* is 15%, what is the present worth of the new product?

SOLUTION

It is easiest to calculate the yearly net revenue per unit before building the spreadsheet shown in Figure 5-5. Those values are the yearly price minus the $120 of costs, which equals $80, $80, $60, $40, and $20.

	A	B	C	D	E
1	12%	i			
2			Net Unit	Cash	
3	Year	Sales (M)	Revenue	Flow ($M)	
4	0			−300	
5	1	1.2	80	96	
6	2	3.5	80	280	
7	3	7	60	420	
8	4	5	40	200	
9	5	3	20	60	
10	D4+NPV(A1,D5:D9) =			$469	Million

FIGURE 5-5 Present worth of a new product.

SUMMARY

Present worth analysis is suitable for almost any economic analysis problem. But it is particularly desirable when we wish to know the present worth of future costs and benefits. And we frequently want to know the value today of such things as income-producing assets, stocks, and bonds.

For present worth analysis, the proper economic criteria are:

Neither input nor output is fixed	Maximize (PW of benefits − PW of costs) or, more simply stated: Maximize NPW
Fixed input	Maximize the PW of benefits
Fixed output	Minimize the PW of costs

To make valid comparisons, we need to analyze each alternative in a problem over the same **analysis period** or **planning horizon.** If the alternatives do not have equal lives, some technique must be used to achieve a common analysis period. One method is to select an analysis period equal to the least common multiple of the alternative lives. Another method is to select an analysis period and estimate end-of-analysis-period salvage values for the alternatives.

Capitalized cost is the present worth of cost for an infinite analysis period ($n = \infty$). When $n = \infty$, the fundamental relationship is $A = iP$. Some form of this equation is used whenever there is a problem with an infinite analysis period.

The numerous assumptions routinely made in solving economic analysis problems include the following.

1. Present sums P are beginning-of-period, and all series receipts or disbursements A and future sums F occur at the end of the interest period. The compound interest tables were derived on this basis.

2. In industrial economic analyses, the point of views for computing the consequences of alternatives is that of the total firm. Narrower views can result in suboptimal solutions.

3. Only the differences between the alternatives are relevant. Past costs are sunk costs and generally do not affect present or future costs. For this reason they are ignored.

4. The investment problem is isolated from the financing problem. We generally assume that all required money is borrowed at interest rate i.

5. For now, stable prices are assumed. The inflation–deflation problem is deferred to Chapter 14. Similarly, our discussion of income taxes is deferred to Chapter 12.

6. Often uniform cash flows or arithmetic gradients are reasonable assumptions. However, spreadsheets simplify the finding of PW in more complicated problems.

PROBLEMS

Present Value of One Alternative

5-1 Compute the present value, P, for the following cash flows.

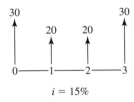

$i = 15\%$

5-3 Compute the present value, P, for the following cash flows.

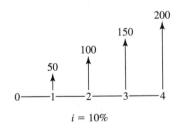

$i = 10\%$

5-2 Compute the present value, P, for the following cash flows.

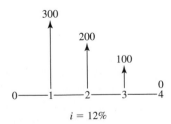

$i = 12\%$

(*Answer*: $P = \$498.50$)

5-4   Find the value of Q so that the present value is 0.

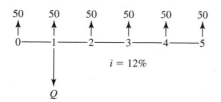

5-5   Compute the present value, P, for the following cash flows.

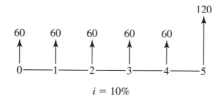

5-6   Compute the present value, P, for the following cash flows.

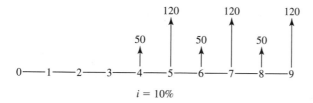

(*Answer:* $P = 324.71)

5-7   Use a geometric gradient formula to compute the present value, P, for the following cash flows.

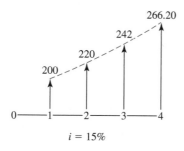

5-8   If $i = 10\%$, compute the present value, P, for the following cash flows.

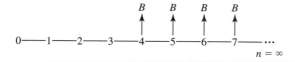

5-9   A stonecutter, assigned to carve the headstone for a well-known engineering economist, began with the following design.

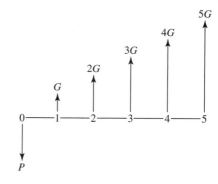

He then started the equation as follows:

$$P = G(P/G, i, 6)$$

He realized he had made a mistake. The equation should have been

$$P = G(P/G, i, 5) + G(P/A, i, 5)$$

The stonecutter does not want to discard the stone and start over. He asks you to help him with his problem. What one compound interest factor can be added to make the equation correct?

$$P = G(P/G, i, 6)(\quad , i, \quad)$$

5-10   Compute the present value, P, for the following cash flows (assume series repeats forever).

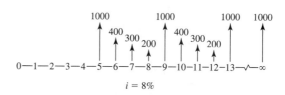

5-11   A project has a net present worth of -140 as of January 1, 2010. If a 10% interest rate is used, what was the project NPW as of December 31, 2007?

5-12   The annual income from a rented house is $12,000. The annual expenses are $3000. If the house can be sold for $145,000 at the end of 10 years, how much could you afford to pay for it now, if you considered 18% to be a suitable interest rate? (*Answer:* $68,155)

5-13 Consider the following cash flow. At a 6% interest rate, what value of P at the end of Year 1 is equivalent to the benefits at the end of Years 2 through 7?

Year	Cash Flow
1	$-P$
2	$100
3	200
4	300
5	400
6	500
7	600

5-14 How much would the owner of a building be justified in paying for a sprinkler system that will save $750 a year in insurance premiums if the system has to be replaced every 20 years and has a salvage value equal to 10% of its initial cost? Assume money is worth 7%.   (*Answer:* $8156)

5-15 A manufacturer is considering purchasing equipment which will have the following financial effects:

Year	Disbursements	Receipts
0	$4400	$ 0
1	660	880
2	660	1980
3	440	2420
4	220	1760

If money is worth 6%, should he invest in the equipment?

5-16 A corporate bond has a face value of $1000 with maturity date 20 years from today. The bond pays interest semiannually at a rate of 8% per year based on the face value. The interest rate paid on similar corporate bonds has decreased to a current rate of 6%. Determine the market value of the bond.

5-17 Calculate the present worth of a 4.5%, $5000 bond with interest paid semiannually. The bond matures in 10 years, and the investor wants to make 8% per year compounded quarterly on the investment.

5-18 In a present worth analysis of certain equipment, one alternative has a net present worth of +420, based on a 6-year analysis period that equals the useful life of the alternative. A 10% interest rate was used in the computations.

The alternative device is to be replaced at the end of the 6 years by an identical item with the same cost, benefits, and useful life. Using a 10% interest rate, compute the net present worth of the alternative equipment for the 12-year analysis period. (*Answer:* NPW = +657.09)

5-19 On February 1, the Miro Company needs to purchase some office equipment. The company is short of cash and expects to be short for several months. The treasurer has said that he could pay for the equipment as follows:

Date	Payment
April 1	$150
June 1	300
Aug. 1	450
Oct. 1	600
Dec. 1	750

A local office supply firm will agree to sell the equipment to Miro now and accept payment according to the treasurer's schedule. If interest will be charged at 3% every 2 months, with compounding once every 2 months, how much office equipment can the Miro Company buy now? (*Answer:* $2020)

5-20 A machine costs $980,000 to purchase and will provide $200,000 a year in benefits. The company plans to use the machine for 13 years and then will sell the machine for scrap, receiving $20,000. The company interest rate is 12%. Should the machine be purchased?

5-21 Annual maintenance costs for a particular section of highway pavement are $2000. The placement of a new surface would reduce the annual maintenance cost to $500 per year for the first 5 years and to $1000 per year for the next 5 years. After 10 years the annual maintenance would again be $2000. If maintenance costs are the only saving, what investment can be justified for the new surface. Assume interest at 4%.

5-22 An investor is considering buying a 20-year corporate bond. The bond has a face value of $1000 and pays 6% interest per year in two semiannual payments. Thus the purchaser of the bond will receive $30 every 6 months in addition to $1000 at the end of 20 years, along with the last $30 interest payment.

If the investor wants to receive 8% interest, compounded semiannually, how much would he or she be willing to pay for the bond?

5-23   A road building contractor has received a major highway construction contract that will require 50,000 m³ of crushed stone each year for 5 years. The stone can be obtained from a quarry for $5.80/m³. As an alternative, the contractor has decided to try to buy the quarry. He believes that if he owned the quarry, the stone would cost him only $4.30/m³. He thinks he could resell the quarry at the end of 5 years for $40,000. If the contractor uses a 10% interest rate, how much would he be willing to pay for the quarry?

5-24   A new office building was constructed 5 years ago by a consulting engineering firm. At that time the firm obtained a bank loan for $100,000 with a 12% annual interest rate, compounded quarterly. The terms of the loan call for equal quarterly payments to repay the loan in 10 years. The loan also allows for its prepayment at any time without penalty.

   As a result of internal changes in the firm, it is now proposed to refinance the loan through an insurance company. The new loan would be for a 20-year term with an interest rate of 8% per year, compounded quarterly. The new equal quarterly payments would repay the loan in the 20-year period. The insurance company requires the payment of a 5% loan initiation charge (often described as a "5-point loan fee"), which will be added to the new loan.

   (a) What is the balance due on the original mortgage if 20 payments have been made in the last five years?

   (b) What is the difference between the equal quarterly payments on the present bank loan and the proposed insurance company loan?

5-25   IBP Inc. is considering establishing a new machine to automate a meatpacking process. The machine will save $50,000 in labor annually. The machine can be purchased for $200,000 today and will be used for 10 years. It has a salvage value of $10,000 at the end of its useful life. The new machine will require an annual maintenance cost of $9000. The corporation has a minimum rate of return of 10%. Do you recommend automating the process?

5-26   Argentina is considering constructing a bridge across the Rio de la Plata to connect its northern coast to the southern coast of Uruguay. If this bridge is constructed, it will reduce the travel time from Buenos Aires, Argentina, to São Paulo, Brazil, by over 10 hours, and there is the potential to significantly improve the flow of manufactured goods between the two countries. The cost of the new bridge, which will be the longest bridge in the world, spanning over 50 miles, will be $700 million. The bridge will require an annual maintenance of $10 million for repairs and upgrades and is estimated to last 80 years. It is estimated that 550,000 vehicles will use the bridge during the first year of operation, and an additional 50,000 vehicles per year until the tenth year. These data are based on a toll charge of $90 per vehicle. The annual traffic for the remainder of the life of the bridge will be 1,000,000 vehicles per year. The Argentine government requires a minimum rate of return of 9% to proceed with the project.

   (a) Does this project provide sufficient revenues to offset its costs?

   (b) What considerations are there besides economic factors in deciding whether to construct the bridge?

5-27   A student has a job that leaves her with $250 per month in disposable income. She decides that she will use the money to buy a car. Before looking for a car, she arranges a 100% loan whose terms are $250 per month for 36 months at 18% annual interest. What is the maximum car purchase price that she can afford with her loan?

5-28   The student in Problem 5-27 finds a car she likes and the dealer offers to arrange financing. His terms are 12% interest for 60 months and no down payment. The car's sticker price is $12,000. Can she afford to buy this car with her $250 monthly disposable income?

5-29   The student in Problem 5-28 really wants this particular car. She decides to try to negotiate a different interest rate. What is the highest interest rate that she can accept, given a 60-month term and $250 per month payments?

5-30   We know a car can be had for 60 monthly payments of $199. The dealer has set us a nominal interest rate of 4.5% compounded daily. What is the purchase price?

5-31 By installing some elaborate inspection equipment on its assembly line, the Robot Corp. can avoid hiring an extra worker who would have earned $26,000 a year in wages and an additional $7500 a year in employee benefits. The inspection equipment has a 6-year useful life and no salvage value. Use a nominal 18% interest rate in your calculations. How much can Robot afford to pay for the equipment if the wages and worker benefits would have been paid

(a) At the end of each year

(b) Monthly

Explain why the answer in (b) is larger. Assume the compounding matches the way the wages and benefits would have been paid (that is, annually and monthly).

5-32 Jerry Stans, a young industrial engineer, prepared an economic analysis for some equipment to replace one production worker. The analysis showed that the present worth of benefits (of employing one less production worker) just equaled the present worth of the equipment costs, assuming a 10-year useful life for the equipment. It was decided not to buy the equipment.

A short time later, the production workers won a new 3-year union contract that granted them an immediate 40¢-per-hour wage increase, plus an additional 25¢-per-hour wage increase in each of the two subsequent years. Assume that in each and every future year, a 25¢-per-hour wage increase will be granted.

Jerry Stans has been asked to revise his earlier economic analysis. The present worth of benefits of replacing one production employee will now increase. Assuming an interest rate of 8%, the justifiable cost of the automation equipment (with a 10-year useful life) will increase by how much? Assume the plant operates a single 8-hour shift, 250 days per year.

Lives Match

5-33 Two alternative courses of action have the following schedules of disbursements:

Year	A	B
0	−$1300	
1	0	−$100
2	0	−200
3	0	−300
4	0	−400
5	0	−500
	−$1300	−$1500

Based on a 6% interest rate, which alternative should be selected?

5-34 Dick Dickerson Construction, Inc. has asked you to help them select a new backhoe. You have a choice between a wheel-mounted version, which costs $50,000 and has an expected life of 5 years and a salvage value of $2000, and a track-mounted one, which costs $80,000, with a 5-year life and an expected salvage value of $10,000. Both machines will achieve the same productivity. Interest is 8%. Which one will you recommend? Use a present worth analysis.

5-35 Walt Wallace Construction Enterprises is investigating the purchase of a new dump truck. Interest is 9%. The cash flows for two likely models are as follows:

Model	First Cost	Annual Operating Cost	Annual Income	Salvage Value	Life (years)
A	$50,000	$2000	$9,000	$10,000	10
B	80,000	1000	12,000	30,000	10

(a) Using present worth analysis, which truck should the firm buy, and why?

(b) Before the construction company can close the deal, the dealer sells out of Model B and cannot get any more. What should the firm do now and why?

5-36 Two different companies are offering a punch press for sale. Company A charges $250,000 to deliver and install the device. Company A has estimated that the machine will have maintenance and operating costs of $4000 a year and will provide an annual benefit of $89,000. Company B charges $205,000 to deliver and install the device. Company B has estimated maintenance and operating costs of the press at $4300 a year, with an annual benefit of $86,000. Both machines will last 5 years and can be sold for $15,000 for the scrap metal. Use an interest rate of 12%. Which machine should your company buy?

5-37 A battery manufacturing plant has been ordered to cease discharging acidic waste liquids containing mercury into the city sewer system. As a result, the firm must now adjust the pH and remove the mercury from its waste liquids. Three firms have provided quotations on the necessary equipment. An analysis of the quotations provided the following table of costs.

Bidder	Installed Cost	Annual Operating Cost	Annual Income from Mercury Recovery	Salvage Value
Foxhill Instrument	$ 35,000	$8000	$2000	$20,000
Quicksilver	40,000	7000	2200	0
Almaden	100,000	2000	3500	0

If the installation can be expected to last 20 years and money is worth 7%, which equipment should be purchased?   (*Answer:* Almaden)

5-38 A new tennis court complex is planned. Both alternatives will last 18 years, and the interest rate is 7%. Use present worth analysis to determine which should be selected.

	Construction Cost	Annual O&M
A	$500,000	$25,000
B	640,000	10,000

Contributed by D. P. Loucks, Cornell University

5-39 Telefono Mexico is expanding its facilities to serve a new manufacturing plant. The new plant will require 2000 telephone lines this year, and another 2000 lines after expansion in 10 years. The plant will be in operation for 30 years. The telephone company is evaluating two options to serve the demand.

Option 1 Provide one cable now with capacity to serve 4000 lines. The cable cost will be $200,000, and annual maintenance costs will be $15,000.

Option 2 Provide a cable with capacity to serve 2000 lines now and a second cable to serve the other 2000 lines in 10 years. The cost of each cable will be $150,000, and each cable will have an annual maintenance of $10,000.

The telephone cables will last at least 30 years, and the cost of removing the cables is offset by their salvage value.

(a) Which alternative should be selected, assuming a 10% interest rate?

(b) Will your answer to (*a*) change if the demand for additional lines occurs in 5 years instead of 10 years?

5-40 A consulting engineer has been engaged to advise a town how best to proceed with the construction of a 200,000 m^3 water supply reservoir. Since only 120,000 m^3 of storage will be required for the next 25 years, an alternative to building the full capacity now is to build the reservoir in two stages. Initially, the reservoir could be built with 120,000 m^3 of capacity and then, 25 years hence, the additional 80,000 m^3 of capacity could be added by increasing the height of the reservoir. Estimated costs are as follows:

	Construction Cost	Annual Maintenance Cost
Build in two stages		
First stage: 120,000 m^3 reservoir	$14,200,000	$75,000
Second stage: Add 80,000 m^3 of capacity, additional construction and maintenance costs	12,600,000	25,000
Build full capacity now 200,000 m^3 reservoir	22,400,000	100,000

If interest is computed at 4%, which construction plan is preferred?

5-41 A city has developed a plan to provide for future municipal water needs. The plan proposes an aqueduct that passes through 500 feet of tunnel in a nearby mountain. Two alternatives are being considered. The first proposes to build a full-capacity tunnel now for $556,000. The second proposes to build a half-capacity tunnel now (cost = $402,000), which should be adequate for 20 years, and then to build a second parallel half-capacity tunnel. The maintenance cost of the tunnel lining for the full-capacity tunnel is $40,000 every 10 years, and for each half-capacity tunnel it is $32,000 every 10 years.

The friction losses in the half-capacity tunnel will be greater than if the full-capacity tunnel were built. The estimated additional pumping costs in the single half-capacity tunnel will be $2000 per year, and for the two half-capacity tunnels it will be $4000 per year. Based on capitalized cost and a 7% interest rate, which alternative should be selected?

Lives Differ

5-42 Use an 8-year analysis period and a 10% interest rate to determine which alternative should be selected:

	A	B
First cost	$5300	$10,700
Uniform annual benefit	$1800	$2,100
Useful life, in years	4	8

5-43 Use capitalized cost to determine which type of road surface is preferred on a particular section of highway. Use 12% interest rate.

	A	B
Initial cost	$500,000	$700,000
Annual maintenance	35,000	25,000
Periodic resurfacing	350,000	450,000
Resurfacing interval	10 years	15 years

5-44 A man had to have the muffler replaced on his 2-year-old car. The repairman offered two alternatives. For $50 he would install a muffler guaranteed for 2 years. But for $65 he would install a muffler guaranteed "for as long as you own the car." Assuming the present owner expects to keep the car for about 3 more years, which muffler would you advise him to have installed if you thought 20% were a suitable interest rate and the less expensive muffler would only last 2 years?

5-45 An engineer has received two bids for an elevator to be installed in a new building. The bids, plus his evaluation of the elevators, are tabulated.

Given a 10% interest rate, which bid should be accepted?

Alternatives	Westinghome	Itis
Installed Cost	$45,000	$54,000
Life (years)	10	15
Annual Cost	$2700	2850
Salvage value	$3000	4500

5-46 A weekly business magazine offers a 1-year subscription for $58 and a 3-year subscription for $116. If you thought you would read the magazine for at least the next 3 years, and consider 20% as a minimum rate of return, which way would you purchase the magazine: With three 1-year subscriptions or a single 3-year subscription? (*Answer:* Choose the 3-year subscription.)

Perpetual Life

5-47 A small dam was constructed for $2 million. The annual maintenance cost is $15,000. If interest is 5%, compute the capitalized cost of the dam, including maintenance.

5-48 A depositor puts $25,000 in a savings account that pays 5% interest, compounded semiannually. Equal annual withdrawals are to be made from the account, beginning one year from now and continuing forever. What is the maximum annual withdrawal?

5-49 What amount of money deposited 50 years ago at 8% interest would provide a perpetual payment of $10,000 a year beginning this year?

5-50 The president of the E. L. Echo Corporation thought it would be appropriate for his firm to "endow a chair" in the Department of Industrial Engineering of the local university. If the professor holding the chair will receive $67,000 per year, and the interest received on the endowment fund is expected to be 8%, what lump sum of money will the Echo Corporation need to provide to establish the endowment fund? (*Answer:* $837,500)

5-51 Dr. Fog E. Professor is retiring and wants to endow a chair of engineering economics at his university. It is expected that he will need to cover an annual cost of $100,000 forever. What lump sum must he donate to the university today if the endowment will earn 10% interest?

5-52 The local botanical society wants to ensure that the gardens in the town park are properly cared for. The group recently spent $100,000 to plant the gardens. The members want to set up a perpetual fund to provide $100,000 for future replantings of the gardens every 10 years. If interest is 5%, how much money would be needed to forever pay the cost of replanting?

5-53 A home builder must construct a sewage treatment plant and deposit sufficient money in a perpetual trust fund to pay the $5000 per year operating cost and to replace the treatment plant every 40 years. The plant will cost $150,000, and future replacement plants will also cost $150,000 each. If the trust fund earns 8% interest, what is the builder's capitalized cost to construct the plant and future replacements, and to pay the operating costs?

5-54 A man who likes cherry blossoms very much wants an urn full of them put on his grave once each year forever after he dies. In his will, he intends to leave a certain sum of money in trust at a local bank to pay the florist's annual bill. How much money should be left for this purpose? Make whatever assumptions you feel are justified by the facts presented. State your assumptions, and compute a solution.

5-55 An elderly lady decided to distribute most of her considerable wealth to charity and to keep for herself only enough money to provide for her living. She feels that $1000 a month will amply provide for her needs. She will establish a trust fund at a bank that pays 6% interest, compounded monthly. At the end of each month she will withdraw $1000. She has arranged that, upon her death, the balance in the account is to be paid to her niece, Susan. If she opens the trust fund and deposits enough money to pay herself $1000 a month in interest as long as she lives, how much will Susan receive when her aunt dies?

5-56 A trust fund is to be established for three purposes: (1) to provide $750,000 for the construction and $250,000 for the initial equipment of a small engineering laboratory; (2) to pay the $150,000 per year laboratory operating cost; and (3) to pay for $100,000 of replacement equipment every 4 years, beginning 4 years from now.

At 6% interest, how much money is required in the trust fund to provide for the laboratory and equipment and its perpetual operation and equipment replacement?

5-57 The local Audubon Society has just put a new bird feeder in the park at a cost of $500. The feeder has a useful life of 5 years and an annual maintenance cost of $50. Our cat, Fred, was very impressed with the project. He wants to establish a fund that will maintain the feeder in perpetuity. Replacement feeders cost $500 every 5 years. If the fund earns 5% interest, what amount must Fred raise for its establishment?

5-58 We want to donate a marble birdbath to the city park as a memorial to our cat, Fred, while he can still enjoy it. We also want to set up a perpetual care fund to cover future expenses "forever." The initial cost of the bath is $5000. Routine annual operating costs are $200 per year, but every fifth year the cost will be $500 to cover major cleaning and maintenance as well as operation.

(a) What is the capitalized cost of this project if the interest rate is 8%?

(b) How much is the present worth of this project if it is to be demolished after 75 years? The final $500 payment in the 75th year will cover the year's operating cost and the site reclamation.

5-59 A local symphony association offers memberships as follows:

Continuing membership, per year	$ 15
Patron lifetime membership	375

The patron membership has been based on the symphony association's belief that it can obtain a 4% rate of return on its investment. If you believed 4% to be an appropriate rate of return, would you be willing to purchase the patron membership? Explain why or why not.

5-60 A rather wealthy man decides to arrange for his descendants to be well educated. He wants each child to have $60,000 for his or her education. He plans to set up a perpetual trust fund so that six children will receive this assistance in each generation. He estimates that there will be four generations per century, spaced 25 years apart. He expects the trust to be able to obtain a 4% rate of return, and the first recipients to receive the money 10 years hence. How much money should he now set aside in the trust? (*Answer:* $389,150)

5-61 A new bridge project is being evaluated at $i = 5\%$. Recommend an alternative based on the capitalized cost for each.

	Construction Cost	Annual O&M	Life (years)
Concrete	$50 million	$ 250,000	70
Steel	40 million	1,000,000	50

5-62 A new stadium is being evaluated at $i = 6\%$. Recommend an alternative for the main structural material based on the capitalized cost for each.

	Construction Cost	Annual O&M	Life (years)
Concrete	$25 million	$200,000	80
Steel	21 million	1,000,000	60

Multiple Alternatives

5-63 A firm is considering three mutually exclusive alternatives as part of a production improvement program. The alternatives are:

	A	B	C
Installed cost	$10,000	$15,000	$20,000
Uniform annual benefit	1,625	1,530	$1,890
Useful life, in years	10	20	20

The salvage value at the end of the useful life of each alternative is zero. At the end of 10 years, Alternative A could be replaced with another A with identical cost and benefits. The maximum attractive rate of return is 6%. Which alternative should be selected?

5-64 A steam boiler is needed as part of the design of a new plant. The boiler can be fired by natural gas, fuel oil, or coal. A decision must be made on which fuel to use. An analysis of the costs shows that the installed cost, with all controls, would be least for natural gas at $30,000; for fuel oil it would be $55,000; and for coal it would be $180,000. If natural gas is used rather than fuel oil, the annual fuel cost will increase by $7500. If coal is used rather than fuel oil, the annual fuel cost will be $15,000 per year less. Assuming 8% interest, a 20-year analysis period, and no salvage value, which is the most economical installation?

5-65 Austin General Hospital is evaluating new office equipment offered by three companies. In each case the interest rate is 15% and the useful life of the equipment is 4 years. Use NPW analysis to determine the company from which you should purchase the equipment.

	Company A	Company B	Company C
First cost	$15,000	$25,000	$20,000
Maintenance and operating costs	1,600	400	900
Annual benefit	8,000	13,000	11,000
Salvage value	3,000	6,000	4,500

5-66 The following costs are associated with three tomato-peeling machines being considered for use in a canning plant.

L.C.M.

	Machine A	Machine B	Machine C
First cost	$52,000	$63,000	$67,000
Maintenance and operating costs	15,000	9,000	12,000
Annual benefit	38,000	31,000	37,000
Salvage value	13,000	19,000	22,000
Useful life, in years	4	6	12

$NPW \, 12 = NPW_4 \, [1+ (P/F, 12, 4) \, (P/F, 12, 8)]$
$26,113 \, [1 + (1.12)-4 + (1.12)-8]$
$53,255$

If the canning company uses an interest rate of 12%, which is the best alternative? Use NPW to make your decision. (*Note:* Consider the least common multiple as the study period.)

5-67 A railroad branch line to a landfill site is to be constructed. It is expected that the railroad line will be used for 15 years, after which the landfill site will be closed and the land turned back to agricultural use. The railroad track and ties will be removed at that time.

In building the railroad line, either treated or untreated wood ties may be used. Treated ties have an installed cost of $6 and a 10-year life; untreated ties are $4.50 with a 6-year life. If at the end of 15 years the ties then in place have a remaining useful life of 4 years or more, they will be used by the railroad elsewhere and have an estimated salvage value of $3 each. Any ties that are removed at the end of their service life, or too close to the end of their service life to be used elsewhere, can be sold for 50¢ each.

Determine the most economical plan for the initial railroad ties and their replacement for the

15-year period. Make a present worth analysis assuming 8% interest.

5-68 A building contractor obtained bids for some asphalt paving, based on a specification. Three paving sub-contractors quoted the following prices and terms of payment:

Paving Co.	Price	Payment Schedule
Quick	$85,000	50% payable immediately 25% payable in 6 months 25% payable at the end of one year
Tartan	82,000	Payable immediately
Faultless	84,000	25% payable immediately 75% payable in 6 months

The building contractor uses a 12% nominal interest rate, compounded monthly, in this type of bid analysis. Which paving subcontractor should be awarded the paving job?

5-69 Given the following data, use present worth analysis to find the best alternative, A, B, or C.

	A	B	C
Initial cost	$10,000	15,000	$12,000
Annual benefit	6,000	10,000	5,000
Salvage value	1,000	−2,000	3,000
Useful life	2 years	3 years	4 years

Use an analysis period of 12 years and 10% interest.

5-70 Consider the following four alternatives. Three are "do something" and one is "do nothing."

	A	B	C	D
Cost	$0	$50	$30	$40
Net annual benefit	0	12	4.5	6
Useful life, in years		5	10	10

At the end of the 5-year useful life of B, a replacement is not made. If a 10-year analysis period and a 10% interest rate are selected, which is the preferred alternative?

5-71 A cost analysis is to be made to determine what, if anything, should be done in a situation offering three "do-something" and one "do-nothing" alternatives. Estimates of the cost and benefits are as follows.

Alternatives	Cost	Uniform Annual Benefit	End-of-Useful-Life Salvage Value	Useful Life (years)
1	$500	$135	$ 0	5
2	600	100	250	5
3	700	100	180	10
4	0	0	0	0

Use a 10-year analysis period for the four mutually exclusive alternatives. At the end of 5 years, Alternatives 1 and 2 may be replaced with identical alternatives (with the same cost, benefits, salvage value, and useful life).

(a) If an 8% interest rate is used, which alternative should be selected?

(b) If a 12% interest rate is used, which alternative should be selected?

5-72 Consider A–E, five mutually exclusive alternatives:

	A	B	C	D	E
Initial cost	$600	$600	$600	$600	$600
Uniform annual benefits					
for first 5 years	100	100	100	150	150
for last 5 years	50	100	110	0	50

The interest rate is 10%. If all the alternatives have a 10-year useful life, and no salvage value, which alternative should be selected?

5-73 An investor has carefully studied a number of companies and their common stock. From his analysis, he has decided that the stocks of six firms are the best of the many he has examined. They represent about the same amount of risk, and so he would like to determine one single stock in which to invest. He plans to keep the stock for 4 years and requires a 10% minimum attractive rate of return.

Which stock from Table P5-73, if any, should the investor consider buying?   (*Answer:* Spartan Products)

TABLE P5-73

Common Stock	Price per Share	Annual End-of-Year Dividend per Share	Estimated Price at End of 4 Years
Western House	$23\frac{3}{4}$	$1.25	$32
Fine Foods	45	4.50	45
Mobile Motors	$30\frac{5}{8}$	0	42
Spartan Products	12	0	20
U.S. Tire	$33\frac{3}{8}$	2.00	40
Wine Products	$52\frac{1}{2}$	3.00	60

TABLE P5-84

Year	0	1	2	3	4	5	6	7
Net cash ($)	0	−120,000	−60,000	20,000	40,000	80,000	100,000	60,000

5-74 Six mutually exclusive alternatives, *A–F*, are being examined. For an 8% interest rate, which alternative should be selected? Each alternative has a 6-year useful life.

	Initial Cost	Uniform Annual Benefit
A	$ 20.00	6.00
B	35.00	9.25
C	55.00	13.38
D	60.00	13.78
E	80.00	24.32
F	100.00	24.32

5-75 The management of an electronics manufacturing firm believes it is desirable to automate its production facility. The automated equipment would have a 10-year life with no salvage value at the end of 10 years. The plant engineering department has surveyed the plant and has suggested there are eight mutually exclusive alternatives.

Plan	Initial Cost (thousands)	Net Annual Benefit (thousands)
1	$265	$51
2	220	39
3	180	26
4	100	15
5	305	57
6	130	23
7	245	47
8	165	33

If the firm expects a 10% rate of return, which plan, if any, should it adopt? (*Answer:* Plan 1)

Spreadsheets

5-76 Assume monthly car payments of $500 per month for 4 years and an interest rate of 1/2% per month. What initial principal or PW will this repay?

5-77 Assume annual car payments of $6000 for 4 years and an interest rate of 6% per year. What initial principal or PW will this repay?

5-78 Assume annual car payments of $6000 for 4 years and an interest rate of 6.168% per year. What initial principal or PW will this repay?

5-79 Why do the values in Problems 5-76, 5-77, and 5-78 differ?

5-80 Assume mortgage payments of $1000 per month for 30 years and an interest rate of 1/2% per month. What initial principal or PW will this repay?

5-81 Assume annual mortgage payments of $12,000 for 30 years and an interest rate of 6% per year. What initial principal or PW will this repay?

5-82 Assume annual mortgage payments of $12,000 for 30 years and an interest rate of 6.168% per year. What initial principal or PW will this repay?

5-83 Why do the values in Problems 5-80, 5-81, and 5-82 differ?

5-84 Ding Bell Imports requires a return of 15% on all projects. If Ding is planning an overseas development project with the cash flows shown in Table P5-84, what is the project's net present value?

5-85   Maverick Enterprises is planning a new product. Annual sales, unit costs, and unit revenues are as tabulated; the first cost of R&D and setting up the assembly line is $42,000. If *i* is 10%, what is the PW?

Year	Annual Sales	Cost/unit	Price/unit
1	$ 5,000	$3.50	$6
2	6,000	3.25	5.75
3	9,000	3.00	5.50
4	10,000	2.75	5.25
5	8,000	2.5	4.5
6	4,000	2.25	3

5-86   Northern Engineering is analyzing a mining project. Annual production, unit costs, and unit revenues are in the table. The first cost of the mine setup is $8 million. If *i* is 15%, what is the PW?

Year	Annual Production (tons)	Cost per ton	Price per ton
1	70,000	$25	$35
2	90,000	20	34
3	120,000	22	33
4	100,000	24	34
5	80,000	26	35
6	60,000	28	36
7	40,000	30	37

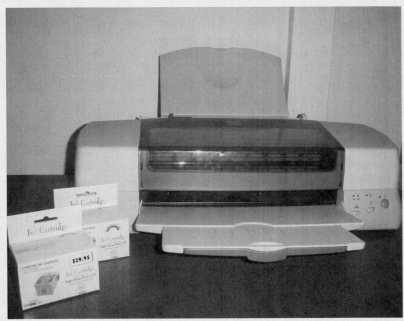

ANNUAL CASH FLOW ANALYSIS

Lowest Prices on the Net! Buy Now!

Next time you review your e-mail and scroll through the spam, take a look at how many unsolicited messages are offering cut-rate ink cartridges for your printer.

Strange, isn't it? Why would spam pests imagine they can capture your attention with ads for cheap printer ink, especially when competing spammers are advertising many other products?

The answer becomes clear when you look at the price of an ink cartridge for a typical inkjet printer. Cartridges often cost $30 or more, and they need to be replaced frequently. By contrast, a good quality inkjet printer can often be purchased for under $100.

QUESTIONS TO CONSIDER

1. Retailers frequently attract buyers by advertising low sale prices for printers, but they almost never mention the cost of the ink cartridges. In analyzing the likely cost of operating a printer over its useful lifetime, how much weight should the buyer give to the price of the printer itself, and how much to the cost of the ink cartridges?
2. King Camp Gillette, inventor of the safety razor, designed his product to work with a specially crafted blade that did not need sharpening and could be disposed of after a few uses. Gillette eventually announced, to everyone's astonishment, that he would be giving his razors away. Many people couldn't imagine how he could possibly make money this way. In fact, Gillette revolutionized marketing, and his business revenue soared. Can you explain why?
3. Would Gillette's strategy work with a product such as a car? Why or why not?
4. What ethical issues do producers and marketers face in designing and selling their products? Is it true that "anything goes in business" so "caveat emptor"?

After Completing This Chapter...

The student should be able to:

- Define *equivalent uniform annual cost (EUAC)* and *equivalent uniform annual benefits (EUAB)*.
- Resolve an engineering economic analysis problem into its annual cash flow equivalent.
- Conduct an *equivalent uniform annual worth (EUAW) analysis* for a single investment.
- Use EUAW, EUAC, and EUAB to compare alternatives with equal, common multiple, or continuous lives, or over some fixed study period.
- Develop and use spreadsheets to analyze loans for purposes of building an amortization table, calculating interest versus principal, finding the balance due, and determining whether to pay off a loan early.

T his chapter is devoted to annual cash flow analysis—the second of the three major analysis techniques. With present worth, analysis, we resolved an alternative into an equivalent cash sum. This might have been an equivalent net present worth, a present worth of cost, or a present worth of benefit. Here we compare alternatives based on their equivalent annual cash flows. Depending on the particular situation, we may wish to compute the equivalent uniform annual cost (EUAC), the equivalent uniform annual benefit (EUAB), or their difference, the equivalent uniform annual worth (EUAW) = (EUAB − EUAC).

To prepare for a discussion of annual cash flow analysis, we will review some annual cash flow calculations, then examine annual cash flow criteria. Following this, we will proceed with annual cash flow analysis.

ANNUAL CASH FLOW CALCULATIONS

Resolving a Present Cost to an Annual Cost

Equivalence techniques were used in prior chapters to convert money, at one point in time, to some equivalent sum or series. In annual cash flow analysis, the goal is to convert money to an equivalent uniform annual cost or benefit. The simplest case is to convert a present sum P to a series of equivalent uniform end-of-period cash flows. This is illustrated in Example 6-1.

EXAMPLE 6-1

A student bought $1000 worth of home furniture. If the items are expected to last 10 years, what will the equivalent uniform annual cost be if interest is 7%?

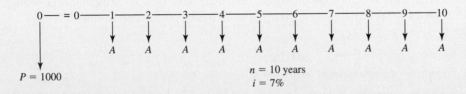

SOLUTION

$$\text{Equivalent uniform annual cost} = P(A/P, i, n)$$
$$= 1000(A/P, 7\%, 10)$$
$$= \$142.40$$

Equivalent uniform annual cost is $142.40.

Treatment of Salvage Value

When there is a salvage value at the end of an asset's useful life, this decreases the equivalent uniform annual cost.

EXAMPLE 6-2

The student in Example 6-1 now believes the furniture can be sold at the end of 10 years for $200. Under these circumstances, what is the equivalent uniform annual cost?

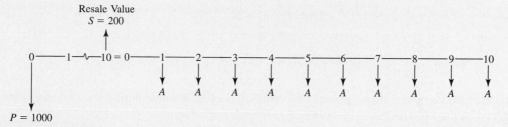

$P = 1000$

SOLUTION

For this situation, the problem may be solved by means of three different calculations.

SOLUTION 1

$$\text{EUAC} = P(A/P, i, n) - S(A/F, i, n) \qquad (6\text{-}1)$$
$$= 1000(A/P, 7\%, 10) - 200(A/F, 7\%, 10)$$
$$= 1000(0.1424) - 200(0.0724)$$
$$= 142.40 - 14.48 = \$127.92$$

This method reflects the annual cost of the cash disbursement minus the annual benefit of the future resale value.

SOLUTION 2

Equation 6-1 describes a relationship that may be modified by an identity presented in Chapter 4:

$$(A/P, i, n) = (A/F, i, n) + i \qquad (6\text{-}2)$$

Substituting this into Equation 6-1 gives:

$$\text{EUAC} = P(A/F, i, n) + Pi - S(A/F, i, n) \qquad (6\text{-}3)$$
$$= (P - S)(A/F, i, n) + Pi$$
$$= (1000 - 200)(A/F, 7\%, 10) + 1000(0.07)$$
$$= 800(0.0724) + 70 = 57.92 + 70$$
$$= \$127.92$$

This method computes the equivalent annual cost due to the unrecovered $800 when the furniture is sold, and adds annual interest on the $1000 investment.

SOLUTION 3

If the value for $(A/F, i, n)$ from Equation 6-2 is substituted into Equation 6-1, we obtain:

$$\text{EUAC} = P(A/P, i, n) - S(A/P, i, n) + Si \qquad (6\text{-}4)$$
$$= (P - S)(A/P, i, n) + Si$$
$$= (1000 - 200)(A/P, 7\%, 10) + 200(0.07)$$
$$= 800(0.1424) + 14 = 113.92 + 14 = \$127.92$$

This method computes the annual cost of the $800 decline in value during the 10 years, plus interest on the $200 tied up in the furniture as the salvage value.

When there is an initial disbursement P followed by a salvage value S, the annual cost may be computed in the three different ways introduced in Example 6-2.

$$\mathbf{EUAC} = P(A/P, i, n) - S(A/F, i, n) \qquad (6\text{-}1)$$

$$\mathbf{EUAC} = (P - S)(A/F, i, n) + Pi \qquad (6\text{-}3)$$

$$\mathbf{EUAC} = (P - S)(A/P, i, n) + Si \qquad (6\text{-}4)$$

Each of the three calculations gives the same results. In practice, the first and third methods are most commonly used. The EUAC calculated in Equations 6-1, 6-3, and 6-4 is also known as the *capital recovery cost* of a project.

EXAMPLE 6-3

Bill owned a car for 5 years. One day he wondered what his uniform annual cost for maintenance and repairs had been. He assembled the following data:

Year	Maintenance and Repair Cost for Year
1	$ 45
2	90
3	180
4	135
5	225

Compute the equivalent uniform annual cost (EUAC) assuming 7% interest and end-of-year disbursements.

SOLUTION

The EUAC may be computed for this irregular series of payments in two steps:

1. Use single payment present worth factors to compute the present worth of cost for the 5 years.
2. With the PW of cost known, use the capital recovery factor to compute EUAC.

$$PW \text{ of cost} = 45(P/F, 7\%, 1) + 90(P/F, 7\%, 2) + 180(P/F, 7\%, 3)$$
$$+ \ 135(P/F, 7\%, 4) + 225(P/F, 7\%, 5)$$
$$= 45(0.9346) + 90(0.8734) + 180(0.8163) + 135(0.7629) + 225(0.7130)$$
$$= \$531$$
$$EUAC = 531(A/P, 7\%, 5) = 531(0.2439) = \$130$$

EXAMPLE 6-4

Bill reexamined his calculations and found that in his table he had reversed the maintenance and repair costs for Years 3 and 4. The correct table is:

Year	Maintenance and Repair Cost for Year
1	$ 45
2	90
3	135
4	180
5	225

Recompute the EUAC.

SOLUTION

This time the schedule of disbursements is an arithmetic gradient series plus a uniform annual cost, as follows:

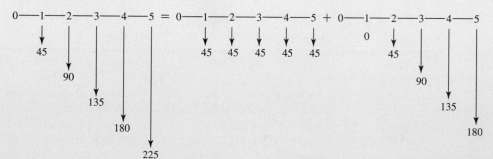

$$\text{EUAC} = 45 + 45(A/G, 7\%, 5)$$

$$= 45 + 45(1.865)$$

$$= \$129$$

Since the timing of the expenditures is different in Examples 6-3 and 6-4, we would not expect to obtain the same EUAC.

The examples have shown four essential points concerning cash flow calculations:

1. There is a direct relationship between the present worth of cost and the equivalent uniform annual cost. It is

$$\text{EUAC} = (\text{PW of cost})(A/P, i, n)$$

2. In a problem, expending money increases the EUAC, while receiving money—for example, from an item is salvage value—decreases the EUAC.

3. When there are irregular cash disbursements over the analysis period, a convenient method of solution is to first determine the PW of cost; then use the equation in Item 1 to calculate the EUAC.

4. Where there is an arithmetic gradient, EUAC may be rapidly computed by using the arithmetic gradient uniform series factor, $(A/G, i, n)$.

ANNUAL CASH FLOW ANALYSIS

The criteria for economic efficiency are presented in Table 6-1. One notices immediately that the table is quite similar to Table 5-1. It is apparent that, if you are maximizing the present worth of benefits, simultaneously you must be maximizing the equivalent uniform annual worth. This is illustrated in Example 6-5.

TABLE 6-1 Annual Cash Flow Analysis

Input/Output	Situation	Criterion
Neither input nor output is fixed	Typical, general situation	Maximize equivalent uniform annual worth (EUAW = EUAB − EUAC)
Fixed input	Amount of money or other input resources is fixed	Maximize equivalent uniform benefits (maximize EUAB)
Fixed output	There is a fixed task, benefit, or other output to be accomplished	Minimize equivalent uniform annual cost (minimize EUAC)

EXAMPLE 6-5

A firm is considering which of two devices to install to reduce costs. Both devices have useful lives of 5 years with no salvage value. Device A costs $1000 and can be expected to result in $300 savings annually. Device B costs $1350 and will provide cost savings of $300 the first year; however, savings will increase $50 annually, making the second year savings $350, the third year savings $400, and so forth. With interest at 7%, which device should the firm purchase?

SOLUTION

Device A

$$\text{EUAW} = -1000(A/P, 7\%, 5) + 300$$
$$= -1000(0.2439) + 300 = \$56.1$$

Device B

$$\text{EUAW} = -1350(A/P, 7\%, 5) + 300 + 50(A/G, 7\%, 5)$$
$$= -1350(0.2439) + 300 + 50(1.865)$$
$$= \$64.0$$

To maximize EUAW, select Device B.

Example 6-5 was presented earlier, as Example 5-1, where we found:

$$\text{PW}_A = -1000 + 300(P/A, 7\%, 5)$$
$$= -1000 + 300(4.100) = \$230$$

This is converted to EUAW by multiplying by the capital recovery factor:

$$\text{EUAW}_A = 230(A/P, 7\%, 5) = 230(0.2439) = \$56.1$$

Similarly, for machine B

$$\text{PW}_B = -1350 + 300(P/A, 7\%, 5) + 50(P/G, 7\%, 5)$$
$$= -1350 + 400(4.100) + 50(7.647) = \$262.4$$

and, hence,

$$\text{EUAW}_B = 262.4(A/P, 7\%, 5) = 262.4(0.2439)$$
$$= \$64.0$$

We see, therefore, that it is easy to convert the present worth analysis results into the annual cash flow analysis results. We could go from annual cash flow to present worth just as easily, by using the series present worth factor. And, of course, both methods show that Device B is the preferred alternative.

EXAMPLE 6-6

Three alternatives are being considered for improving an operation on the assembly line along with the "do-nothing" alternative. Equipment costs vary, as do the annual benefits of each in comparison to the present situation. Each of Plans A, B, and C has a 10-year life and a salvage value equal to 10% of its original cost.

	Plan A	Plan B	Plan C
Installed cost of equipment	$15,000	$25,000	$33,000
Material and labor savings per year	14,000	9,000	14,000
Annual operating expenses	8,000	6,000	6,000
End-of-useful life salvage value	1,500	2,500	3,300

If interest is 8%, which plan, if any, should be adopted?

SOLUTION

Since neither installed cost nor output benefits are fixed, the economic criterion is to maximize EUAW = EUAB − EUAC.

	Plan A	Plan B	Plan C
Equivalent uniform annual benefit (EUAB)			
Material and labor per year	$14,000	$9,000	$14,000
Salvage value $(A/F, 8\%, 10)$	104	172	228
EUAB =	$14,104	$9,172	$14,228
Equivalent uniform annual cost (EUAC)			
Installed cost $(A/P, 8\%, 10)$	$ 2,235	$3,725	$ 4,917
Annual operating expenses	8,000	6,000	6,000
EUAC =	$10,235	$9,725	$10,917
EUAW = EUAB − EUAC =	$ 3,869	−$ 553	$ 3,311

Based on our criterion of maximizing EUAW, Plan A is the best of the four alternatives. Since the do-nothing alternative has EUAW = 0, it is a more desirable alternative than Plan B.

ANALYSIS PERIOD

In Chapter 5, we saw that the analysis period is an important consideration in computing present worth comparisons. In such problems, a common analysis period must be used for all alternatives. In annual cash flow comparisons, we again have the analysis period question. Example 6-7 will help in examining the problem.

EXAMPLE 6-7

Two pumps are being considered for purchase. If interest is 7%, which pump should be bought?

	Pump A	Pump B
Initial cost	$7000	$5000
End-of-useful-life salvage value	1500	1000
Useful life, in years	12	6

SOLUTION

The annual cost for 12 years of Pump A can be found by using Equation 6-4:

$$EUAC = (P - S)(A/P, i, n) + Si$$

$$= (7000 - 1500)(A/P, 7\%, 12) + 1500(0.07)$$

$$= 5500(0.1259) + 105 = \$797$$

Now compute the annual cost for 6 years of Pump B:

$$EUAC = (5000 - 1000)(A/P, 7\%, 6) + 1000(0.07)$$

$$= 4000(0.2098) + 70 = \$909$$

For a common analysis period of 12 years, we need to replace Pump B at the end of its 6-year useful life. If we assume that another pump B' can be obtained, having the same $5000 initial cost, $1000 salvage value, and 6-year life, the cash flow will be as follows:

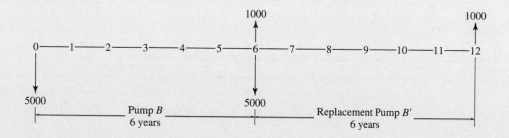

For the 12-year analysis period, the annual cost for Pump B is

$$EUAC = [5000 - 1000(P/F, 7\%, 6) + 5000(P/F, 7\%, 6)$$

$$- 1000(P/F, 7\%, 12)] \times (A/P, 7\%, 12)$$

$$= [5000 - 1000(0.6663) + 5000(0.6663) - 1000(0.4440)] \times (0.1259)$$

$$= (5000 - 666 + 3331 - 444)(0.1259)$$

$$= (7211)(0.1259) = \$909$$

The annual cost of B for the 6-year analysis period is the same as the annual cost for the 12-year analysis period. This is not a surprising conclusion when one recognizes that the annual cost of the first 6-year period is repeated in the second 6-year period. Thus the lengthy calculation of EUAC for 12 years of Pump B and B′ was not needed. By assuming that the shorter-life equipment is replaced by equipment with identical economic consequences, we have avoided a lot of calculations. Select Pump A.

Analysis Period Equal to Alternative Lives

If the analysis period for an economy study coincides with the useful life for each alternative, then the economy study is based on this analysis period.

Analysis Period a Common Multiple of Alternative Lives

When the analysis period is a common multiple of the alternative lives (for example, in Example 6-7, the analysis period was 12 years with 6- and 12-year alternative lives), a "replacement with an identical item with the same costs, performance, and so forth" is frequently assumed. This means that when an alternative has reached the end of its useful life, we assume that it will be replaced with an identical item. As shown in Example 6-7, the result is that the EUAC for Pump B with a 6-year useful life is equal to the EUAC for the entire analysis period based on Pump B *plus* the replacement unit, Pump B′.

Under these circumstances of identical replacement, we can compare the annual cash flows computed for alternatives based on their own service lives. In Example 6-7, the annual cost for Pump A, based on its 12-year service life, was compared with the annual cost for Pump B, based on its 6-year service life.

Analysis Period for a Continuing Requirement

Many times an economic analysis is undertaken to determine how to provide for a more or less continuing requirement. One might need to pump water from a well as a continuing requirement. There is no distinct analysis period. In this situation, the analysis period is assumed to be long but undefined.

If, for example, we had a continuing requirement to pump water and alternative Pumps A and B had useful lives of 7 and 11 years, respectively, what should we do? The customary assumption is that Pump A's annual cash flow (based on a 7-year life) may be compared to Pump B's annual cash flow (based on an 11-year life). This is done without much concern that the least common multiple of the 7- and 11-year lives is 77 years. This comparison of "different-life" alternatives assumes identical replacement (with identical costs, performance, etc.) when an alternative reaches the end of its useful life.

This continuing requirement, which can also be described as an *indefinitely long horizon*, is illustrated in Example 6.8. Since this is longer than the lives of the alternatives, we can make the best decision possible given current information by minimizing EUAC or maximizing EUAW or EUAB. At a later time, we will make another replacement and there will be more information on costs *at that time*.

EXAMPLE 6-8

Pump B in Example 6-7 is now believed to have a 9-year useful life. Assuming the same initial cost and salvage value, compare it with Pump A using the same 7% interest rate.

SOLUTION

If we assume that the need for A or B will exist for some continuing period, the comparison of costs per year for the unequal lives is an acceptable technique. For 12 years of Pump A:

$$EUAC = (7000 - 1500)(A/P, 7\%, 12) + 1500(0.07) = \$797$$

For 9 years of Pump B:

$$EUAC = (5000 - 1000)(A/P, 7\%, 9) + 1000(0.07) = \$684$$

For minimum EUAC, select Pump B.

Infinite Analysis Period

At times we have an alternative with a limited (finite) useful life in an infinite analysis period situation. The equivalent uniform annual cost may be computed for the limited life. The assumption of identical replacement (replacements have identical costs, performance, etc.) is often appropriate. Based on this assumption, the same EUAC occurs for each replacement of the limited-life alternative. The EUAC for the infinite analysis period is therefore equal to the EUAC computed for the limited life. With identical replacement,

$$EUAC_{\text{infinite analysis period}} = EUAC_{\text{for limited life } n}$$

A somewhat different situation occurs when there is an alternative with an infinite life in a problem with an infinite analysis period:

$$EUAC_{\text{infinite analysis period}} = P(A/P, i, \infty) + \text{Any other annual costs}$$

When $n = \infty$, we have $A = Pi$ and, hence, $(A/P, i, \infty)$ equals i.

$$EUAC_{\text{infinite analysis period}} = Pi + \text{Any other annual costs}$$

EXAMPLE 6-9

In the construction of an aqueduct to expand the water supply of a city, there are two alternatives for a particular portion of the aqueduct. Either a tunnel can be constructed through a mountain, or a pipeline can be laid to go around the mountain. If there is a permanent need for the aqueduct, should the tunnel or the pipeline be selected for this particular portion of the aqueduct? Assume a 6% interest rate.

SOLUTION

	Tunnel Through Mountain	**Pipeline Around Mountain**
Initial cost	$5.5 million	$5 million
Maintenance	0	0
Useful life	Permanent	50 years
Salvage value	0	0

Tunnel

For the tunnel, with its permanent life, we want $(A/P, 6\%, \infty)$. For an infinite life, the capital recovery is simply interest on the invested capital. So $(A/P, 6\%, \infty) = i$, and we write

$$\text{EUAC} = Pi = \$5.5 \text{ million}(0.06)$$

$$= \$330,000$$

Pipeline

$$\text{EUAC} = \$5 \text{ million}(A/P, 6\%, 50)$$

$$= \$5 \text{ million}(0.0634) = \$317,000$$

For fixed output, minimize EUAC. Select the pipeline.

The difference in annual cost between a long life and an infinite life is small unless an unusually low interest rate is used. In Example 6-9 the tunnel is assumed to be permanent. For comparison, compute the annual cost if an 85-year life is assumed for the tunnel.

$$\text{EUAC} = \$5.5 \text{ million}(A/P, 6\%, 85)$$

$$= \$5.5 \text{ million}(0.0604)$$

$$= \$332,000$$

The difference in time between 85 years and infinity is great indeed; yet the difference in annual costs in Example 6-9 is very small.

Some Other Analysis Period

The analysis period in a particular problem may be something other than one of the four we have so far described. It may be equal to the life of the shorter-life alternative, the longer-life alternative, or something entirely different. One must carefully examine the consequences of each alternative throughout the analysis period and, in addition, see what differences there might be in salvage values, and so forth, at the end of the analysis period.

EXAMPLE 6-10

Suppose that Alternative 1 has a 7-year life and a salvage value at the end of that time. The replacement cost at the end of 7 years may be more or less than the original cost. If the replacement is retired prior to 7 years, it will have a terminal value that exceeds the end-of-life salvage value. Alternative 2 has a 13-year life and a terminal value whenever it is retired. If the situation indicates that 10 years is the proper analysis period, set up the equations to compute the EUAC for each alternative. Use results from Example 5-5 to compute the results.

SOLUTION

Alternative 1

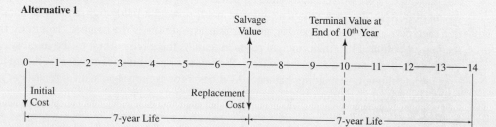

Alternative 2

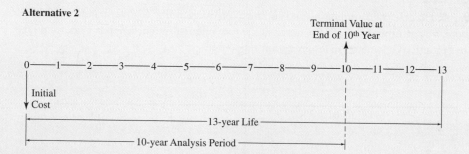

Alternative 1

$$EUAC_1 = [\text{Initial cost} + (\text{Replacement cost} - \text{Salvage value})(P/F, i, 7)$$
$$- (\text{Terminal value})(P/F, i, 10)](A/P, i, 10)$$
$$= 64,076(A/P, 8\%, 10) \quad \text{using results from Example 5-5}$$
$$= 64,076(8.1490) = \$9547$$

Alternative 2

$$EUAC_2 = [\text{Initial cost} - (\text{Terminal value})(P/F, i, 10)](A/P, i, 10)$$
$$= 69,442(A/P, 8\%, 10) \quad \text{using results from Example 5-5}$$
$$= 69,442(0.1490) = \$10,347$$

Select Alternative 1.

USING SPREADSHEETS TO ANALYZE LOANS

Loan and bond payments are made by firms, agencies, and individual engineers. Usually, the payments in each period are constant. Spreadsheets make it easy to:

- Calculate the loan's amortization schedule
- Decide how a payment is to be split between principal and interest
- Find the balance due on a loan
- Calculate the number of payments remaining on a loan.

Building an Amortization Schedule

As illustrated in previous chapters and Appendix 1, an amortization schedule lists for each payment period: the loan payment, interest paid, principal paid, and remaining balance. For each period the interest paid equals the interest rate times the balance remaining from the period before. Then the principal payment equals the payment minus the interest paid. Finally, this principal payment is applied to the balance remaining from the preceding period to calculate the new remaining balance. As a basis for comparison with spreadsheet loan functions, Figure 6-1 shows this calculation for Example 6-11.

EXAMPLE 6-11

An engineer wanted to celebrate graduating and getting a job by spending $2400 on new furniture. Luckily the store was offering 6-month financing at the low interest rate of 6% per year nominal (really $1/2$% per month). Calculate the amortization schedule.

SOLUTION

	A	B	C	D	E
1	2400	Initial balance			
2	0.50%	i			
3	6	N			
4	$407.03	Payment	$= -PMT(A2,A3,A1)$		
5					
6			Principal	Ending	
7	Month	Interest	Payment	Balance	
8	0			2400.00	=A1
9	1	12.00	395.03	2004.97	=D8−C9
10	2	10.02	397.00	1607.97	
11	3	8.04	398.99	1208.98	
12	4	6.04	400.98	807.99	
13	5	4.04	402.99	405.00	
14	6	2.03	405.00	0.00	
15				=$A$4−B14	
16				=Payment−Interest	
17			=$A$2*D13		
18			=rate*previous balance		

FIGURE 6-1 Amortization schedule for furniture loan.

The first step is to calculate the monthly payment:

$$A = 2400(A/P, \frac{1}{2}\%, 6) = 2400(0.1696) = \$407.0$$
$$= \text{PMT}(0.005, 6, -2400) = \$407.03$$

With this information the engineer can use the spreadsheet of Figure 6-1 to obtain the amortization schedule.

How Much to Interest? How Much to Principal?

For a loan with constant payments, we can answer these questions for any period without the full amortization schedule. For a loan with constant payments, the functions IPMT and PPMT directly answer these questions. For simple problems, both functions have four arguments $(i, t, n, -P)$, where t is the time period being calculated. Both functions have optional arguments that permit adding a balloon payment (an F) and changing from end-of-period payments to beginning-of-period payments.

For example, consider Period 4 of Example 6-11. The spreadsheet formulas give the same answer as shown in Figure 6-1.

$$\text{Interest}_4 = \text{IPMT}(0.5\%, 4, 6, -2400) = \$6.04$$
$$\text{Principal payment}_4 = \text{PPMT}(0.5\%, 4, 6, -2400) = \$400.98$$

Finding the Balance Due on a Loan

An amortization schedule is one used to calculate the balance due on a loan. A second, easier way is to remember that the balance due equals the present worth of the remaining payments. Interest is paid in full after each payment, so later payments are simply based on the balance due.

EXAMPLE 6-12

A car is purchased with a 48-month, 9% nominal loan with an initial balance of $15,000. What is the balance due halfway through the 4 years?

SOLUTION

The first step is to calculate the monthly payment, at a monthly interest rate of $\frac{3}{4}\%$. This equals

$$\text{Payment} = 15,000(A/P, 0.75\%, 48) \quad \text{or} \quad = \text{PMT}(0.75\%, 48, -15000)$$
$$= (15,000)(0.0249) = \$373.50 \quad \text{or} \quad = \$373.28$$

The next step will use the spreadsheet answer, because it is more accurate (there are only three significant digits in the tabulated factor).

After 24 payments and with 24 left, the remaining balance equals $(P/A, i, N_{remaining})$ payment

$$\text{Balance} = (P/A, 0.75\%, 24)\$373.28 \qquad \text{or} \quad = \text{PV}(0.75\%, 24, 373.28)$$

$$= (21.889)(373.28) = \$8170.73 \quad \text{or} \quad = \$8170.78$$

Thus halfway through the repayment schedule, 54.5% of the original balance is still owed.

Pay Off Debt Sooner by Increasing Payments

Paying off debt can be a good investment because the investment earns the rate of interest on the loan. For example, this could be 8% for a mortgage, 10% for a car loan, or 19% for a credit card. When one is making extra payments on a loan, the common question is: How much sooner will the debt be paid off? Until the debt is paid off, any early payments are essentially locked up, since the same payment amount is owed each month.

The first reason that spreadsheets are convenient is fractional interest rates. For example, an auto loan might be at a nominal rate of 13% with monthly compounding or 1.08333% per month. The second reason is that the function NPER calculates the number of periods remaining on a loan.

NPER can be used to calculate how much difference is made by one extra payment or by increasing all payments by $x\%$. Extra payments are applied entirely to principal, so the interest rate, remaining balance, and payment amounts are all known. $N_{remaining}$ equals NPER(i, payment, remaining balance) with optional arguments for beginning-of-period cash flows and balloon payments. The signs of the payment and the remaining balance must be different.

EXAMPLE 6-13

Maria has a 7.5% mortgage with monthly payments for 30 years. Her original balance was $100,000, and she just made her twelfth payment. Each month she also pays into a reserve account, which the bank uses to pay her fire and liability insurance ($900 annually) and property taxes ($1500 annually). By how much does she shorten the loan if she makes an extra *loan* payment today? If she makes an extra *total* payment? If she increases each total payment to 110% of her current total payment?

SOLUTION

The first step is to calculate Maria's *loan* payment for the 360 months. Rather than calculating a six-significant digit monthly interest rate, it is easier to use 0.075/12 in the spreadsheet formulas.

$$\text{Payment} = \text{PMT}(0.075/12, 360, -100000) = \$699.21$$

The remaining balance after 12 such payments is the present worth of the remaining 348 payments.

$$\text{Balance}_{12} = \text{PV}(0.075/12, 348, 699.21) = \$99,077.53$$
$$\text{(after 12 payments, she has paid off \$922!)}$$

If she pays an extra $699.21, then the number of periods remaining is

$$\text{NPER}(0.075/12, -699.21, 99077.53 - 699.21) = 339.5$$

This is 8.5 payments less than the 348 periods left before the extra payment. If she makes an extra total payment, then

$$\text{Total payment} = 699.21 + 900/12 + 1500/12 = \$899.21/\text{month}$$

$$\text{NPER}(0.075/12, -699.21, 99244 - 899.21) = 337.1$$

or 2.4 more payments saved. If she makes an extra 10% payment on the total payment of $899.21, then

$$\text{NPER}(0.075/12, -(1.1 * 899.21 - 200), 99077.53) = 246.5 \text{ payments}$$

or 101.5 payments saved.

Note that $200 of the total payment goes to pay for insurance and taxes.

SUMMARY

Annual cash flow analysis is the second of the three major methods of resolving alternatives into comparable values. When an alternative has an initial cost P and salvage value S, there are three ways of computing the equivalent uniform annual cost:

- $\text{EUAC} = P(A/P, i, n) - S(A/F, i, n)$ (6-1)
- $\text{EUAC} = (P - S)(A/F, i, n) + Pi$ (6-3)
- $\text{EUAC} = (P - S)(A/P, i, n) + Si$ (6-4)

All three equations give the same answer. This quantity is also known as the *capital recovery cost* of the project.

The relationship between the present worth of cost and the equivalent uniform annual cost is:

- $\text{EUAC} = (\text{PW of cost})(A/P, i, n)$

The three annual cash flow criteria are:

Neither input nor output fixed	Maximize EUAW = EUAB − EUAC
For fixed input	Maximize EUAB
For fixed output	Minimize EUAC

In present worth analysis there must be a common analysis period. Annual cash flow analysis, however, allows some flexibility provided the necessary assumptions are suitable in the situation being studied. The analysis period may be different from the lives of the alternatives, and provided the following criteria are met, a valid cash flow analysis may be made.

1. When an alternative has reached the end of its useful life, it is assumed to be replaced by an identical replacement (with the same costs, performance, etc.).

2. The analysis period is a common multiple of the useful lives of the alternatives, or there is a continuing or perpetual requirement for the selected alternative.

If neither condition applies, it is necessary to make a detailed study of the consequences of the various alternatives over the entire analysis period with particular attention to the difference between the alternatives at the end of the analysis period.

There is very little numerical difference between a long-life alternative and a perpetual alternative. As the value of n increases, the capital recovery factor approaches i. At the limit, $(A/P, i, \infty) = i$.

One of the most common uniform payment series is the repayment of loans. Spreadsheets are useful in analyzing loans (balance due, interest paid, etc.) for several reasons: they have specialized functions, many periods are easy, and any interest rate can be used.

PROBLEMS

Annual Calculations

6-1 Compute the EUAB for these cash flows based on a 10% interest rate.

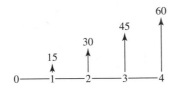

(*Answer:* EUAB= $35.72)

6-2 Compute the EUAC for these cash flows.

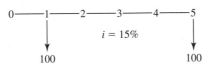

6-3 Compute the EUAB for these cash flows.

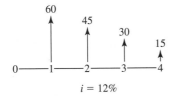

6-4 If $i = 6\%$, compute the EUAB over 6 years that is equivalent to the two receipts shown.

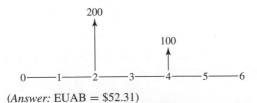

(*Answer:* EUAB = $52.31)

6-5 A loan of $100 is to be repaid in three equal semiannual (every 6 months) payments. If the annual interest rate is 7% compounded semiannually, how much is each payment? (*Answer:* $35.69)

6-6 When he started work on his twenty-second birthday, D. B. Cooper decided to invest money each month with the objective of becoming a millionaire by the time he reaches age 65. If he expects his investments to yield 18% per annum, compounded monthly, how much should he invest each month? (*Answer:* $6.92 a month.)

6-7 The average age of engineering students at graduation is a little over 23 years. This means that the working career of most engineers is almost exactly 500 months. How much would an engineer need to save each month to become a millionaire by the end of his working career? Assume a 15% interest rate, compounded monthly.

6-8 An engineer wishes to have $5 million by the time he retires in 40 years. Assuming 15% nominal interest, compounded continuously, what annual sum must he set aside? (*Answer:* $2011)

6-9 Art Arfons, a K-State-educated engineer, has made a considerable fortune. He wishes to start a perpetual scholarship for engineering students at K-State. The scholarship will provide a student with an annual stipend of $2500 for each of 4 years (freshman through senior), plus an additional $5000 during the senior year to cover entertainment expenses. Assume that students graduate in 4 years, a new award is given every 4 years, and the money is paid at the beginning of each year with the first award at the beginning of Year 1. The interest rate is 8%.

(*a*) Determine the equivalent uniform annual cost (EUAC) of providing the scholarship.

(*b*) How much money must Art donate to K-State?

6-10 For the diagram, compute the value of D that results in a net equivalent uniform annual worth (EUAW) of 0.

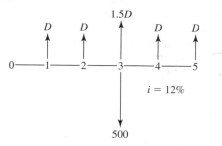

6-11 An electronics firm invested $60,000 in a precision inspection device. It cost $4000 to operate and maintain in the first year and $3000 in each of the subsequent years. At the end of 4 years, the firm changed their inspection procedure, eliminating the need for the device. The purchasing agent was very fortunate in being able to sell the inspection device for $60,000, the original price. The plant manager asks you to compute the equivalent uniform annual cost of the device during the 4 years it was used. Assume interest at 10% per year. (*Answer:* $9287)

6-12 A firm is about to begin pilot plant operation on a process it has developed. One item of optional equipment that could be obtained is a heat exchanger unit. The company finds that a unit now available for $30,000 could be used in other company operations. It is estimated that the heat exchanger unit will be worth $35,000 at the end of 8 years. This seemingly high salvage value is due primarily to the fact that the $30,000 purchase price is really a rare bargain. If the firm believes 15% is an appropriate rate of return, what annual benefit is needed to justify the purchase of the heat exchanger unit? (*Answer:* $4135)

6-13 A firm purchased some equipment at a very favorable price of $30,000. The equipment resulted in an annual net saving of $1000 per year during the 8 years it was used. At the end of 8 years, the equipment was sold for $40,000. Assuming interest at 8%, did the equipment purchase prove to be desirable?

6-14 A machine costs $20,000 and has a 5-year useful life. At the end of the 5 years, it can be sold for $4000. If annual interest is 8%, compounded semiannually, what is the equivalent uniform annual cost of the machine?

6-15 Mr. Wiggley wants to buy a new house. It will cost $178,000. The bank will loan 90% of the purchase price at a nominal interest rate of 10.75% compounded weekly, and Mr. Wiggley will make monthly payments. What is the amount of the monthly payments if he intends to pay the house off in 25 years?

6-16 Steve Lowe must pay his property taxes in two equal installments on December 1 and April 1. The two payments are for taxes for the fiscal year that begins on July 1 and ends the following June 30. Steve purchased a home on September 1. Assuming the annual property taxes remain at $850 per year for the next several years, Steve plans to open a savings account and to make uniform monthly deposits the first of each month. The account is to be used to pay the taxes when they are due.

To open the account, Steve deposits a lump sum equivalent to the monthly payments that will not have been made for the first year's taxes. The savings account pays 9% interest, compounded monthly and payable quarterly (March 31, June 30, September 30, and December 31). How much money should Steve put into the account when he opens it on September 1? What uniform monthly deposit should he make from that time on? (*Answers:* Initial deposit $350.28; monthly deposit $69.02)

6-17 A motorcycle is for sale for $2600. The dealer is willing to sell it on the following terms:

> No down payment; pay $44 at the end of each of the first 4 months; pay $84 at the end of each month after that until the loan has been paid in full.

At a 12% annual interest rate compounded monthly, how many $84 payments will be required?

6-18 Your company must make a $500,000 balloon payment on a lease 2 years and 9 months from today. You have been directed to deposit an amount of money quarterly, beginning today, to provide for the $500,000 payment. The account pays 4% per year, compounded quarterly. What is the required quarterly deposit? *Note:* Lease payments are due at the beginning of the quarter.

6-19 For the diagram, compute the value of C that results in a net equivalent annual worth (EUAW) of 0.

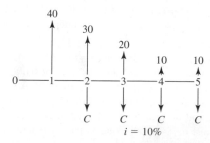

6-20 If interest is 10%, what is the EUAB?

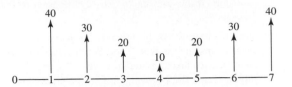

6-21 The maintenance foreman of a plant in reviewing his records found that a large press had the following maintenance cost record:

5 years ago	$ 600
4 years ago	700
3 years ago	800
2 years ago	900
Last year	1000

After consulting with a lubrication specialist, he changed the preventive maintenance schedule. He believes that maintenance will be $900 this year and will decrease by $100 a year in each of the following 4 years. If his estimate of the future is correct, what will be the equivalent uniform annual maintenance cost for the 10-year period? Assume interest at 8%. (*Answer:* $756)

6-22 Linda O'Shay deposited $30,000 in a savings account as a perpetual trust. She believes the account will earn 7% annual interest during the first 10 years and 5% interest thereafter. The trust is to provide a uniform end-of-year scholarship at the university. What uniform amount could be used for the student scholarship each year, beginning at the end of the first year and continuing forever?

6-23 An engineer has a fluctuating future budget for the maintenance of a particular machine. During each of the first 5 years, $1000 per year will be budgeted. During the second 5 years, the annual budget will be $1500 per year. In addition, $3500 will be budgeted for an overhaul of the machine at the end of the fourth year, and another $3500 for an overhaul at the end of the eighth year.

The engineer asks you to compute the uniform annual expenditure that would be equivalent to these fluctuating amounts, assuming interest at 6% per year.

6-24 A machine has a first cost of $150,000, an annual operation and maintenance cost of $2500, a life of 10 years, and a salvage value of $30,000. At the end of Years 4 and 8, it requires a major service, which costs $20,000 and $10,000, respectively. At the end of Year 5, it will need to be overhauled at a cost of $45,000. What is the equivalent uniform annual cost of owning and operating this particular machine?

6-25 There is an annual receipt of money that varies from $100 to $300 in a fixed pattern that repeats forever. If interest is 10%, compute the EUAB, also continuing forever, that is equivalent to the fluctuating disbursements.

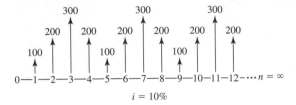

$i = 10\%$

Annual Comparisons

6-26 Alice White has arranged to buy some home recording equipment. She estimates that it will have a 5-year useful life and no salvage value. The dealer has offered Alice two alternative ways to pay for the equipment:

   A. Pay $2000 immediately and $500 at the end of one year.

   B. Pay nothing until the end of 4 years, when a single payment of $3000 must be made.

   Alice believes 12% is a suitable interest rate. Use an annual cash flow analysis to determine which method of payment she should select. (*Answer:* Select B)

6-27 The Johnson Company pays $200 a month to a trucker to haul wastepaper and cardboard to the city dump. The material could be recycled if the company were to buy a $6000 hydraulic press baler and spend $3000 a year for labor to operate the baler. The baler has an estimated useful life of 30 years and no salvage value. Strapping material would cost $200 per year for the estimated 500 bales a year that would be produced. A wastepaper company will pick up the bales at the plant and pay Johnson $2.30 per bale for them. Use an annual cash flow analysis in working this problem.

   (*a*) If interest is 8%, is it economical to install and operate the baler?

   (*b*) Would you recommend that the baler be installed?

6-28 Jenny McCarthy is an engineer for a municipal power plant. The plant uses natural gas, which is currently provided from an existing pipeline at an annual cost of $10,000 per year. Jenny is considering a project to construct a new pipeline. The initial cost of the new pipeline would be $35,000, but it would reduce the annual cost to $5000 per year. Assume an analysis period of 20 years and no salvage value for either the existing or new pipeline. The interest rate is 6%.

(a) Determine the equivalent uniform annual cost (EUAC) for the new pipeline.

(b) Should the new pipeline be built?

6-29 Claude James, a salesman, needs a new car for business use. He expects to be promoted to a supervisory job at the end of 3 years, and he will no longer be "on the road." The company reimburses salesmen each month at the rate of 25¢ per mile driven. Claude finds that there are three different ways of obtaining his chosen car:

A. Pay cash: the price is $13,000.

B. Lease the car: the monthly charge is $350 on a 36-month lease, payable at the end of each month; at the end of the 3-year period, the car is returned to the leasing company.

C. Lease the car with an option to buy at the end of the lease: pay $360 a month for 36 months; at the end of that time, Claude could buy the car, if he chooses, for $3500.

Claude believes he should use a 12% interest rate. If the car could be sold for $4000 at the end of 3 years, which method should he use to obtain it?

6-30 When he purchased his home, Al Silva borrowed $80,000 at 10% interest to be repaid in 25 equal annual end-of-year payments. After making 10 payments, Al found he could refinance the balance due on his loan at 9% interest for the remaining 15 years.

To refinance the loan, Al must pay the original lender the balance due on the loan, plus a penalty charge of 2% of the balance due; to the new lender he also must pay a $1000 service charge to obtain the loan. The new loan would be made equal to the balance due on the old loan, plus the 2% penalty charge, and the $1000 service charge. Should Al refinance the loan, assuming that he will keep the house for the next 15 years? Use an annual cash flow analysis in working this problem.

6-31 A firm must decide whether to provide their salespeople with firm-owned cars or to pay a mileage allowance for their own cars. New cars would cost about $18,000 each and could be resold 4 years later for about $7000 each. Annual operating costs would be $600 per year plus 12¢ per mile. If the salespeople drove their own cars, the firm would pay 30¢ per mile. How many miles must each salesperson drive each year for it to be economically

practical for the firm to provide the cars. Assume a 10% annual interest rate. Use an annual cash flow analysis.

6-32 The town of Dry Gulch needs more water from Pine Creek. The town engineer has selected two plans for comparison: a *gravity plan* (divert water at a point 10 miles up Pine Creek and pipe it by gravity to the town) and a *pumping plan* (divert water at a point closer to town). The pumping plant would be built in two stages, with half-capacity installed initially and the other half installed 10 years later.

The analysis will assume a 40-year life, 10% interest, and no salvage value. Use an annual cash flow analysis to find which plan is more economical.

	Gravity	Pumping
Initial investment	$2,800,000	$1,400,000
Additional investment in 10th year	None	200,000
Operation and maintenance	10,000/yr	25,000/yr
Power cost		
Average first 10 years	None	50,000/yr
Average next 30 years	None	100,000/yr

6-33 A manufacturer is considering replacing a production machine tool. The new machine, costing $3700, would have a life of 4 years and no salvage value, but would save the firm $500 per year in direct labor costs and $200 per year in indirect labor costs. The existing machine tool was purchased 4 years ago at a cost of $4000. It will last 4 more years and will have no salvage value. It could be sold now for $1000 cash. Assume that money is worth 8% and that differences in taxes, insurance, and so forth are negligible. Use an annual cash flow analysis to determine whether the new machine should be purchased.

6-34 Two possible routes for a power line are under study. Data on the routes are as follows:

	Around the Lake	Under the Lake
Length	15 km	5 km
First cost	$5000/km	$25,000/km
Maintenance	$200/km/yr	$400/km/yr
Useful life, in years	15	15
Salvage value	$3000/km	$5000/km
Yearly power loss	$500/km	$500/km
Annual property taxes	2% of first cost	2% of first cost

If 7% interest is used, should the power line be routed around the lake or under the lake?   (*Answer:* Around the lake.)

6-35 An oil refinery must now begin sending its waste liquids through a costly treatment process before discharging them into a nearby stream. The engineering department estimates costs at $30,000 for the first year. It is estimated that if process and plant alterations are made, the waste treatment cost will decline $3000 each year. As an alternate, a specialized firm, Hydro-Clean, has offered a contract to process the waste liquids for 10 years for a fixed price of $15,000 per year, payable at the end of each year. Either way, there should be no need for waste treatment after 10 years. The refinery manager considers 8% to be a suitable interest rate. Use an annual cash flow analysis to determine whether the Hydro-Clean offer should be accepted.

6-36 Bill Anderson buys a car every 2 years as follows: initially he makes a down payment of $6000 on a $15,000 car. The balance is paid in 24 equal monthly payments with annual interest at 12%. When he has made the last payment on the loan, he trades in the 2-year-old car for $6000 on a new $15,000 car, and the cycle begins over again.

Doug Jones decided on a different purchase plan. He thought he would be better off if he paid $15,000 cash for a new car. Then he would make a monthly deposit in a savings account so that, at the end of 2 years, he would have $9000 in the account. The $9000 plus the $6000 trade-in value of the car will allow Doug to replace his 2-year-old car by paying $9000 for a new one. The bank pays 6% interest, compounded quarterly.

(*a*) What is Bill Anderson's monthly payment to pay off the loan on the car?

(*b*) After he purchased the new car for cash, how much per month should Doug Jones deposit in his savings account to have enough money for the next car 2 years hence?

(*c*) Why is Doug's monthly savings account deposit smaller than Bill's payment?

6-37 Two mutually exclusive alternatives are being considered.

Year	A	B
0	−$3000	−$5000
1	+845	+1400
2	+845	+1400
3	+845	+1400
4	+845	+1400
5	+845	+1400

One of the alternatives must be selected. Using a 15% nominal interest rate, compounded continuously, determine which one. Solve by annual cash flow analysis.

Different Lives

6-38 A certain industrial firm desires an economic analysis to determine which of two different machines should be purchased. Each machine is capable of performing the same task in a given amount of time. Assume the minimum attractive return is 8%.

	Machine X	Machine Y
First cost	$5000	$8000
Estimated life, in years	5	12
Salvage value	0	$2000
Annual maintenance cost	0	150

Which machine would you choose? Base your answer on annual cost. (*Answers:* X = $1252; Y = $1106)

6-39 A company must decide whether to buy Machine *A* or Machine *B*:

	Machine A	Machine B
Initial cost	$10,000	$20,000
Useful life, in years	4	10
End-of-useful-life salvage value	$10,000	$10,000
Annual maintenance	1,000	0

At a 10% interest rate, which machine should be installed? Use an annual cash flow analysis in working this problem.   (*Answer:* Machine *A*)

6-40 Consider the following two mutually exclusive alternatives:

	A	B
Cost	$100	$150
Uniform annual benefit	16	24
Useful life, in years	∞	20

Alternative B may be replaced with an identical item every 20 years at the same $150 cost and will have the same $24 uniform annual benefit. Using a 10% interest rate and an annual cash flow analysis, which alternative should be selected?

6-41 A pump is needed for 10 years at a remote location. The pump can be driven by an electric motor if a power line is extended to the site. Otherwise, a gasoline engine will be used. Use an annual cash flow analysis and a 10% interest rate. How should the pump be powered?

	Gasoline	Electric
First cost	$2400	$6000
Annual operating cost	1200	750
Annual maintenance	300	50
Salvage value	300	600
Life, in years	5	10

6-42 A suburban taxi company is considering buying taxis with diesel engines instead of gasoline engines. The cars average 50,000 km a year.

	Diesel	Gasoline
Vehicle cost	$13,000	$12,000
Useful life, in years	3	4
Fuel cost per liter	48¢	51¢
Mileage, in km/liter	35	28
Annual repairs	$ 300	$ 200
Annual insurance premium	500	500
End-of-useful-life resale value	2,000	3,000

Use an annual cash flow analysis to determine the more economical choice if interest is 6%.

6-43 The manager in a canned food processing plant is trying to decide between two labeling machines.

Find A

	Machine A	Machine B
First cost	$15,000	$25,000
Maintenance and operating costs	1,600	400
Annual benefit	8,000	13,000
Salvage value	3,000	6,000
Useful life, in years	7	10

Eua

Assume an interest rate of 12%. Use annual cash flow analysis to determine which machine should be chosen.

6-44 Consider the following three mutually exclusive alternatives:

	A	B	C
Cost	$100	$150.00	$200.00
Uniform annual benefit	10	17.62	55.48
Useful life, in years	∞	20	5

Assuming that Alternatives B and C are replaced with identical units at the end of their useful lives, and an 8% interest rate, which alternative should be selected? Use an annual cash flow analysis in working this problem.   (*Answer:* Select C)

6-45 Carp, Inc. wants to evaluate two methods of shipping their products. Use an interest rate of 15% and annual cash flow analysis to decide which is the most desirable alternative.

	A	B
First cost	$700,000	$1,700,000
Maintenance and operating costs	18,000	29,000
+ Cost gradient (begin Year 1)	+900/yr	+750/yr
Annual benefit	154,000	303,000
Salvage value	142,000	210,000
Useful life, in years	10	20

6-46 A college student has been looking for new tires and has found the following alternatives:

Tire Warranty (months)	Price per Tire
12	$39.95
24	59.95
36	69.95
48	90.00

The student feels that the warranty period is a good estimate of the tire life and that a 10% interest rate is appropriate. Using an annual cash flow analysis, which tire should be purchased?

6-47 Consider the following alternatives:

	A	B
Cost	$50	$180
Uniform annual benefit	15	60
Useful life, in years	10	5

The analysis period is 10 years, but there will be no replacement for Alternative *B* after 5 years. Based on a 15% interest rate, which alternative should be selected? Use an annual cash flow analysis.

6-48 Some equipment will be installed in a warehouse that a firm has leased for 7 years. There are two alternatives:

	A	*B*
Cost	$100	$150
Uniform annual benefit	55	61
Useful life, in years	3	4

At any time after the equipment is installed, it has no salvage value. Assume that Alternatives *A* and *B* will be replaced at the end of their useful lives by identical equipment with the same costs and benefits. For a 7-year analysis period and a 10% interest rate, use an annual cash flow analysis to determine which alternative should be selected.

6-49 Uncle Elmo needs to replace the family privy. The local sanitary engineering firm has submitted two alternative structural proposals with respective cost estimates as shown. Which construction should Uncle Elmo choose if his minimum attractive rate of return is 6%? Use both a present worth and annual cost approach in your comparison.

	Masonite	**Brick**
First cost	$250	$1000
Annual maintenance	20	10
Salvage value	10	100
Service life, in years	4	20

Spreadsheets and Loans

6-50 A student loan totals $18,000 at graduation. The interest rate is 6%, and there will be 60 payments beginning 1 month after graduation. What is the monthly payment? What is owed after the first 2 years of payments? (*Answer*: Payment = $347.99, balance due = $11,439)

6-51 The student in Problem 6-50 received $1500 as a graduation present. If an extra $1500 is paid at Month 1, when is the final payment made? How much is it. (*Answer*: $98 in Month 55)

6-52 A new car is purchased for $12,000 with a 0% down, 9% loan. The loan is for 4 years. After making 30 payments, the owner wants to pay off the loan's remaining balance. How much is owed?

6-53 A year after buying her car, Anita has been offered a job in Europe. Her car loan is for $15,000 at a 9% nominal interest rate for 60 months. If she can sell the car for $12,000, how much does she get to keep after paying off the loan?

6-54 A $78,000 mortgage has a 30-year term and a 9% nominal interest rate.

 (*a*) What is the monthly payment?

 (*b*) After the first year of payments, what is the outstanding balance?

 (*c*) How much interest is paid in Month 13? How much principal?

6-55 A $92,000 mortgage has a 30-year term and a 9% nominal interest rate.

 (*a*) What is the monthly payment?

 (*b*) After the first year of payments, what fraction of the loan has been repaid?

 (*c*) After the first 10 years of payments, what is the outstanding balance?

 (*d*) How much interest is paid in Month 25? How much principal?

6-56 A 30-year mortgage for $95,000 is issued at a 9% nominal interest rate.

 (*a*) What is the monthly payment?

 (*b*) How long does it take to pay off the mortgage, if $1000 per month is paid?

 (*c*) How long does it take to pay off the mortgage, if double payments are made?

6-57 A 30-year mortgage for $145,000 is issued at a 6% nominal interest rate.

 (*a*) What is the monthly payment?

 (*b*) How long does it take to pay off the mortgage, if $1000 per month is paid?

 (*c*) How long does it take to pay off the mortgage, if 20% extra is paid each month?

6-58 Solve Problem 6-34 for the breakeven first cost per kilometer of going under the lake.

6-59 Redo Problem 6-42 to calculate the EUAW of the alternatives as a function of miles driven per year to see if there is a crossover point in the decision process. Graph your results.

6-60 Set up Problem 6-27 on a spreadsheet to make all the input data variable and determine various scenarios which would make the baler economical.

6-61 Develop a spreadsheet to solve Problem 6-32. What is the breakeven cost of the additional pumping investment in Year 10?

CHAPTER 7

Rate of Return Analysis

Bar Codes Give a Number; RFID Codes Tell a Story

Both radio frequency identification (RFID) and bar code technology encode information, but RFID uses radio waves rather than optical scanning to read that information. This means that the RFID tag can be read without a line of sight. The shift from manual price tags, shipping labels, and baggage tags to bar codes revolutionized operations by reducing costs and errors. Now the shift to RFID is revolutionizing operations again. Today's smart tag is a paper sandwich of a bar code and an RFID tag.

A bar code speaks five or six times in its lifetime; RFID codes can speak 200 to 1000 times each second. Your car keys are an RFID chip allowing you to open or start your car remotely. RFID is in the EZ Pass on your windshield, a rail pass in China, your new Smart Passport, a subway pass in Washington, D.C., some employee IDs, and garment labels. RFID chips are used to tightly control the inventory of packs of razor blades—including the control of pilferage. They are used in boxes of Alaska wild seafood to track temperatures from packaging to restaurant delivery. They are required on many deliveries to the military and major retailers.

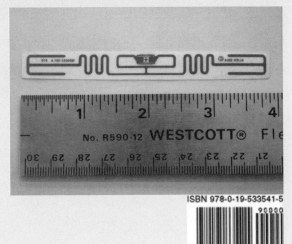

Because RFID tags carry so much more information, organizations need to measure how they improve performance in many ways. RFID is changing the old ways of operating in logistics and supply chains and business in general. Today's wholesalers, retailers, manufacturers, and military decision makers trust RFID technology to create real-time inventory tracking, while eliminating the jobs of many warehouse workers and cashiers.

As with any technology that allows change, there are some who dislike it. In addition, some raise ethical questions about the ability to track items after purchase and people as they move about a facility. Some think of RFID as spy chips. However, before you challenge RFID, understand where it is now and think where it will lead you.

Contributed by Oliver Hedgepeth, University of Alaska Anchorage, author of RFID Metrics, *CRC Press (2007).*

QUESTIONS TO CONSIDER

1. The growth rate for the market in RFID systems from 2004 to 2010 is estimated to be about 50% annually. This is built on many individual projects—each of which must be economically justified. Does the growth rate imply anything about the rate of return on the individual projects?

2. The cost of RFID tags could be as low as 5¢ in 2008, much higher than the bar code tag cost of less than a penny. What are the cost and benefit factors that you would use to make a decision to replace bar codes with RFID tags? Or to incur the cost of both in a smart tag?

3. The Smart Passport and its information may not be meaningful without access to a database file on the person being scanned. How would you advise U.S. citizens or defense department shippers and logisticians who use Smart Passports when they are traveling outside the country?

4. What would you suggest to management about the 24/7 stream of data each second that could come from an inventory counted by RFID tags?

5. Go to the Internet and find five different RFID applications. List the problems these users think they are solving and why they seem to think the chips will work. Do you think they have thought through the problem sufficiently? Are there ethical or social concerns for each application? Explain your answer.

6. The RFID computer chip sewn in the label of your new garment can still be active or readable when you get home, when you wear it to work, and when you return to the store where you bought it. Some consumers have already expressed concern about RFID chips, calling them an invasion of privacy. What ethical and social concerns are raised by RFID chips?

After Completing This Chapter...

The student should be able to:

- Evaluate project cash flows with the *internal rate of return* (IRR) measure.
- Plot a project's present worth (PW) against the interest rate.
- Use an *incremental* rate of return analysis to evaluate competing alternatives.
- Develop and use spreadsheets to make IRR and incremental rate of return calculations.

I n this chapter we will examine four aspects of rate of return, the third major analysis method. First, the meaning of "rate of return" is explained; second, the calculation of rate of return is illustrated; third, rate of return analysis problems are presented; and fourth, incremental analysis is presented. In an appendix to the chapter, we describe difficulties sometimes encountered in attempting to compute an interest rate for cash flows of certain kinds.

Rate of return analysis is the most frequently used exact analysis technique in industry. Problems in computing the rate of return sometimes occur, but its major advantage is that it is a single figure of merit that is readily understood.

Consider these statements:

- The net present worth on a project is $32,000.
- The equivalent uniform annual net benefit is $2800.
- The project will produce a 23% rate of return.

While none of these statements tells the complete story, the third one measures the project's desirability in terms that are widely and easily understood. Thus, this measure is accepted by engineers and business leaders alike.

There is another advantage to rate of return analysis. In both present worth and annual cash flow calculations, one must select an interest rate—and this may be a difficult and controversial item. In rate of return analysis, no interest rate is introduced into the calculations (except as described in the appendix to this chapter). Instead, we compute a rate of return (more accurately called *internal rate of return*) from the cash flow. To decide how to proceed, the calculated rate of return is compared with a preselected **minimum attractive rate of return,** or simply MARR. This is the same value of *i* used for present worth and annual cash flow analysis.

INTERNAL RATE OF RETURN

A project's rate of return or internal rate of return can be defined by using the calculation of a present worth or an equivalent uniform annual worth.

Internal rate of return is the interest rate at which the present worth and equivalent uniform annual worth are equal to 0.

This definition is easy to remember, and it also gives us a clear way to solve for the rate of return. We've already done this in earlier chapters when we solved for the interest rate on a loan or investment.

Other definitions based on the unpaid balance of a loan or the unrecovered investment can help clarify why this rate of return is also called the *internal rate of return* or IRR. In Chapter 3 we examined four plans to repay $5000 in 5 years with interest at 8% (Table 3-1). In each case the amount loaned ($5000) and the loan duration (5 years) were the same. Yet the total interest paid to the lender varied from $1200 to $2347. We saw, however, that the lender received 8% interest each year on the amount of money actually owed. And, at the end of 5 years, the principal and interest payments exactly repaid the $5000 debt with interest at 8%. We say the lender received an "8% rate of return."

Internal rate of return is defined as the interest rate paid on the unpaid balance of a *loan* such that the payment schedule makes the unpaid loan balance equal to zero when the final payment is made.

Instead of lending money, we might invest $5000 in a machine tool with a 5-year useful life and an equivalent uniform annual benefit of $1252. The question becomes, What rate of return would we receive on this investment? The cash flows would be:

Year	Cash Flow
0	−$5000
1	+1252
2	+1252
3	+1252
4	+1252
5	+1252

We recognize the cash flow as reversed from Plan 3 of Table 3-1. We know that five payments of $1252 are equivalent to a present sum of $5000 when interest is 8%. Therefore, the rate of return on this investment is 8%. For an investment, we may define internal rate of return as follows:

> **Internal rate of return** is the interest rate earned on the unrecovered *investment* such that the payment schedule makes the unrecovered investment equal to zero at the end of the investment's life.

It must be understood that the 8% rate of return does not mean an annual return of 8% on the $5000 investment, or $400 in each of the 5 years with $5000 returned at the end of Year 5. Instead, each $1252 payment represents an 8% return on the unrecovered investment *plus* a partial return of the investment. This may be tabulated as follows:

Year	Cash Flow	Unrecovered Investment at Beginning of Year	8% Return on Unrecovered Investment	Investment Repayment at End of Year	Unrecovered Investment at End of Year
0	−$5000				
1	+1252	$5000	$ 400	$ 852	$4148
2	+1252	4148	331	921	3227
3	+1252	3227	258	994	2233
4	+1252	2233	178	1074	1159
5	+1252	1159	93	1159	0
			$1260	$5000	

This cash flow represents a $5000 investment with benefits that produce an 8% rate of return on the unrecovered investment.

Although the definitions of internal rate of return are stated differently, for loan and for an investment, there is only one fundamental concept. It is that **the internal rate of return is the interest rate at which the benefits are equivalent to the costs** or the present worth (PW) is 0. Since we are describing the funds that remain within the investment throughout its life, the resulting rate of return is described as the internal rate of return, i.

CALCULATING RATE OF RETURN

To calculate a rate of return on an investment, we must convert the various conse-quences of the investment into a cash flow. Then we solve the cash flow for the unknown value of the internal rate of return (IRR). Five forms of the cash flow equation are as follows:

$$PW \text{ of benefits} - PW \text{ of costs} = 0 \qquad (7\text{-}1)$$

$$\frac{PW \text{ of benefits}}{PW \text{ of costs}} = 1 \qquad (7\text{-}2)$$

$$\text{Present worth} = \text{Net present worth}[1] = 0 \qquad (7\text{-}3)$$

$$EUAW = EUAB - EUAC = 0 \qquad (7\text{-}4)$$

$$PW \text{ of costs} = PW \text{ of benefits} \qquad (7\text{-}5)$$

The five equations represent the same concept in different forms. They relate costs and benefits with the IRR as the only unknown. The calculation of rate of return is illustrated by the following examples.

EXAMPLE 7-1

An engineer invests $5000 at the end of every year for a 40-year career. If the engineer wants $1 million in savings at retirement, what interest rate must the investment earn?

SOLUTION

Using Equation 7-3, we write

$$\text{Net PW} = 0 = -\$5000(F/A, i, 40) + \$1,000,000$$

Rewriting, we see that

$$(F/A, i, 40) = \$1,000,000/\$5000 = 200$$

We then look at the compound interest tables for the value of i where $(F/A, i, 40) = 200$. If no tabulated value of i gives this value, then we will interpolate, using the two closest values or solve exactly with a calculator or spreadsheet. In this case $(F/A, 0.07, 40) = 199.636$, which to three significant digits equals 200. Thus the required rate of return for the investment is 7%. (The exact value is 7.007%.)

While this example has been defined in terms of an engineer's personal finances, we could just as easily have said, "a mining firm makes annual deposits of $50,000 into a reclamation fund for 40 years. If the firm must have $10 million when the mine is closed, what interest rate must the investment earn?" Since the annual deposit and required final amount are 10 times larger in each case, the *F/A* factor and the answer are obviously the same.

---

[1]Remember that present value (PV), present worth (PW), net present value (NPV), and net present worth (NPW) are synonyms in practice and in this text.

EXAMPLE 7-2

An investment resulted in the following cash flow. Compute the rate of return.

Year	Cash Flow
0	−$700
1	+100
2	+175
3	+250
4	+325

SOLUTION

$$EUAW = EUAB - EUAC = 0$$

$$100 + 75(A/G, i, 4) - 700(A/P, i, 4) = 0$$

Here, we have two different interest factors in the equation. We will not be able to solve it as easily as Example 7-1. Since there is no convenient direct method of solution, we will solve the equation by trial and error. Try $i = 5\%$ first:

$$EUAB - EUAC = 0$$

$$100 + 75(A/G, 5\%, 4) - 700(A/P, 5\%, 4) = 0$$

$$100 + 75(1.439) - 700(0.2820) = 0$$

$$EUAB - EUAC = 208 - 197 = +11$$

The EUAC is too low. If the interest rate is increased, EUAC will increase. Try $i = 8\%$:

$$EUAB - EUAC = 0$$

$$100 + 75(A/G, 8\%, 4) - 700(A/P, 8\%, 4) = 0$$

$$100 + 75(1.404) - 700(0.3019) = 0$$

$$EUAB - EUAC = 205 - 211 = -6$$

This time the EUAC is too large. We see that the true rate of return is between 5% and 8%. Try $i = 7\%$:

$$EUAB - EUAC = 0$$

$$100 + 75(A/G, 7\%, 4) - 700(A/P, 7\%, 4) = 0$$

$$100 + 75(1.416) - 700(0.2952) = 0$$

$$EUAB - EUAC = 206 - 206 = 0$$

The IRR is 7%.

EXAMPLE 7-3

A local firm sponsors a student loan program for the children of employees. No interest is charged until graduation, and then the interest rate is 5%. Maria borrows $9000 per year, and she graduates after 4 years. Since tuition must be paid ahead of time, assume that she borrows the money at the start of each year.

If Maria makes five equal annual payments, what is each payment? Use the cash flow from when she started borrowing the money to when it is all paid back, and then calculate the internal rate of return for Maria's loan. Is this arrangement attractive to Maria?

SOLUTION

Maria owes $36,000 at graduation. The first step is to calculate the five equal annual payments to repay this loan at 5%.

$$\text{Loan payment} = \$36,000(A/P, 5\%, 5) = 36,000(0.2310) = \$8316$$

Maria receives $9000 by borrowing at the start of each year. She graduates at the end of Year 4. At the end of Year 4, which is also the beginning of Year 5, interest starts to accrue. She makes her first payment at the end of Year 5, which is one year after graduation.

Year	0	1	2	3	4	5	6	7	8	9
Cash Flow	9000	9000	9000	9000	0	−8316	−8316	−8316	−8316	−8316

The next step is to write the present worth equation in factor form, so that we can apply Equation 7-3 and set it equal to 0. This equation has three factors, so we will have to solve the problem by picking interest rates and substituting values. The present worth value is a function of i.

$$\text{PW}(i) = 9000[1 + (P/A, i, 3)] - 8316(P/A, i, 5)(P/F, i, 4)$$

The first two interest rates used are 0% (because it is easy) and 3% because the subsidized rate will be below the 5% that is charged after graduation.

At 0% any P/A factor equals n, and any P/F factor equals 1.

$$\text{PW}(0\%) = 9000 \times 4 - 8316 \times 5 = -5180$$

$$\text{PW}(3\%) = 9000 \times (1 + 2.829) - 8316 \times 4.580 \times 0.8885 = 620.5$$

Since PW(i) has opposite signs for 0 and 3, there is a value of i between 0 and 3% which is the IRR. Because the value for 3% is closer to 0, the IRR will be closer to 3%. Try 2% next.

$$\text{PW}(2\%) = 9000 \times (1 + 2.884) - 8316 \times 4.713 \times 0.9238 = -1251$$

As shown in Figure 7-1, interpolating between 2 and 3% leads to

$$\text{IRR} = 2\% + (3\% - 2\%)[1251/(1251 + 620.8)] = 2.67\%.$$

This rate is quite low, and it makes the loan look like a good choice.

FIGURE 7-1 Plot of PW versus interest rate i.

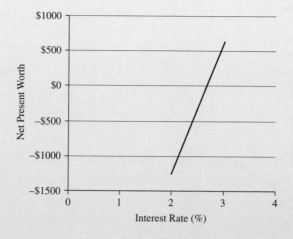

We can prove that the rate of return in Example 7-3 is very close to 2.66% by tabulating how much Maria owes each year, the interest that accrues, and her borrowings and payments.

Year	Amount Owed at Start of Year	Interest at 2.66%	Cash Flow at End of Year	Amount Owed at End of Year
1	$9,000	$239	$9000	$18,239
2	18,239	485	9000	27,725
3	27,725	737	9000	37,462
4	37,462	996	0	38,459
5	38,459	1023	−8315	31,166
6	31,166	829	−8315	23,680
7	23,680	630	−8315	15,995
8	15,995	425	−8315	8,106
9	8,106	216	−8315	6*

*This small amount owed indicates that the interest rate is slightly less than 2.66%.

If in Figure 7-1 net present worth (NPW) had been computed for a broader range of values of i, Figure 7-2 would have been obtained. From this figure it is apparent that the error resulting from linear interpolation increases as the interpolation width increases.

Plot of NPW Versus Interest Rate i

Figure 7-2—the second plot of NPW versus interest rate i—is an important source of information. For a cash flow where borrowed money is repaid, the NPW plot would appear as in Figure 7-3. The borrowed money is received early in the time period with a later repayment of an equal sum, plus payment of interest on the borrowed money. In all cases in which interest is charged and the amount borrowed is fully repaid, the NPW at 0% will be negative.

For a cash flow representing an investment followed by benefits from the investment, the plot of NPW versus i (we will call it an **NPW plot** for convenience) would have the form of Figure 7-4. As the interest rate increases, future benefits are discounted more heavily and the NPW decreases.

FIGURE 7-2 Replot of NPW Versus interest rate i over a larger range of values.

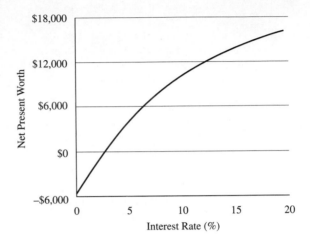

Year	Cash Flow
0	$+P$
1	$-A$
2	$-A$
3	$-A$
4	$-A$
.	.
.	.
.	.

FIGURE 7-3 Typical NPW plot for borrowed money.

Year	Cash Flow
0	$-P$
1	$+$Benefit A
2	$+A$
3	$+A$
4	$+A$
.	.
.	.
.	.

FIGURE 7-4 NPW plot for a typical investment.

Thus, interest is a charge for the use of someone else's money or a receipt for letting others use our money. The interest rate is positive, and it is either a charge for borrowing money or a receipt for lending money. Negative interest rates are more. A loan with a forgiveness provision (means not all principal is repaid) can have a negative rate. Some investments perform poorly and have negative rates.

EXAMPLE 7-4

A new corporate bond was initially sold by a stockbroker to an investor for $1000. The issuing corporation promised to pay the bondholder $40 interest on the $1000 face value of the bond every 6 months, and to repay the $1000 at the end of 10 years. After one year the bond was sold by the original buyer for $950.

(a) What rate of return did the original buyer receive on his investment?

(b) What rate of return can the new buyer (paying $950) expect to receive if he keeps the bond for its remaining 9-year life?

SOLUTION TO PART a

The original bondholder sold the bond for less ($950) than the purchase price ($1000). So the semiannual rate of return is less than the 4% semiannual interest rate on the bond.

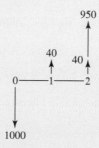

Since $40 is received each 6 months, we will solve the problem using a 6-month interest period. Let PW of cost = PW of benefits, and write

$$1000 = 40(P/A, i, 2) + 950(P/F, i, 2)$$

Try $i = 1\frac{1}{2}\%$:

$$1000 \neq 40(1.956) + 950(0.9707) = 78.24 + 922.17 \neq 1000.41$$

The interest rate per 6 months, $\text{IRR}_{6 \text{ mon}}$, is very close to $1\frac{1}{2}\%$. This means the nominal (annual) interest rate is $2 \times 1.5\% = 3\%$. The effective (annual) interest rate $= \text{IRR} = (1 + 0.015)^2 - 1 = 3.02\%$.

SOLUTION TO PART b

The new buyer will redeem the bond for more ($1000) than the purchase price ($950). So the semiannual rate of return is more than the 4% semiannual interest rate on the bond.

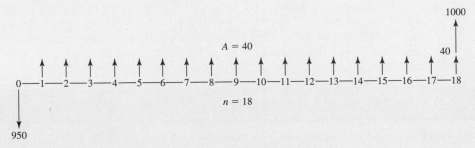

Given the same $40 semiannual interest payments, for 6-month interest periods we write

$$950 = 40(P/A, i, 18) + 1000(P/F, i, 18)$$

Try $i = 5\%$:

$$950 \neq 40(11.690) + 1000(0.4155) = 467.60 + 415.50 \neq 883.10$$

The PW of benefits is too low. Try a lower interest rate, say, $i = 4\%$:

$$950 \neq 40(12.659) + 1000(0.4936) = 506.36 + 493.60$$
$$\neq 999.96$$

The value of the IRR is between 4 and 5%. By interpolation,

$$\text{IRR} = 4\% + (1\%)\left(\frac{999.96 - 950.00}{999.96 - 883.10}\right) = 4.43\%$$

The nominal interest rate is $2 \times 4.43\% = 8.86\%$. The effective interest rate is $(1 + 0.0443)^2 - 1 = 9.05\%$.

INTEREST RATES WHEN THERE ARE FEES OR DISCOUNTS

Often when firms and individuals borrow money, there are fees charged in addition to the interest. This can be as simple as the underwriting fee that a firm is charged when it sells a bond. In Example 7-5 we add that underwriting fee to the bond in Example 7-4, and look at the bond from the firm's perspective rather than the investor's.

Example 7-6 shows how a cash discount is "unstated interest." When one is buying a new car, this may be stated as a choice between a "cash" rebate and a low-interest loan. Fees and cash discounts raise the interest rate on a loan, and the true cost of the loan is its internal rate of return.

EXAMPLE 7-5

The corporate bond in Example 7-4 was part of a much larger offering that the firm arranged with the underwriter. Each of the bonds had a face value of $1000 and a life of 10 years. Since $40 or

4% of the face value was paid in interest every 6 months, the bond had a nominal interest rate of 8% per year. If the firm paid the underwriter a 1% fee to sell the bond, what is the effective interest rate that the firm is paying on the bond?

SOLUTION

From the firm's perspective, it receives $1000 minus the fee at time 0, then it pays interest every 6 months for 10 years, and then it pays $1000 to redeem the bond. The 1% fee reduces what the firm receives when the bond is sold to $990. The interest payments are $40 every 6 months. This is easiest to model using twenty 6-month periods.

$$PW(i) = 990 - 40(P/A, i, 20) - 1000(P/F, i, 20)$$

Since the nominal interest rate is 4% every 6 months, we know that the fee will raise this some. So let us use the next higher table of 4.5%.

$$PW(4.5\%) = 990 - 40(P/A, 4.5\%, 20) - 1000(P/F, 4.5\%, 20)$$
$$= 990 - 40 \times 13.008 - 1000 \times 0.4146 = \$55.08$$

We know that the PW of the interest and final bond payoff is $1000 at 4%.

$$PW(4\%) = 990 - 1000 = -10$$

Now we interpolate to find the interest rate for each 6-month period.

$$i = 4\% + (4.5\% - 4\%) \times 10/(10 + 55.08) = 4.077\%$$

The effective annual rate is

$$i_{eff} = 1.04077^2 - 1 = 0.0832 = 8.32\%$$

EXAMPLE 7-6

A manufacturing firm may decide to buy an adjacent property so that it can expand its warehouse. If financed through the seller, the property's price is $300,000, with 20% down and the balance due in five annual payments at 12%. The seller will accept 10% less if cash is paid. The firm does not have $270,000 in cash, but it can borrow this amount from a bank. What is the rate of return or IRR for the loan offered by the seller?

SOLUTION

One choice is to pay $270,000 in cash. The other choice is to pay $60,000 down and five annual payments computed at 12% on a principal of $240,000. The annual payments equal

$$240,000(A/P, 12\%, 5) = 240,000 \times 0.2774 = \$66,576$$

The two ways of borrowing the money and the *incremental* difference between them can be summarized in a cash flow table. The incremental difference is the result of subtracting the second set of cash flows from the first.

Year	Pay Cash	Borrow from Property Owner	Incremental Difference
0	−$270,000	−$60,000	−$210,000
1		−66,576	66,576
2		−66,576	66,576
3		−66,576	66,576
4		−66,576	66,576
5		−66,576	66,576

This choice between (1) −$270,000 now and (2) −$60,000 now and −$66,576 annually for 5 years can be stated by setting them equal in PW terms, as follows:

$$-270,000 = -60,000 - 66,576(P/A, \text{IRR}, 5)$$

The true amount borrowed is $210,000, which is the difference between the $60,000 down payment and the $270,000 cash price. This is the incremental amount paid or invested by the firm instead of borrowing from the seller. Collecting the terms, the PW equation is the same as the PW equation for the incremental difference.

$$0 = -210,000 + 66,576(P/A, \text{IRR}, 5)$$

In either case, the final equation for finding the IRR is:

$$PW = 0 = -210,000 + 66,576(P/A, \text{IRR}, 5)$$

$$(P/A, IRR, 5) = 3.154$$

Looking in the tables we see that the IRR is between 15% and 18%. Interpolating gives us the following:

$$\text{IRR} = 15\% + (18\% - 15\%) \times [(3.352 - 3.154)/(3.352 - 3.127)] = 17.6\%$$

on a borrowed amount of $210,000. This is a relatively high rate of interest, so that borrowing from a bank and paying cash to the property owner is better. Note that using a spreadsheet or a financial calculator gives an exact answer of 17.62%.

LOANS AND INVESTMENTS ARE EVERYWHERE

Examples 7-4, 7-5, and 7-6 were about borrowing money through bonds and loans, but many applications for the rate of return are stated in other ways. Example 7-7 is a common problem on university campuses—buying parking permits for an academic year or for a quarter or a semester at a time.

Buying a year's parking permit is investing more money now to avoid paying for another shorter permit later. Choosing to buy a shorter permit is a loan, where the money saved by not buying the annual permit is borrowed to be repaid with the cost of the second semester permit.

EXAMPLE 7-7

An engineering student is deciding whether to buy two 1-semester parking permits or an annual permit. The annual parking permit costs \$180 due August 15th; the semester permits are \$130 due August 15th and January 15th. These values are for yellow permits at the University of Alaska Anchorage for 2006–07. What is the rate of return for buying the annual permit?

SOLUTION

Before we solve this mathematically, let us describe it in words. We are equating the \$50 cost difference now between the two permits with the \$130 cost to buy another semester permit in 5 months. Since the \$130 is 2.6 times the \$50, it is clear that we will get a high interest rate.

This is most easily solved by using, monthly periods, and the payment for the second semester is 5 months later. The cash flow table adds a column for the incremental difference to the information in the cash flow diagram.

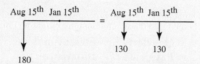

Time	Annual Pass	Two Semester Passes	Incremental Difference
Aug. 15th	−\$180	−\$130	−\$50
Jan. 15th		−130	130

Setting the two PWs equal to each other we have

$$-\$180 = -\$130[1 + (P/F, i_{mon}, 5)]$$

$$(P/F, i_{mon}, 5) = -\$50/-\$130 = 5/13 = 0.3846$$

Rather than interpolating, we can use the formula for the P/F factor.

$$1/(1 + i_{mon})^5 = 0.3846$$

$$(1 + i_{mon})^5 = 2.600$$

$$1 + i_{mon} = 1.2106$$

$$i_{mon} = 21.06\%, \text{which is an extremely high rate per month}$$

On an annual basis, the effective interest rate is $(1.2106^{12} - 1) = 891\%$. Unless the student is planning on graduating in January, it is clearly better to buy the permit a year at a time.

Examples 7-8 and 7-9 are about two common situations faced by firms and by individuals—buying insurance with a choice of payment plans and deciding whether to buy or lease a vehicle or equipment. Like the parking permit case, these situations can be described as investing more now to save money later, or borrowing money now to be paid later.

EXAMPLE 7-8

An engineering firm can pay for its liability insurance for errors and omissions on an annual or quarterly basis. If paid quarterly, the insurance costs $10,000. If paid annually, the insurance costs $35,000. What is the rate of return for paying on annual basis?

SOLUTION

This is most easily solved by using quarterly periods. As shown in the cash flow diagram and table, insurance must be paid at the start of each quarter. The cash flow table adds a column for the incremental difference to the information in the cash flow diagram.

Quarter	Annual Payment	Quarterly Payments	Incremental Difference
0	−35,000	−10,000	−25,000
1		−10,000	10,000
2		−10,000	10,000
3		−10,000	10,000

Setting the two PWs equal to each other we have

$$-\$35,000 = -\$10,000(1 + (P/A, i_{qtr}, 3))$$

$$(P/A, i_{qtr}, 3) = -\$25,000/ - \$10,000 = 2.5$$

$$(P/A, 9\%, 3) = 2.531$$

$$(P/A, 10\%, 3) = 2.487$$

$$i_{qtr} = 0.09 + (0.10 - 0.09)(2.531 - 2.5)/(2.531 - 2.487)$$

$$= 0.09 + 0.01(0.7045) = 9.705\%$$

On an annual basis the effective interest rate is $(1.09705^4 - 1) = 44.8\%$.

Unless the firm is planning on going out of business, it is clearly better to buy the insurance a year at a time. Note that a calculator or spreadsheet gives the exact quarterly interest rate: 9.7010%.

EXAMPLE 7-9

Mountain Environmental Consulting may buy some field equipment for $40,000 or lease it for $2500 per month. In either case, the equipment will be replaced in 2 years. The salvage value of the equipment after 2 years would be $60,000. What is the IRR or cost of the lease?

SOLUTION

The first step is to summarize the cash flows in a table. We must use monthly periods and remember that lease fees are paid at the start of the period.

Month	Buy	Lease	Incremental Difference
0	−$40,000	−$2500	−$37,500
1–23		−2500	2,500
24	6,000	0	6,000

The easiest way to analyze this example is to simply set the present worths of buying and leasing equal.

$$-\$40,000 + \$6000(P/F, i_m, 24) = -\$2500 - \$2500(P/A, i_m, 23)$$

$$0 = \$37,500 - \$2500(P/A, i_m, 23) - \$6000(P/F, i_m, 24)$$

In other words, if Mountain Environmental leases the equipment, it avoids the expenditure of $37,500 at time 0. It incurs a cost of $2500 at the ends of Months 1 through 23, and it gives up the salvage value of $6000 at the end of Month 24. This cash flow pattern (positive at time 0 and negative in later years) occurs because leasing is a way to borrow money.

The column of incremental difference describes the decision to buy rather than lease, which invests more $37,500 now to avoid the later lease payments.

For calculators and spreadsheets, this equation is easier to solve if $n = 24$ for both factors. There is a total cash flow of −$6000 at the end of Month 24. This cash flow can be split into −$2500 and −$3500. Then both the uniform and single payment periods continue to the end of Month 24.

$$0 = \$37,500 - \$2500(P/A, i_m, 24) - \$3500(P/F, i_m, 24)$$

$$PW(0\%) = \$37,500 - \$2500 \times 24 - \$3500 \times 1 = -\$26,000$$

$$PW(5\%) = \$37,500 - \$2500 \times 13.799 - \$3500 \times 0.3101 = \$1917$$

The answer is clearly between 0% and 5%, and closer to 5%. Try 4% next.

$$PW(4\%) = \$37,500 - \$2500 \times 15.247 - \$3500 \times 0.3901 = -\$1983$$

$$i_{mon} = 4\% + (5\% - 4\%) \times 1983/(1983 + 1917) = 4.51\%$$

The effective annual rate is

$$i_{eff} = 1.0451^{12} - 1 = 0.6975 = 69.8\%$$

In this case leasing is an extremely expensive way to obtain the equipment.

INCREMENTAL ANALYSIS

When there are two alternatives, rate of return analysis is performed by computing the *incremental rate of return*—ΔIRR—on the difference between the alternatives. In Examples 7-6 through 7-9 this was done by setting the present worths of the two alternatives equal.

Now we will calculate the increment more formally. Since we want to look at increments of investment, the cash flow for the difference between the alternatives is computed by taking the higher initial-cost alternative *minus* the lower initial-cost alternative. If ΔIRR is the same or greater than the MARR, choose the higher-cost alternative. If ΔIRR is less than the MARR, choose the lower-cost alternative.

Two-Alternative Situation	**Decision**
ΔIRR ≥ MARR	Choose the higher-cost alternative
ΔIRR < MARR	Choose the lower-cost alternative

Rate of return and incremental rate of return analysis are illustrated by Examples 7-10 through 7-13. Example 7-10 illustrates a *very* important point that was not part of our earlier examples of parking permits, insurance, and equipment to be bought or leased. In those cases, the decision that we needed the insurance, permit, or equipment had already been made. We only needed to decide the best way to obtain it. In Example 7-10, there is a *do-nothing* alternative that must be considered.

EXAMPLE 7-10

If an electromagnet is installed on the input conveyor of a coal-processing plant, it will pick up scrap metal in the coal. Removing this scrap will save an estimated $1200 per year in costs associated with machinery damage. The electromagnetic equipment has an estimated useful life of 5 years and no salvage value. Two suppliers have been contacted: Leaseco will provide the equipment in return for three beginning-of-year annual payments of $1000 each; Saleco will provide the equipment for $2783. If the MARR is 10%, should the project be done, and if so, which supplier should be selected?

SOLUTION

Before we analyze which supplier should be selected, we must decide whether the do-nothing alternative would be a better choice. Since the first cost of Leaseco is lower than the first cost with Saleco, let us compare Leaseco with doing nothing. This is the same as finding the present worth or IRR of the Leaseco investment.

The cash flow at Time 0 for Leaseco is –$1000. At the end of Years 1 and 2, the firm spends $1000 on the lease and saves $1200 in machinery damage for a net of $200. Then for 3 years the firm saves $1200 annually. Since the firm is investing $1000 to save $4000 spread over 5 years, the arrangement is clearly worthwhile. In fact, at the 10% MARR, the PW is $1813. (Problem 7-20 asks you to find the 49% rate of return.)

To compare Saleco and Leaseco, we first recognize that both will provide equipment with the same useful life and benefits. Thus, the comparison is a fixed-output situation. In rate of return analysis, the method of solution is to examine the differences between the alternatives. By taking Saleco – Leaseco, we obtain an increment of investment.

Year	Leaseco	Saleco	Difference Between Alternatives: Saleco − Leaseco
0	−$1000	−$2783	−$1783
1	$\begin{cases} -1000 \\ +1200 \end{cases}$	+1200	+1000
2	$\begin{cases} -1000 \\ +1200 \end{cases}$	+1200	+1000
3	+1200	+1200	0
4	+1200	+1200	0
5	+1200	+1200	0

Compute the NPW at various interest rates on the increment of investment represented by the difference between the alternatives.

Year n	Cash Flow: Saleco − Leaseco	PW* At 0%	At 8%	At 20%	At ∞ %
0	−$1783	−$1783	−$1783	−$1783	−$1783
1	+1000	+1000	+926	+833	0
2	+1000	+1000	+857	+694	0
3	0	0	0	0	0
4	0	0	0	0	0
5	0	0	0	0	0
	NPW =	+217	0	−256	−1783

*Each year the cash flow is multiplied by $(P/F, i, n)$.

At 0%: $(P/F, 0\%, n) = 1$ for all values of n

At ∞%: $(P/F, \infty\%, 0) = 1$

$(P/F, \infty\%, n) = 0$ for all other values of n

From the plot of these data in Figure 7-5, we see that NPW = 0 at $i = 8\%$.

FIGURE 7-5 NPW plot for Example 7-10.

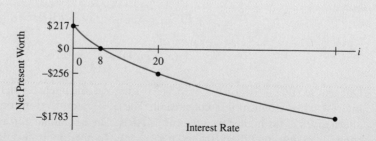

Thus, the incremental rate of return—ΔIRR—of selecting Saleco rather than Leaseco is 8%. This is less than the 10% MARR. Select Leaseco.

EXAMPLE 7-11

You must select one of two mutually exclusive alternatives. (*Note*: Engineering economists often use the term "mutually exclusive alternatives" to emphasize that selecting one alternative precludes selecting any other.) The alternatives are as follows:

Year	Alt. 1	Alt. 2
0	−$10	−$20
1	+15	+28

Any money not invested here may be invested elsewhere at the MARR of 6%. If you can choose only one alternative one time, which one would you select using the internal rate of return (IRR) analysis method?

SOLUTION

Using the IRR analysis method, we will select the lesser-cost alternative (Alt. 1), unless we find that the additional cost of Alt. 2 produces enough additional benefits to make it preferable instead. If we consider Alt. 2 in relation to Alt. 1, then

$$\begin{bmatrix}\text{Higher-cost}\\\text{Alt. 2}\end{bmatrix} = \begin{bmatrix}\text{Lower-cost}\\\text{Alt. 1}\end{bmatrix} + \begin{bmatrix}\text{Differences between}\\\text{Alt. 1 and Alt. 2}\end{bmatrix}$$

or

$$\text{Differences between Alt. 1 and Alt. 2} = \begin{bmatrix}\text{Higher-cost}\\\text{Alt. 2}\end{bmatrix} - \begin{bmatrix}\text{Lower-cost}\\\text{Alt. 1}\end{bmatrix}$$

The choice between the two alternatives reduces to examining the differences between them. We can compute the rate of return on the differences between the alternatives.

Year	Alt. 1	Alt. 2	Alt. 2 − Alt. 1
0	−$10	−$20	−$20 − (−$10) = −$10
1	+15	+28	+28 − (+15) = +13

$$0 = PW_{\text{Alt.2−Alt.1}} = -10 + 13(P/F, i, 1)$$

$$(P/F, i, 1) = \frac{10}{13} = 0.7692$$

One can see that if $10 increases to $13 in one year, the interest rate must be 30%. The compound interest tables confirm this conclusion. The 30% rate of return on the difference between the alternatives is far higher than the 6% MARR. The additional $10 investment to obtain Alt. 2 is superior to investing the $10 elsewhere at 6%. To obtain this desirable increment of investment, with its 30% rate of return, Alt. 2 is selected.

To understand more about Example 7-11, compute the rate of return for each alternative.

Alternative 1

$$0 = PW = -\$10 + \$15(P/F, i, 1)$$

$$(P/F, i, 1) = \frac{10}{15} = 0.6667 \Longrightarrow i = 50\%$$

Alternative 2

$$0 = PW = -\$20 + \$28(P/F, i, 1)$$

$$(P/F, i, 1) = \frac{20}{28} = 0.7143 \Longrightarrow i = 40\%$$

The higher rate of return of Alt. 1 is attractive, but Alt. 1 is not the correct solution. Solve the problem again, this time using present worth analysis.

Present Worth Analysis

Alternative 1

$$NPW = -10 + 15(P/F, 6\%, 1) = -10 + 15(0.9434) = +\$4.15$$

Alternative 2

$$NPW = -20 + 28(P/F, 6\%, 1) = -20 + 28(0.9434) = +\$6.42$$

Alternative 1 has a 50% rate of return and an NPW (at the 6% MARR) of +$4.15. Alternative 2 has a 40% rate of return on a larger investment, with the result that its NPW (at the 6% MARR) is +$6.42. Our economic criterion is to maximize the return, rather than the rate of return. To maximize NPW, select Alt. 2. This agrees with the incremental rate of return analysis.

EXAMPLE 7-12

If the computations for Example 7-11 do not convince you, and you still think Alternative 1 would be preferable, try this problem.

You have $20 in your wallet and two alternative ways of lending Bill some money.

(a) Lend Bill $10 with his promise of a 50% return. That is, he will pay you back $15 at the agreed time.

(b) Lend Bill $20 with his promise of a 40% return. He will pay you back $28 at the same agreed time.

You can select whether to lend Bill $10 or $20. This is a one-time situation, and any money not lent to Bill will remain in your wallet. Which alternative do you choose?

SOLUTION

So you see that a 50% return on the smaller sum is less rewarding to you than 40% on the larger sum? Since you would prefer to have $28 than $25 ($15 from Bill plus $10 remaining in your wallet) after the loan is paid, lend Bill $20.

EXAMPLE 7-13

Solve Example 7-11 again, but this time compute the interest rate on the increment (Alt. 1 − Alt. 2 instead of (Alt. 2 − Alt. 1)). How do you interpret the results?

SOLUTION

This time the problem is being viewed as follows:

$$\text{Alt. 1} = \text{Alt. 2} + [\text{Alt. 1} - \text{Alt. 2}]$$

Year	Alt. 1	Alt. 2	[Alt. 1 − Alt. 2]
0	−$10	−$20	−$10 − (−$20) = +$10
1	+15	+28	+15 − (+28) = −13

We can write one equation in one unknown:

$$\text{NPW} = 0 = +10 - 13(P/F, i, 1)$$

$$(P/F, i, 1) = \frac{10}{13} = 0.7692 \implies i = 30\%$$

Once again the interest rate is found to be 30%. The critical question is, What does the 30% represent? Looking at the increment again:

Year	Alt. 1 − Alt. 2
0	+$10
1	−13

The cash flow does *not* represent an investment; instead, it represents a loan. It is as if we borrowed $10 in Year 0 (+$10 represents a receipt of money) and repaid it in Year 1 (−$13 represents a disbursement). The 30% interest rate means this is the amount *we would pay* for the loan.

Is this a desirable borrowing scenario? Since the MARR on investments is 6%, it is reasonable to assume our maximum interest rate on borrowing would also be 6%. Here the interest rate is 30%, which means the borrowing is undesirable. Since Alt. 1 = Alt. 2 plus the undesirable (Alt. 1−Alt. 2) increment, we should reject Alternative 1. This means we should select Alternative 2—the same conclusion reached in Example 7-11.

Example 7-13 illustrated that one can analyze either **increments of investment** or **increments of borrowing.** When looking at increments of investment, we accept the increment when the incremental rate of return equals or exceeds the minimum attractive rate of return ($\Delta\text{IRR} \geq \text{MARR}$). When looking at increments of borrowing, we accept the increment when the incremental interest rate is less than or equal to the *minimum* attractive rate of return ($\Delta\text{IRR} \leq \text{MARR}$). One way to avoid much of the possible confusion is to organize the solution to any problem so that one is examining increments of investment. This is illustrated in the next example.

EXAMPLE 7-14

A firm is considering which of two devices to install to reduce costs. Both devices have useful lives of 5 years and no salvage value. Device A costs $1000 and can be expected to result in $300 savings annually. Device B costs $1350 and will provide cost savings of $300 the first year but will increase $50 annually, making the second-year savings $350, the third-year savings $700, and so forth. For a 7% MARR, which device should the firm purchase?

SOLUTION

This problem has been solved by present worth analysis (Example 5-1) and annual cost analysis (Example 6-5). This time we will use rate of return analysis, which must be done on the incremental investment.

			Difference Between Alternatives:
Year	Device A	Device B	Device B − Device A
0	−$1000	−$1350	−$350
1	+300	+300	0
2	+300	+350	+50
3	+300	+400	+100
4	+300	+450	+150
5	+300	+500	+200

For the difference between the alternatives, write a single equation with i as the only unknown.

$$PW(i) = 0 = -350 + 50(P/G, i, 5)$$

$$(P/G, i, 5) = 7 \text{ so } i \text{ is between } 9\%, (P/G, 9\%, 5)$$

$$= 7.111, \text{ and } 10\%, (P/G, 10\%, 5) = 6.862.$$

$$i = 9\% + (10\% - 9\%)(7.111 - 7)/(7.111 - 6.862) = 9.45\%$$

This is more than the 7% MARR; therefore, the increment is desirable. Reject Device A and choose Device B.

Analysis Period

In discussing present worth analysis and annual cash flow analysis, an important consideration is the analysis period. This is also true in rate of return analysis. The solution method for two alternatives is to examine the differences between the alternatives. Clearly, the examination must cover the selected analysis period. For now, we can only suggest that the assumptions made should reflect one's perception of the future as accurately as possible.

In Example 7-15 the analysis period is a common multiple of the alternative service lives and identical replacement is assumed. This problem illustrates an analysis of the differences between the alternatives over the analysis period.

EXAMPLE 7-15

Two machines are being considered for purchase. If the MARR is 10%, which machine should be bought? Use an IRR analysis comparison.

	Machine X	Machine Y
Initial cost	$200	$700
Uniform annual benefit	95	120
End-of-useful-life salvage value	50	150
Useful life, in years	6	12

SOLUTION

The solution is based on a 12-year analysis period and a replacement machine X that is identical to the present machine X. The cash flow for the differences between the alternatives, is as follows:

Year	Machine X	Machine Y	Machine Y − Machine X
0	−$200	−$700	−$500
1	+95	+120	+25
2	+95	+120	+25
3	+95	+120	+25
4	+95	+120	+25
5	+95	+120	+25
6	+95 +50 −200	+120	+25 +150
7	+95	+120	+25
8	+95	+120	+25
9	+95	+120	+25
10	+95	+120	+25
11	+95	+120	+25
12	+95 +50	+120 +150	+25 +100

PW of cost of differences = PW of benefits of differences

$$500 = 25(P/A, i, 12) + 150(P/F, i, 6) + 100(P/F, i, 12)$$

The sum of the benefits over the 12 years is $550, which is only a little greater than the $500 additional cost. This indicates that the rate of return is quite low. Try $i = 1\%$.

$$500 \doteq 25(11.255) + 150(0.942) + 100(0.887)$$
$$\doteq 281 + 141 + 89 = 511$$

The interest rate is too low. Try $i = 1\frac{1}{2}\%$:

$$500 \stackrel{?}{=} 25(10.908) + 150(0.914) + 100(0.836)$$
$$\stackrel{?}{=} 273 + 137 + 84 = 494$$

The internal rate of return on the $Y - X$ increment, IRR_{Y-X}, is about 1.3%, far below the 10% minimum attractive rate of return. The additional investment to obtain Machine Y yields an unsatisfactory rate of return, therefore X is the preferred alternative.

SPREADSHEETS AND RATE OF RETURN ANALYSIS

The spreadsheet functions covered in earlier chapters are particularly useful in calculating internal rates of returns (IRRs). If a cash flow diagram can be reduced to at most one P, one A, and/or one F, then the RATE *investment function* can be used. Otherwise the IRR *block function* is used with a cash flow in each period.

The Excel investment function is RATE(n,A,P,F,type,guess). The A, P, and F cannot all be the same sign. The F, type, and guess are optional arguments. The "type" is end- or beginning-of-period cash flow (for A, but not F), and the "guess" is the starting value in the search for the IRR.

Considering Example 7-1, where $P = -8200$, $A = 2000$, and $n = 5$, the RATE function would be:

$$\text{RATE}(5,2000,-8200)$$

which gives an answer of 7.00%, which matches that found in Example 7-1.

For Example 7-2, where $P = -700$, $A = 100$, $G = 75$, and $n = 4$, the RATE function cannot be used, since it has no provisions for the arithmetic gradient, G. Suppose the years (row 1) and the cash flows (row 2) are specified in columns B through E. The internal rate of return calculated using IRR(B2:F2) is 6.91%.

	A	B	C	D	E	F
1	Year	0	1	2	3	4
2	Cash flow	-700	100	175	250	325

Figure 7-6 illustrates the use of spreadsheet to graph the present worth of a cash flow series versus the interest rate. The interest rate with present worth equal to 0 is the IRR. The y axis on this graph has been modified so that the x axis intersects at a present worth of -50 rather than at 0. To do this click on the y axis, then right-click to bring up the "format axis" option. Select this and then select the tab for the *scale* of the axis. This has a selection for the intersection of the x axis. This process ensures that the x-axis labels are outside the graph.

SUMMARY

Rate of return may be defined as the interest rate paid on the unpaid balance of a loan such that the loan is exactly repaid if the schedule of payments is followed. On an investment, rate

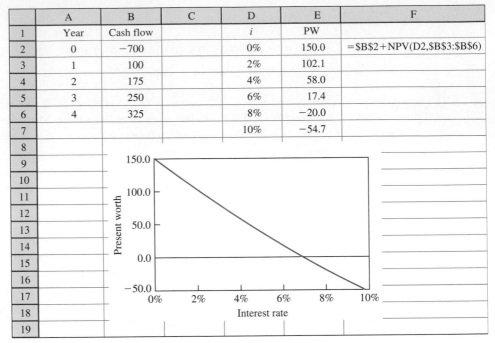

	A	B	C	D	E	F
1	Year	Cash flow		i	PW	
2	0	−700		0%	150.0	=$B$2+NPV(D2,$B$3:$B$6)
3	1	100		2%	102.1	
4	2	175		4%	58.0	
5	3	250		6%	17.4	
6	4	325		8%	−20.0	
7				10%	−54.7	
8						
9						
10						
11						
12						
13						
14						
15						
16						
17						
18						
19						

FIGURE 7-6 Graphing present worth versus i.

of return is the interest rate earned on the unrecovered investment such that the payment schedule makes the unrecovered investment equal to zero at the end of the life of the investment. Although the two definitions of rate of return are stated differently, there is only one fundamental concept being described. It is that the rate of return is the interest rate i at which the benefits are equivalent to the costs.

There are a variety of ways of writing the cash flow equation in which the rate of return i may be the single unknown. Five of them are as follows:

$$\text{PW of benefits} - \text{PW of costs} = 0$$

$$\frac{\text{PW of benefits}}{\text{PW of costs}} = 1$$

$$\text{NPW} = 0$$

$$\text{EUAB} - \text{EUAC} = 0$$

$$\text{PW of costs} = \text{PW of benefits}$$

Rate of return analysis: Rate of return analysis is the most frequently used method in industry, as the resulting rate of return is readily understood. Also, the difficulties in selecting a suitable interest rate to use in present worth and annual cash flow analysis are avoided.

Criteria

Two Alternatives

Compute the incremental rate of return—ΔIRR—on the increment of *investment* between the alternatives. Then,

- if ΔIRR $\geq$ MARR, choose the higher-cost alternative, or,
- if ΔIRR $<$ MARR, choose the lower-cost alternative

When an increment of *borrowing* is examined, where ΔIRR is the incremental interest rate,

- if ΔIRR $\leq$ MARR, the increment is acceptable, or
- if ΔIRR $>$ MARR, the increment is not acceptable

Three or More Alternatives

See Chapter 8.

Looking Ahead

Rate of return is further described in Appendix 7A. This material concentrates on the difficulties that occur with some cash flows that may yield more than one root for the rate of return equation.

PROBLEMS

Rate of Return

7-1 Compute the rate of return for the following cash flow to within $\frac{1}{2}$%.

Year	Cash Flow
0	−$100
1–10	+27

(*Answer:* 23.9%)

7-2 The Diagonal Stamp Company, which sells used postage stamps to collectors, advertises that its average price has increased from $1 to $5 in the last 5 years. Thus, management states, investors who had purchased stamps from Diagonal 5 years ago would have received a 100% rate of return each year.

(*a*) To check their calculations, compute the annual rate of return.

(*b*) Why is your computed rate of return less than 100%?

7-3 A table saw costs $175 at a local store. You may either pay cash for it or pay $35 now and $12.64 a month

for 12 months beginning 30 days hence. If you choose the time payment plan, what nominal annual interest rate will you be charged?   (*Answer:* 15%)

7-4 An investment of $5000 in Biotech common stock proved to be very profitable. At the end of 3 years the stock was sold for $25,000. What was the rate of return on the investment?

7-5 Helen is buying a $12,375 car with a $3000 down payment, followed by 36 monthly payments of $325 each. The down payment is paid immediately, and the monthly payments are due at the end of each month. What nominal annual interest rate is Helen paying? What effective interest rate?   (*Answers:* 15%; 16.08%)

7-6 Peter Minuit bought an island from the Manhattoes Indians in 1626 for $24 worth of glass beads and trinkets. The 1991 estimate of the value of land on this island was $12 billion. What rate of return would the Indians have received if they had retained title to the island rather than selling it for $24?

7-7 An engineer invests $5800 at the end of every year for a 35-year career. If the engineer wants $1 million in savings at retirement, what interest rate must the investment earn?

7-8 A mining firm makes annual deposits of $250,000 into a reclamation fund for 20 years. If the firm must have $10 million when the mine is closed, what interest rate must the investment earn?

7-9 You spend $1000 and in return receive two payments of $1094.60—one at the end of 3 years and the other at the end of 6 years. Calculate the resulting rate of return.

7-10 Fred, our cat, just won the local feline lottery to the tune of 3000 cans of "9-Lives" cat food (assorted flavors). A local grocer offers to take the 3000 cans and in return, supply Fred with 30 cans a month for the next 10 years. What rate of return, in terms of nominal annual rate, will Fred realize on this deal? (Compute to nearest 0.01%.)

7-11 A woman went to the Beneficial Loan Company and borrowed $3000. She must pay $119.67 at the end of each month for the next 30 months.

(*a*) Calculate the nominal annual interest rate she is paying to within ±0.15%.

(*b*) What effective annual interest rate is she paying?

7-12 For the following diagram, compute the IRR to within $1/2$%.

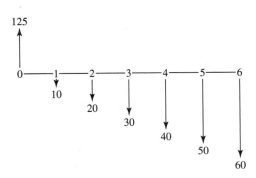

7-13 For the following diagram, compute the rate of return.

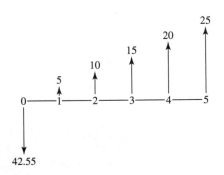

7-14 Consider the following cash flow:

Year	Cash Flow
0	−$500
1	+200
2	+150
3	+100
4	+50

Compute the rate of return represented by the cash flow.

7-15 Consider the following cash flow:

Year	Cash Flow
0	−$1000
1	0
2	+300
3	+300
4	+300
5	+300

Compute the rate of return on the $1000 investment to within 0.1%. (*Answer:* 5.4%)

7-16 Compute the rate of return for the following cash flow.

Year	Cash Flow
0	−$500
1–3	0
4	+4500

7-17 A bank proudly announces that it has changed its interest computation method to continuous compounding. Now $2000 left in the bank for 9 years will double to $4000.

(*a*) What nominal interest rate, compounded continuously, is the bank paying?

(*b*) What effective interest rate is it paying?

7-18 Compute the rate of return for the following cash flow to within 0.5%.

Year	Cash Flow
0	−$640
1	0
2	100
3	200
4	300
5	300

(*Answer:* 9.3%)

7-19 Compute the rate of return for the following cash flow.

Year	Cash Flow
1–5	−$233
6–10	+1000

7-20 For Example 7-10, find the rate of return on the Leaseco proposal compared with doing nothing. (*Answer*: 49%)

7-21 For the following diagram, compute the interest rate at which the costs are equivalent to the benefits.

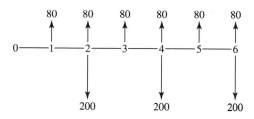

(*Answer:* 50%)

7-22 For the following diagram, compute the rate of return on the $3810 investment.

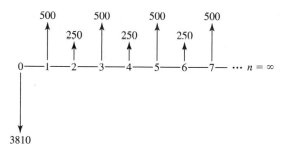

7-23 For the following diagram, compute the rate of return.

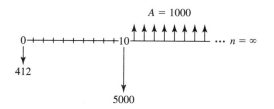

7-24 Consider the following cash flow:

Year	Cash Flow
0	−$400
1	0
2	+200
3	+150
4	+100
5	+50

Write one equation, with i as the only unknown, for the cash flow. In the equation you are not to use more

than two single payment compound interest factors. (You may use as many other factors as you wish.) Then solve your equation for i.

7-25 You have just been elected into the "Society of Honorable Engineers." First-year dues are waived in honor of your election. Thus, your first-payment of $200 is due at the end of the year, and annual dues are expected to increase 3% annually. After 40 years you become a life member and no further dues are owed. Instead of paying annual dues, however, you can pay a one-time $2000 life membership fee.

(*a*) Show the equation for determining the rate of return for buying a life membership.

Contributed by D.P. Loucks, Cornell University

(*b*) What is the rate of return?

Net Present Worth vs. *i* and Bonds

7-26 For Problem 7-1, graph the PW versus the interest rate for values from 0% to 50%. Is this the typical PW graph for an investment?

7-27 For Problem 7-18, graph the PW versus the interest rate for values from 0% to 50%. Is this the typical PW graph for an investment?

7-28 A man buys a corporate bond from a bond brokerage house for $925. The bond has a face value of $1000 and pays 4% of its face value each year. If the bond will be paid off at the end of 10 years, what rate of return will the man receive?   (*Answer:* 4.97%)

7-29 A well-known industrial firm has issued $1000 bonds that carry a 4% nominal annual interest, rate paid semiannually. The bonds mature 20 years from now, at which time the industrial firm will redeem them for $1000 plus the terminal semiannual interest payment. From the financial pages of your newspaper you learn that the bonds may be purchased for $715 each ($710 for the bond plus a $5 sales commission). What nominal annual rate of return would you receive if you purchased the bond now and held it to maturity 20 years hence?   (*Answer:* 6.6%)

7-30 On April 2, 1988, an engineer bought a $1000 bond of an American airline for $875. The bond paid 6% on its principal amount of $1000, half in each of its April 1 and October 1 semiannual payments; it repaid the $1000 principal sum on October 1, 2001. What nominal rate of return did the engineer receive from the bond if he held it to its maturity (on October 1, 2001)?   (*Answer:* 7.5%)

7-31 ABC Corporation's recently issued bonds pay interest semiannually and mature in 10 years. The face value of each bond is $1000, and 6.8% is the nominal interest rate.

(a) What is the effective interest rate an investor receives?

(b) If a 0.75% fee is deducted by the brokerage firm from the initial $1000, what is the effective annual interest rate paid by ABC Corporation?

7-32 ABC Corporation is issuing some *zero coupon bonds*, which pay no interest. At maturity in 20 years they pay a face value of $10,000. The bonds are expected to sell for $3118 when issued.

(a) What is the effective interest rate an investor receives?

(b) A 1% fee (based on the face value) is deducted by the brokerage firm from the initial sales revenue. What is the effective annual interest rate paid by ABC Corporation?

Discounts and Fees

7-33 The cash price of a machine tool is $3500. The dealer is willing to accept a $1200 down payment and 24 end-of-month monthly payments of $110 each. At what effective interest rate are these terms equivalent? (*Answer:* 14.4%)

7-34 A local bank makes auto loans. It charges 4% per year in the following manner: if $3600 is borrowed to be repaid over a 3-year period, the bank interest charge is $3600 × 0.04 × 3 years = $432. The bank deducts the $432 of interest from the $3600 loan and gives the customer $3168 in cash. The customer must repay the loan by paying $1/36$ of $3600, or $100, at the end of each month for 36 months. What nominal annual interest rate is the bank actually charging for this loan?

7-35 Jan purchased 100 shares of Peach Computer stock for $18 per share, plus a $45 brokerage commission. Every 6 months she received a dividend from Peach of 50 cents per share. At the end of 2 years, just after receiving the fourth dividend, she sold the stock for $23 per share and paid a $58 brokerage commission from the proceeds. What annual rate of return did she receive on her investment?

7-36 A used car dealer advertises financing at 0% interest over 3 years with monthly payments. You must pay a processing fee of $250 at signing. The car you like costs $6000.

(a) What is your effective annual interest rate?

(b) You believe that the dealer would accept $5200 if you paid cash. What effective annual interest rate would you be paying, if you financed with the dealer?

7-37 A new car dealer advertises financing at 0% interest over 4 years with monthly payments or a $3000 rebate if you pay cash.

(a) The car you like costs $12,000. What effective annual interest rate would you be paying if you financed with the dealer?

(b) The car you like costs $18,000. What effective annual interest rate would you be paying if you financed with the dealer?

(c) The car you like costs $24,000. What effective annual interest rate would you be paying if you financed with the dealer?

7-38 A used car dealer advertises financing at 4% interest over 3 years with monthly payments. You must pay a processing fee of $250 at signing. The car you like costs $6000.

(a) What is your effective annual interest rate?

(b) You believe that the dealer would accept $5200 if you paid cash. What effective annual interest rate would be paying if you financed with the dealer?

(c) Compare these answers with those for Problem 7-36. What can you say about what matters the most for determining the effective interest rate?

7-39 Some laboratory equipment sells for $75,000. The manufacturer offers financing at 8% with annual payments for 4 years for up to $50,000 of the cost. The salesman is willing to cut the price by 10% if you pay cash. What is the interest rate you would pay by financing?

7-40 A home mortgage with monthly payments for 30 years is available at 6% interest. The home you are buying costs $120,000, and you have saved $12,000 to meet the requirement for a 10% down payment. The lender charges "points" of 2% of the loan value as a loan origination and processing fee. This fee is added to the initial balance of the loan.

(a) What is your monthly payment?

(b) If you keep the mortgage until it is paid off in 30 years, what is your effective annual interest rate?

(c) If you move to a larger house in 10 years and pay off the loan, what is your effective annual interest rate?

Money by Mail	Non-Negotiable INE 1/96	
For the Amount of **$3000 or $2000 or $1000** **Dollars** Pay to the Order of **I Feel Rich** **Limited Time Offer**		To borrow $3000, $2000, or $1000

| For the Amount of $3000 Dollars Pay to the Order of _____ **I Feel Rich** Total of Payments **$4,280.40** Number of Monthly Payments 36 Months Amount Financed $3,246.25 | APR 18.95% Finance Charge $1034.29 Monthly Payment $118.90 Credit Line Premium $83.46* Credit Disability Premium $162.65* | $3000 loan terms |

| For the Amount of **$2000 Dollars** Pay to the Order of _____ **I Feel Rich** Total of Payments $2,731.50 Number of Monthly Payments 30 Months Amount Financed $2,137.25 | APR 19.95% Finance Charge $594.25 Monthly Payment $91.05 Credit Line Premium $44.38* Credit Disability Premium $92.87* | $2000 loan terms |

| For the Amount of **$1000 Dollars** Pay to the Order of _____ **I Feel Rich** Total of Payments $1,300.80 Number of Monthly Payments 24 Months Amount Financed $1055.26 | APR 20.95% Finance Charge $245.54 Monthly Payment $54.20 Credit Line Premium $16.91* Credit Disability Premium $39.02* | $1000 loan terms |

*Credit insurance. If selected, premium will be paid from amount financed. If not selected, cash advance is total amount financed.

(d) If you are transferred in 3 years and pay off the loan, what is your effective annual interest rate?

7-41 A finance company is using the above "Money by Mail" offer. Calculate the yearly nominal IRR received by the company if a customer chooses the loan of $2000 and accepts the credit insurance (Life and Disability).

7-42 A finance company is using the "Money by Mail" offer shown in Problem 7-41. Calculate the yearly nominal IRR received by the company if a customer chooses the $3000 loan but declines the credit insurance.

Investments and Loans

7-43 An investor bought a one-acre lot on the outskirts of a city for $9000 cash. Each year he paid $80 of property taxes. At the end of 4 years, he sold the lot for a net value of $15,000. What rate of return did he receive on his investment?   (*Answer:* 12.92%)

7-44 A mine is for sale for $240,000. It is believed the mine will produce a profit of $65,000 the first year, but the profit will decline $5000 a year after that, eventually reaching zero, whereupon the mine will be worthless. What rate of return would this $240,000 investment produce for the purchaser of the mine?

7-45 An apartment building in your neighborhood is for sale for $140,000. The building has four units, which are rented at $500 per month each. The tenants have long-term leases that expire in 5 years. Maintenance and other expenses for care and upkeep are $8000 annually. A new university is being built in the vicinity and it is expected that the building could be sold for $160,000 after 5 years.

(a) What is the internal rate of return for this investment?

(b) Should this investment be accepted if the other options have a rate of return of 12%?

7-46   A new machine can be purchased today for $300,000. The annual revenue from the machine is calculated to be $67,000, and the equipment will last 10 years. Expect the maintenance and operating costs to be $3000 a year and to increase $600 per year. The salvage value of the machine will be $20,000. What is the rate of return for this machine?

7-47   An insurance company is offering to sell an annuity for $20,000 cash. In return the firm will guarantee to pay the purchaser 20 annual end-of-year payments, with the first payment amounting to $1100. Subsequent payments will increase at a uniform 10% rate each year (second payment is $1210; third payment is $1331, etc.). What rate of return would someone who buys the annuity receive?

7-48   Fifteen families live in Willow Canyon. Although several water wells have been drilled, none has produced water. The residents take turns driving a water truck to a fire hydrant in a nearby town. They fill the truck with water and then haul it to a storage tank in Willow Canyon. Last year truck fuel and maintenance cost $3180. This year the residents are seriously considering spending $100,000 to install a pipeline from the nearby town to their storage tank. What rate of return would the Willow Canyon residents receive on their new water supply pipeline if the pipeline is considered to last

(a) Forever?

(b) 100 years?

(c) 50 years?

(d) Would you recommend that the pipeline be installed? Explain.

7-49   An investor bought 100 shares of Omega common stock for $9000. He held the stock for 9 years. For the first 4 years he received annual end-of-year dividends of $800. For the next 4 years he received annual dividends of $400. He received no dividend for the ninth year. At the end of the ninth year he sold his stock for $6000. What rate of return did he receive on his investment?

7-50   One aspect of obtaining a college education is the prospect of improved future earnings in comparison to non-college graduates. Sharon Shay estimates that a college education has a $28,000 equivalent cost at graduation. She believes the benefits of her education will occur throughout 40 years of employment. She thinks that during the first 10 years out of college, her income will be higher than that of a non-college graduate by $3000 per year. During the subsequent 10 years, she projects an annual income that is $6000 per year higher. During the last 20 years of employment, she estimates an annual salary that is $12,000 above the level of the non-college graduate. If her estimates are correct, what rate of return will she receive as a result of her investment in a college education?

7-51   Upon graduation, every engineer must decide whether to go on to graduate school. Estimate the costs of going full time to the university to obtain a master of science degree. Then estimate the resulting costs and benefits. Combine the various consequences into a cash flow table and compute the rate of return. Nonfinancial benefits are probably relevant here too.

7-52   A popular magazine offers a lifetime subscription for $200. Such a subscription may be given as a gift to an infant at birth (the parents can read it in those early years), or taken out by an individual for himself. Normally, the magazine costs $12.90 per year. Knowledgeable people say it probably will continue indefinitely at this $12.90 rate. What rate of return would be obtained if a life subscription were purchased for an infant, rather than paying $12.90 per year beginning immediately? You may make any reasonable assumptions, but the compound interest factors must be *correctly* used.

7-53   The following advertisement appeared in the *Wall Street Journal* on Thursday, February 9, 1995:

> "There's nothing quite like the Seville SmartLease. Seville SLS $0 down, $599 a month/36 months."

Terms are first month's lease payment of $599 plus $625 refundable security deposit and a consumer down payment of $0 for a total of $1224 due at lease signing. Monthly payment is based on a net capitalized cost of $39,264 for total monthly payments of $21,564. Payment examples based on a 1995 Seville SLS: $43,658 MSRP including destination charge.

Tax, license, title fees, and insurance extra. Option to purchase at lease end for $27,854. Mileage charge of $0.15 per mile over 36,000 miles.

(*a*) Set up the cash flows.

(*b*) Determine the interest rate (nominal and effective) for the lease.

7-54 An engineering student is deciding whether to buy two 1-semester parking permits or an annual permit. The annual parking permit costs $100 due August 15th, and the semester permits are $65 due August 15th and January 15th. These values are for evening permits at the University of Alaska Anchorage for 2006–07. What is the rate of return for buying the annual permit?

7-55 An engineering student is deciding whether to buy two 1-semester parking permits or an annual permit. Using the dates and costs for your university, find the rate of return for the incremental cost of the annual permit.

7-56 An engineering firm can pay for its liability insurance on an annual or quarterly basis. If paid quarterly, the insurance costs $18,000. If paid annually, the insurance costs $65,000. What are the quarterly rate of return and the nominal and effective interest rates for paying on an annual basis?

7-57 An engineering firm can pay for its liability insurance on an annual or quarterly basis. If paid quarterly, the insurance costs $18,000. If paid annually, the insurance costs $65,000. Use a spreadsheet to calculate the exact quarterly interest rate for paying on an annual basis.

7-58 An engineering student must decide whether to pay for auto insurance on a monthly or an annual basis. If paid annually, the cost is $1650. If paid monthly, the cost is $150 at the start of each month. What is the rate of return for buying the insurance on an annual basis?

7-59 For your auto or home insurance, find out the cost of paying annually or on a shorter term. What is the rate of return for buying the insurance on an annual basis?

Incremental Analysis

7-60 Two alternatives are as follows:

Year	A	B
0	−$2000	−$2800
1	+800	+1100
2	+800	+1100
3	+800	+1100

If 5% is considered the minimum attractive rate of return, which alternative should be selected?

7-61 Consider two mutually exclusive alternatives:

Year	X	Y
0	−$100	−$50.0
1	35	16.5
2	35	16.5
3	35	16.5
4	35	16.5

If the minimum attractive rate of return is 10%, which alternative should be selected?

7-62 Two mutually exclusive alternatives are being considered. Both have a 10-year useful life. If the MARR is 8%, which alternative is preferred?

	A	B
Initial cost	$100.00	$50.00
Uniform annual benefit	19.93	11.93

7-63 Consider two mutually exclusive alternatives:

Year	X	Y
0	−$5000	−$5000
1	−3000	+2000
2	+4000	+2000
3	+4000	+2000
4	+4000	+2000

If the MARR is 8%, which alternative should be selected?

7-64 Two mutually exclusive alternatives are being considered. Both have lives of 5 years. Alternative *A* has a first cost of $2500 and annual benefits of $746. Alternative *B* costs $6000 and has annual benefits of $1664

If the minimum attractive rate of return is 8%, which alternative should be selected? Solve the problem by

(*a*) Present worth analysis

(*b*) Annual cash flow analysis

(*c*) Rate of return analysis

7-65 A contractor is considering whether to buy or lease a new machine for his layout site work. Buying a new machine will cost $12,000 with a salvage value of $1200 after the machine's useful life of 8 years. On the other hand, leasing requires an annual lease payment of $3000. Assuming that the MARR is 15% and on the basis of an internal rate of return analysis, which alternative should the contractor be advised to accept. The cash flows are as follows:

Year (*n*)	Alt. *A* (buy)	Alt. *B* (lease)
0	−$12,000	−$3000
1		−3000
2		−3000
3		−3000
4		−3000
5		−3000
6		−3000
7		−3000
8	+1200	0

7-66   Two hazardous environment facilities are being evaluated, with the projected life of each facility being 10 years. The cash flows are as follows:

	Alt. *A*	Alt. *B*
First cost	$615,000	$300,000
Maintenance and operating cost	10,000	25,000
Annual benefits	158,000	92,000
Salvage value	65,000	−5,000

The company uses a MARR of 15%. Using rate of return analysis, which alternative should be selected?

7-67   The owner of a corner lot wants to find a use that will yield a desirable return on his investment. After much study and calculation, the owner decides that the two best alternatives are:

	Build Gas Station	Build Soft Ice Cream Stand
First cost	$80,000	$120,000
Annual property taxes	3,000	5,000
Annual income	11,000	16,000
Life of building, in years	20	20
Salvage value	0	0

If the owner wants a minimum attractive rate of return on his investment of 6%, which of the two alternatives would you recommend?

7-68   After 15 years of working for one employer, you transfer to a new job. During each of these years your employer contributed (that is, she diverted from your salary) $1500 to an account for your retirement (a fringe benefit), and you contributed a matching amount each year. The whole fund was invested at 5% during that time, and the value of the account now stands at $30,000. You are now faced with two

alternatives. (1) You may leave both contributions in the fund until retirement in 35 years, during which you will get the future value of this amount at 5% interest per year. (2) You may take out the total value of "your" contributions, which is $15,000 (one-half of the total $30,000). You can do as you wish with the money you take out, but the other half will be lost as far as you are concerned. In other words, you can give up $15,000 today for the sake of getting now the other $15,000. Otherwise, you must wait 35 years more to get the accumulated value of the entire fund. Which alternative is more attractive? Explain your choice.

7-69   In his will, Frank's uncle has given Frank the choice between two alternatives:

Alternative 1 $2000 cash
Alternative 2 $150 cash now plus $100 per month for 20 months beginning the first day of next month

(*a*) At what rate of return are the two alternatives equivalent?

(*b*) If Frank thinks the rate of return in (*a*) is too low, which alternative should he select?

7-70   A bulldozer can be purchased for $380,000 and used for 6 years, when its salvage value is 15% of the first cost. Alternatively, it can be leased for $60,000 a year. (Remember that lease payments occur at the start of the year.) The firm's interest rate is 12%.

(*a*) What is the interest rate for buying versus leasing? Which is the better choice?

(*b*) If the firm will receive $65,000 more each year than it spends on operating and maintenance costs, should the firm obtain the bulldozer? What is the rate of return for the bulldozer using the best financing plan?

7-71   A diesel generator for electrical power can be purchased by a remote community for $480,000 and used for 10 years, when its salvage value is $50,000. Alternatively, it can be leased for $70,000 a year. (Remember that lease payments occur at the start of the year.) The community's interest rate is 8%.

(*a*) What is the interest rate for buying versus leasing? Which is the better choice?

(*b*) The community will spend $80,000 less each year for fuel and maintenance, than it currently spends on buying power. Should it obtain the generator? What is the rate of return for the generator using the best financing plan?

Analysis Period

7-72   Two alternatives are being considered:

	A	B
First cost	$9200	$5000
Uniform annual benefit	$1850	$1750
Useful life, in years	8	4

If the minimum attractive rate of return is 7%, which alternative should be selected?

7-73   Jean has decided it is time to buy a new battery for her car. Her choices are:

	Zappo	Kicko
First cost	$56	$90
Guarantee period, in months	12	24

Jean believes the batteries can be expected to last only for the guarantee period. She does not want to invest extra money in a battery unless she can expect a 50% rate of return. If she plans to keep her present car another 2 years, which battery should she buy?

7-74   Two alternatives are being considered:

	A	B
Initial cost	$9200	$5000
Uniform annual benefit	1850	1750
Useful life, in years	8	4

Base your computations on a MARR of 7% and an 8-year analysis period. If identical replacement is assumed, which alternative should be selected?

7-75   Two investment opportunities are as follows:

	A	B
First cost	$150	$100
Uniform annual benefit	25	22.25
End-of-useful-life salvage value	20	0
Useful life, in years	15	10

At the end of 10 years, Alt. B is not replaced. Thus, the comparison is 15 years of A versus 10 years of

B. If the MARR is 10%, which alternative should be selected?

Spreadsheets

7-76   The Southern Guru Copper Company operates a large mine in a South American country. A legislator in the National Assembly said in a speech that most of the capital for the mining operation was provided by loans from the World Bank; in fact, Southern Guru has only $500,000 of its own money actually invested in the property. The cash flow for the mine is:

Year	Cash Flow
0	$0.5 million investment
1	3.5 million profit
2	0.9 million profit
3	3.9 million profit
4	8.6 million profit
5	4.3 million profit
6	3.1 million profit
7	6.1 million profit

The legislator divided the $30.4 million total profit by the $0.5 million investment. This produced, he said, a 6080% rate of return on the investment. Southern Guru, claiming the actual rate of return is much lower, asks you to compute it.

7-77   A young engineer's starting salary is $52,000. The engineer expects annual raises of 3%. The engineer will deposit 10% of the annual salary at the end of each year in a savings account that earns 4%. How much will the engineer have saved for starting a business after 15 years? We suggest that the spreadsheet include at least columns for the year, the year's salary, the year's deposit, and the year's cumulative savings. (*Answer:* $126,348)

7-78   A young engineer's starting salary is $55,000. The engineer expects annual raises of 2%. The engineer will deposit 10% of the annual salary at the end of each year in a savings account that earns 5%. How much will the engineer have saved for retirement after 40 years?

7-79   A young engineer's starting salary is $55,000. The engineer expects annual raises of 2%. The engineer will deposit a constant percentage of the annual salary at the end of each year in a savings account that earns 5%. What percentage must be saved so that there will be $1 million in savings for retirement after 40 years?

(*Hint*: Use GOAL SEEK: see last section of Chapter 8.)

7-80 Find the average starting engineer's salary for your discipline. Find and reference a source for the average annual raise you can expect. If you deposit 10% of your annual salary at the end of each year in a savings account that earns 4%, how much will you have saved for retirement after 40 years?

APPENDIX 7A
Difficulties in Solving for an Interest Rate

After completing this chapter appendix, students should be able to:

- Describe why some project's cash flows cannot be solved for a single positive interest rate.
- Identify when multiple roots can occur.
- Evaluate how many potential roots exist for a particular project.
- Use the *modified internal rate of return (MIRR)* methodology in multiple-root cases.

Example 7A-1 illustrates the situation.

EXAMPLE 7A-1

The Going Aircraft Company has an opportunity to supply a large airplane to Interair, a foreign airline. Interair will pay $19 million when the contract is signed and $10 million one year later. Going estimates its second- and third-year costs at $50 million each. Interair will take delivery of the airplane during Year 4 and agrees to pay $20 million at the end of that year and the $60 million balance at the end of Year 5. Compute the rate of return on this project.

SOLUTION

The PW of each cash flow can be computed at various interest rates. For example, for Year 2 and $i = 10\%$: PW $= -50(P/F, 10\%, 2) = -50(0.826) = -41.3$.

Year	Cash Flow	0%	10%	20%	40%	50%
0	+$19	+$19	+$19	+$19	+$19	+$19
1	+10	+10	+9.1	+8.3	+7.1	+6.7
2	−50	−50	−41.3	−34.7	−25.5	−22.2
3	−50	−50	−37.6	−28.9	−18.2	−14.8
4	+20	+20	+13.7	+9.6	+5.2	+3.9
5	+60	+60	+37.3	+24.1	+11.2	+7.9
	PW =	+$9	+$0.1	−$2.6	−$1.2	+$0.5

FIGURE 7A-1
PW plot.

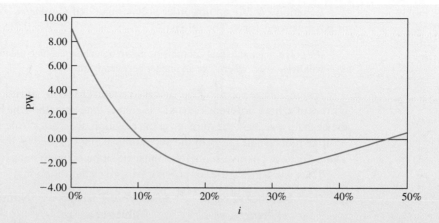

The PW plot for this cash flow is represented in Figure 7A-1. We see that this cash flow produces *two* points at which PW = 0: one at 10.24% and the other at 47.30%.

Why Multiple Solutions Can Occur

Example 7A-1 produced unexpected results. This happens because there were two changes in the signs of the cash flows. Years 0 and 1 were positive, Years 2 and 3 were negative, and Years 4 and 5 were positive. The cash flow series went from positive cash flows to negative cash flows to positive cash flows.

Most cash flow series have only one shift in sign. Investments start with one or more years of negative cash flows followed by many years of positive cash flows. Loans begin with a positive cash flow that is repaid with later negative cash flows. These problems have a unique rate of return because they have a single change in the sign of the cash flows.

Having more than one sign change in the cash flow series can produce multiple points or roots at which the PW equals 0. To see how many roots are possible, we link solving an economic analysis problem to solving a mathematical equation.

A project's cash flows are the values from CF_0 to CF_n.

Year	Cash Flow
0	CF_0
1	CF_1
2	CF_2
⋮	⋮
n	CF_n

The equation to find the internal rate of return, where PW = 0, is written as follows:

$$PW = 0 = CF_0 + CF_1(1+i)^{-1} + CF_2(1+i)^{-2} + \cdots + CF_n(1+i)^{-n} \qquad (7A\text{-}1)$$

If we let $x = (1+i)^{-1}$, then Equation 7A-1 may be written

$$0 = CF_0 + CF_1 x^1 + CF_2 x^2 + \cdots + CF_n x^n \qquad (7A\text{-}2)$$

Equation 7A-2 is an *n*th-order polynomial, and *Descartes' rule* describes the number of positive roots for *x*. The rule is:

> If a polynomial with real coefficients has *m* sign changes, then the number of positive roots will be $m - 2k$, where *k* is an integer between 0 and $m/2$.

A sign change exists when successive nonzero terms, written according to ascending powers of *x*, have different signs. If *x* is greater than zero, then the number of sign changes in the cash flows equals the number in the equation. Descartes' rule means that the number of positive roots (values of *x*) of the polynomial cannot exceed *m*, the number of sign changes. The number of positive roots for *x* must either be equal to *m* or less by an even integer.

Thus, Descartes' rule for polynomials gives the following:

Number of Sign Changes, *m*	Number of Positive Values of *x*	Number of Positive Values of *i*
0	0	0
1	1	1 or 0
2	2 or 0	2, 1, or 0
3	3 or 1	3, 2, 1, or 0
4	4, 2, or 0	4, 3, 2, 1, or 0

If *x* is greater than 1, the corresponding value of *i* is negative. If there is only one root and it is negative, then it is a valid IRR. One example of a valid negative IRR comes from a bad outcome, such as the IRR for a failed research and development project. Another example would be a privately sponsored student loan set up so that if a student volunteers for the Peace Corps, only half of the principal need be repaid.

If a project has a negative and a positive root for *i*, then for most projects only the positive root of *i* is used.

Projects with Multiple Sign Changes

Example 7A-1 is representative of projects for which there are initial payments (e.g., when the order for the ship, airplane, or building is signed), then the bulk of the costs occur, and then there are more payments on completion.

Projects of other types often have two or more sign changes in their cash flows. Projects with a salvage cost typically have two sign changes. This salvage cost can be large for environmental restoration at termination. Examples include pipelines, open-pit mines, and nuclear power plants. The following cash flow diagram is representative.

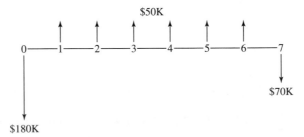

Many enhancement projects for existing mines and deposits have a pattern of two sign changes. Example 7A-2 describes an oil well in an existing field. The initial investment recovers more of the resource and speeds recovery of resources that would have been recovered eventually. The resources shifted for earlier recovery can lead to two sign changes.

In Example 7A-3, we consider staged construction, where three sign changes are common.

Other examples can be found in incremental comparisons of alternatives with unequal lives. In the next section we learn how to determine whether each of these examples has multiple roots.

EXAMPLE 7A-2

Adding an oil well to an existing field costs $4 million (4M). It will increase recovered oil by $3.5M, and it shifts $4.5M worth of production from Years 5, 6, and 7 to earlier years. Thus, the cash flows for Years 1 through 4 total $8M and Years 5 through 7 total −$4.5M. If the well is justified, one reason is that the oil is recovered sooner. How many roots for the PW equation are possible?

SOLUTION

The first step is to draw the cash flow diagram and count the number of sign changes. The following pattern is representative, although most wells have a longer life.

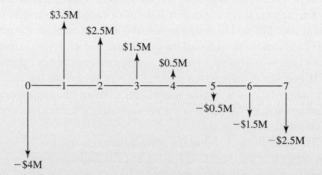

There are two sign changes, thus there may be 0, 1, or 2 positive roots for the PW = 0 equation. The additional recovery corresponds to an investment, and the shifting of recovery to earlier years corresponds to a loan (positive cash flow now and negative later). Thus, the oil wells are neither an investment nor a loan; they are a combination of both.

EXAMPLE 7A-3

A project has a first cost of $120,000. Net revenues begin at $30,000 in Year 1 and then increase by $2000 per year. In Year 5 the facility is expanded at a cost of $60,000 so that demand can continue to expand at $2000 per year. How many roots for the PW equation are possible?

SOLUTION

The first step is to draw the cash flow diagram. Then counting the three sign changes is easy.

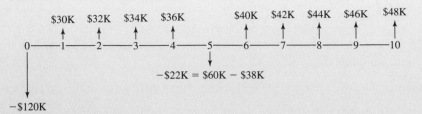

With three sign changes, there may be 0, 1, 2, or 3 positive roots for the PW = 0 equation.

Evaluating How Many Roots There Are

The number of sign changes tells us how many roots are possible—not how many roots there are. Rather than covering the many mathematical approaches that *may* tell us if the root will be unique, it is more useful to employ the power of the spreadsheet. A spreadsheet can show us if a root is unique and the value of each root that exists.

The approach is simple. For any set of cash flows, graph the PW as a function of the interest rate. We are interested in positive interest rates, so the graph usually starts at $i = 0$. Since *Descartes' rule* is based on $x > 0$, and $x = (1 + i)^{-1}$, sometimes the graph is started close to $i = -1$. This will identify any negative values of i that solve the equation.

If the root is unique, we're done. We've found the internal rate of return. If there are multiple roots, then we know the project's PW at all interest rates—including the one our organization uses. We can also use the approach of the next section to find a *modified* internal rate of return.

The easiest way to build the spreadsheet is to use the NPV(i,values) function in Excel. Remember that this function applies to cash flows in periods 1 to n, so that the cash flow at time 0 must be added in separately. The following examples use a spreadsheet to answer the question of how many roots each cash flow diagram in the last section has.

EXAMPLE 7A-4

This project is representative of ones with a salvage cost. How many roots for the PW equation exist?

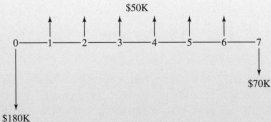

Figure 7A-2 shows the spreadsheet calculations and the graph of PW versus *i*.

	A	B	C	D	E	F	G
1	Year	Cash Flow		*i*	PW		
2	0	−180		−40%	−126.39	=$B$2+NPV(D2,$B$3:$B$9)	
3	1	50		−30%	219.99		
4	2	50		−20%	189.89		
5	3	50		−10%	114.49		
6	4	50		0%	50.00		
7	5	50		10%	1.84		
8	6	50		20%	−33.26		
9	7	−70		30%	−59.02		
10				40%	−78.24		
11	IRR	10.45%		50%	−92.88		
12	root	−38.29%					
13							
14							
15							
16							
17							
18							
19							
20							
21							
22							

FIGURE 7A-2 PW versus *i* for project with salvage cost.

In this case, there is 1 positive root of 10.45%. The value can be used as an IRR. There is also a negative root of $i = −38.29\%$. This root is not useful. With two sign changes, other similar diagrams may have 0, 1, or 2 positive roots for the PW = 0 equation. The larger the value of the final salvage cost, the more likely it is that 0 or 2 positive roots will occur.

EXAMPLE 7A-5   (7A-2 revisited)

Adding an oil well to an existing field had the following cash flow diagram. How many roots for the PW equation exist and what are they?

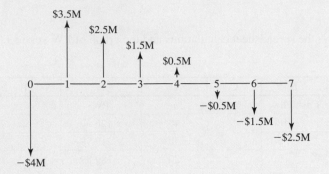

Figure 7A-3 shows the spreadsheet calculations and the graph of PW versus i.

	A	B	C	D	E	F	G
1	Year	Cash Flow		i	PW		
2	0	−4.00		0%	−0.50	=$B$2+NPV(D2,$B$3:$B$9)	
3	1	3.50		5%	0.02		
4	2	2.50		10%	0.28		
5	3	1.50		15%	0.37		
6	4	0.50		20%	0.36		
7	5	−0.50		25%	0.29		
8	6	−1.50		30%	0.19		
9	7	−2.50		35%	0.06		
10				40%	−0.08		
11	root	4.73%		45%	−0.22		
12	root	37.20%		50%	−0.36		

FIGURE 7A-3 PW versus i for oil well.

In this case, there are positive roots at 4.73 and 37.20%. These roots are not useful. This project is a combination of an investment and a loan, so we don't even know whether we want a high rate or a low rate. If our interest rate is about 20%, then the project has a positive PW. However, small changes in the data can make for large changes in these results.

It is useful to apply the modified internal rate of return described in the next section.

EXAMPLE 7A-6 (7A-3 revisited)

A project has a first cost of $120,000. Net revenues begin at $30,000 in Year 1, and then increase by $2000 per year. In Year 5 the facility is expanded at a cost of $60,000 so that demand can continue to expand at $2000 per year. How many roots for the PW equation have a positive value for i?

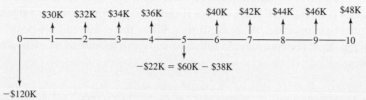

$$-\$22K = \$60K - \$38K$$

SOLUTION

Figure 7A-4 shows the spreadsheet calculations and the graph of PW versus i.

	A	B	C	D	E	F	G
1	Year	Cash Flow		i	PW		
2	0	−120		−80%	579646330	=$B$2+NPV(D2,$B$3:$B$12)	
3	1	30		−60%	735723		
4	2	32		−40%	17651		
5	3	34		−30%	1460		
6	4	36		0%	210		
7	5	−22		10%	73		
8	6	40		20%	7		
9	7	42		30%	−28		
10	8	44		40%	−48		
11	9	46		50%	−62		
12	10	48		60%	−71		
13				70%	−78		
14	IRR	21.69%		80%	−83		
15				90%	−87		
16							

FIGURE 7A-4 PW versus i for project with staged construction.

In this case, there is 1 positive root of 21.69%. The value can be used as an IRR, and the project is very attractive.

When there are two or more sign changes in the cash flow, we know that there are several possibilities concerning the number of positive values of i. Probably the greatest danger in this situation is to fail to recognize the multiple possibilities and to solve for a value of i. The approach of constructing the PW plot both establishes whether there are multiple roots and what their values are. This may be tedious to do by hand but is very easy with a spreadsheet (or a graphing calculator).

If there is a single positive value of i, we have no problem. On the other hand, if there is no positive value of i, or if there are multiple positive values, the situation may be attractive, unattractive, or confusing. When there are multiple positive values of i, none of them should be considered a suitable measure of the rate of return or attractiveness of the cash flow.

Modified Internal Rate of Return (MIRR)

Two external rates of return can be used to ensure that the resulting equation is solvable for a unique rate of return—the MIRR. The MIRR is a measure of the attractiveness of the cash flows, but it is also a function of the two external rates of return.

The rates that are *external* to the project's cash flows are (1) the rate at which the organization normally invests and (2) the rate at which it normally borrows. These are external rates for investing, e_{inv}, and for financing, e_{fin}. Because profitable firms invest at higher rates than they borrow at, the rate for financing is generally higher than the rate for financing. Sometimes a single external rate is used for both, but this requires the questionable assumption that investing and financing happen at the same rate.

The approach is:

1. Combine cash flows in each period into a single net receipt, R_t, or net expense, E_t.
2. Find the present worth of the expenses with the financing rate.
3. Find the future worth of the receipts with the investing rate.
4. Find the MIRR which makes the present and future worths equivalent.

The result is Equation 7A-3. This equation will have a unique root, since it has a single negative present worth and a single positive future worth. There is only one sign change in the resulting series.

$$(F/P, \text{MIRR}, n) \sum_t E_t(P/F, e_{fin}, t) = \sum_t R_t(F/P, e_{inv}, n - t) \qquad (7A\text{-}3)$$

There are other external rates of return, but the MIRR has historically been the most clearly defined. All of the external rates of return are affected by the assumed values for the investing and financing rates, so none are a *true* rate of return on the project's cash flow. The MIRR also has an Excel function, so it now can very easily be used. Example 7A-7 illustrates the calculation, which is also summarized in Figure 7A-5.

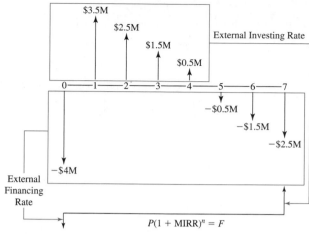

FIGURE 7A-5 MIRR for the oil well.

EXAMPLE 7A-7   (7A-2 and 7A-5 revisited)

Adding an oil well to an existing field had the cash flows summarized in Figure 7A-5. If the firm normally borrows money at 8% and invests at 15%, find the modified internal rate of return (MIRR).

SOLUTION

Figure 7A-6 shows the spreadsheet calculations.

	A	B	C	D	E	F	G	H	I
1	8%	external financing rate							
2	15%	external investing rate							
3	Year	0	1	2	3	4	5	6	7
4	Cash Flow	−4.00	3.50	2.50	1.50	0.50	−0.50	−1.50	2.50
5	13.64%		Cell A5 contains = MIRR(A4:I4,A1,A2)						

FIGURE 7A-6 MIRR for oil well.

It is also possible to calculate the MIRR by hand. While more work, the process does clarify what the MIRR function is doing.

1. Each period's cash flow is already a single net receipt or expenditure.
2. Find the present worth of the expenses with the financing rate.

$$PW = -4M - 0.5M(P/F, 8\%, 5) - 1.5M(P/F, 8\%, 6) - 2.5M(P/F, 8\%, 7)$$

$$= -4M - 0.5M(0.6806) - 1.5M(0.6302) - 2.5M(0.5835) = -6.744M$$

3. Find the future worth of the receipts with the investing rate.

$$FW = 3.5M(F/P, 15\%, 6) + 2.5M(F/P, 15\%, 5) + 1.5M(F/P, 15\%, 4) + 0.5M(F/P, 15\%, 3)$$

$$= 3.5M(2.313) + 2.5M(2.011) + 1.5M(1.749) + 0.5M(1.521) = 16.507M$$

4. Find the MIRR that makes the present and future worths equivalent.

$$0 = (1 + MIRR)^n \cdot PW + FW$$

$$0 = (1 + MIRR)^7(-6.744M) + 16.507M$$

$$(1 + MIRR)^7 = 16.507M/6.744M = 2.448$$

$$(1 + MIRR) = 2.448^{1/7} = 1.1364$$

$$MIRR = 13.64\%$$

The MIRR does allow calculation of a rate of return for *any* set of cash flows. However, the result is only as realistic as the external rates that are used. The MIRR value can depend as much on the external rates that are used, as it does on the cash flows that it is describing.

SUMMARY

In cash flows with more than one sign change, we find that solving the cash flow equation can result in more than one positive rate of return. Typical situations include a new oil well in an existing field, a project with a significant salvage cost, and staged construction.

In a sign change, successive nonzero values in the cash flow have different signs (that is, they change from + to −, or vice versa). Zero sign changes indicates there is no rate of return, as the cash flow is either all disbursements or all receipts.

One sign change is the usual situation, and a single positive rate of return generally results. There will be a negative rate of return whenever loan repayments are less than the loan or an investment fails to return benefits at least equal to the investment.

Multiple sign changes may result in multiple positive roots for i. When they occur, none of the positive multiple roots is a suitable measure of the project's economic desirability. If multiple roots are identified by graphing the present worth versus the interest rate, then the modified internal rate of return can be used to evaluate the project.

Graphing the present worth versus the interest rate ensures that the analyst recognizes that the cash flow has multiple sign changes. Otherwise a rate could be found and used that is not in fact a meaningful descriptor of the project.

The modified internal rate of return (MIRR) relies on rates for investing and borrowing that are external to the project. The number of sign changes are reduced to one, ensuring that the MIRR can be found.

PROBLEMS

Unless the problem asks a different question or provides different data: (1) determine how many roots are possible, and (2) graph the PW versus the interest rate to see whether multiple roots occur. If the root is a unique IRR, it is the project's rate of return. If there are multiple roots, then use an *external investing rate* of 12% and an *external borrowing rate* of 6%. Compute and use the MIRR as the project's rate of return.

7A-1 Find the rate of return for the following cash flow:

Year	Cash Flow
0	−$15,000
1	10,000
2	−8,000
3	11,000
4	13,000

(*Answer:* 21.2% IRR)

7A-2 A group of businessmen formed a partnership to buy and race an Indianapolis-type racing car. They agreed to pay an individual $50,000 for the car and associated equipment. The payment was to be in a lump sum at the end of the year. In what must have been "beginner's luck," the group won a major race the first week and $80,000. The rest of the first year, however, was not as good: at the end of the first year, the group had to pay out $35,000 for expenses plus the $50,000 for the car and equipment. The second year was a poor one: the group had to pay $70,000 just to clear up the racing debts at the end of the year. During the third and fourth years, racing income just equaled costs. When the group was approached by a prospective buyer for the car, they readily accepted $80,000 cash, which was paid at the end of the fourth year. What rate of return did the businessmen obtain from their racing venture?   (*Answer:* 9.6% MIRR)

7A-3 A student organization, at the beginning of the fall quarter, bought and operated a soft-drink vending machine as a means of helping finance its activities. The vending machine cost $75 and was installed at a gasoline station near the university. The student organization pays $75 every 3 months to the station owner for the right to keep the vending machine at the station. During the year the student organization owned the machine, they received the following quarterly income from it, before making the $75 quarterly payment to the station owner:

Quarter	Income
Fall	$150
Winter	25
Spring	125
Summer	150

At the end of one year, the student group resold the machine for $50. Determine the quarterly cash flow. Then determine a quarterly rate of return, a nominal annual rate, and an effective annual rate.

7A-4 Given the following cash flow, determine the rate of return on the project.

Year	Cash Flow
0	−$500
1	+2000
2	−1200
3	−300

(*Answer:* 11.3% MIRR)

7A-5 Given the following cash flow, determine the rate of return on the investment.

Year	Cash Flow
0	−$500
1	200
2	−500
3	1200

(*Answer:* IRR = 21.1% IRR)

7A-6 Given the following cash flow, determine the rate of return on the investment.

Year	Cash Flow
0	−$100
1	360
2	−570
3	360

7A-7 Compute the rate of return on the investment characterized by the following cash flow.

Year	Cash Flow
0	−$110
1	−500
2	300
3	−100
4	400
5	500

7A-8 Compute the rate of return on the investment characterized by the following cash flow.

Year	Cash Flow
0	−$50.0
1	+20.0
2	−40
3	+36.8
4	+36.8
5	+36.8

(*Answer:* IRR = 15.4%)

7A-9 A firm invested $15,000 in a project that appeared to have excellent potential. Unfortunately, a lengthy labor dispute in Year 3 resulted in costs that exceeded benefits by $8000. The cash flow for the project is as follows:

Year	Cash Flow
0	−$15,000
1	+10,000
2	+6,000
3	−8,000
4	+4,000
5	+4,000
6	+4,000

Compute the rate of return for the project.

7A-10 The following cash flow has no positive interest rate. The project, which had a projected life of 5 years, was terminated early.

Year	Cash Flow
0	−$50
1	+20
2	+20

There is, however, a negative interest rate. Compute its value. (*Answer:* −13.7%)

7A-11 For the following cash flow, compute the rate of return.

Year	Cash Flow
0	−$20
1	0
2	−10
3	+20
4	−10
5	+100

7A-12 Given the following cash flow:

Year	Cash Flow
0	−$800
1	+500
2	+500
3	−300
4	+400
5	+275

What is the rate of return on the investment? (*Answer:* 26.55% IRR)

7A-13 Consider the following cash flow.

Year	Cash Flow
0	−$100
1	240
2	−143

If the minimum attractive rate of return is 12%, should the project be undertaken?

7A-14 Refer to the strip-mining project in Example 5-10. Compute the rate of return for the project.

7A-15 Consider the following cash flow.

Year	Cash Flow
0	−$500
1	800
2	170
3	−550

Compute the rate of return on the investment.

7A-16 Determine the rate of return on the investment for the following cash flow.

Year	Cash Flow
0	−$100
1	+360
2	−428
3	168

7A-17 Compute the rate of return on the investment on the following cash flow.

Year	Cash Flow
0	−$1200
1	+358
2	+358
3	+358
4	+358
5	+358
6	−394

7A-18 Determine the rate of return on the investment on the following cash flow.

Year	Cash Flow
0	−$3570
1–3	+1000
4	−3170
5–8	+1500

7A-19 Bill bought a vacation lot he saw advertised on television for $800 down and monthly payments of $55. When he visited the lot, he found it was not something he wanted to own. After 40 months he was finally able to sell the lot. The new owner assumed the balance of the loan on the lot and paid Bill $2500. What rate of return did Bill receive on his investment?

7A-20 Compute the rate of return on an investment having the following cash flow.

Year	Cash Flow
0	−$850
0	+600
2–9	+200
10	−1800

7A-21 Assume that the following cash flows are associated with a project.

Year	Cash Flow
0	−$16,000
1	−8,000
2	11,000
3	13,000
4	−7,000
5	8,950

Compute the rate of return for this project.

7A-22 Compute the rate of return for the following cash flow.

Year	Cash Flow
0	−$200
1	+100
2	+100
3	+100
4	−300
5	+100
6	+200
7	+200
8	−124.5

(*Answer:* 20%)

7A-23 Following are the annual cost data for a tomato press.

Year	Cash Flow
0	−$210,000
1	88,000
2	68,000
3	62,000
4	−31,000
5	30,000
6	55,000
7	65,000

What is the rate of return associated with this project?

7A-24 A project has been in operation for 5 years, yielding the following annual cash flows:

Year	Cash Flow
0	−$103,000
1	102,700
2	−87,000
3	94,500
4	−8,300
5	38,500

Calculate the rate of return and state whether it has been an acceptable rate of return.

7A-25 Consider the following situation.

Year	Cash Flow
0	−$200
1	+400
2	−100

What is the rate of return?

7A-26 An investor is considering two mutually exclusive projects. He can obtain a 6% before-tax rate of return on external investments, but he requires a minimum attractive rate of return of 7% for these projects. Use a 10-year analysis period to compute the incremental rate of return from investing in Project *A* rather than Project *B*.

	Project *A*: Build Drive-Up Photo Shop	Project *B*: Buy Land in Hawaii
Initial capital investment	$58,500	$ 48,500
Net uniform annual income	6,648	0
Salvage value 10 years hence	30,000	138,000
Computed rate of return	8%	11%

7A-27 In January 2003, an investor bought a convertible debenture bond issued by the XLA Corporation. The bond cost $1000 and paid $60 per year interest in annual payments on December 31. Under the convertible feature of the bond, it could be converted into 20 shares of common stock by tendering the bond, together with $400 cash. The next business day after the investor received the December 31, 2005, interest payment, he submitted the bond together with $400 to the XLA Corporation. In return, he received the 20 shares of common stock. The common stock paid no dividends. On December 31, 2007, the investor sold the stock for $1740, terminating his 5-year investment in XLA Corporation. What rate of return did he receive?

7A-28 A problem often discussed in the engineering economy literature is the "oil-well pump problem." Pump 1 is a small pump; Pump 2 is a larger pump that costs more, will produce slightly more oil, and will produce it more rapidly. If the MARR is 20%, which pump should be selected? Assume that any temporary external investment of money earns 10% per year and that any temporary financing is done at 6%.

Year	Pump 1 ($000s)	Pump 2 ($000s)
0	−$100	−$110
1	+70	+115
3	+70	+30

CHOOSING THE BEST ALTERNATIVE

Coauthored with John Whittaker

Tapping into Controversy

When New York's Tappan Zee Bridge opened in 1955, it seemed big enough to handle any amount of traffic that was likely to come its way. The bridge spanned the Hudson River at a crossing in Westchester County, well north of New York City, near towns seemingly so removed from Manhattan that they were referred to as "satellite communities." Invoking visions of westward expansion and unpopulated vistas, state government officials proudly told voters that the $81 million bridge project would open up "new frontiers."

During its first full year of operation, the bridge carried only 18,000 vehicles per day, far below its design capacity of 100,000. By 2000, the satellite communities were called suburbs, and the Tappan Zee Bridge was being used by 135,000 vehicles every day. That number was expected to increase to 175,000 by 2020.

Clearly, something must be done to ease the growing congestion on the bridge. But what? The bridge's usable space has already been expanded once by converting a large median strip into a reversible traffic lane. And the bridge itself is nearing the end of its useful life, which originally was intended to be 50 years.

Upgrading the existing bridge would likely cost over $1 billion and would not adequately alleviate traffic congestion. Building a new bridge could cost as much as $4 billion dollars—quite a jump from the original price tag.

Why so costly? Inflation explains part of it, of course. The original Tappan Zee Bridge would cost almost $600 million in today's dollars. The new bridge would also be bigger. It would have eight lanes instead of six, a 33% increase.

But much of the cost would come from additional expenses that could not have been contemplated 50 years ago. Backers of the original bridge happily predicted that it would increase property values and attract new homebuyers to "outlying areas." And so it did. Today, however, many local residents fear that enlarging the bridge would simply encourage even more traffic—soon leading back to the same quandary they are now in.

To address these concerns, some government officials have suggested incorporating light-rail tracks and bus lanes into the new structure, in an attempt to encourage greater use of mass transit. Any such additions to the bridge would considerably drive up the cost of construction. Others have suggested installation of a more sophisticated toll-collection system that would give discounts to off-peak travelers.

QUESTIONS TO CONSIDER

1. In today's more environmentally aware decision-making climate, costly features such as the proposed light-rail lanes may have to be included in any proposal for new bridge construction just to get the project off the drawing board. How does this affect the minimum cost of public works projects?
2. Electronic toll collection permits vehicles to pay without stopping, which can dramatically improve traffic flow. The setting of tolls has always included considerations of fairness. Electronic toll collection introduces new possibilities of ethical issues that may be an important part of setting tolls. For example, should some individuals or vehicles pay a different toll based on income or number of people in the vehicle? Should tolls change with time of day or amount of congestion, and how predictable does the toll need to be? What is ethical on these and other toll-related issues?
3. Expansion and growth bring many challenges for society. What do you view as the primary ethical issues surrounding concepts like urban sprawl, eminent domain, and sustainable development?

After Completing This Chapter...
The student should be able to:

- Use a *graphical technique* to visualize and solve problems involving mutually exclusive choices.
- Define *incremental analysis* and differentiate it from a standard present worth, annual worth, and internal rate of return analyses.
- Use spreadsheets to solve incremental analysis problems.

INCREMENTAL ANALYSIS

In previous editions this chapter was titled, *Incremental Analysis,* and it extended the incremental analysis presented in Chapter 7 to multiple alternatives—solely for the rate of return (IRR) measure. The approach relied on a series of numerical comparisons of challengers and defenders.

In this edition we use the more powerful and easier-to-understand approach of graphical present worth (PW), equivalent uniform annual cost (EUAC), or equivalent uniform annual worth (EUAW) comparisons. This also supports calculation of incremental rates of return.

This graphical approach has the added benefit of focusing on the difference between alternatives. Often the difference is much smaller than the uncertainty in our estimated data. For ease of grading and instruction, we may assume in this text and in your exams that answers are exact, but in the real world the uncertainty in the data must always be considered.

Engineering design continually selects one of a set of alternatives to be part of the design. In engineering economy the words *mutually exclusive* alternatives are often used to emphasize that only one alternative may be implemented. Thus our problem is selecting the best of these mutually exclusive alternatives.

In earlier chapters we did this by maximizing PW, minimizing EUAC, or maximizing EUAW. We will do the same here. In Chapter 7 we compared two alternatives incrementally to decide whether the IRR on the increment was acceptable. Thus any two alternatives can be compared by recognizing that:

[Higher-cost alternative] = [Lower-cost alternative] + [Increment between them]

When there are two alternatives, only a single incremental analysis is required. With more alternatives, a series of "challenger–defender" comparisons is required. Also, only by doing the analysis step by step can we determine which pairs must be compared. For example, if there are 4 alternatives, then 3 of 6 possible comparisons must be made. For 5 alternatives then 4 of 10 possible comparisons must be made. For N alternatives, $N-1$ comparisons must be made from $N(N-1)/2$ possibilities.

The graphical approach is best implemented with spreadsheets, but it provides more information, and it is easier to understand and to present to others.

GRAPHICAL SOLUTIONS

Examples 8-1 and 8-2 illustrate why incremental or graphical analyses for the IRR criterion are required. They show that graphing makes it easy to choose the best alternative. Chapter 4 presented *xy* plots done with spreadsheets, and in Chapter 7 spreadsheets were used to graph the present worth versus the rate of return. In this chapter the present worth of each alternative is one *y* variable for graphs with multiple alternatives (or variables). In Chapter 9, one of the spreadsheet sections will present some of the ways that graphs can be customized for a better appearance.

EXAMPLE 8-1

The student engineering society is building a snack cart to raise money. The members must decide what capacity the cart should be able to serve. To serve 100 customers per hour will cost $10,310, and to serve 150 customers per hour will cost $13,400. The 50% increase in capacity is less than 50% of $10,310 because of economies of scale, but the increase in net revenue will be less than 50%, since the cart will not always be serving 150 customers per hour. The estimated net annual income for the lower capacity is $3300, and for the higher capacity it is $4000. After 5 years the cart is expected to have no salvage value. The engineering society is unsure of what interest rate to use to decide on the capacity. Make a recommendation.

SOLUTION

Since we do not know the interest rate, the easiest way to analyze the problem is to graph the PW of each alternative versus the interest rate, as in Figure 8-1. The Excel function[1]

$$= -\text{cost} + \text{PV}(\text{interest rate}, \text{life}, -\text{annual benefit})$$

is used to graph these two equations:

$$\text{PW}_{\text{low}} = -\$10,310 + \$3300(P/A, i, 5)$$
$$\text{PW}_{\text{high}} = -\$13,400 + \$4000(P/A, i, 5)$$

FIGURE 8-1 Maximizing PW to choose best alternative.

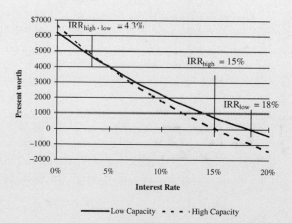

We want to maximize the PW, so the choice between these alternatives is defined by where their curves intersect (the incremental rate of return of 4.3%), not by where those curves intersect the PW = 0 axis. Those internal rates of return of 15 and 18% are irrelevant to the choice between the low- and high-capacity carts. This is why we did not analyze more than two alternatives in Chapter 7, leaving that for this chapter.

---

[1]The cost term in the equation has a minus sign because it is a cost and we are calculating a present worth. The annual benefit term has a minus sign because of the sign convention for the PV function.

From Figure 8-1, we see that if the interest rate is below 4.3%, the high-capacity cart has a higher PW. From 4.3 to 18% the lower-capacity cart has a higher PW. Above 18%, the third alternative of doing nothing is a better choice than building a low-capacity cart. This can be summarized in a choice table:

Interest Rate	Best Choice
$0\% \leq i \leq 4.3\%$	High capacity
$4.3\% \leq i \leq 18.0\%$	Low capacity
$18.0\% \leq i$	Do nothing

The interest rate where the low- and high-capacity curves intersect can be found by analyzing the *incremental* investment. The following table summarizes the added cost, capacity, and benefit for that incremental investment.

	Low	High	High – Low
Cost	$10,310	$13,400	$3,090
Capacity	100	150	50
Annual benefit	$3,300	$4,000	$700
Life, in years	5	5	5

If the equation for the PW of the incremental investment is solved for its IRR, the answer is 4.3%. The Excel function is = RATE(life, annual benefit, –first cost).

$$PW_{high-low} = -\$3090 + \$700(P/A, i, 5)$$

This example has been defined in terms of a student engineering society. But very similar problems are faced by a civil engineer sizing the weighing station for trucks carrying fill to a new earth dam, by an industrial engineer designing a package-handling station, and by a mechanical engineer sizing energy conservation equipment.

EXAMPLE 8-2

Solve Example 7-15 by means of an NPW graph. Two machines are being considered for purchase. If the minimum attractive rate of return (MARR) is 10%, which machine should be bought?

	Machine X	Machine Y
Initial cost	$200	$700
Uniform annual benefit	$95	$120
End-of-useful-life salvage value	$50	$150
Useful life, in years	6	12

SOLUTION

Since the useful lives of the two alternatives are different, for an NPW analysis we must adjust them to the same analysis period. If the need seems continuous, then the like-for-like assumption is reasonable and a 12-year analysis period can be used. The annual cash flows and the incremental cash flows are as follows:

End of Year	Cash Flows		$Y - X$
	Machine X	Machine Y	
0	−$200	−$700	−$500
1	95	120	25
2	95	120	25
3	95	120	25
4	95	120	25
5	95	120	25
6	−55	120	175
7	95	120	25
8	95	120	25
9	95	120	25
10	95	120	25
11	95	120	25
12	145	270	125

By means of a spreadsheet program, the NPWs are calculated for a range of interest rates and then plotted on an NPW graph (Figure 8-2).

For a MARR of 10%, Machine X is clearly the superior choice. In fact, as the graph clearly illustrates, Machine X is the correct choice for most values of MARR. The intersection point of the two graphs can be found by calculating ΔIRR, the rate of return on the incremental investment. From the spreadsheet IRR function applied to the incremental cash flows for $Y - X$:

$$\Delta\text{IRR} = 1.3\%$$

FIGURE 8-2 NPW graph.

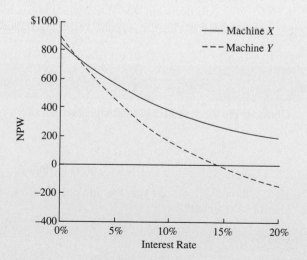

Then we can see from the graph that for MARR greater than 1.3%, Machine X is the right choice, and for MARR values less than 1.3%, Machine Y is the right choice.

	NPW	
Rate	Machine X	Machine Y
0%	$840.00	$890.00
1.322	752.24	752.24
2	710.89	687.31
4	604.26	519.90
6	515.57	380.61
8	441.26	263.90
10	378.56	165.44
12	325.30	81.83
14	279.77	10.37
16	240.58	−51.08
18	206.65	−104.23
20	177.10	−150.47

Examples 8-3 through 8-6 increase the number of alternatives being considered, but the same approach is used. Graph the PW, EUAW, or EUAC, and for each possible interest rate choose the best alternative. Generally this means maximizing the PW or EUAW, but for Example 8-4 the EUAC is minimized.

Example 8-4 is a typical design problem where the most cost-effective solution must be chosen, but the dollar value of the benefit is not defined. For example, every building must have a roof, but deciding whether it should be metal, shingles, or a built-up membrane is a cost decision. No value is placed on a dry building; it is simply a requirement. In Example 8-4 having the pressure vessel is a requirement.

EXAMPLE 8-3

Consider the three mutually exclusive alternatives:

	A	B	C
Initial cost	$2000	$4000	$5000
Uniform annual benefit	410	639	700

Each alternative has a 20-year life and no salvage value. If the MARR is 6%, which alternative should be selected?

SOLUTION

Using a spreadsheet to plot the NPW of the three alternatives produces Figure 8-3.

FIGURE 8-3 NPW graph of Alternatives A, B, and C.

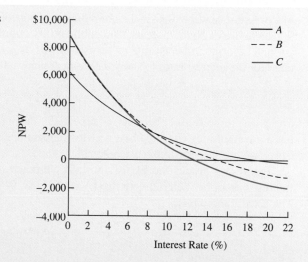

The highest line at any interest rate shows which alternative will provide the maximum NPW at that rate. From the graph we see that Alternative C has the maximum, then B, and finally A. The precise breakpoints can be found by calculating the ΔIRR of the corresponding increment.

$$NPW(C - B) = -\$5000 + \$4000 + (\$700 - \$639)(P/A, i, 20) = 0$$

$$\Delta IRR(C - B) = 2\% \qquad \text{using} = RATE(20, 700 - 639, 4000 - 5000)$$

$$NPW(B - A) = -\$4000 + \$2000 + (\$639 - \$410)(P/A, i, 20) = 0$$

$$\Delta IRR(B - A) = 9.6\% \qquad \text{using} = RATE(20, 639 - 410, 2000 - 4000)$$

The information from the graph can be presented in a choice table as follows:

If		MARR ≥	9.6%	Choose Alt. A
If	9.66%	≥ MARR ≥	2%	Choose Alt. B
If	2%	≥ MARR ≥	0%	Choose Alt. C

and the answer to the original question is select Alternative B if MARR is 6%.

EXAMPLE 8-4

A pressure vessel can be made out of brass, stainless steel, or titanium. The first cost and expected life for each material are:

	Brass	**Stainless Steel**	**Titanium**
Cost	\$100,000	\$175,000	\$300,000
Life, in years	4	10	25

The pressure vessel will be in the nonradioactive portion of a nuclear power plant that is expected to have a life of 50 to 75 years. The public utility commission and the power company have not yet agreed on the interest rate to be used for decision making and rate setting. Build a choice table for the interest rates to show where each material is the best.

SOLUTION

The pressure vessel will be replaced repeatedly during the life of the facility, and each material has a different life. Thus, the best way to compare the materials is using EUAC (see Chapter 6). This assumes like-for-like replacements.

Figure 8-4 graphs the EUAC for each alternative. In this case the best alternative at each interest rate is the material with the *lowest* EUAC. (We maximize worths, but minimize costs.) The Excel function is

$$= \text{PMT(interest rate, life, } -\text{first cost)}$$

The factor equation is

$$\text{EUAC} = \text{first cost}(A/P, i, \text{life})$$

FIGURE 8-4 EUAC comparison of alternatives.

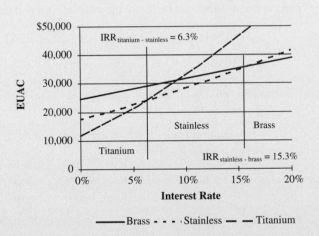

Because these alternatives have different length lives, calculating the incremental IRRs is best done using the spreadsheet function GOAL SEEK (see the last section of this chapter). The choice table for each material is:

Interest Rate	Best Choice
$0\% \ \leq i \leq 6.3\%$	Titanium
$6.3\% \ \leq i \leq 15.3\%$	Stainless steel
$15.3\% \leq i$	Brass

EXAMPLE 8-5

The following information refers to three mutually exclusive alternatives. The decision maker wishes to choose the right machine but is uncertain what MARR to use. Create a choice table that will help the decision maker to make the correct economic decision.

	Machine X	Machine Y	Machine Z
Initial cost	$200	$700	$425
Uniform annual benefit	65	110	100
Useful life, in years	6	12	8

SOLUTION

In this question, the lives of the three alternatives are different. Thus we cannot directly use the present worth criterion but must first make some assumptions about the period of use and the analysis period. As we saw in Chapter 6, when the service period is expected to be continuous and the assumption of like-for-like replacement reasonable, we can assume a series of replacements and compare annual worth values just as we did with present worth values. There is a direct relationship between the present worth and the equivalent uniform annual worth. It is

$$EUAW = NPW(A/P, i, n).$$

We will make these assumptions in this instance, and instead of plotting NPW, we shall plot EUAW. This was done on a spreadsheet using the Excel function = benefit + PMT (interest rate, life, initial cost), and the result is Figure 8-5.

The EUAW graph shows that for low values of MARR, Y is the correct choice. Then, as MARR increases, Z and finally X become the preferred machines. This can be expressed in a choice table read directly from the graph. If the "do-nothing" alternative is available, the table is the following:

FIGURE 8-5 EUAW graph.

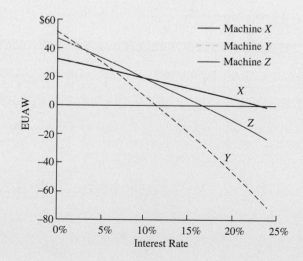

If		MARR ≥	23%	do nothing
If	23% ≥	MARR ≥	10%	choose X
If	10% ≥	MARR ≥	3.5%	choose Z
If	3.5% ≥	MARR ≥	0%	choose Y

If the "do-nothing" alternative is not available, then the table reads

If		MARR ≥	10%	choose X
If	10% ≥	MARR ≥	3.5%	choose Z
If	3.5% ≥	MARR ≥	0%	choose Y

The choice now is back in the decision maker's hands. There is still the need to determine MARR, but if the uncertainty was, for example, that MARR was some value in the range 12 to 18%, it can be seen that for this problem it doesn't matter. The answer is Machine X in any event. If, however, the uncertainty of MARR were in the 7 to 13% range, it is clear the decision maker would have to determine MARR with greater accuracy before solving this problem.

EXAMPLE 8-6

The following information is for five mutually exclusive alternatives that have 20-year useful lives. The decision maker may choose any one of the options or reject them all. Prepare a choice table.

	Alternatives				
	A	B	C	D	E
Cost	$4,000	$2,000	$6,000	$1,000	$9,000
Uniform annual benefit	639	410	761	117	785

SOLUTION

Figure 8-6 is an NPW graph of the alternatives constructed by means of a spreadsheet.

FIGURE 8-6 NPW graph.

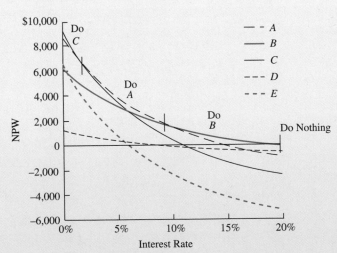

The graph clearly shows that Alternatives D and E are never part of the solution. They are dominated by the other three. The crossover points can either be read from the graph (if you have plotted it at a large enough scale) or found by calculating the ΔIRR of the intersecting graphs.

Calculating the incremental interest rates:

$$\Delta IRR(C - A)$$

$$\$6000 - \$4000 = (\$761 - \$639)(P/A, i, 20) \quad i = 2\%$$

$$\Delta IRR(A - B)$$

$$\$4000 - \$2000 = (\$639 - \$410)(P/A, i, 20) \quad i = 9.6\%$$

and to find where the NPW of A crosses the 0 axis

$$IRR(A)$$

$$\$4000 = \$639 \quad (P/A, i, 20) \quad i = 20\%$$

Placing these numbers in a choice table:

If		MARR $\geq$	20%	do nothing
If	20%	> MARR >	9.6%	select B
If	9.6%	$\geq$ MARR $\geq$	2%	select A
If	2%	$\geq$ MARR $\geq$	0%	select C

A final point to note on this example is that if we view the IRRs of the five alternatives, the only information we can glean from them is that if the

	A	B	C	D	E
IRR	15%	20%	11%	10%	6%

"do-nothing" alternative is available, it will be chosen if MARR is greater than that of the largest IRR. There is nothing in the numbers to tell us which alternatives will be in the final solution set; only the NPW graph will show us where the change points will be.[2]

ELEMENTS IN COMPARING MUTUALLY EXCLUSIVE ALTERNATIVES

1. Be sure all the alternatives are identified. In textbook problems the alternatives will be well defined, but real-life problems may be less clear. Before proceeding, one must

---

[2]There are analytical techniques for determining which incremental investments enter the solution set, but they are beyond the scope of this book and, in any case, are redundant in this era of spreadsheets. Interested readers can consult *Economic Analysis for Engineers and Managers* by Sprague and Whittaker (Prentice-Hall 1987).

have all the mutually exclusive alternatives tabulated, including the "do-nothing" or "keep doing the same thing" alternative, if appropriate.

2. Construct an NPW or EUAW graph showing all alternatives plotted on the same axes. This would be a difficult task were it not for spreadsheets.

3. Examine the line of maximum values and determine which alternatives create it, and over what range.

4. Determine the changeover points where the line of maximum NPW changes from one alternative to another. These can either be read directly off the graph or calculated, since they are the intersection points of the two curves and, what is more important and meaningful for engineering economy, they are **the ∆IRR of the incremental investment** between the two alternatives.

5. Create a choice table to present the information in compact and easily understandable form.

DOING A SET OF INCREMENTAL CHALLENGER–DEFENDER COMPARISONS

This chapter has focused on doing a graphical comparison of the PW of each alternative over a range of interest rates. We've calculated incremental interest rates, but we've looked at the curves to see which pair of intersecting PW curves we were analyzing. Before spreadsheets, this type of problem was solved by a series of challenger–defender comparisons, where defender ≡ best alternative identified at this stage of the analysis, and challenger ≡ next alternative being evaluated. Example 8-7 illustrates this approach.

The numerically based pairwise incremental comparisons is done at a single interest rate, and it does not show that "near" their intersection, two PW curves have "essentially" the same value. It also does not produce a choice table, since it is at a single interest rate. A choice table can be constructed, but it requires even more incremental comparisons.

EXAMPLE 8-7

For the alternatives in Example 8-6, conduct a pairwise incremental comparison. Which alternative is the best at an interest rate of 10%? Each of the five alternatives has no salvage value at the end of a 20-year useful life.

	Alternatives				
	A	*B*	*C*	*D*	*E*
Cost	$4000	$2000	$6000	$1000	$9000
Annual benefit	639	410	761	117	785

SOLUTION

1. The first step is to reorder the alternatives in order of increasing cost. This ensures that for each pairwise comparison the increment is an investment.

Alternatives

	D	B	A	C	E
Cost	$1000	$2000	$4000	$6000	$9000
Annual benefit	117	410	639	761	785

2. Calculate the IRR of the least expensive alternative to see if it is better than doing nothing at a MARR of 10%.

$$PW_D = 0 = -\$1000 + \$117(P/A, IRR_D, 20)$$

Solving this without a spreadsheet may take several tries and interpolation, but using the Excel function =RATE(life, annual benefit, –first cost) makes it easy and the answer exact:

$$IRR_D = 9.94\%$$

Because investment D earns less than 10%, doing nothing is preferred to doing D.

3. Doing nothing is still our *defender,* and the next Alternative, B, becomes the *challenger.* Calculate the IRR of Alternative B to see if it is better than doing nothing at a MARR of 10%.

$$PW_B = 0 = -\$2000 + \$410(P/A, IRR_B, 20)$$

Using the Excel function = RATE(life, annual benefit, –first cost), we find

$$IRR_B = 19.96\%$$

Because investment B earns more than 10%, Alternative B is preferred to doing nothing.

4. Alternative B is now our *defender,* and the next Alternative, A, becomes the *challenger.* This comparison must be made incrementally.

$$PW_{A-B} = 0 = -(\$4000 - \$2000) + (\$639 - \$410)(P/A, IRR_{A-B}, 20)$$

Using the Excel function =RATE(life, annual benefit, –first cost), we have

$$IRR_{A-B} = 9.63\%$$

Because the incremental investment $A - B$ earns less than 10%, Alternative B is preferred to doing Alternative A.

5. Alternative B is still our *defender,* and the next Alternative, C, becomes the *challenger.* This comparison must be made incrementally.

$$PW_{C-B} = 0 = -(\$6000 - \$2000) + (\$761 - \$410)(P/A, IRR_{C-B}, 20)$$

Using the Excel function =RATE(life, annual benefit, –first cost), we have

$$IRR_{C-B} = 6.08\%$$

Because the incremental investment $C - B$ earns less than 10%, Alternative B is preferred to doing Alternative C.

6. Alternative B is still our *defender,* and the final Alternative, E, becomes the *challenger.* This comparison must be made incrementally.

$$PW_{E-B} = 0 = -(\$9000 - \$2000) + (\$785 - \$410)(P/A, IRR_{E-B}, 20)$$

Using the Excel function =RATE(life, annual benefit, –first cost)

$$IRR_{E-B} = 0.67\%$$

Because the incremental investment $E - B$ earns less than 10%, Alternative B is preferred to doing Alternative E.

At an interest rate of 10%, Alternative B is the best choice. But only by referring to Figure 8-6 can we see that Alternative B is the best choice for all interest rates between 9.63% (where Alternative A is equally attractive) and 19.96% (where doing nothing is equally attractive). With other data, the curve for Alternative B might intersect at an incremental comparison we hadn't done. Also this pairwise comparison has not even compared Alternatives A and C so we cannot complete the choice table.

CHOOSING AN ANALYSIS METHOD

At this point, we have examined in detail the three major economic analysis techniques: present worth analysis, annual cash flow analysis, and rate of return analysis. A practical question is, which method should be used for a particular problem?

While the obvious answer is to use the method requiring the least computations, a number of factors may affect the decision.

1. Unless the MARR—minimum attractive rate of return (or minimum required interest rate for invested money)—is known, neither present worth analysis nor annual cash flow analysis is possible.

2. Present worth analysis and annual cash flow analysis often require far less computation than rate of return analysis.

3. In some situations, a rate of return analysis is easier to explain to people unfamiliar with economic analysis. At other times, an annual cash flow analysis may be easier to explain.

4. Business enterprises generally adopt one, or at most two, analysis techniques for broad categories of problems. If you work for a corporation and the policy manual specifies rate of return analysis, you would appear to have no choice in the matter.

Since one may not always be able to choose the analysis technique computationally best suited to the problem, this book illustrates how to use each of the three methods in all feasible situations. Ironically, the most difficult method—rate of return analysis—is the one used most frequently by engineers in industry!

SPREADSHEETS AND GOAL SEEK

An incremental analysis of two alternatives is easily done with the RATE or IRR functions when the lives of the alternatives are the same. The problem is more difficult, however, when the lives are different. As was discussed in Chapters 5 and 6, alternatives having lives of different lengths are usually compared by assuming that the alternatives are repeated until the least common multiple of their lives. This can be done with a spreadsheet, but Excel supports an easier approach.

Excel has a tool called GOAL SEEK that identifies a formula cell, a target value, and a variable cell. This tool causes the variable cell to be changed automatically until the formula cell equals the target value. To find an IRR for an incremental analysis, the formula cell can be the difference between two equivalent annual worths with a target value of 0. Then if the variable cell is the interest rate, GOAL SEEK will find the IRR.

In Excel this tool is accessed by selecting T(ools) on the main toolbar or menu and G(oal seek) on the sub-menu. As shown in Example 8-8, the variable cell (with the interest rate) must somehow affect the formula cell (difference in equivalent uniform annual worths or costs (EUAWs or EUACs)), although it need not appear directly in the formula cell. In Figures 8-7a and 8-7b the interest rate (cell A1) appears in the EUAC formulas (cells D3 and D4) but not in the formula cell (cell D5).

EXAMPLE 8-8

Two different asphalt mixes can be used on a highway. The good mix will last 6 years and will cost $600,000 to buy and lay down. The better mix will last 10 years and will cost $800,000 to buy and lay down. Find the incremental IRR for using the more expensive mix.

SOLUTION

This example would be difficult to solve without the GOAL SEEK tool. The least common multiple of 6 and 10 is 30 years, which is the comparison period. With the GOAL SEEK tool a very simple spreadsheet does the job. In Figure 8-7a the spreadsheet is shown before GOAL SEEK. Figure 8-7b shows the result after the goal (D5) has been set = 0 and A1 has been chosen as the variable cell to change.

	A	B	C	D	E	F
1	8.00%	Interest rate				
2	Alternative	Cost	Life	EUAC		
3	Good	600000	6	129,789	= −PMT($A$1,C3,B3)	
4	Better	800000	10	119,224	= −PMT($A$1,C4,B4)	
5			difference=	10,566	= D3−D4	

FIGURE 8-7 Spreadsheet before GOAL SEEK.

	A	B	C	D	E	F
1	14.52%	Interest rate				
2	Alternative	Cost	Life	EUAC		
3	Good	600000	6	156,503	= −PMT($A$1,C3,B3)	
4	Better	800000	10	156,503	= −PMT($A$1,C4,B4)	
5			difference=	0	= D3−D4	

FIGURE 8-8 Spreadsheet after GOAL SEEK.

The incremental IRR found by GOAL SEEK is 14.52%.

SUMMARY

For choosing from a set of mutually exclusive alternatives, the rate of return technique is more complex than the present worth or annual cash flow techniques because in the last two techniques the numbers can be compared directly, whereas with the rate of return it is necessary to consider the *increment of investment*. This is fairly straightforward if there are only two alternatives, but it becomes more and more complex as the number of alternatives increases.

A visual display of the problem can be created by using a spreadsheet to to graph the economic value of the alternatives. The steps are as follows:

1. Be sure all the alternatives are identified.
2. Construct an NPW or EUAW (or EUAC) graph showing all alternatives plotted on the same axes.
3. Examine the line of maximum values (or minimum for the EUAC) and determine which alternatives create it, and over what range.
4. Determine the changeover points (ΔIRRs).
5. Create a choice table.

This graphical approach, although more values are calculated, is a more powerful one because, by allowing the decision maker to see the range over which the choices are valid, it provides a form of sensitivity analysis. It also makes it clear that "close" to the changeover point the alternatives are very similar in value.

PROBLEMS

These problems are organized such that the (*a*) parts require graphical analysis and as such are much more easily done with spreadsheets, and the (*b*) parts require numerical incremental analysis. Some problems include only one approach.

Two Action Alternatives

8-1 Given the following, construct a choice table for interest rates from 0% to 100%.

Year	X	Y
0	−$10	−$20
1	+15	+28

8-2 Consider two mutually exclusive alternatives and the do-nothing alternative:

Year	Buy X	Buy Y
0	−$100.0	−$50.0
1	+31.5	+16.5
2	+31.5	+16.5
3	+31.5	+16.5
4	+31.5	+16.5

Construct a choice table for interest rates from 0% to 100%.

8-3 Consider three alternatives *A*, *B*, and "do nothing": Construct a choice table for interest rates from 0% to 100%.

Year	A	B
0	−$100	−$150
1	+30	+43
2	+30	+43
3	+30	+43
4	+30	+43
5	+30	+43

8-4 A firm is considering two alternatives that have no salvage value.

	A	B
Initial cost	$10,700	$5500
Uniform annual benefits	2,100	1800
Useful life, in years	8	4

At the end of 4 years, another *B* may be purchased with the same cost, benefits, and so forth.

(*a*) Construct a choice table for interest rates from 0% to 100%.

(*b*) If the MARR is 10%, which alternative should be selected?

8-5 Our cat Fred's summer kitty-cottage needs a new roof. He's considering the following two proposals and feels a 15-year analysis period is in line with his remaining lives. (There is no salvage value for old roofs.)

	Thatch	Slate
First cost	$20	$40
Annual upkeep	5	2
Service life, in years	3	5

(a) Construct a choice table for interest rates from 0% to 100%.

(b) Which roof should he choose if his MARR is 12%? What is the actual value of the IRR on the incremental cost?

8-6 Don Garlits is a landscaper. He is considering the purchase of a new commercial lawn mower, either the Atlas or the Zippy. Construct a choice table for interest rates from 0% to 100%.

	Atlas	Zippy
Initial cost	$6700	$16,900
Annual operation and maintenance cost	1500	1,200
Annual benefit	4000	4,500
Salvage value	1000	3,500
Useful life, in years	3	6

8-7 The South End bookstore has an annual profit of $170,000. The owner is considering opening a second bookstore on the north side of the campus. He can lease an existing building for 5 years with an option to continue the lease for a second 5-year period. If he opens the second bookstore, he expects the existing store will lose some business that will he gained by "The North End," the new bookstore. It will take $500,000 of store fixtures and inventory to open The North End. He believes that the two stores will have a combined profit of $260,000 a year after all the expenses of both stores have been paid.

The owner's economic analysis is based on a 5-year period. He will be able to recover this $500,000 investment at the end of 5 years by selling the store fixtures and inventory.

(a) Construct a choice table for interest rates from 0% to 100%.

(b) The owner will not open The North End unless he can expect a 15% rate of return. What should he do?

8-8 A paper mill is considering two types of pollution control equipment.

	Neutralization	Precipitation
Initial cost	$700,000	$500,000
Annual chemical cost	40,000	110,000
Salvage value	175,000	125,000
Useful life, in years	5	5

(a) Construct a choice table for interest rates from 0% to 100%.

(b) The firm wants a 12% rate of return on any avoidable increments of investment. Which equipment should be purchased?

8-9 A stockbroker has proposed two investments in low-rated corporate bonds paying high interest rates and selling below their stated value (in other words, junk bonds). Both bonds are rated as equally risky.

Bond	Stated Value	Annual Interest Payment	Current Market Price, Including Buying Commission	Bond Maturity*
Gen Dev	$1000	$ 94	$480	15 years
RJR	1000	140	630	15

*At maturity the bondholder receives the last interest payment plus the bond stated value.

(a) Construct a choice table for interest rates from 0% to 100%.

(b) Which, if any, of the bonds should you buy if your MARR is 25%?

Multiple Alternatives

8-10 Consider the following alternatives:

	A	B	C
Initial cost	$300	$600	$200
Uniform annual benefits	41	98	35

Each alternative has a 10-year useful life and no salvage value.

(a) Construct a choice table for interest rates from 0% to 100%.

(b) If the MARR is 8%, which alternative should be selected?

8-11 The following four mutually exclusive alternatives have no salvage value after 10 years.

	A	B	C	D
First cost	$75	$50	$50	$85
Uniform annual benefit	16	12	10	17
Computed rate of return	16.8%	20.2%	15.1%	15.1%

(a) Construct a choice table for interest rates from 0% to 100%.

(b) Using 8% for the MARR, which alternative should be selected?   (*Answer: A*)

8-12 The following three mutually exclusive alternatives have no salvage value after 5 years:

	A	B	C
First cost	$200	$300	$600
Uniform annual benefit	59.7	77.1	165.2
Computed rate of return	15%	9%	11.7%

Construct a choice table for interest rates from 0% to 100%.

8-13 Consider four mutually exclusive alternatives, each having an 8-year useful life:

	A	B	C	D
First cost	$1000	$800	$600	$500
Uniform annual benefit	122	120	97	122
Salvage value	750	500	500	0

(a) Construct a choice table for interest rates from 0% to 100%.

(b) If the minimum attractive rate of return is 8%, which alternative should be selected?

8-14 Three mutually exclusive projects are being considered:

	A	B	C
First cost	$1000	$2000	$3000
Uniform annual benefit	150	150	0
Salvage value	1000	2700	5600
Useful life, in years	5	6	7

When each project reached the end of its useful life, it would be sold for its salvage value and there would be no replacement.

(a) Construct a choice table for interest rates from 0% to 100%.

(b) If 8% is the desired rate of return, which project should be selected?

8-15 Consider three mutually exclusive alternatives and the do-nothing alternative.

	A	B	C
Initial cost	$770.00	$1406.30	$2563.30
Uniform annual benefit	420.00	420.00	420.00
Useful life, in years	2	4	8
Computed rate of return	6.0%	7.5%	6.4%

The analysis period is 8 years. At the end of 2, 4, and 6 years, Alt. A will have an identical replacement. Alternative B will have a single identical replacement at the end of 4 years. Construct a choice table for interest rates from 0% to 100%.

8-16 Consider the three alternatives:

	A	B	C
Initial cost	$1500	$1000	$2035
Annual benefit in each of first 5 years	250	250	650
Annual benefit in each of subsequent 5 years	450	250	145

Each alternative has a 10-year useful life and no salvage value.

Construct a choice table for interest rates from 0% to 100%.

8-17 Three mutually exclusive alternatives are being considered.

	A	B	C
Initial investment	$50,000	$22,000	$15,000
Annual net income	5,093	2,077	1,643
Computed rate of return	8%	7%	9%

Each alternative has a 20-year useful life with no salvage value.

(a) Construct a choice table for interest rates from 0% to 100%.

(b) If the minimum attractive rate of return is 7%, which alternative should be selected?

8-18 Consider the following alternatives:

	A	B	C
Initial cost	$100.00	$150.00	$200.00
Uniform annual benefit	10.00	17.62	55.48
Useful life, in years	Infinite	20	5

Alternatives B and C are replaced at the end of their useful lives with identical replacements. Use an infinite analysis period.

(a) Construct a choice table for interest rates from 0% to 100%.

(b) At an 8% interest rate, which alternative is better?

8-19 Three mutually exclusive alternatives are being studied.

Year	A	B	C
0	−$20,000	−$20,000	−$20,000
1	+10,000	+10,000	+5,000
2	+5,000	+10,000	+5,000
3	+10,000	+10,000	+5,000
4	+6,000	0	+15,000

(a) Construct a choice table for interest rates from 0% to 100%.

(b) If the MARR is 12%, which alternative should be selected?

8-20 A firm is considering three mutually exclusive alternatives as part of a production improvement program. The alternatives are as follows:

	A	B	C
Installed cost	$10,000	$15,000	$20,000
Uniform annual benefit	1,625	1,625	1,890
Useful life, in years	10	20	20

For each alternative, the salvage value at the end of useful life is zero. At the end of 10 years, Alt. A could be replaced by another A with identical cost and benefits.

(a) Construct a choice table for interest rates from 0% to 100%.

(b) The MARR is 6%. If the analysis period is 20 years, which alternative should be selected?

8-21 A new 10,000-square-meter warehouse next door to the Tyre Corporation is for sale for $450,000. The terms offered are $100,000 down with the balance being paid in 60 equal monthly payments based on 15% interest. It is estimated that the warehouse would have a resale value of $600,000 at the end of 5 years.

Tyre has the cash and could buy the warehouse but does not need all the warehouse space at this time. The Johnson Company has offered to lease half the new warehouse for $2500 a month.

Tyre presently rents and uses 7000 square meters of warehouse space for $2700 a month. It has the option of reducing the rented space to 2000 square meters, in which case the monthly rent would be $1000 a month. Further, Tyre could cease renting warehouse space entirely. Tom Clay, the Tyre Corp. plant engineer, is considering three alternatives:

1. Buy the new warehouse and lease the Johnson Company half the space. In turn, the Tyre-rented space would be reduced to 2000 square meters.

2. Buy the new warehouse and cease renting any warehouse space.

3. Continue as is, with 7000 square meters of rented warehouse space.

Construct a choice table for interest rates from 0% to 100%.

8-22 QZY, Inc. is evaluating new widget machines offered by three companies. The machines have the following characteristics:

	Company A	Company B	Company C
First cost	$15,000	$25,000	$20,000
Maintenance and operating	1,600	400	900
Annual benefit	8,000	13,000	9,000
Salvage value	3,000	6,000	4,500
Useful life, in years	4	4	4

(a) Construct a choice table for interest rates from 0% to 100%.

(b) MARR = 15%. From which company, if any, should you buy the widget machine? Use rate of return analysis.

8-23 The Croc Co. is considering a new milling machine from among three alternatives: each has life of 10 years:

	Alternative		
	Deluxe	Regular	Economy
First cost	$220,000	$125,000	$75,000
Annual benefit	79,000	43,000	28,000
Maintenance and operating costs	38,000	13,000	8,000
Salvage value	16,000	6,900	3,000

(a) Construct a choice table for interest rates from 0% to 100%.

(b) MARR = 15%. Using incremental rate of return analysis, which alternative, if any, should the company choose?

8-24 Wayward Airfreight, Inc. has asked you to recommend a new automatic parcel sorter. You have obtained the following bids:

	SHIP-R	SORT-Of	U-SORT-M
First cost	$184,000	$235,000	$180,000
Salvage value	38,300	44,000	14,400
Annual benefit	75,300	89,000	68,000
Yearly maintenance and operating cost	21,000	21,000	12,000
Useful life, in years	7	7	7

(a) Construct a choice table for interest rates from 0% to 100%.

(b) Using an MARR of 15% and a rate of return analysis, which alternative, if any, should be selected?

8-25 A firm is considering the following alternatives, as well as a fifth choice: do nothing. Each alternative has a 5-year useful life.

	1	2	3	4
Initial cost	$100,000	$130,000	$200,000	$330,000
Uniform annual net income ($000s)	26.38	38.78	47.48	91.55
Computed rate of return	10%	15%	6%	12%

(a) Construct a choice table for interest rates from 0% to 100%.

(b) The firm's minimum attractive rate of return is 8%. Which alternative should be selected?

8-26 Construct a choice table for interest rates from 0% to 100%.

	Alternatives			
	A	B	C	D
Initial cost	$2000	5000	4000	3000
Annual benefit	800	500	400	1300
Salvage value	2000	1500	1400	3000
Life, in years	5	6	7	4

8-27 In a particular situation, four mutually exclusive alternatives are being considered. Each of the alternatives costs $1300 and has no salvage value after a 10-year life.

Alternative	Annual Benefit	Calculated Rate of Return
A	$100 at end of first year; increasing $30 per year thereafter	10.0%
B	$10 at end of first year; increasing $50 per year thereafter	8.8%
C	Annual end-of-year benefit = $260	15.0%
D	$450 at end of first year; declining $50 per year thereafter	18.1%

(a) Construct a choice table for interest rates from 0% to 100%.

(b) If the MARR is 8%, which alternative should be selected? (*Answer:* Alt. *C*)

8-28 A more detailed examination of the situation in Problem 8-27 reveals that there are two additional mutually exclusive alternatives to be considered. Both cost more than the $1300 for the four original alternatives. Both have no salvage value after a useful live of 10 years.

Alternative	Cost	Annual End-of-Years Benefit	Calculated Rate of Return
E	$3000	$ 488	10.0%
F	5850	1000	11.2%

(a) Construct a choice table for interest rates from 0% to 100%.

(b) If the MARR remains at 8%, which one of the six alternatives should be selected? (*Answer:* Alt. *F*)

8-29 The owner of a downtown parking lot has employed a civil engineering consulting firm to advise him on the economic feasibility of constructing an office building on the site. Bill Samuels, a newly hired civil engineer, has been assigned to make the analysis. He has assembled the following data:

Alternative	Total Investment*	Total Net Annual Revenue from Property
Sell parking lot	$ 0	$ 0
Keep parking lot	200,000	22,000
Build 1-story building	400,000	60,000
Build 2-story building	555,000	72,000
Build 3-story building	750,000	100,000
Build 4-story building	875,000	105,000
Build 5-story building	1,000,000	120,000

* Includes the value of the land.

The analysis period is to be 15 years. For all alternatives, the property has an estimated resale (salvage) value at the end of 15 years equal to the present total investment.

(a) Construct a choice table for interest rates from 0% to 100%.

(b) If the MARR is 10%, what recommendation should Bill make?

8-30　A firm is considering moving its manufacturing plant from Chicago to a new location. The industrial engineering department was asked to identify the various alternatives together with the costs to relocate the plant, and the benefits. The engineers examined six likely sites, together with the do-nothing alternative of keeping the plant at its present location. Their findings are summarized as follows:

Plant Location	First Cost ($000s)	Uniform Annual Benefit ($000s)
Denver	$300	$ 52
Dallas	550	137
San Antonio	450	117
Los Angeles	750	167
Cleveland	150	18
Atlanta	200	49
Chicago	0	0

The annual benefits are expected to be constant over the 8-year analysis period.

(a) Construct a choice table for interest rates from 0% to 100%.

(b) If the firm uses 10% annual interest in its economic analysis, where should the manufacturing plant be located?   (*Answer:* Dallas)

8-31　An oil company plans to purchase a piece of vacant land on the corner of two busy streets for $70,000. On properties of this type, the company installs businesses of four different types.

Plan	Cost	Type of Business
A	$ 75,000	Conventional gas station with service facilities for lubrication, oil changes, etc.
B	230,000	Automatic carwash facility with gasoline pump island in front
C	30,000	Discount gas station (no service bays)
D	130,000	Gas station with low-cost, quick-carwash facility

* Cost of improvements does not include the $70,000 cost of land.

In each case, the estimated useful life of the improvements is 15 years. The salvage value for each is estimated to be the $70,000 cost of the land. The net annual income, after paying all operating expenses, is projected as follows:

Plan	Net Annual Income
A	$23,300
B	44,300
C	10,000
D	27,500

(a) Construct a choice table for interest rates from 0% to 100%.

(b) If the oil company expects a 10% rate of return on its investments, which plan (if any) should be selected?

8-32　*The Financial Advisor* is a weekly column in the local newspaper. Assume you must answer the following question. "I recently retired at age 65, and I have a tax-free retirement annuity coming due soon. I have three options. I can receive (A) $30,976 now, (B) $359.60 per month for the rest of my life (assume 20 years), or (C) $513.80 per month for the next 10 years. What should I do?" Ignore the timing of the monthly cash flows and assume that the payments are received at the end of year.
Contributed by D.P. Loucks, Cornell University

(a) Develop a choice table for interest rates from 0% to 50%. (You do not know what the reader's interest rate is.)

(b) If $i = 9\%$, use an incremental rate of return analysis to recommend which option should be chosen.

8-33　A firm must decide which of three alternatives to adopt to expand its capacity. The firm wishes a minimum annual profit of 20% of the initial cost of each separable increment of investment. Any money not invested in capacity expansion can be invested elsewhere for an annual yield of 20% of initial cost.

Alt.	Initial Cost	Annual Profit	Profit Rate
A	$100,000	$30,000	30%
B	300,000	66,000	22%
C	500,000	80,000	16%

Which alternative should be selected? Use a challenger–defender rate of return analysis.

8-34 The New England Soap Company is considering adding some processing equipment to the plant to aid in the removal of impurities from some raw materials. By adding the processing equipment, the firm can purchase lower-grade raw material at reduced cost and upgrade it for use in its products.

Four different pieces of processing equipment are being considered:

	A	B	C	D
Initial investment	$10,000	$18,000	25,000	$30,000
Annual saving in materials costs	4,000	6,000	7,500	9,000
Annual operating cost	2,000	3,000	3,000	4,000

The company can obtain a 15% annual return on its investment in other projects and is willing to invest money on the processing equipment only as long as it can obtain 15% annual return on each increment of money invested. Which one, if any, of the alternatives should be selected? Use a challenger–defender rate of return analysis assuming an indefinite or perpetual life.

Cash vs. Loan vs. Lease

8-35 Frequently we read in the newspaper that one should lease a car rather than buying it. For a typical 24-month lease on a car costing $9400, the monthly lease charge is about $267. At the end of the 24 months, the car is returned to the lease company (which owns the car). As an alternative, the same car could be bought with no down payment and 24 equal monthly payments, with interest at a 12% nominal annual percentage rate. At the end of 24 months the car is fully paid for. The car would then be worth about half its original cost.

(a) Over what range of nominal before-tax interest rates is leasing the preferred alternative?

(b) What are some of the reasons that would make leasing more desirable than is indicated in (a)?

8-36 *The Financial Advisor* is a weekly column in the local newspaper. Assume you must answer the following question. "I need a new car that I will keep for 5 years. I have three options. I can (A) pay $19,999 now, (B) make monthly payments for a 6% 5-year loan with 0% down, or (C) make lease payments of $299.00 per month for the next 5 years. The lease option also requires an up-front payment of $1000. What should I do?"

Assume that the number of miles driven matches the assumptions for the lease, and the vehicle's value after 5 years is $6500. Remember that lease payments are made at the beginning of the month, and the salvage value is only received if you own the vehicle.

(a) Develop a choice table for nominal interest rates from 0% to 50%. (You do not know what the reader's interest rate is.)

(b) If $i = 9\%$, use an incremental rate of return analysis to recommend which option should be chosen.

(*Answer:* For nominal interest rates of 0 to 6% pay cash, of 6 to 30.0% use the loan, and of over 30.0% use the lease.)

8-37 *The Financial Advisor* is a weekly column in the local newspaper. Assume you must answer the following question. "I need a new car that I will keep for 5 years. I have three options. I can (A) pay $15,999 now, (B) make monthly payments for a 9% 5-year loan with 0% down, or (C) make lease payments of $269.00 per month for the next 5 years. The lease option also requires an up-front payment of $500. What should I do?"

Assume that the number of miles driven matches the assumptions for the lease, and the vehicle's value after 5 years is $4500. Remember that lease payments are made at the beginning of the month, and the salvage value is only received if you own the vehicle.

(a) Develop a choice table for nominal interest rates from 0% to 50%. (You do not know what the reader's interest rate is.)

(b) If $i = 9\%$, use an incremental rate of return analysis to recommend which option should be chosen.

8-38 Assume that the vehicle described in Problem 8-36 will be kept for 10 rather than 5 years. If leased, the vehicle can be purchased for its value at the end of 5 years. At the end of 10 years, the vehicle will be worth $2000.

(a) Develop a choice table for nominal interest rates from 0% to 50%. (You do not know what the reader's interest rate is.)

(b) If $i = 9\%$, use an incremental rate of return analysis to recommend which option should be chosen.

(c) Are your answers different than in Problem 8-36? Why or why not?

8-39   Contact a car dealer and choose a car (keep it reasonable—no Lamborghinis). Tell the people at the dealership that you are a student working on an assignment. Be truthful and don't argue; if they don't want to help you, leave and find a friendlier dealer. For both buying and leasing, show all assumptions, costs, and calculations. Do not include the cost of maintenance, gasoline, oil, water, fluids, and other routine expenses in your calculations.

Determine the car's sales price (no need to negotiate) and the costs for sales tax, license, and fees. Estimate the "Blue Book value" in 5 years. Determine the monthly payment based on a 5-year loan at 9% interest. Assume that your down payment is large enough to cover only the sales tax, license, and fees. Calculate the equivalent uniform monthly cost of owning the car.

Identify the costs to lease the car (if available assume a 5-year lease). This includes the monthly lease payment, required down payments, and any return fees that are required. Calculate the equivalent uniform monthly cost of leasing the car.

The salesperson probably does not have the answers to many of these questions. Write a one- to two-page memo detailing the costs. Make a recommendation: Should you own or lease your car? Include nonfinancial items and potential financial items in your conclusions, such as driving habits and whether you are likely to drive more than the allowed number of miles dictated in the lease.

Contributed by Neal Lewis, University of Bridgeport

OTHER ANALYSIS TECHNIQUES

Clean, Green, and Far Between

Designers and engineers have made great progress in developing environment-friendly construction materials and building techniques. And "green" office buildings offer numerous advantages. They can improve worker productivity, reduce health and safety costs, improve the indoor environmental quality, and reduce energy and maintenance costs.

There is a widely held belief, however, that green structures cost more to build, especially with respect to first costs. In the commercial sector, barely 1% of office construction projects have bothered to apply for certification as green buildings because managers believe that the application process is too complicated and requires too much paperwork. Despite this, the U.S. Green Building Council reports that the number of square feet of LEED projects has increased eightfold, from 80 million in 2002 to 642 million in 2006. LEED, or Leadership in Energy and Environmental Design, is the nationally accepted benchmark for design, construction, and operation of high-performance green buildings.

Commercial developers have found that environment-friendly features can add to the expense of construction. Even though additional costs may increase overall construction expenses by only 1 or 2%, they can make it harder for the builder to recoup its investment and break even on the project in a timely manner. Furthermore, there is a general reluctance by many in this industry to embrace new ideas or new technologies.

Renters may like the idea of being cleaner and greener—but they may be unwilling to pay extra for it unless they can see a tangible impact to their bottom line.

Revised by Kim Lascola Needy, University of Pittsburgh

QUESTIONS TO CONSIDER

1. Office leases frequently require building owners, rather than tenants, to pay heating and cooling costs. What effect might this have on the decision making of potential tenants who are considering renting space in green buildings?
2. The green building movement has had more success among developers who hold onto their buildings for years and rent them out, rather than selling them as soon as they are constructed. What factors might influence their views?
3. Many environment-friendly buildings are architecturally distinctive and feature better-quality materials and workmanship than traditional commercial structures. Advocates for the environment hope these characteristics will help green buildings attract a rent "premium." How might these features make the buildings more attractive to tenants? Is it the green features or the higher-quality materials and workmanship that add significantly to the perceived higher first costs of green buildings?
4. How can the costs and benefits of green buildings be economically validated by an independent party so that designers and engineers can make a fair assessment?
5. Does the economic attractiveness of green buildings depend on which measure is used for evaluation?
6. What ethical questions arise from state or municipal regulations intended to promote or require green building practices?

After Completing This Chapter...

You should be able to:

- Use future worth, benefit–cost ratio, payback period, and sensitivity analysis methods to solve engineering economy problems.
- Link the use of the *future worth* analysis to the present worth and annual worth methods developed earlier.
- Mathematically develop the *benefit–cost ratio,* and use this model to select alternatives and make economic choices.
- Understand the concept of the *payback period* of an investment, and be able to calculate this quantity for prospective projects.
- Demonstrate a basic understanding of *sensitivity* and *breakeven analyses* and the use of these tools in an engineering economic analysis.
- Use a spreadsheet to perform *sensitivity* and *breakeven analyses.*

Chapter 9 examines four topics:

- Future worth analysis
- Benefit–cost ratio or present worth index analysis
- Payback period
- Sensitivity, breakeven, and what-if analysis

Future worth analysis is very much like present worth analysis, dealing with *then* (future worth) rather than with *now* (present worth) situations.

Previously, we have written economic analysis relationships based on either:

$$\text{PW of cost} = \text{PW of benefit} \qquad \text{or} \qquad \text{EUAC} = \text{EUAB}$$

Instead of writing it in this form, we could define these relationships as

$$\frac{\text{PW of benefit}}{\text{PW of cost}} = 1 \qquad \text{or} \qquad \frac{\text{EUAB}}{\text{EUAC}} = 1$$

When economic analysis is based on these ratios, the calculations are called benefit–cost ratio analysis. The PW ratio is also known as a present worth index.

Payback period is an approximate analysis technique, generally defined as the time required for cumulative benefits to equal cumulative costs.

Sensitivity describes how much a problem element must change to change a particular decision. Closely related is breakeven analysis, which determines the conditions under which two alternatives are equivalent. What-if analysis changes one or all vaiables to see how the economic value and recommended decision change. Thus, breakeven and what-if analysis are forms of sensitivity analysis.

FUTURE WORTH ANALYSIS

In present worth analysis, alternatives are compared in terms of their present consequences. In annual cash flow analysis, the comparison was in terms of equivalent uniform annual costs (or benefits). But the concept of resolving alternatives into comparable units is not restricted to a present or annual comparison. The comparison may be made at any point in time. In many situations we would like to know what the *future* situation will be, if we take some particular course of action *now*. This is called **future worth analysis.**

EXAMPLE 9-1

Ron Jamison, a 20-year-old college student, smokes about a carton of cigarettes a week. He wonders how much money he could accumulate by age 65 if he quit smoking now and put his cigarette money into a savings account. Cigarettes cost $35 per carton. Ron expects that a savings account would earn 5% interest, compounded semiannually. Compute Ron's future worth at age 65.

SOLUTION

$$\text{Semiannual saving } \$35/\text{carton} \times 26 \text{ weeks} = \$910$$
$$\text{Future worth (FW)} = A(F/A, 2\tfrac{1}{2}\%, 90) = 910(329.2) = \$299,572$$

EXAMPLE 9-2

An east coast firm has decided to establish a second plant in Kansas City. There is a factory for sale for \$850,000 that could be remodeled and used. As an alternative, the firm could buy vacant land for \$85,000 and have a new plant constructed there. Either way, it will be 3 years before the firm will be able to get a plant into production. The timing and cost of the various components for the factory are:

Year	Construct New Plant		Remodel Available Factory	
0	Buy land	\$ 85,000	Purchase factory	\$850,000
1	Design and initial construction costs	200,000	Design and remodeling costs	250,000
2	Balance of construction costs	1,200,000	Additional remodeling costs	250,000
3	Setup of production equipment	200,000	Setup of production equipment	250,000

If interest is 8%, which alternative has the lower equivalent cost when the firm begins production at the end of Year 3?

SOLUTION

New Plant

$$\text{Future worth of cost (FW)} = 85,000(F/P, 8\%, 3) + 200,000(F/A, 8\%, 3)$$
$$+ 1,000,000(F/P, 8\%, 1) = \$1,836,000$$

Remodel Available Factory

$$\text{Future worth of cost (FW)} = 850{,}000(F/P, 8\%, 3) + 250{,}000(F/A, 8\%, 3)$$
$$= \$1{,}882{,}000$$

The total cost of remodeling the available factory ($1,600,000) is smaller than the total cost of a new plant ($1,685,000). However, the timing of the expenditures, is better with the new plant. The new plant is projected to have the smaller future worth of cost and thus is the preferred alternative.

BENEFIT–COST RATIO ANALYSIS

At a given minimum attractive rate of return (MARR), we would consider an alternative acceptable, provided

$$\text{PW of benefits} - \text{PW of costs} \geq 0 \qquad \text{or} \qquad \text{EUAB} - \text{EUAC} \geq 0$$

These could also be stated as a ratio of benefits to costs, or

$$\text{Benefit–cost ratio } \frac{B}{C} = \frac{\text{PW of benefit}}{\text{PW of costs}} = \frac{\text{EUAB}}{\text{EUAC}} \geq 1$$

Rather than using present worth or annual cash flow analysis to solve problems, we can base the calculations on the benefit–cost ratio, B/C. The criteria are presented in Table 9-1. In Table 9-1 the two special cases where maximizing the B/C ratio is correct are listed below the more common situation where incremental analysis is required. In Chapter 16 we will detail how this measure is applied in the public sector. Its use there is so pervasive, that the term *present worth index* is sometimes used to distinguish private sector applications.

TABLE 9-1 Benefit–Cost Ratio Analysis

	Situation	Criterion
Neither input nor output fixed	Neither amount of money or other inputs nor amount of benefits or other outputs are fixed	*Two alternatives:* Compute incremental benefit–cost ratio ($\Delta B/\Delta C$) on the increment of investment between the alternatives. If $\Delta B/\Delta C \geq 1$, choose higher-cost alternative; otherwise, choose lower-cost alternative. *Three or more alternatives:* Solve by benefit–cost ratio incremental analysis
Fixed input	Amount of money or other input resources are fixed	Maximize B/C
Fixed output	Fixed task, benefit, or other output to be accomplished	Maximize B/C

EXAMPLE 9-3

A firm is trying to decide which of two devices to install to reduce costs. Both devices have useful lives of 5 years and no salvage value. Device *A* costs $1000 and can be expected to result in $300 savings annually. Device *B* costs $1350 and will provide cost savings of $300 the first year, but savings will increase by $50 annually, making the second-year savings $350, the third-year savings $400, and so forth. With interest at 7%, which device should the firm purchase?

SOLUTION

We have used three types of analysis thus far to solve this problem: present worth in Example 5-1, annual cash flow in Example 6-5, and rate of return in Example 7-14. First we correctly analyze this incrementally, then we look at each device's benefit–cost ratio.

Incremental *B–A*

$$\text{PW of cost} = \$350$$

$$\text{PW of benefits} = 50(P/G, 7\%, 5)$$

$$= 50(7.647) = \$382$$

$$\frac{\text{B}}{\text{C}} = \frac{\text{PW of benefit}}{\text{PW of costs}} = \frac{382}{350} = 1.09$$

The increment is justified at the MARR of 7%. Device *B* should be purchased.

Device *A*

$$\text{PW of cost} = \$1000$$

$$\text{PW of benefits} = 300(P/A, 7\%, 5)$$

$$= 300(4.100) = \$1230$$

$$\frac{\text{B}}{\text{C}} = \frac{\text{PW of benefit}}{\text{PW of costs}} = \frac{1230}{1000} = 1.23$$

Device *B*

$$\text{PW of cost} = \$1350$$

$$\text{PW of benefit} = 300(P/A, 7\%, 5) + 50(P/G, 7\%, 5)$$

$$= 300(4.100) + 50(7.647) = 1230 + 382 = 1612$$

$$\frac{\text{B}}{\text{C}} = \frac{\text{PW of benefit}}{\text{PW of costs}} = \frac{1612}{1350} = 1.19$$

Maximizing the benefit–cost ratio indicates the wrong choice, Device *A*. Incremental analysis must be used.

EXAMPLE 9-4

In Example 7-15 we analyzed two machines that were being considered for purchase. Assuming 10% interest, which machine should be bought?

	Machine X	Machine Y
Initial cost	$200	$700
Uniform, annual benefit	95	120
End-of-useful-life salvage value	50	150
Useful life, in years	6	12

SOLUTION

Assuming a 12-year analysis period, the cash flow table is:

Year	Machine X	Machine Y
0	−$200	−$700
1–5	+95	+120
6	+95 / −200 / +50	+120
7–11	+95	+120
12	+95 / +50	+120 / +150

We will solve the problem using

$$\frac{B}{C} = \frac{EUAB}{EUAC}$$

and considering the salvage value of the machines to be reductions in cost, rather than increases in benefits. This choice affects the ratio value, but not the decision.

Machine X

$$EUAC = 200(A/P, 10\%, 6) - 50(A/F, 10\%, 6)$$

$$= 200(0.2296) - 50(0.1296) = 46 - 6 = \$40$$

$$EUAB = \$95$$

Note that this assumes the replacement for the last 6 years has identical costs. Under these circumstances, the EUAC for the first 6 years equals the EUAC for all 12 years.

Machine Y

$$EUAC = 700(A/P, 10\%, 12) - 150(A/F, 10\%, 12)$$

$$= 700(0.1468) - 150(0.0468) = 103 - 7 = \$96$$

$$EUAB = \$120$$

Machine Y − Machine X

$$\frac{\Delta B}{\Delta C} = \frac{120 - 95}{96 - 40} = \frac{25}{56} = 0.45$$

The incremental benefit–cost ratio of less than 1 represents an undesirable increment of investment. We therefore choose the lower-cost alternative—Machine X. If we had computed benefit–cost ratios for each machine, they would have been:

Machine X   **Machine Y**

$$\frac{B}{C} = \frac{95}{40} = 2.38 \qquad \frac{B}{C} = \frac{120}{96} = 1.25$$

Although $B/C = 1.25$ for Machine Y (the higher-cost alternative), we must not use this fact as the basis for selecting the more expensive alternative. It only indicates that Y would be acceptable if X were unavailable. The incremental benefit–cost ratio, $\Delta B/\Delta C$, clearly shows that Y is a less desirable alternative than X. Also, we must not jump to the conclusion that the best alternative is always the one with the largest B/C ratio. This, too, may lead to incorrect decisions—as we saw in Example 9-3, and we shall see again when we examine problems with three or more alternatives.

EXAMPLE 9-5

Consider the five mutually exclusive alternatives from Examples 8-6 and 8-7 plus an additional alternative, F. They have 20-year useful lives and no salvage value. If the minimum attractive rate of return is 6%, which alternative should be selected?

	A	B	C	D	E	F
Cost	$4000	$2000	$6000	$1000	$9000	$10,000
PW of benefit	7330	4700	8730	1340	9000	9,500
$\dfrac{B}{C} = \dfrac{\text{PW of benefits}}{\text{PW of cost}}$	1.83	2.35	1.46	1.34	1.00	0.95

SOLUTION

Incremental analysis is needed to solve the problem. The steps in the solution are the same as the ones presented in Example 8-7 for incremental rate of return, except here the criterion is $\Delta B/\Delta C$, and the cutoff is 1, rather than ΔIRR with a cutoff of MARR.

1. Be sure all the alternatives are identified.
2. (Optional) Compute the B/C ratio for each alternative. Since there are alternatives for which $B/C \geq 1$, we will discard any with $B/C < 1$. Discard Alt. F.

3. Arrange the remaining alternatives in ascending order of investment.

	D	B	A	C	E
Cost (= PW of cost)	$1000	$2000	$4000	$6000	$9000
PW of benefits	1340	4700	7330	8730	9000
B/C	1.34	2.35	1.83	1.46	1.00

	Increment $B - D$	Increment $A - B$	Increment $C - A$
ΔCost	$1000	$2000	$2000
ΔBenefit	3360	2630	1400
$\Delta B/\Delta C$	3.36	1.32	0.70

4. For each increment of investment, if $\Delta B/\Delta C \geq 1$ the increment is attractive. If $\Delta B/\Delta C < 1$ the increment of investment is not desirable. The increment $B - D$ is desirable, so B is preferred to D. The increment $A - B$ is desirable. Thus, Alt. A is preferred. Increment $C - A$ is not attractive since $\Delta B/\Delta C = 0.70$. Now we compare A and E:

	Increment $E - A$
ΔCost	$5000
ΔBenefit	1670
$\Delta B/\Delta C$	0.33

The increment is undesirable. We choose Alt. A as the best of the six alternatives. *Note:* The best alternative does not have the highest B/C ratio.

Benefit–cost ratio analysis may be graphically represented. Figure 9-1 is a graph of Example 9-5. We see that F has a B/C < 1 and can be discarded. Alternative D is the starting point for examining the separable increments of investment. The slope of line $B - D$ indicates a $\Delta B/\Delta C$ ratio of >1. This is also true for line $A - B$. Increment $C - A$ has a slope much flatter than B/C = 1, indicating an undesirable increment of investment. Alternative C is therefore discarded and A retained. Increment $E - A$ is similarly unattractive. Alternative A is therefore the best of the six alternatives.

FIGURE 9-1 Benefit–cost ratio graph of Example 9-5.

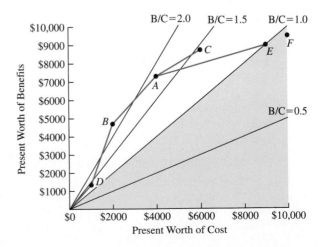

Note two additional things about Figure 9-1: first, even if alternatives with B/C ratio < 1 had not been initially excluded, they would have been systematically eliminated in the incremental analysis. Second, Alt. *B* had the highest B/C ratio (B/C = 2.35), but it is not the best of the six alternatives. We saw the same situation in rate of return analysis of three or more alternatives. The reason is the same in both analysis situations. We seek to maximize the *total* profit, not the profit rate.

Variations on the Theme of the Benefit–Cost Ratio

Two of the most common applications of the benefit–cost ratio typically use modified versions of the ratio. The basic ratio has been defined as placing *all* benefits in the numerator of the ratio and *all* costs in the denominator. Example 9-4 modified this slightly by considering the salvage values as reducing the costs rather than as increasing the benefits.

In the public sector, it is common to define the benefit–cost ratio so that the numerator includes all consequences to the users or the public and the denominator includes all consequences to the sponsor or government. For example, the numerator might include the positive benefits of improved highway traffic flow and the disbenefits of congestion during construction, since both accrue to the public or users. The denominator in this case includes consequences to the government, such as the costs of construction and the reduced maintenance cost for the new highway.

Example 9-6 illustrates this for a public project. Then in Example 9-7 the same numbers are put in a private context. Here the benefit–cost ratio is typically called a present worth index. The calculation is modified so that the denominator is the project's first cost, and all other consequences are placed in the numerator. This formulation of the benefit–cost ratio emphasizes the "bang for buck" of how much return for each dollar of investment.

We will examine the public sector application of the benefit–cost ratio in more detail in Chapter 16, and the present worth index will be used in Chapter 15. Here and in those chapters, the same standard applies for all versions of the ratio. Is the ratio ≥ 1? More importantly, if one version of the ratio is ≥ 1, then all versions are ≥ 1. As shown in Examples 9-6 and 9-7, the values of the ratios may differ, but whether they are above or below 1 and the recommended decisions do not change.

EXAMPLE 9-6

Traffic congestion on Riverview Boulevard has reached a point where something must be done. Two suggested plans have a life of 15 years, because that is the scheduled time for completion of the new Skyway Highway. After that time traffic will fall well below current levels.

Adding right-turn lanes at key intersections will cost $8.9M (million) with annual maintenance costs for signals and lane painting of $150,000. Added congestion during construction is a disbenefit of $900,000, but the reduced congestion after construction is an annual benefit of $1.6M. This benefit actually starts lower and increases over time, but for a simple initial analysis we are assuming a uniform annual benefit.

Adding a second left-turn lane at a few key intersections will cost an additional $3M with an added annual maintenance cost of $75,000. This construction is more disruptive, and the total disbenefit for congestion during construction is $2.1M. Upon completion, the total benefit for reduced congestion will be $2.2M annually.

Which alternatives is preferred if the interest rate is 10%? Analyze using a government B/C ratio (public in numerator and government in denominator).

SOLUTION

Since something *must* be done and we have only two identified alternatives, we could simply analyze the difference between the alternatives to see which is better. But we are going to start by analyzing the less expensive "right-turns" alternative to check that this is a reasonable choice for what *must* be done.

The user consequences include an annual benefit for reduced congestion and a first "cost" that is the disbenefit of increased congestion during construction.

$$PW_{\text{right turns}} = -900,000 + 1,600,000(P/A, 10\%, 15)$$
$$= -900,000 + 1,600,000(7.606) = \$11.27M$$

The government costs include a first cost for construction and annual maintenance costs. Note that these are calculated as present *costs*.

$$PC_{\text{right turns}} = 8,900,000 + 150,000(P/A, 10\%, 15)$$
$$= 8,900,000 + 150,000(7.606) = \$10.04M$$

The benefit–cost ratio for public divided by government consequences is:

$$B/C \text{ ratio} = \$11.27M/\$10.04M = 1.122$$

Thus, the right-turns-only alternative is better than doing nothing.

Now we evaluate the incremental investment for also doing the left-turn improvements. Because we are using a benefit–cost ratio, this evaluation must be done incrementally. The user consequences include an incremental annual benefit for reduced congestion and an incremental first "cost" that is the disbenefit of increased congestion during construction.

$$PW_{\text{left turns} - \text{right turns}} = -1,200,000 + 600,000(P/A, 10\%, 15)$$
$$= -1,200,000 + 600,000(7.606) = \$3.364M$$

The government costs include a first cost for construction and annual maintenance costs.

$$PC_{\text{left turns} - \text{right turns}} = 3,000,000 + 75,000(P/A, 10\%, 15)$$
$$= 3,000,000 + 75,000(7.606) = \$3.570M$$

The benefit–cost ratio for public divided by government consequences is:

$$B/C \text{ ratio} = \$3.364M/\$3.570M = 0.942$$

Thus, the right-turns-only alternative is better than adding the left-turn increment.

EXAMPLE 9-7

The industrial engineering department of Amalgamated Widgets is considering two alternatives for improving material flow in its factory. Both plans have a life of 15 years, because that is the estimated remaining life for the factory.

A minimal reconfiguration will cost \$8.9M (million) with annual maintenance costs of \$150,000. During construction there is a cost of \$900,000 for extra material movements and overtime, but more efficient movement of materials will save \$1.6M annually. The cost savings actually start lower and increase over time, but for a simple initial analysis we are assuming a uniform annual cost savings.

Reconfiguring a second part of the plant will cost an additional \$3M with an added annual maintenance cost of \$75,000. This construction is more disruptive, and the total cost for material movement and overtime congestion during construction is \$2.1M. Once complete the total cost savings for more efficient movement of materials is \$2.2M annually.

Which alternatives is preferred if the interest rate is 10%? Analyze using a present worth index (all consequences in Years 1 to n in numerator and all first costs in denominator).

SOLUTION

Since something *must* be done and we have only two identified alternatives, we could simply analyze the difference between the alternatives to see which is better. But we are going to start by analyzing the less expensive minimal reconfiguration alternative to check that this is a reasonable choice for what *must* be done.

The consequences in Years 1 to n include an annual cost savings for more efficient flow and annual maintenance costs.

$$PW_{\text{Years 1 to } n} = (1,600,000 - 150,000)(P/A, 10\%, 15)$$

$$= (1,600,000 - 150,000)(7.606) = \$11.03M$$

The first costs include a first cost for construction and the cost for disruption during construction.

$$PC = 8,900,000 + 900,000 = \$9.8M$$

The present worth index is:

$$PW \text{ index} = \$11.03M/\$9.8M = 1.125$$

Thus, the minimal reconfiguration is better than doing nothing.

Now we evaluate the incremental investment for also reconfiguring the second part of the plant. Because we are using a present worth index, this evaluation must be done incrementally. The annual consequences include an incremental annual cost savings and incremental maintenance costs.

$$PW_{\text{Years 1 to } n} = (600,000 - 75,000)(P/A, 10\%, 15)$$

$$= 525,000(7.606) = \$3.993M$$

There is a first cost for construction and for the associated disruption.

$$PC = 3,000,000 + 1,200,000 = \$4.2M$$

The present worth index is:

$$PW \text{ index} = \$3.993M/\$4.2M = 0.951$$

Thus, the minimal reconfiguration is better than also reconfiguring the second part of the plant.

In Examples 9-6 and 9-7, which numbers in the numerator and denominator were changed, and so did the exact values of the B/C ratio and present worth index. However, the conclusions did not. The ratios were above 1.0 for the minimal investment choice. The ratios were below 1.0 for the incremental investment. It was always best to make the minimal investment.

These examples demonstrate that present worth analysis and incremental benefit–cost ratio analysis lead to the same optimal decision. We saw in Chapter 8 that rate of return and present worth analysis led to identical decisions. Any of the exact analysis methods—present worth, annual cash flow, rate of return, or benefit–cost ratio—will lead to the same decision. Benefit–cost ratio analysis is extensively used in economic analysis at all levels of government.

PAYBACK PERIOD

Payback period is the period of time required for the profit or other benefits from an investment to equal the cost of the investment. This is the general definition for payback period. Other definitions consider depreciation of the investment, interest, and income taxes; they, too, are simply called "payback period." We will limit our discussion to the simplest form.

> **Payback period** is the period of time required for the project's profit or other benefits to equal the project's cost.

The criterion in all situations is to minimize the payback period. The computation of payback period is illustrated in Examples 9-8 and 9-9.

EXAMPLE 9-8

The cash flows for two alternatives are as follows:

Year	A	B
0	−$1000	−$2783
1	+200	+1200
2	+200	+1200
3	+1200	+1200
4	+1200	+1200
5	+1200	+1200

You may assume the benefits occur throughout the year rather than just at the end of the year. Based on payback period, which alternative should be selected?

SOLUTION

Alternative *A*

Payback period is how long it takes for the profit or other benefits to equal the cost of the investment. In the first 2 years, only $400 of the $1000 cost is recovered. The remaining $600 cost is recovered in the first half of Year 3. Thus the payback period for Alt. *A* is 2.5 years.

Alternative *B*

Since the annual benefits are uniform, the payback period is simply

$$\$2783/\$1200 \text{ per year} = 2.3 \text{ years}$$

To minimize the payback period, choose Alt. *B*.

EXAMPLE 9-9   (Example 5-4 revisited)

A firm is trying to decide which of two weighing scales it should install to check a package-filling operation in the plant. If both scales have a 6-year life, which one should be selected? Assume an 8% interest rate.

Alternative	Cost	Uniform Annual Benefit	End-of-Useful-Life Salvage Value
Atlas scale	$2000	$450	$100
Tom Thumb scale	3000	600	700

SOLUTION

Atlas Scale

$$\text{Payback period} = \frac{\text{Cost}}{\text{Uniform annual benefit}}$$

$$= \frac{2000}{450} = 4.4 \text{ years}$$

Tom Thumb Scale

$$\text{Payback period} = \frac{\text{Cost}}{\text{Uniform annual benefit}}$$

$$= \frac{3000}{600} = 5 \text{ years}$$

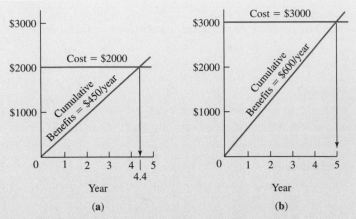

FIGURE 9-2 Payback period plots for Example 9-9: (**a**) Atlas scale and (**b**) Tom Thumb scale.

Figure 9-2 illustrates the situation. To minimize payback period, select the Atlas scale.

There are four important points to be understood about payback period calculations:

1. This is an approximate, rather than an exact, economic analysis calculation.
2. All costs and all profits, or savings of the investment before payback are included *without* considering differences in their timing.
3. All the economic consequences beyond the payback period are completely ignored.
4. Being an approximate calculation, payback period may or may not select the correct alternative.

This last point—that payback period may select the *wrong* alternative—was illustrated by Example 9-9. When payback period is used, the Atlas scale appears to be the more attractive alternative. Yet, when the same problem was solved earlier by the present worth method (Example 5-4), the Tom Thumb scale was the chosen alternative. The reason for the different conclusions is the $700 salvage value at the end of 6 years. The salvage value occurs after the payback period; so it was ignored in the payback calculation. It *was* considered in the present worth analysis, which correctly showed that the Tom Thumb scale was in fact more desirable.

But if payback period calculations are approximate and may lead to selecting the wrong alternative, why are they used? First, the calculations can be readily made by people unfamiliar with economic analysis. Second, payback period is easily understood. Earlier we pointed out that this is also an advantage to rate of return.

Moreover, payback period *does* measure how long it will take for the cost of the investment to be recovered from its benefits. Firms are often very interested in this time period: a rapid return of invested capital means that the funds can be reused sooner for other purposes. But one must not confuse the *speed* of the return of the investment, as measured by the payback period, with economic *efficiency*. They are two distinctly separate concepts. The former emphasizes the quickness with which invested funds return to a firm; the latter considers the overall profitability of the investment.

Example 9-10 illustrates how using the payback period criterion may result in an unwise decision.

EXAMPLE 9-10

A firm is buying production equipment for a new plant. Two alternative machines are being considered for a particular operation.

	Tempo Machine	**Dura Machine**
Installed cost	$30,000	$35,000
Net annual benefit after all annual expenses have been deducted	12,000 the first year, *declining* $3000 per year thereafter	1000 the first year, increasing $3000 per year thereafter
Useful life, in years	4	8

Neither machine has any salvage value. Compute the payback period for each machine.

SOLUTION BASED ON PAYBACK PERIOD

FIGURE 9-3 Payback period plots for Example 9-10: (**a**) Tempo machine and (**b**) Dura machine.

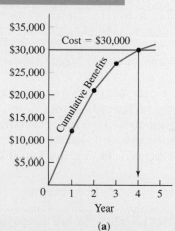
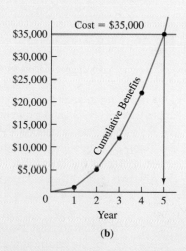

The Tempo machine has a declining annual benefit, while the Dura has an increasing annual benefit. Figure 9-3 shows the Tempo has a 4-year payback period and the Dura has a 5-year payback period. To minimize the payback period, the Tempo is selected.

Now, as a check on the payback period analysis, compute the rate of return for each alternative. Assume the minimum attractive rate of return is 10%.

SOLUTION BASED ON RATE OF RETURN

The cash flows for the two alternatives are as follows:

Year	Tempo Machine	Dura Machine
0	−$30,000	−$35,000
1	+12,000	+1,000
2	+9,000	+4,000

3	+6,000	+7,000
4	+3,000	+10,000
5	0	+13,000
6	0	+16,000
7	0	+19,000
8	0	+22,000
	0	+57,000

Tempo Machine

Since the sum of the cash flows for the Tempo machine is zero, we see immediately that the $30,000 investment just equals the subsequent benefits. The resulting rate of return is 0%.

Dura Machine

$$PW(i) = -35,000 + 1000(P/A, i, 8) + 3000(P/G, i, 8)$$

$$PW(20\%) = -35,000 + 1000(3.837) + 3000(9.883) = -1514$$

The 20% interest rate has discounted future benefits too much; it is too high. Try $i = 15\%$:

$$PW(15\%) = -35,000 + 1000(4.487) + 3000(12.481) = 6930$$

This time, the interest rate is too low. Linear interpolation would show that the rate of return is approximately 19%; and a spreadsheet gives the exact answer.

Using an exact calculation—rate of return—it is clear that the Dura machine is far superior to the Tempo. On the other hand, the shorter payback period for the Tempo does measure the speed of the return of the investment. The conclusion to be drawn is that **liquidity** and **profitability** may be two quite different criteria.

From the discussion and the examples, we see that payback period can measure the speed of the return of the investment. This might be quite important, for example, for a company that is short of working capital or for a firm in an industry experiencing rapid changes in technology. Calculation of payback period alone, however, must not be confused with a careful economic analysis. Ignoring all cash flows after the payback period is seldom wise. We have shown that a short payback period does not always mean that the associated investment is desirable. Thus, payback period is not a suitable replacement for accurate economic analysis calculations.

SENSITIVITY AND BREAKEVEN ANALYSIS

Since many data gathered in solving a problem represent *projections* of future consequences, there may be considerable uncertainty regarding the data's accuracy. Since the goal is to make good decisions, an appropriate question is: To what extent do variations in the data affect my decision? When small variations in a particular estimate would change which alternative is selected, the decision is said to be **sensitive to the estimate.** To better evaluate the impact of any particular estimate, we compute "how much a particular estimate would need to change in order to change a particular decision." This is called **sensitivity analysis.**

An analysis of the sensitivity of a problem's decision to its various parameters highlights the important aspects of that problem. For example, estimated annual maintenance and salvage values may vary substantially. Sensitivity analysis might indicate that a certain decision is insensitive to the salvage-value estimate over the full range of possible values. But, at the same time, we might find that the decision is sensitive to changes in the annual maintenance estimate. Under these circumstances, one should place greater emphasis on improving the annual maintenance estimate and less on the salvage-value estimate.

As indicated at the beginning of this chapter, breakeven analysis is a form of sensitivity analysis that is often presented as a **breakeven chart**. Another nomenclature that is some-times used for the breakeven point is *point of indifference*. One application of these tools is **stage construction.** Should a facility be constructed now to meet its future full-scale requirement, Or should it be constructed in stages as the need for the increased capacity arises? Three examples are:

- Should we install a cable with 400 circuits now or a 200-circuit cable now and another 200-circuit cable later?
- A 10 cm water main is needed to serve a new area of homes. Should it be installed now, or should a 15 cm main be installed to ensure an adequate water supply to adjoining areas later, when other homes have been built?
- An industrial firm needs a new warehouse now and estimates that it will need to double its size in 4 years. The firm could have a warehouse built now and later enlarged, or have the warehouse with capacity for expanded operations built right away.

Examples 9-11 and 9-13 illustrate sensitivity and breakeven analysis.

EXAMPLE 9-11

Consider a project that may be constructed to full capacity now or may be constructed in two stages.

Construction Alternative	Costs
Two-stage construction	
Construct first stage now	$100,000
Construct second stage	120,000
n years from now	
Full-capacity construction	
Construct full capacity now	140,000

Other Factors

1. All facilities will last for 40 years regardless of when they are installed; after 40 years, they will have zero salvage value.

2. The annual cost of operation and maintenance is the same for both two-stage construction and full-capacity construction.

3. Assume an 8% interest rate.

Plot "age when second stage is constructed" versus "costs for both alternatives." Mark the breakeven point on your graph. What is the sensitivity of the decision to second-stage construction 16 or more years in the future?

SOLUTION

Since we are dealing with a common analysis period, the calculations may be either annual cost or present worth. Present worth calculations appear simpler and are used here.

Construct Full Capacity Now

$$\text{PW of cost} = \$140,000$$

Two-Stage Construction

If the first stage is to be constructed now and the second stage n years hence, compute the PW of cost for several values of n (years).

$$\text{PW of cost} = 100,000 + 120,000(P/F, 8\%, n)$$

$$n = 5 \quad \text{PW} = 100,000 + 120,000(0.6806) = \$181,700$$
$$n = 10 \quad \text{PW} = 100,000 + 120,000(0.4632) = \ \ 155,600$$
$$n = 20 \quad \text{PW} = 100,000 + 120,000(0.2145) = \ \ 125,700$$
$$n = 30 \quad \text{PW} = 100,000 + 120,000(0.0994) = \ \ 111,900$$

These data are plotted in the form of a breakeven chart in Figure 9-4.

FIGURE 9-4 Breakeven chart for Example 9-11.

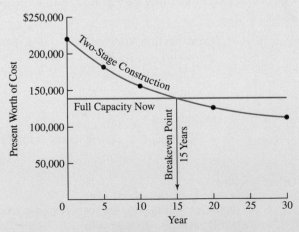

In Figure 9-4 we see that the PW of cost for two-stage construction naturally decreases as the second stage is deferred. The one-stage construction (full capacity now) is unaffected by the x-axis variable and, hence, is a horizontal line.

The breakeven point on the graph is where both alternatives have the same PW. We see that if the second stage is deferred for 15 years, then the PW of cost of two-stage construction is equal to one-stage construction; Year 15 is the breakeven point. The graph also shows that if the second stage were to be needed prior to Year 15, then one-stage construction, with its smaller PW of cost, would be preferred. On the other hand, if the second stage would not be required until after 15 years, two-stage construction is preferred.

This breakeven point can also be calculated by setting the two alternatives equal to each other.

$$PW = 140{,}000 = 100{,}000 + 120{,}000(P/F, 8\%, n)$$

$$(P/F, 8\%, n) = \frac{40{,}000}{120{,}000} = 0.3333$$

From the tables

$$n = 14 + (15 - 14)(0.3405 - 0.3333)/(0.3405 - 0.3152)$$

$$n = 14.3 \text{ years}$$

From Excel's NPER function or GOAL SEEK, we find that the answer is 14.27 years.

The decision on how to construct the project is sensitive to the age at which the second stage is needed *only* if the range of estimates includes 15 years. For example, if one estimated that the second-stage capacity would be needed between 5 and 10 years hence, the decision is insensitive to that estimate. For any value within that range, the decision does not change. But, if the second-stage capacity were to be needed sometime between, say, 12 and 18 years, the decision would be sensitive to the estimate of when the full capacity would be needed.

One question posed by this example is *how* sensitive the decision is to the need for the second stage at 16 years or beyond. The graph shows that the decision is insensitive. In all cases for construction on or after 16 years, two-stage construction has a lower PW of cost.

EXAMPLE 9-12

Example 8-3 posed the following situation. Three mutually exclusive alternatives are given, each with a 20-year life and no salvage value. The minimum attractive rate of return is 6%.

	A	B	C
Initial cost	$2000	$4000	$5000
Uniform annual benefit	410	639	700

In Example 8-3 we found that Alt. *B* was the preferred alternative at 6%. Here we would like to know how sensitive the decision is to our estimate of the initial cost of *B*. If *B* is preferred at an initial cost of $4000, it will continue to be preferred at any smaller initial cost. But *how much* higher than $4000 can the initial cost be and still have *B* the preferred alternative? With neither input nor output fixed, maximizing net present worth is a suitable criterion.

Alternative A

$$\text{NPW}_A = 410(P/A, 6\%, 20) - 2000$$
$$= 410(11.470) - 2000 = \$2703$$

Alternative B

Let $x = $ initial cost of B.

$$\text{NPW}_B = 639(P/A, 6\%, 20) - x$$
$$= 639(11.470) - x$$
$$= 7329 - x$$

Alternative C

$$\text{NPW}_C = 700(P/A, 6\%, 20) - 5000$$
$$= 700(11.470) - 5000 = \$3029$$

For the three alternatives, we see that B will only maximize NPW as long as its NPW is greater than 3029.

$$3029 = 7329 - x$$
$$x = 7329 - 3029 = \$4300$$

Therefore, B is the preferred alternative if its initial cost does not exceed \$4300.

Figure 9-5 is a breakeven chart for the three alternatives. Here the criterion is to maximize NPW; as a result, the graph shows that B is preferred if its initial cost is less than \$4300. At an initial cost above \$4300, C is preferred. We have a breakeven point at \$4300. When B has an initial cost of \$4300, B and C are equally desirable.

FIGURE 9-5 Breakeven chart for Example 9-12.

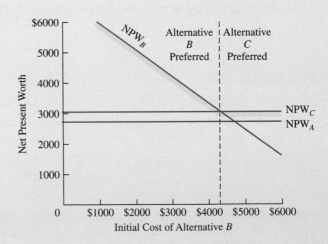

EXAMPLE 9-13

In both Examples 9-6 (traffic congestion on Riverview Boulevard) and 9-7 (reconfiguring the plant of Amalgamated Widgets), the life of 15 years is clearly subject to some uncertainty. While holding the other data constant, analyze the sensitivity of the recommended decisions to the project life. Use the present worth measure since it is the same for both examples.

SOLUTION

The two alternatives have the following present worth values.

$$PW_{\text{right turns or minimal}} = -900{,}000 - 8{,}900{,}000 + (1{,}600{,}000 - 150{,}000)(P/A, 10\%, n)$$
$$= -9{,}800{,}000 + 1{,}450{,}000(P/A, 10\%, n)$$

$$PW_{\text{left turns or 2nd part of plant}} = -2{,}100{,}000 - 8{,}900{,}000 - 3{,}000{,}000$$
$$+ (2{,}200{,}000 - 225{,}000)(P/A, 10\%, n)$$
$$= \;\; 14{,}000{,}000 + 1{,}975{,}000(P/A, 10\%, n)$$

These could be analyzed for breakeven values of n. However, it is easier to use the graphing technique for multiple alternatives that was presented in Chapter 8. Instead of using the interest rate for the x axis, use n.

As shown in Figure 9-6, the right-turn or minimal alternative is the best one for lives of 12 to 16 years. The left-turn increment or 2nd part of plant, is the best choice for lives of 17 or more years. If the life is 11 years or less, doing nothing is better. To keep the graph readable, Figure 9-6 includes only years 10 through 20.

FIGURE 9-6 Breakeven chart for Example 9-13.

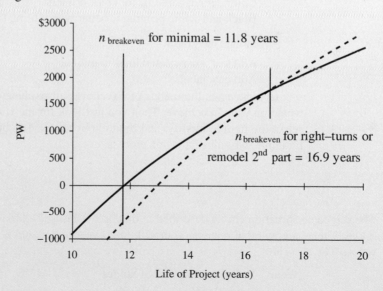

Breakeven points can be estimated from graphs or calculated with formulas or GOAL SEEK. Sensitivity analysis and breakeven point calculations can be very useful in identifying how different estimates affect the calculations. It must be recognized that these calculations assume that all parameters except one are held constant, and the sensitivity of the decision to that one variable is evaluated. The next section presents ways to modify your Excel chart to make it more effective.

GRAPHING WITH SPREADSHEETS FOR SENSITIVITY AND BREAKEVEN ANALYSIS

Chapter 4 introduced drawing xy plots with spreadsheets, and Chapter 7 reviewed this procedure for plotting present worth versus i. The Chapter 7 plot (Figure 7-6) is an example of breakeven analysis, because it is used to determine at what interest rate does the project break even or have a present worth of 0. This section will present some of the spreadsheet tools and options that can make the xy plots more effective and attractive.

The spreadsheet tools and options can be used to:

- Modify the x or y axes

  Specify the minimum or maximum value
  Specify at what value the other axis intersects (default is 0)

- Match line types to data

  Use line types to distinguish one curve from another
  Use markers to show real data
  Use lines without markers to plot curves (straight segments or smooth curves)

- Match chart colors to how displayed

  Color defaults are fine for color computer screen
  Color defaults are OK for color printers
  Black-and-white printing is better with editing (use line types not colors)

- Annotate the graph

  Add text, arrows, and lines to graphs
  Add data labels

In most cases the menus of Excel are self-explanatory, so the main step is deciding what you want to achieve. Then you just look for the way to do it. Left clicks are used to select the item to modify, and right clicks are used to bring up the options for that item. Example 9-14 illustrates this process.

EXAMPLE 9-14

The staged construction choice described in Example 9-11 used a broad range of x values for the x axis. Create a graph that focuses on the 10- to 20-year period and is designed for printing in a report. The costs are:

Year	Full Capacity	Two Stages
0	$140,000	$100,000
n	0	120,000

SOLUTION

The first step is to create a table of values that shows the present worth of the costs for different values of n — the length of time until the second stage or full capacity is needed. Notice that the full capacity is calculated at $n = 0$. The only reason to calculate the corresponding value for staged construction is to see if the formula is properly entered, since building both stages at the same time will not really cost $220,000. The values for staged construction at 5, 10, 20, and 30 years check with the values in Example 9-11.

The next step is to select cells A8:C13, which includes the x values and two series of y values. Then the ChartWizard tool is selected. In the first step, the xy (scatter) plot is selected with the option of smoothed lines without markers. In Step 2 no action is required, since the cells A8:C13 were selected first. In Step 3 labels are added for the x and y axes. In Step 4 the chart is moved around on the worksheet page, so that it does not overlap with the data. The result is shown in Figure 9-7 (except that this is the color screen version printed to a black-and-white printer).

Our first step in cleaning up the graph is to delete the formula in cell C9, since two-stage construction will not be done at Time 0. We also delete the label in the adjacent cell, which explains the formula. Then we create a new label for cell C10. As shown in Appendix A, the easy way to create that label is to insert an apostrophe, or space, as the first entry in cell C10. This converts the formula to a label which we can copy to D10. Then we delete the apostrophe in cell C10.

FIGURE 9-7 Automatic graph from spreadsheet.

	A	B	C	D
1	Time	Full Capacity	Two Stage	
2	0	140,000	100,000	
3	n		120,000	
4				
5	Life	40		
6	i	8%		
7			Present Worth of	
8	n	Full Capacity	Two Stage	
9	0	140,000	220,000	=$C$2+PV($B$6,A9,0,−$C$3)
10	5	140,000	181,670	
11	10	140,000	155,583	
12	20	140,000	125,746	
13	30	140,000	111,925	

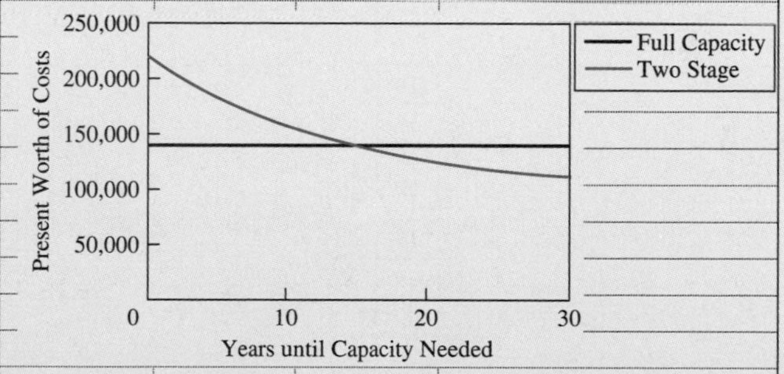

The axis scales must be modified to focus on the area of concern. Select the x axis and change the minimum from automatic to 10 and the maximum to 20. Select the y axis and change the minimum to 125,000 and the maximum to 160,000.

Left-click on the plot area to select it. Then right-click to bring up the options. Select Format Plot Area and change the area pattern to "none." This will eliminate the gray fill that made Figure 9-7 difficult to read.

Left-click on the two-stage curve to select it. Then right-click for the options. Format the data series using the "patterns" tab. Change the line style from solid to dashed, the line color from automatic to black, and increase the line weight. Similarly, increase the line weight for the full-capacity line. Finally, select a grid line and change the line style to dotted. The result is far easier to read in black and white.

To further improve the graph, we can replace the legend with annotations on the graph. Left-click somewhere in the white area around the graph to select "chart area." Right-click and then choose the chart options on the menu. The legends tab will let us delete the legend by turning "show legend" off. Similarly, we can turn the x-axis gridlines on. The line style for these gridlines should be changed to match the y-axis gridlines. This allows us to see that the breakeven time is between 14 and 15 years.

To make the graph less busy, change the scale on the x axis so that the interval is 5 years rather than automatic. Also eliminate the gridlines for the y axis (by selecting the chart area, chart

FIGURE 9-8 Spreadsheet of Figure 9-7 with improved graph.

	A	B	C	D
1	Time	Full Capacity	Two Stage	
2	0	140,000	100,000	
3	n		120,000	
4				
5	life	40		
6	i	8%		
7		Present worth of		
8	n	Full Capacity	Two Stage	
9	0	140,000		
10	5	140,000	181,670	=$C$2+PV($B$6,A10,0,−$C$3)
11	10	140,000	155,583	
12	20	140,000	125,746	
13	30	140,000	111,925	
14				

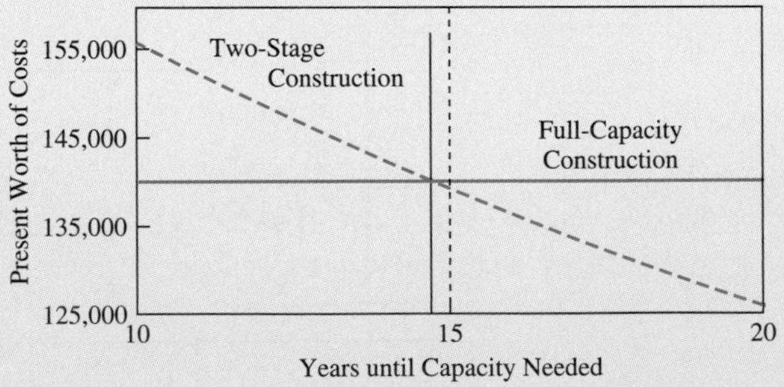

options, and gridlines tabs). The graph size can be increased for easier reading, as well. This may require specifying an interval of 10,000 for the scale of the *y* axis.

Finally to add the labels for the full-capacity curve and the two-stage curve, find the toolbar for graphics, which is open when the chart is selected (probably along the bottom of the spreadsheet). Select the text box icon, and click on a location close to the two-stage chart. Type in the label for two-stage construction. Notice how including a return and a few spaces can shape the label to fit the slanted line. Add the label for full construction. Figure 9-8 is the result.

DOING WHAT-IF ANALYSIS WITH SPREADSHEETS

Breakeven charts change one variable at a time, while what-if analysis may change many of the variables in a problem. However, spreadsheets remain a very powerful tool for this form of sensitivity analysis. In Example 9-15 a project appears to be very promising. However, what-if analysis indicates that a believable scenario raises some questions about whether the project should be done.

EXAMPLE 9-15

You are an assistant to the vice president for manufacturing. The staff at one of the plants has recommended approval for a new product with a new assembly line to produce it. The VP believes that the numbers presented are too optimistic, and she has added a set of adjustments to the original estimates. Analyze the project's benefit–cost ratio or present worth index as originally submitted. Reanalyze the project, asking "What if the VP's adjustments are correct?"

	Initial Estimate	Adjustment
First cost	$70,000	+10%
Units/year	1,200	−20%
Net unit revenue	$ 25	−15%
Life, in years	8	−3
Interest rate	12%	none

SOLUTION

Figure 9-9 shows that the project has a 2.13 benefit–cost ratio with the initial estimates, but only a value of 0.96 with the what-if adjustments. Thus we need to determine which set of numbers is more realistic. Real-world experience suggests that in many organizations the initial estimates are too optimistic. Auditing of past projects is the best way to develop adjustments for future projects.

FIGURE 9-9 Spreadsheet for what-if analysis.

	A	B	C	D
1		Initial Estimate	Adjust-ment	Adjusted Values
2	First cost	$70,000	10%	$77,000
3	Units/year	1,200	−20%	960
4	Net unit revenue	$25	−15%	$21
5	Life (years)	8	−3	5
6	Interest rate	12%	none	12%
7				
8	Benefits	149,029		73,537
9	Cost	70,000		77,000
10	B/C Ratio	2.13		0.96
11				
12		=PV(B6,B5-B3*B4)		

SUMMARY

In this chapter, we have looked at four new analysis techniques.

Future worth: When the comparison between alternatives will be made in the future, the calculation is called future worth. This is very similar to present worth, which is based on the present, rather than a future point in time.

Benefit–cost ratio analysis: This technique is based on the ratio of benefits to costs using either present worth or annual cash flow calculations. The method is graphically similar to present worth analysis. When neither input nor output is fixed, incremental benefit–cost ratios ($\Delta B/\Delta C$) are required. The method is similar in this respect to rate of return analysis. Benefit–cost ratio analysis is often used at the various levels of government.

Payback period: Here we define payback as the period of time required for the profit or other benefits of an investment to equal the cost of the investment. Although simple to use and understand, payback is a poor analysis technique for ranking alternatives. While it provides a measure of the speed of the return of the investment, it is not an accurate measure of the profitability of an investment.

Sensitivity, breakeven, and what-if analysis: These techniques are used to see how sensitive a decision is to estimates for the various parameters. Breakeven analysis is done to locate conditions under which the alternatives are equivalent. This is often presented in the form of breakeven charts. Sensitivity analysis is an examination of a range of values for some parameters to determine their effect on a particular decision. What-if analysis changes one to many estimates to see what the result is.

PROBLEMS

P. 284

Future Worth

9-1 Compute the future worth for the following cash flows.

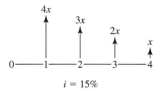

$i = 12\%$

(*Answer: F* $= \$1199$)

9-2 For the following cash flows, compute the future worth.

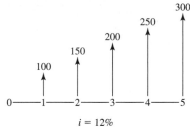

$i = 15\%$

9-3 For the following cash flows, compute the future worth.

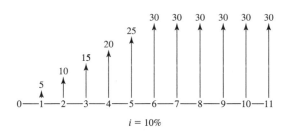

$i = 10\%$

9-4 For a 12% interest rate, compute the value of *F* so the following cash flows have a future worth of 0.

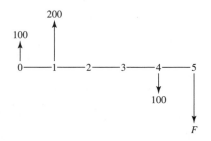

9-5 Compute *F* so the following cash flows have a future worth of 0.

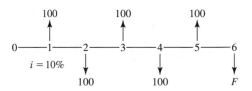

9-6 Calculate the present worth and the future worth of a series of 10 annual cash flows with the first cash flow equal to $15,000 and each successive cash flow increasing by $1200. The interest rate is 12%.

9-7 Sally deposited $100 a month in her savings account for 24 months. For the next 5 years she made no deposits. What is the future worth in Sally's savings account at the end of the 7 years, if the account earned 6% annual interest, compounded monthly? (*Answer:* $3430.78)

9-8 A 20-year-old student decided to set aside $100 on his 21st birthday for investment. Each subsequent year through his 55th birthday, he plans to increase the investment on a $100 arithmetic gradient. He will not set aside additional money after his 55th birthday. If the student can achieve a 12% rate of return, what is the future worth of the investments on his 65th birthday? (*Answer:* $1,160,700)

9-9 You can buy a piece of vacant land for $30,000 cash. You plan to hold it for 15 years and then sell it at a profit. During this period, you would pay annual property taxes of $600. You would have no income from the property. Assuming that you want a 10% rate of return, at what net price would you have to sell the land 15 years hence? (*Answer:* $144,373)

9-10 An individual who makes $32,000 per year anticipates retiring in 30 years. If his salary is increased by $600 each year and he deposits 10% of his yearly salary into a fund that earns 7% interest, what is the future worth at retirement?

9-11 Stamp collecting has become an increasingly popular—and expensive—hobby. One favorite method is to save plate blocks (usually four stamps with the printing plate number in the margin) of each new stamp as it is issued. With rising postage rates and increased numbers of new stamps being issued, this collecting plan costs more each year.

Stamps have been a good place to invest money over the last 10 years, as the demand for stamps previously issued has caused resale prices to increase 18% each year. Suppose a collector purchased $100 worth of stamps 10 years ago, and increased his purchases by $50 per year in each subsequent year. After 10 years of stamp collecting, what is the future worth of the stamp collection?

9-12 The interest rate is 16% per year and there are 48 compounding periods per year. The principal is $50,000. What is the future worth in 5 years?

9-13 In the early 1980s, planners were examining alternate sites for a new London airport. The economic analysis included the value of structures that would need to be removed from various airport sites. At one potential site, the twelfth-century Norman church of St. Michaels, in the village of Stewkley, would have had to be demolished. The planners used the value of the fire insurance policy on the church—a few thousand pounds sterling—as the church's value.

An outraged antiquarian wrote to the London *Times* that an equally plausible computation would be to assume that the original cost of the church (estimated at 100 pounds sterling) be increased at the rate of 10% per year for 800 years. Based on his proposal, what would be the future worth of St. Michaels? (*Note:* There was great public objection to tearing down the church, and it was spared.)

9-14 Bill made a budget and planned to deposit $150 a month in a savings account, beginning September 1. He did this, but on the following January 1, he reduced the monthly deposits to $100. He made 18 deposits, four at $150 and 14 at $100. If the savings account paid 6% interest, compounded monthly, what was the future worth of his savings immediately after the last deposit?   (*Answer:* $2094.42)

9-15 A company deposits $1000 in a bank at the beginning of each year for 6 years. The account earns 8% interest, compounded every 6 months. What is in the account at the end of 6 years?

9-16 Don Ball is a 55-year-old engineer. According to mortality tables, a male at age 55 has an average life expectancy of 21 more years. Don has accumulated $48,500 toward his retirement. He is now adding $5000 per year to his retirement fund. The fund earns 12% interest. Don will retire when he can obtain an annual income from his retirement fund of $20,000, assuming he lives to age 76. He will make no provision for a retirement income after age 76. What is the youngest age at which Don can retire?

9-17 Jean invests $100 in Year 1 and doubles the amount each year after that (so the investment is $100, 200, 400, 800, ...). If she does this for 10 years, and the investment pays 10% annual interest, what is the future worth of her investment?

9-18 If you invested $2500 in a 24-month bank certificate of deposit (CD) paying 8.65%, compounded monthly, what is the future worth of the CD when it matures?

9-19 After receiving an inheritance of $25,000 on her 21st birthday, Ayn deposited the inheritance in a savings account with an effective annual interest rate of 6%. She decided to make regular deposits, beginning with $1000 on her 22nd birthday and increasing by $200 each year (i.e., $1200 on her 23rd birthday, $1400 on her 24th birthday, etc.). What was the future worth of Ayn's deposits on her 56th birthday?

9-20 The Association of General Contractors (AGC) is endowing a fund of $1 million for the Construction Engineering Technology Program at Grambling State University. The AGC established an escrow account in which 10 equal end-of-year deposits that earn 7% compound interest were to be made. After seven deposits, the Louisiana legislature revised laws relating to the licensing fees AGC can charge its members, and there was no deposit at the end of Year 8. What must the amount of the remaining equal end-of-year deposits be, to ensure that the $1 million is available on schedule for the Construction Engineering Technology Program?

9-21 A new engineer is considering investing in an individual retirement account (IRA) with a mutual fund that has an average annual return of 10%. What is the future worth of her IRA at age 65 if she makes annual investments of $2000 into the fund beginning on her 25th birthday? Assume that the fund continues to earn an annual return of 10%.

9-22 IPS Corp. will upgrade its package-labeling machinery. It costs $150,000 to buy the machinery and have it installed. Operation and maintenance costs, which are $1500 per year for the first 3 years, increase by 500 per year for the machine's 10-year life. The machinery has a salvage value of 5% of its initial cost. Interest is 10%. What is the future worth of cost of the machinery?

9-23 A company is considering buying a new bottle-capping machine. The initial cost of the machine is $325,000 and it has a 10-year life. Monthly maintenance costs are expected to be $1200 per month for the first 7 years and $2000 per month for the remaining years. The machine requires a major overhaul costing $55,000 at the end of the fifth year of service. Assume that all these costs occur at the end of the appropriate period. What is the future value of all the costs of owning and operating this machine if the nominal interest rate is 7.2%?

9-24 A family starts an education fund for their son Patrick when he is 8 years old, investing $150 on his eighth birthday, and increasing the yearly investment by $150 per year until Patrick is 18 years old. The fund pays 9% annual interest. What is the fund's future worth when Patrick is 18?

9-25 A bank account pays 19.2% interest with monthly compounding. A series of deposits started with a deposit of $5000 on January 1, 2007. Deposits in the series were to occur each 6 months. Each deposit in the series is for $150 less than the one before it. The last deposit in the series will be due on January 1, 2022. What is the future worth of the account on July 1, 2024, if the balance was zero before the first deposit and no withdrawals are made?

9-26 A recent college graduate got a good job and began a savings account. He authorized the bank to automatically transfer $75 each month from his checking account to the savings account. The bank made the first withdrawal on July 1, 2007 and is instructed to make the last withdrawal on January 1, 2025. The bank pays a nominal interest rate of 4.5% and compounds twice a month. What is the future worth of the account on January 1, 2025?

9-27 Bob, an engineer, decided to start a college fund for his son. Bob will deposit a series of equal, semiannual cash flows with each deposit equal to $1500. Bob made the first deposit on July 1, 2008 and will make the last deposit on July 1, 2028. Joe, a friend of Bob's, received an inheritance on April 1, 2013, and has decided to begin a college fund for his daughter. Joe wants to send his daughter to the same college as Bob's son. Therefore, Joe needs to accumulate the same amount of money on July 1, 2028, as Bob will have accumulated from his semiannual deposits. Joe never took engineering economics and had no idea how to determine the amount that should be deposited. He decided to deposit $40,000 on July 1, 2013. Will Joe's deposit be sufficient? If not, how much should he have put in? Use a nominal interest of 7% with semiannual compounding on all accounts.

9-28 A business executive is offered a management job at Generous Electric Company, which offers him a 5-year contract that calls for a salary of $62,000 per year, plus 600 shares of GE stock at the end of the 5 years. This executive is currently employed by Fearless Bus Company, which also has offered him a 5-year contract. It calls for a salary of $65,000, plus 100 shares of Fearless stock each year. The Fearless stock is currently worth $60 per share and pays an annual dividend of $2 per share. Assume end-of-year payments of salary and stock. Stock dividends begin one year after the stock is received. The executive believes that the value of the stock and the dividend will remain constant. If the executive considers 9% a suitable rate of return in this situation, what must the Generous Electric stock be worth per share to make the two offers equally attractive? Use the future worth analysis method in your comparison. (*Answer:* $83.76)

9-29 Pick a discretionary expense that you incur on a regular basis, such as buying cigarettes weekly, buying fashion items monthly, buying sports tickets monthly, or going to movies weekly. Assume that you instead place the money in an investment account that earns 9% annually. After 40 years, how much is in the account?

Benefit–Cost Ratio *p. 286*

9-30 Each of the three alternatives shown has a 5-year useful life. If the MARR is 10%, which alternative should be selected? Solve the problem by benefit–cost ratio analysis.

	A	B	C
Cost	$600.0	$500.0	$200.0
Uniform annual benefit	158.3	138.7	58.3

(*Answer: B*)

9-31 Consider three alternatives, each with a 10-year useful life. If the MARR is 10%, which alternative should be selected? Solve the problem by benefit–cost ratio analysis.

	A	B	C
Cost	$800	$300	$150
Uniform annual benefit	142	60	33.5

9-32 An investor is considering buying some land for $100,000 and constructing an office building on it. Three different buildings are being analyzed.

	Building Height		
	2 Stories	**5 Stories**	**10 Stories**
Cost of building (excluding cost of land)	$400,000	$800,000	$2,100,000
Resale value* of land + building after 20-year horizon	200,000	300,000	400,000
Annual net rental income	70,000	105,000	256,000

*Resale value to be considered a reduction in cost, rather than a benefit.

Using benefit–cost ratio analysis and an 8% MARR, determine which alternative, if any, should be selected.

9-33   Using benefit–cost ratio analysis, determine which one of the three mutually exclusive alternatives should be selected. Each alternative has a 6-year useful life. Assume a 10% MARR.

	A	**B**	**C**
First cost	$560	$340	$120
Uniform annual benefit	140	100	40
Salvage value	40	0	0

9-34   Consider four alternatives, each of which has an 8-year useful life:

	A	**B**	**C**	**D**
Cost	$100.0	$80.0	$60.0	$50.0
Uniform annual benefit	12.2	12.0	9.7	12.2
Salvage value	75.0	50.0	50.0	0

If the MARR is 8%, which alternative should be selected? Solve the problem by benefit–cost ratio analysis.

9-35   Using benefit–cost ratio analysis, a 5-year useful life, and a 15% MARR, determine which of the following alternatives should be selected.

	A	**B**	**C**	**D**	**E**
Cost	$100	$200	$300	$400	$500
Uniform annual benefit	37	69	83	126	150

9-36   Five mutually exclusive investment alternatives have been proposed. Based on benefit–cost ratio analysis, and a MARR of 15%, which alternative should be selected?

Year	A	B	C	D	E
0	−$200	−$100	−$125	−$150	−$225
1–5	+68	+25	+42	+52	+68

9-37   A project will cost $50,000. The benefits at the end of the first year are estimated to be $10,000, increasing at a 10% uniform rate in subsequent years. Using an 8-year analysis period and a 10% interest rate, compute the benefit–cost ratio.

9-38   A do-nothing and two mutually exclusive alternatives are being considered for reducing traffic congestion. User benefits come from reduced congestion once the project is complete, while user disbenefits are due to increased congestion during construction. The interest rate is 9%, and the life of each alternative is 15 years. Which alternative should be chosen?

	A	**B**
User benefits ($M/yr)	2.1	2.6
User disbenefits ($M)	1.2	2.1
First cost ($M)	6.9	9.9
Operations and maintenance ($M/yr)	0.75	0.825

(a) Use the benefit–cost ratio.

(b) Use the modified benefit–cost ratio.

(c) Use the public/government version of the B/C ratio.

(d) Assume these numbers apply to a private firm and use a present worth index.

(e) Are your recommendations for (a) through (d) consistent? Which measure gives the largest value? Why?

9-39   A school is overcrowded and there are three options. The do-nothing alternative corresponds to continuing to use modular classrooms. The school can be expanded, or a new school can be built to "split the load" between the schools. User benefits come from improvements in school performance for the expanded or new schools. If a new school is built, there are more benefits because more students will be able to walk to school, the average distance for those who ride the school buses will be shorter, and the schools will be smaller and more "student friendly." The disbenefits for the expanded school are due to the impact of the construction process during the school year. The interest rate is 8%, and the life of each alternative is 20 years. Which alternative should be chosen? What is the incremental ratio for the preferred alternative?

	A	**B**
User benefits ($M/yr)	2.1	3.1
User disbenefits ($M)	0.8	0
First cost ($M)	8.8	10.4
Operations and maintenance ($M/yr)	0.95	1.7

(a) Use the benefit–cost ratio.

(b) Use the modified benefit–cost ratio.

(c) Use the public/government version of the B/C ratio.

(d) Assume these numbers apply to a private firm and use a present worth index.

(e) Are your recommendations for (a) through (d) consistent? Which measure gives the largest value? Why?

Payback Period and Exact Methods

9-40   Able Plastics, an injection-molding firm, has negotiated a contract with a national chain of department stores. Plastic pencil boxes are to be produced for a 2-year period. Able Plastics has never produced the item before and requires all new dies. If the firm invests $67,000 for special removal equipment to unload the completed pencil boxes from the molding machine, one machine operator can be eliminated. This would save $26,000 per year. The removal equipment has no salvage value and is not expected to be used after the 2-year production contract is completed. The equipment, although useless to Able, would be serviceable for about 15 years. What is the payback period? Should Able Plastics buy the removal equipment?

9-41   A cannery is considering installing an automatic case-sealing machine to replace current hand methods. If they purchase the machine for $3800 in June, at the beginning of the canning season, they will save $400 per month for the 4 months each year that the plant is in operation. Maintenance costs of the case-sealing machine are expected to be negligible. The case-sealing machine is expected to be useful for five annual canning seasons and then will have no salvage value. What is the payback period? What is the nominal annual rate of return?

9-42   A project has the following costs and benefits. What is the payback period?

Year	Costs	Benefits
0	$1400	
1	500	
2	300	$400
3–10		300 in each year

9-43   A car dealer leases a small computer with software for $5000 per year. As an alternative he could buy the computer for $7000 and lease the software for $3500 per year. Any time he would decide to switch to some other computer system he could cancel the software lease and sell the computer for $500. If he buys the computer and leases the software,

(a) What is the payback period?

(b) If he kept the computer and software for 6 years, what would be the benefit–cost ratio, based on a 10% interest rate?

9-44   A large project requires an investment of $200 million. The construction will take 3 years: $30 million will be spent during the first year, $100 million during the second year, and $70 million during the third year of construction. Two project operation periods are being considered: 10 years with the expected net profit of $40 million per year and 20 years with the expected net profit of $32.5 million per year. For simplicity of calculations it is assumed that all cash flows occur at end of year. The company minimum required return on investment is 10%.

Calculate for each alternative:

(a) The payback periods

(b) The total equivalent investment cost at the end of the construction period

(c) The equivalent uniform annual worth of the project (use the operation period of each alternative)

Which operation period should be chosen?

9-45   Two alternatives with identical benefits are being considered:

	A	**B**
Initial cost	$500	$800
Uniform annual cost	200	150
Useful life, in years	8	8

(a) Compute the payback period if Alt. B is purchased rather than Alt. A.

(b) Use a MARR of 12% and benefit–cost ratio analysis to identify the alternative that should be selected.

9-46   Tom Sewel has gathered data on the relative costs of a solar water heater system and a conventional electric water heater. The data are based on statistics for a mid-American city and assume that during cloudy days an electric heating element in the solar heating system will provide the necessary heat.

The installed cost of a conventional electric water tank and heater is $200. A family of four uses an average of 300 liters of hot water a day, which takes $230 of electricity per year. The glass-lined tank has a 20-year guarantee. This is probably a reasonable estimate of its actual useful life.

The installed cost of two solar panels, a small electric pump, and a storage tank with auxiliary electric heating element is $1400. It will cost $60 a year for electricity to run the pump and heat water on cloudy days. The solar system will require $180 of maintenance work every 4 years. Neither the conventional electric water heater nor the solar water heater will have any salvage value at the end of its useful life.

Using Tom's data, what is the payback period if the solar water heater system is installed, rather than the conventional electric water heater?

9-47 Consider four mutually exclusive alternatives:

	A	B	C	D
Cost	$75.0	$50.0	$15.0	$90.0
Uniform annual benefit	18.8	13.9	4.5	23.8

Each alternative has a 5-year useful life and no salvage value. The MARR is 10%. Which alternative should be selected, based on

(a) Future worth analysis
(b) Benefit–cost ratio analysis
(c) The payback period

9-48 Consider three alternatives:

	A	B	C
First cost	$50	$150	$110
Uniform annual benefit	28.8	39.6	39.6
Useful life, in years*	2	6	4
Computed rate of return	10%	15%	16.4%

*At the end of its useful life, an identical alternative (with the same cost, benefits, and useful life) may be installed.

All the alternatives have no salvage value. If the MARR is 12%, which alternative should be selected?

(a) Solve the problem by future worth analysis.
(b) Solve the problem by benefit–cost ratio analysis.
(c) Solve the problem by payback period.

(d) If the answers in parts (a), (b), and (c) differ, explain why this is the case.

9-49 Consider three mutually exclusive alternatives. The MARR is 10%.

Year	X	Y	Z
0	−$100	−$50	−$50
1	25	16	21
2	25	16	21
3	25	16	21
4	25	16	21

(a) For Alt. X, compute the benefit–cost ratio.
(b) Based on the payback period, which alternative should be selected?
(c) Determine the preferred alternative based on an exact economic analysis method.

9-50 The cash flows for three alternatives are as follows:

Year	A	B	C
0	−$500	−$600	−$900
1	−400	−300	0
2	200	350	200
3	250	300	200
4	300	250	200
5	350	200	200
6	400	150	200

(a) Based on payback period, which alternative should be selected?
(b) Using future worth analysis, and a 12% interest rate, determine which alternative should be selected.

9-51 Three mutually exclusive alternatives are being considered:

	A	B	C
Initial cost	$500	$400	$300
Benefit at end of the first year	200	200	200
Uniform benefit at end of subsequent years	100	125	100
Useful life, in years	6	5	4

At the end of its useful life, an alternative is *not* replaced. If the MARR is 10%, which alternative should be selected

(a) Based on the payback period?
(b) Based on benefit–cost ratio analysis?

9-52

Year	E	F	G	H
0	−$90	−$110	−$100	−$120
1	20	35	0	0
2	20	35	10	0
3	20	35	20	0
4	20	35	30	0
5	20	0	40	0
6	20	0	50	180

(a) Based on future worth analysis, which of the four alternatives is preferred at 6% interest?

(b) Based on future worth analysis, which alternative is preferred at 15% interest?

(a) Based on the payback period, which alternative is preferred?

(d) At 7% interest, what is the benefit–cost ratio for Alt. G?

Sensitivity

9-53 Tom Jackson is buying a new car. Alternative A is an American-built compact. It has an initial cost of $8900 and operating costs of 9 ¢/km, excluding depreciation. From resale statistics, Tom estimates the American car can be resold at the end of 3 years for $1700. Alternative B is a foreign built Fiasco. Its initial cost is $8000, the operating cost, also excluding depreciation, is 8 ¢/km. How low could the resale value of the Fiasco be to provide equally economical transportation? Assume Tom will drive 12,000 km/year and considers 8% as an appropriate interest rate. (*Answer:* $175)

9-54 A newspaper is considering buying locked vending machines to replace open newspaper racks in the downtown area. The vending machines cost $45 each. It is expected that the annual revenue from selling the same quantity of newspapers will increase $12 per vending machine. The useful life of the vending machine is unknown.

(a) To determine the sensitivity of rate of return to useful life, prepare a graph for rate of return versus useful life for lives up to 8 years.

(b) If the newspaper requires a 12% rate of return, what minimum useful life must it obtain from the vending machines?

(c) What would be the rate of return if the vending machines were to last indefinitely?

9-55 If the MARR is 12%, compute the value of X that makes the two alternatives equally desirable.

	A	B
Cost	$800	$1000
Uniform annual benefit	230	230
Useful life, in years	5	X

9-56 If the MARR is 12%, compute the value of X that makes the two alternatives equally desirable.

	A	B
Cost	$150	$ X
Uniform annual benefit	40	65
Salvage value	100	200
Useful life, in years	6	6

9-57 Consider two alternatives:

	A	B
Cost	$500	$300
Uniform annual benefit	75	75
Useful life, in years	Infinity	X

Assume that Alt. B is not replaced at the end of its useful life. If the MARR is 10%, what must be the useful life of B to make Alternatives A and B equally desirable?

9-58 Chris Cook studied the situation described in Problem 9-46 and decided that the solar system will *not* require the $180 of maintenance every 4 years. Chris believes future replacements of either the conventional electric water heater, or the solar water heater system can be made at the same costs and useful lives as the initial installation. Based on a 10% interest rate, what must be the useful life of the solar system to make it no more expensive than the electric water heater system?

9-59 Jane Chang is making plans for a summer vacation. She will take $1000 with her in the form of traveler's checks. From the newspaper, she finds that if she purchases the checks by May 31, she will not have to pay a service charge. That is, she will obtain $1000 worth of traveler's checks for $1000. But if she waits to buy the checks until just before starting her summer trip, she must pay a 1% service charge. (It will cost her $1010 for $1000 of traveler's checks.)

Jane can obtain a 13% interest rate, compounded weekly, on her money. How many weeks after May 31 can she begin her trip and still justify buying the traveler's checks on May 31?

9-60 Fence posts for a particular job cost $10.50 each to install, including the labor cost. They will last 10 years. If the posts are treated with a wood preservative, they can be expected to have a 15-year life. Assuming a 10% interest rate, how much could one afford to pay for the wood preservative treatment?

9-61 A piece of property is purchased for $10,000 and yields a $1000 yearly net profit. The property is sold after 5 years. What is its minimum price to breakeven with interest at 10%?

9-62 Rental equipment is for sale for $110,000. A prospective buyer estimates he would keep the equipment for 12 years and spend $6000 a year on maintaining it. Estimated annual net receipts from equipment rentals would be $14,400. It is estimated the rental equipment could be sold for $80,000 at the end of 12 years. If the buyer wants a 7% rate of return on his investment, what is the maximum price he should pay for the equipment?

9-63 *The Financial Advisor* is a weekly column in the local newspaper. Assume you must answer the following question. "I recently retired at age 65, and I have a tax-free retirement annuity coming due soon. I have three options. I can receive (*A*) $30,976 now, (*B*) $359.60 per month for the rest of my life, or (*C*) $513.80 per month for the next 10 years. What should I do?" Ignore the timing of the monthly cash flows and assume that the payments are received at the end of year. Assume the 10-year annuity will continue to be paid to loved heirs if the person dies before the 10-year period is over.

(*a*) If $i = 6\%$, develop a choice table for lives from 5 to 30 years. (You do not know how long this person or other readers may live.)

(*b*) If $i = 10\%$, develop a choice table for lives from 5 to 30 years. (You do not know how long this person or other readers may live.)

(*c*) How does increasing the interest rate change your recommendations?

Contributed by D. P. Loucks, Cornell University

9-64 A motor with a 200-horsepower output is needed in the factory for intermittent use. A Graybar motor costs $7000, and has an electrical efficiency of 89%. A Blueball motor costs $6000 and has an 85% efficiency. Neither motor would have any salvage value, since the cost to remove it would equal its scrap value. The annual maintenance cost for either motor is estimated at $300 per year. Electric power costs $0.072/kWh (1 hp = 0.746 kW). If a 10% interest rate is used in the calculations, what is the minimum number of hours the higher initial cost Graybar motor must be used each year to justify its purchase?

9-65 Plan *A* requires a $100,000 investment now. Plan *B* requires an $80,000 investment now and an additional $40,000 investment at a later time. At 8% interest, compute the breakeven point for the timing of the $40,000 investment.

9-66 A low-carbon-steel machine part, operating in a corrosive atmosphere, lasts 6 years, and costs $350 installed. If the part is treated for corrosion resistance, it will cost $500 installed. How long must the treated part last to be the preferred alternative, assuming 10% interest?

9-67 Neither of the following machines has any net salvage value.

	A	*B*
Original cost	$55,000	$75,000
Annual expenses		
Operation	9,500	7,200
Maintenance	5,000	3,000
Taxes and insurance	1,700	2,250

At what useful life are the machines equivalent if

(*a*) 10% interest is used in the computations?

(*b*) 0% interest is used in the computations?

9-68 A machine costs $5240 and produces benefits of $1000 at the end of each year for 8 years. Assume an annual interest rate of 10%.

(*a*) What is the payback period (in years)?

(*b*) What is the breakeven point (in years)?

(*c*) Since the answers in (*a*) and (*b*) are different, which one is "correct"?

9-69 Analyze Problem 9-61 again with the following changes:

(*a*) What if the property is purchased for $12,000?

(*b*) What if the yearly net profit is $925?

(*c*) What if it is sold after 7 years?

(*d*) What if (*a*), (*b*), and (*c*) happen simultaneously?

9-70 Analyze Problem 9-53 again with the following changes:

(*a*) What if the Fiasco is more reliable than expected, so that its operating cost is $0.075/km?

(*b*) What if Tom drives only 9000 km/year?

(*c*) What if Tom's interest rate is 6% annually?

(*d*) What if (*a*), (*b*), and (*c*) happen simultaneously?

9-71 Analyze Problem 9-55 again with the following changes:

(*a*) What if B's first cost is $1200?

(*b*) What if B's annual benefit is $280?

(*c*) What if the MARR is 10% annually?

(*d*) What if (*a*), (*b*), and (*c*) happen simultaneously?

9-72 Analyze Problem 9-66 again with the following changes:

(*a*) What if the installed cost of the corrosion-treated part is $600?

(*b*) What if the untreated part will last only 4 years?

(*c*) What if the MARR is 12% annually?

(*d*) What if (*a*), (*b*), and (*c*) happen simultaneously?

CHAPTER 10

UNCERTAINTY IN FUTURE EVENTS

They Only Thought They Were Done

Since the early 1980s, under the federal Superfund program, companies have had to clean up property sites that are contaminated by hazardous waste. So far, hundreds of sites have been cleaned up (or "remediated," in Superfund terminology), at a collective cost that runs to the billions of dollars; hundreds of other sites are still in the process of being decontaminated. Superfund cleanups usually take years to complete, and often involve expensive legal wrangling among the parties involved.

A large number of Superfund sites—including some where cleanup has been completed—have been found to contain trichloroethylene (TCE), a widely used solvent. To complicate matters further, the U.S. Environmental Protection Agency (EPA) has discovered that TCE may be far more toxic than originally thought.

Moreover, the cleanup techniques used in the past to remove the substance from groundwater may actually have made the problem worse by causing TCE to be emitted in vapor form. This vapor can concentrate in buildings and other enclosed spaces, increasing the potential toxicity even further.

EPA is now studying whether previously cleaned Superfund sites should be "reopened" for additional remediation of TCE. The decision is made more difficult by the fact that some previously cleaned sites were sold after the Superfund remediation was finished and may have had buildings constructed on them.

Questions to Consider

1. How can a company quantify the risk of an event such as discovery of TCE on its property?
2. What are some ways a company might estimate the cost of such an event, should it occur?
3. Prior to the mid-1970s, it was common (and legal) for manufacturers to landfill chemicals and other potentially hazardous materials. The Superfund program begun 1980 required companies that had disposed of these materials to clean up the landfill sites, even if disposal had taken place years before, at a time when the practice was not against the law. How can companies anticipate and prepare for future laws that might penalize activities that are legal today but may become illegal (or at least socially unacceptable) in the future?

After Completing This Chapter...

You should be able to:

- Use a range of estimated variables to evaluate a project.
- Describe possible outcomes with probability distributions.
- Combine probability distributions for individual variables into joint probability distributions.
- Use expected values for economic decision making.
- Use economic decision trees to describe and solve more complex problems.
- Measure and consider risk when making economic decisions.
- Understand how simulation can be used to evaluate economic decisions.

An assembly line is built after the engineering economic analysis has shown that the anticipated product demand will generate profits. A new motor, heat exchanger, or filtration unit is installed after analysis has shown that future cost savings will economically justify current costs. A new road, school, or other public facility is built after analysis has shown that the future demand and benefits justify the present cost to build. However, future performance of the assembly line, motor, and so on is uncertain, and demand for the product or public facility is more uncertain.

Engineering economic analysis is used to evaluate projects with long-term consequences when the time value of money matters. Thus, it must concern itself with future consequences; but describing the future accurately is not easy. In this chapter we consider the problem of evaluating the future. The easiest way to begin is to make a careful estimate. Then we examine the possibility of predicting a range of possible outcomes. Finally, we consider what happens when the probabilities of the various outcomes are known or may be estimated. We will show that the tools of probability are quite useful for economic decision making.

ESTIMATES AND THEIR USE IN ECONOMIC ANALYSIS

Economic analysis requires evaluating the future consequences of an alternative. In practically every chapter of this book, there are cash flow tables and diagrams that describe precisely the costs and benefits for future years. We don't really believe that we can exactly foretell a future cost or benefit. Instead, our goal is to select a single value representing the *best* estimate that can be made.

We recognize that estimated future consequences are not precise and that the actual values will be somewhat different from our estimates. Even so, it is likely we have made the tacit assumption that these estimates *are* correct. We know estimates will not always turn out to be correct; yet we treat them like facts once they are *in* the economic analysis. We do the analysis as though the values were exact. This can lead to trouble. If actual costs and benefits are different from the estimates, an undesirable alternative may be selected. This is because the variability of future consequences is concealed by assuming that the best estimates will actually occur. The problem is illustrated by Example 10-1.

EXAMPLE 10-1

Two alternatives are being considered. The best estimates for the various consequences are as follows:

	A	B
Cost	$1000	$2000
Net annual benefit	$150	$250
Useful life, in years	10	10
End-of-useful-life salvage value	$100	$400

If interest is 3½%, which alternative has the better net present worth (NPW)?

SOLUTION

Alternative A

$$\text{NPW} = -1000 + 150(P/A, 3\tfrac{1}{2}\%, 10) + 100(P/F, 3\tfrac{1}{2}\%, 10)$$

$$= -1000 + 150(8.317) + 100(0.7089)$$

$$= -1000 + 1248 + 71$$

$$= +\$319$$

Alternative B

$$\text{NPW} = -2000 + 250(P/A, 3\tfrac{1}{2}\%, 10) + 400(P/F, 3\tfrac{1}{2}\%, 10)$$

$$= -2000 + 250(8.317) + 400(0.7089)$$

$$= -2000 + 2079 + 284$$

$$= +\$363$$

Alternative B, with its larger NPW, would be selected.

Alternate Formation of Example 10-1

Suppose that at the end of 10 years, the actual salvage value for B were \$300 instead of the \$400, the best estimate. If all the other estimates were correct, is B still the preferred alternative?

SOLUTION

Revised *B*

$$\text{NPW} = -2000 + 250(P/A, 3\tfrac{1}{2}\%, 10) + 300(P/F, 3\tfrac{1}{2}\%, 10)$$

$$= -2000 + 250(8.317) + 300(0.7089)$$

$$= -2000 + 2079 + 213$$

$$= +\$292 \to A \text{ is now the preferred alternative.}$$

Example 10-1 shows that the change in the salvage value of Alternative B actually results in a change of preferred alternative. Thus, a more thorough analysis of Example 10-1 would consider (1) which values are uncertain, (2) whether the uncertainty is ±5% or −50 to +80%, and (3) which uncertain values lead to different decisions. A more thorough analysis, which is done with the tools of this chapter, determines which decision is better over the range of possibilities. Explicitly considering uncertainty lets us make better decisions. The tool of breakeven analysis is illustrated in Example 10-2.

EXAMPLE 10-2

Use the data of Example 10-1 to compute the sensitivity of the decision to the Alt. *B* salvage value by computing the breakeven value. For Alt. *A*, NPW = +319. For breakeven between the alternatives,

$$NPW_A = NPW_B$$

$$+319 = -2000 + 250(P/A, 3^1/2\%, 10) + \text{Salvage value}_B(P/F, 3^1/2\%, 10)$$

$$= -2000 + 250(8.317) + \text{Salvage value}_B(0.7089)$$

At the breakeven point

$$\text{Salvage value}_B = \frac{319 + 2000 - 2079}{0.7089} = \frac{240}{0.7089} = \$339$$

When Alt. *B* salvage value > \$339, *B* is preferred; when < \$339, *A* is preferred.

Breakeven analysis, as shown in Example 10-2, is one means of examining the impact of the variability of some estimate on the outcome. It helps by answering the question, How much variability can a parameter have before the decision will be affected? While the preferred decision depends on whether the salvage value is above or below the breakeven value, the economic difference between the alternatives is small when the salvage value is "close" to breakeven. Breakeven analysis does not solve the basic problem of how to take the inherent variability of parameters into account in an economic analysis. This will be considered next.

A RANGE OF ESTIMATES

It is usually more realistic to describe parameters with a range of possible values, rather than a single value. A range could include an **optimistic** estimate, the **most likely** estimate, and a **pessimistic** estimate. Then, the economic analysis can determine whether the decision is sensitive to the range of projected values.

EXAMPLE 10-3

A firm is considering an investment. The most likely data values were found during the feasibility study. Analyzing past data of similar projects shows that optimistic values for the first cost and the annual benefit are 5% better than most likely values. Pessimistic values are 15% worse.

The firm's most experienced project analyst has estimated the values for the useful life and salvage value.

	Optimistic	Most Likely	Pessimistic
Cost	$950	$1000	$1150
Net annual benefit	210	200	170
Useful life, in years	12	10	8
Salvage value	100	0	0

Compute the rate of return for each estimate. If a 10% before-tax minimum attractive rate of return is required, is the investment justified under all three estimates? If it is justified only under some estimates, how can these results be used?

SOLUTION

Optimistic Estimate

$$\text{PW of cost} = \text{PW of benefit}$$

$$\$950 = 210(P/A, \text{IRR}_{opt}, 12) + 100(P/F, \text{IRR}_{opt}, 12)$$

$$\text{IRR}_{opt} = 19.8\%$$

Most Likely Estimate

$$\$1000 = 200(P/A, \text{IRR}_{\text{most likely}}, 10)$$

$$(P/A, \text{IRR}_{\text{most likely}}, 10) = 1000/200 = 5 \rightarrow \text{IRR}_{\text{most likely}} = 15.1\%$$

Pessimistic Estimate

$$\$1150 = 170(P/A, \text{IRR}_{pess}, 8)$$

$$(P/A, \text{IRR}_{pess}, 8) = 1150/170 = 6.76 \rightarrow \text{IRR}_{pess} = 3.9\%$$

From the calculations we conclude that the rate of return for this investment is most likely to be 15.1%, but might range from 3.9% to 19.8%. The investment meets the 10% MARR criterion for two of the estimates. These estimates can be considered to be scenarios of what may happen with this project. Since one scenario indicates that the project is not attractive, we need to have a method of weighting the scenarios or considering how likely each is.

Example 10-3 made separate calculations for the sets of optimistic, most likely, and pessimistic values. The range of scenarios is useful. However, if there are more than a few uncertain variables, it is unlikely that all will prove to be optimistic (best case) or most likely or pessimistic (worst case). It is more likely that many parameters are the most likely values, while some are optimistic and some are pessimistic.

This can be addressed by using Equation 10-1 to calculate average or mean values for each parameter. Equation 10-1 puts four times the weight on the most likely value than on the other two. This equation was developed as an approximation with the beta distribution.

$$\text{Mean value} = \frac{\text{Optimistic value} + 4(\text{Most likely value}) + \text{Pessimistic value}}{6} \quad (10\text{-}1)$$

This approach is illustrated in Example 10-4.

EXAMPLE 10-4

Solve Example 10-3 by using Equation 10-1. Compute the resulting mean rate of return.

SOLUTION

Compute the mean for each parameter:

$$\text{Mean cost} = [950 + 4 \times 1000 + 1150]/6 = 1016.7$$

$$\text{Mean net annual benefit} = [210 + 4 \times 200 + 170]/6 = 196.7$$

$$\text{Mean useful life} = [12 + 4 \times 10 + 8]/6 = 10.0$$

$$\text{Mean salvage life} = 100/6 = 16.7$$

Compute the mean rate of return:

$$\text{PW of cost} = \text{PW of benefit}$$

$$\$1016.7 = 196.7(P/A, \text{IRR}_{\text{beta}}, 10) + 16.7(P/F, \text{IRR}_{\text{beta}}, 10)$$

$$\text{IRR}_{\text{beta}} = 14.2\%$$

Example 10-3 gave a most likely rate of return (15.1%) that differed from the mean rate of return (14.2%) computed in Example 10-4. These values are different because the former is based exclusively on the most likely values and the latter takes into account the variability of the parameters.

In examining the data, we see that the pessimistic values are further away from the most likely values than are the optimistic values. This is a common occurrence. For example, a savings of 10 to 20% may be the maximum possible, but a cost overrun can be 50%, 100%, or even more. This causes the resulting weighted mean values to be less favorable than the most likely values. As a result, the mean rate of return, in this example, is less than the rate of return based on the most likely values.

PROBABILITY

We all have used probabilities. For example, what is the probability of getting a "head" when flipping a coin? Using a model that assumes that the coin is fair, both the head and tail outcomes occur with a probability of 50%, or $1/2$. This probability is the likelihood of an event in a single trial. It also describes the long-run relative frequency of getting heads in many trials (out of 50 coin flips, we expect to average 25 heads).

Probabilities can also be based on data, expert judgment, or a combination of both. Past data on weather and climate, on project completion times and costs, and on highway traffic are combined with expert judgment to forecast future events. These examples can be important in engineering economy.

Another example based on long-run relative frequency is the PW of a flood-protection dam that depends on the probabilities of different-sized floods over many years. This would be based on data about past floods and would include many years of potential flooding. An example of a single event that may be estimated by expert judgment is the probability of a successful outcome for a research and development project, which will determine its PW.

All the data in an engineering economy problem may have some level of uncertainty. However, small uncertainties may be ignored, so that more analysis can be done with the large uncertainties. For example, the price of an off-the-shelf piece of equipment may vary by only ±5%. The price could be treated as a known or deterministic value. On the other hand, demand over the next 20 years will have more uncertainty. Demand should be analyzed as a random or stochastic variable. We should establish probabilities for different values of demand.

There are also logical or mathematical rules for probabilities. If an outcome can never happen, then the probability is 0. If an outcome will certainly happen, then the probability is 1, or 100%. This means that probabilities cannot be negative or greater than 1; in other words, they must be within the interval [0, 1], as indicated shortly in Equation 10-2.

Probabilities are defined so that the sum of probabilities for all possible outcomes is 1 or 100% (Equation 10-3). Summing the probability of 0.5 for a head and 0.5 for a tail leads to a total of 1 for the possible outcomes from the coin flip. An exploration well drilled in a potential oil field will have three outcomes (dry hole, noncommercial quantities, or commercial quantities) whose probabilities will sum to one.

Equations 10-2 and 10-3 can be used to check that probabilities are valid. If the probabilities for all but one outcome are known, the equations can be used to find the unknown probability for that outcome (see Example 10-5).

$$0 \leq \text{Probability} \leq 1 \tag{10-2}$$

$$\sum_{j=1 \text{ to } K} P(\text{outcome}_j) = 1, \quad \text{where there are } K \text{ outcomes} \tag{10-3}$$

In a probability course many probability distributions, such as the normal, uniform, and beta are presented. These continuous distributions describe a large population of data. However, for engineering economy it is more common to use 2 to 5 outcomes with discrete probabilities—even though the 2 to 5 outcomes only represent or approximate the range of possibilities.

This is done for two reasons. First, the data often are estimated by expert judgment, so that using 7 to 10 outcomes would be false accuracy. Second, each outcome requires more analysis. In most cases the 2 to 5 outcomes represents the best trade-off between representing the range of possibilities and the amount of calculation required. Example 10-5 illustrates these calculations.

EXAMPLE 10-5

What are the probability distributions for the annual benefit and life for the following project?

The annual benefit's most likely value is $8000 with a probability of 60%. There is a 30% probability that it will be $5000, and the highest value that is likely $10,000. A life of 6 years is twice as likely as a life of 9 years.

SOLUTION

Probabilities are given for only two of the possible outcomes for the annual benefit. The third value is found from the fact that the probabilities for the three outcomes must sum to 1 (Equation 10-3).

$$1 = P(\text{Benefit is } \$5000) + P(\text{Benefit is } \$8000) + P(\text{Benefit is } \$10,000)$$

$$P(\text{Benefit is } \$10,000) = 1 - 0.6 - 0.3 = 0.1$$

The probability distribution can then be summarized in a table. The histogram or relative frequency diagram is Figure 10-1.

Annual benefit	$5000	$8000	$10,000
Probability	0.3	0.6	0.1

FIGURE 10-1 Probability distribution for annual benefit.

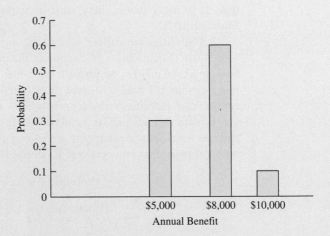

The problem statement tells us:

$$P(\text{life is 6 years}) = 2P(\text{life is 9 years})$$

Equation 10-3 can be applied to write a second equation for the two unknown probabilities:

$$P(6) + P(9) = 1$$

Combining these, we write

$$2P(9) + P(9) = 1$$
$$P(9) = 1/3$$
$$P(6) = 2/3$$

The probability distribution for the life is $P(6) = 66.7\%$ and $P(9) = 33.3\%$.

JOINT PROBABILITY DISTRIBUTIONS

Example 10-5 constructed the probability distributions for a project's annual benefit and life. These examples show how likely each value is for the input data of the problem. We would like to construct a similar probability distribution for the project's present worth. This is the distribution that we can use to evaluate the project. That present worth depends on both input probability distributions, so we need to construct the *joint* probability distribution for the different combinations of their values.

For this introductory text, we assume that two random variables such as the annual benefit and life are unrelated or statistically independent. This means that the *joint* probability of a combined event (Event A defined on the first variable and Event B on the second variable) is the product of the probabilities for the two events. This is Equation 10-4:

$$\text{If A and B are independent, then } P(A \text{ and } B) = P(A) \times P(B) \qquad (10\text{-}4)$$

For example, flipping a coin and rolling a die are statistically independent. Thus, the probability of {flipping a head and rolling a 4} equals the probability of a {heads} = $^1/_2\%$ times the probability of a {4} = 1/6, for a joint probability = 1/12.

The number of outcomes in the joint distribution is the product of the number of outcomes in each variable's distribution. Thus, for the coin and the die, there are 2 times 6, or 12 combinations. Each of the 2 outcomes for the coin is combined with each of the 6 outcomes for the die.

Some variables are not statistically independent, and the calculation of their joint probability distribution is more complex. For example, a project with low revenues may be terminated early and one with high revenues may be kept operating as long as possible. In these cases annual cash flow and project life are not independent. While this type of relationship can sometimes be modeled with economic decision trees (covered later in this chapter), we will limit our coverage in this text to the simpler case of independent variables.

Example 10-6 uses the three values and probabilities for the annual benefit and the two values and probabilities for the life to construct the six possible combinations. Then the values and probabilities are constructed for the project's PW.

EXAMPLE 10-6

The project described in Example 10-5 has a first cost of $25,000. The firm uses an interest rate of 10%. Assume that the probability distributions for annual benefit and life are unrelated or statistically independent. Calculate the probability distribution for the PW.

SOLUTION

Since there are three outcomes for the annual benefit and two outcomes for the life, there are six combinations. The first four columns of the following table show the six combinations of life and annual benefit. The probabilities in columns 2 and 4 are multiplied to calculate the joint probabilities in column 5. For example, the probability of a low annual benefit and a short life is $0.3 \times 2/3$, which equals 0.2 or 20%.

The PW values include the $25,000 first cost and the results of each pair of annual benefit and life. For example, the PW for the combination of high benefit and long life is:

$$PW_{\$10,000,9} = -25,000 + 10,000(P/A, 10\%, 9) = -25,000 + 10,000(5.759) = \$32,590$$

Annual Benefit	Probability	Life (years)	Probability	Joint Probability	PW
$ 5,000	30%	6	66.7%	20.0%	−$ 3,224
8,000	60	6	66.7	40.0	+9,842
10,000	10	6	66.7	6.7	18,553
5,000	30	9	33.3	10.0	3,795
8,000	60	9	33.3	20.0	21,072
10,000	10	9	33.3	3.3	32,590
				100.0%	

Figure 10-2 shows the probabilities for the PW in the form of a histogram for relative frequency distribution, or probability distribution function.

FIGURE 10-2 Probability distribution function for PW.

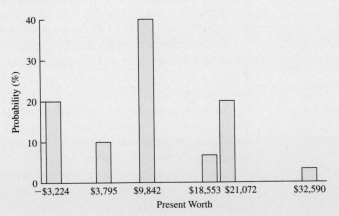

This probability distribution function shows that there is a 20% chance of having a negative PW. It also shows that there is a small (3.3%) chance of the PW being $32,590. The three values used to describe possible annual benefits for the project and the two values for life have been combined to describe the uncertainty in the project's PW.

Creating a distribution, as in Example 10-6, gives us a much better understanding of the possible PW values along with their probabilities. The three possibilities for the annual benefit and the two for the life are representative of the much broader set of possibilities that really exist. Optimistic, most likely, and pessimistic values are a good way to represent the uncertainty about a variable.

Similarly the six values for the PW represent the much broader set of possibilities. The 20% probability of a negative PW is one measure of risk that we will talk about later in the chapter.

Some problems, such as Examples 10-3 and 10-4, have so many variables or different outcomes that constructing the joint probability distribution is arithmetically burdensome. If the values in Equation 10-1 are treated as a discrete probability distribution function, the probabilities are $1/6, 2/3, 1/6$. With an optimistic, most likely, and pessimistic outcome for each of 4 variables, there are $3^4 = 81$ combinations. In Examples 10-3 and 10-4, the salvage value has only two distinct values, so there are still $3 \times 3 \times 3 \times 2 = 54$ combinations.

When the problem is important enough, the effort to construct the joint probability distribution is worthwhile. It gives the analyst and the decision maker a better understanding of what may happen. It is also needed to calculate measures of a project's risk. While spreadsheets can automate the arithmetic, simulation (described at the end of the chapter) can be a better choice when there are a large number of variables and combinations.

EXPECTED VALUE

For any probability distribution we can compute the **expected value (EV)** or arithmetic average (mean). To calculate the EV, each outcome is weighted by its probability, and the results are summed. This is NOT the simple average or unweighted mean. When the class average on a test is computed, this is an unweighted mean. Each student's test has the same weight. This simple "average" is the one that is shown by the button $\bar{x}$ on many calculators.

The expected value is a weighted average, like a student's grade point average (GPA). To calculate a GPA, the grade in each class is weighted by the number of credits. For the expected value of a probability distribution, the weights are the probabilities.

This is described in Equation 10-5. We saw in Example 10-4, that these expected values can be used to compute a rate of return. They can also be used to calculate a present worth as in Example 10-7.

$$\text{Expected value} = \text{Outcome}_A \times P(A) + \text{Outcome}_B \times P(B) + \cdots \tag{10-5}$$

EXAMPLE 10-7

The first cost of the project in Example 10-5 is $25,000. Use the expected values for annual benefits and life to estimate the present worth. Use an interest rate of 10%.

SOLUTION

$$\text{EV}_{\text{benefit}} = 5000(0.3) + 8000(0.6) + 10{,}000(0.1) = \$7300$$

$$\text{EV}_{\text{life}} = 6(2/3) + 9(1/3) = 7 \text{ years}$$

The PW using these values is

$$\text{PW(EV)} = -25{,}000 + 7300(P/A, 10\%, 7) = -25{,}000 + 6500(4.868) = \$10{,}536$$

[*Note:* This is the present worth of the expected values, PW(EV), not the expected value of the present worth, EV(PW). It is an easy value to calculate that approximates the EV(PW), which will be computed from the joint probability distribution found in Example 10-6.]

Example 10-7 is a simple way to approximate the project's expected PW. But the true expected value of the PW is somewhat different. To find it, we must use the joint probability distribution for benefit and life, and the resulting probability distribution function for PW that was derived in Example 10-6. Example 10-8 shows the expected value of the PW or the EV(PW).

EXAMPLE 10-8

Use the probability distribution function of the PW that was derived in Example 10–6 to calculate the EV(PW). Does this indicate an attractive project?

SOLUTION

The table from Example 10-6 can be reused with one additional column for the weighted values of the PW (= PW × probability). Then, the expected value of the PW is calculated by summing the column of present worth values that have been weighted by their probabilities.

Annual Benefit	Probability	Life (years)	Probability	Joint Probability	PW	PW × Joint Probability
$ 5,000	30%	6	66.7%	20.0%	−$ 3,224	−$ 645
8,000	60	6	66.7	40.0	+9,842	+3,937
10,000	10	6	66.7	6.7	18,553	1,237
5,000	30	9	33.3	10.0	3,795	380
8,000	60	9	33.3	20.0	21,072	4,214
10,000	10	9	33.3	3.3	32,590	1,086
				100.0%	EV(PW) =	$10,209

With an expected PW of $10,209, this is an attractive project. While there is a 20% chance of a negative PW, the possible positive outcomes are larger and more likely. Having analyzed the project under uncertainty, we are much more knowledgeable about the potential result of the decision to proceed.

The $10,209 value is more accurate than the approximate value calculated in Example 10-7. The values differ because PW is a nonlinear function of the life. The more accurate value of $10,209 is lower because the annual benefit values for the longer life are discounted by $1/(1+i)$ for more years.

In Examples 10-7 and 10-8, the question was whether the project had a positive PW. With two or more alternatives, the criterion would have been to maximize the PW. With equivalent uniform annual costs (EUACs) the goal is to minimize the EUAC. Example 10-9 uses the criterion of minimizing the EV of the EUAC to choose the best height for a dam.

EXAMPLE 10-9

A dam is being considered to reduce river flooding. But if a dam is built, what height should it be? Increasing the dam's height will (1) reduce a flood's probability, (2) reduce the damage when floods occur, and (3) cost more. Which dam height minimizes the expected total annual cost? The state uses an interest rate of 5% for flood protection projects, and all the dams should last 50 years.

Dam Height (ft)	First Cost	Annual P (flood) > Height	Damages if Flood Occurs
No dam	$ 0	0.25	$800,000
20	700,000	0.05	500,000
30	800,000	0.01	300,000
40	900,000	0.002	200,000

SOLUTION

The easiest way to solve this problem is to choose the dam height with the lowest equivalent uniform annual cost (EUAC). Calculating the EUAC of the first cost requires multiplying the first cost by $(A/P, 5\%, 50)$. For example, for the dam 20 ft high, this is $700,000(A/P, 5\%, 50) = \$38,344$.

Calculating the annual expected flood damage cost for each alternative is simplified because the term for the P(no flood) is zero, because the damages for no flood are $0. Thus we need to calculate only the term for flooding. This is done by multiplying the P(flood) times the damages if a flood happens. For example, the expected annual flood damage cost with no levee is $0.25 \times \$800,000$, or $\$200,000$.

Then the EUAC of the first cost and the expected annual flood damage are added together to find the total EUAC for each height. The 30 ft dam is somewhat cheaper than the 40 ft dam.

Dam Height (ft)	EUAC of First Cost	Expected Annual Flood Damages	Total Expected EUAC
No dam	$ 0	$200,000	$200,000
20	38,344	25,000	63,344
30	43,821	3000	46,821
40	49,299	400	49,699

ECONOMIC DECISION TREES

Some engineering projects are more complex, and evaluating them properly is correspondingly more complex. For example, consider a new product with potential sales volumes ranging from low to high. If the sales volume is low, then the product may be discontinued early in its potential life. On the other hand, if sales volume is high, additional capacity may be added to the assembly line and new product variations may be added. This can be modeled with a decision tree.

The following symbols are used to model decisions with decision trees:

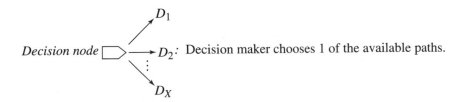

Decision node $\square \longrightarrow D_2$: Decision maker chooses 1 of the available paths.

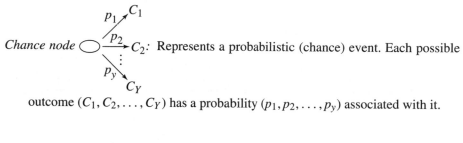

Chance node : Represents a probabilistic (chance) event. Each possible outcome $(C_1, C_2, \ldots, C_Y)$ has a probability $(p_1, p_2, \ldots, p_y)$ associated with it.

Outcome node ⟶▢: Shows result for a particular path through the decision tree.

Pruned branch ─╫►: The double hash mark indicates that a branch has been pruned because another branch has been chosen. This can happen only at decision nodes, not at chance nodes. The term "pruned" is chosen to correspond to the gardener's practice of trimming or pruning off branches to make a tree or bush healthier.

Figure 10-3 illustrates how decision nodes ▷, chance nodes ◯, and outcome nodes ▢ can be used to describe a problem's structure. Details such as the probabilities and costs can be added on the branches that link the nodes. With the branches from decision and chance nodes, the model becomes a decision tree.

Figure 10-3 illustrates that decision trees describe the problem by starting at the decision that must be made and then adding chance and decision nodes in the proper logical sequence. Thus describing the problem starts at the first step and goes forward in time with sequences of decision and chance nodes.

To make the decision, calculations begin with the final nodes in the tree. Since they are the final nodes, enough information is available to evaluate them. At decision nodes the criterion is either to maximize PW or to minimize EUAC. At chance nodes an expected value for PW or EUAC is calculated.

Once all nodes that branch from a node have been evaluated, the originating node can be evaluated. If the originating node is a decision node, choose the branch with the best PW or EUAC and place that value in the node. If the originating node is a chance node, calculate the expected value and place that value in the node. This process "rolls back" values from the terminal nodes in the tree to the initial decision. Example 10-10 illustrates this.

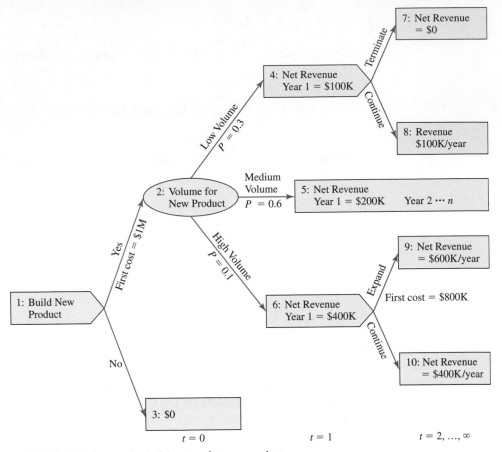

FIGURE 10-3 Economic decision tree for new product.

EXAMPLE 10-10

What decision should be made on the new product summarized in Figure 10-3? What is the expected value of the product's PW? The firm uses an interest rate of 10% to evaluate projects. If the product is terminated after one year, the capital equipment has a salvage value of $550,000 for use with other new products. If the equipment is used for 8 years, the salvage value is $0.

SOLUTION

Evaluating decision trees is done by starting with the end outcome nodes and the decisions that lead to them. In this case the decisions are whether to terminate after 1 year if sales volume is low and whether to expand after 1 year if sales volume is high.

The decision to terminate the product depends on which is more valuable, the equipment's salvage value of $550,000 or the revenue of $100,000 per year for 7 more years. The worth (PW_1) of the salvage value is $550,000. The worth (PW_1) of the revenue stream at the end of Year 1 shown in node 8 is:

$$PW_1 \text{ for node } 8 = 100,000(P/A, 10\%, 7)$$

$$= 100,000(4.868) = \$486,800$$

Thus, terminating the product and using the equipment for other products is better. We enter the two "present worth" values at the end of Year 1 in nodes 7 and 8. We make the *arc to node 7 bold* to indicate that it is our preferred choice at node 4. We use a *double hash mark* to show that we're *pruning the arc to node 8* to indicate that it has been rejected as an inferior choice at node 4.

The decision to expand at node 6 could be based on whether the \$800,000 first cost for expansion can be justified based on increasing annual revenues for 7 years by \$200,000 per year. However, this is difficult to show on the tree. It is easier to calculate the "present worth" values at the end of Year 1 for each of the two choices. The worth (PW_1) of node 9 (expand) is:

$$PW_1 \text{ for node } 9 = -800,000 + 600,000(P/A, 10\%, 7)$$

$$= -800,000 + 600,000(4.868)$$

$$= \$2,120,800$$

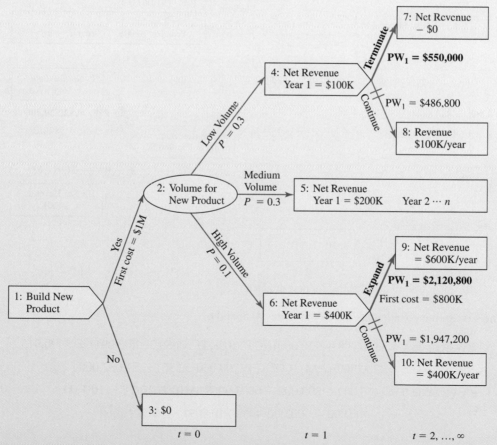

FIGURE 10-4 Partially solved decision tree for new product.

The value of node 10 (continue without expanding) is:

$$\text{PW}_1 \text{ for node } 10 = 400,000(P/A, 10\%, 7)$$
$$= 400,000(4.868)$$
$$= \$1,947,200$$

This is $173,600 less than the expansion node, so the expansion should happen if volume is high. Figure 10-4 summarizes what we know at this stage of the process.

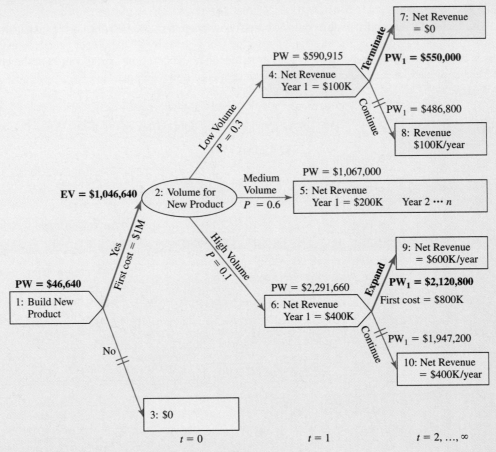

FIGURE 10-5 Solved decision tree for new product.

The next step is to calculate the PW at nodes 4, 5, and 6.

PW at node 4 $= (100,000 + 550,000)(P/F, 10\%, 1) = 650,000(0.9091) = \$590,915$

PW at node 5 $= (200,000)(P/A, 10\%, 8) = 200,000(5.335) = \$1,067,000$

PW at node 6 $= [400,000 - 800,000 + 600,000(P/A, 10\%, 7)](P/F, 10\%, 1)$

$\qquad = [-400,000 + 600,000(4.868)](0.9091) = \$2,291,660$

Now the expected value at node 2 can be calculated

$$\text{EV at node 2} = 0.3(590,915) + 0.6(1,067,000) + 0.1(2,291,660) = \$1,046,640$$

The cost of selecting node 2 is \$1,000,000, so proceeding with the product has an expected PW of \$46,640. This is greater than the \$0 for not building the project. So the decision is to build. Figure 10-5 is the decision tree at the final stage.

Example 10-10 is representative of many problems in engineering economy. The main criterion is maximizing PW or minimizing EUAC. However, as shown in Example 10-11, other criteria, such as risk, are used in addition to expected value.

EXAMPLE 10-11

Consider the economic evaluation of collision and comprehensive (fire, theft, etc.) insurance for a car. This insurance is typically required by lenders, but once the car has been paid for, this insurance is not required. (Liability insurance *is* a legal requirement.)

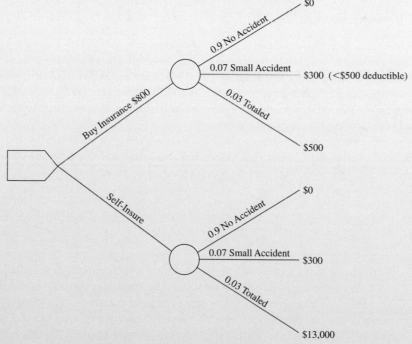

FIGURE 10-6 Decision tree for buying auto collision insurance.

Figure 10-6 begins with a decision node with two alternatives for the next year. Insurance will cost \$800 per year with a \$500 deductible if a loss occurs. The other option is to self-insure, which means to go without buying collision and comprehensive insurance. Then if a loss occurs, the owner must replace the vehicle with money from savings or a loan, or do without a vehicle until the owner can afford to replace it.

Three accident severities are used to represent the range of possibilities: a 90% chance of no accident, a 7% chance of a small accident (at a cost of $300, which is less than the deductible), and a 3% chance of totaling the $13,000 vehicle. Since our driving habits are likely to be the same with and without insurance, the accident probabilities are the same for both chance nodes.

Even though this is a text on engineering economy, we have simplified the problem and ignored the difference in timing of the cash flows. Insurance payments are made at the beginning of the covered period, and accident costs occur during the covered period. Since car insurance is usually paid semiannually, the results of the economic analysis are not changed significantly by the simplification. We focus on the new concepts of expected value, economic decision trees, and risk.

What are the expected values for each alternative, and what decision is recommended?

SOLUTION

The expected values are computed by using Equation 10-5. If insured, the maximum cost equals the deductible of $500. If self-insured, the cost is the cost of the accident.

$$\text{EV}_{\text{accident w/ins.}} = (0.9 \times 0) + (0.07 \times 300) + (0.03 \times 500) = \$36$$

$$\text{EV}_{\text{accident w/o ins.}} = (0.9 \times 0) + (0.07 \times 300) + (0.03 \times 13,000) = \$411$$

Thus, buying insurance lowers the expected cost of an accident by $375. To evaluate whether we should buy insurance, we must also account for the cost of the insurance. Thus, these expected costs are combined with the $0 for self-insuring (total $411) and the $800 for insuring (total $836). Thus self-insuring has an expected value cost that is $425 less per year ($= \$836 - \$411$). This is not surprising, since the premiums collected must cover both the costs of operating the insurance company and the expected value of the payouts.

This is also an example of *expected values alone not determining the decision*. Buying insurance has an expected cost that is $425 per year higher, but that insurance limits the maximum loss to $500 rather than $13,000. The $425 may be worth spending to avoid that risk.

RISK

Risk can be thought of as the chance of getting an outcome other than the expected value—with an emphasis on something negative. One common measure of risk is the probability of a loss (see Example 10-6). The other common measure is the **standard deviation** (σ), which measures the dispersion of outcomes about the expected value. For example, many students have used the normal distribution in other classes. The normal distribution has 68% of its probable outcomes within ± 1 standard deviation of the mean and 95% within ± 2 standard deviations of the mean.

Mathematically, the standard deviation is defined as the square root of the variance. This term is defined as the weighted average of the squared difference between the outcomes of the random variable X and its mean. Thus the larger the difference between the mean and the values, the larger are the standard deviation and the variance. This is Equation 10-6:

$$\text{Standard deviation } (\sigma) = \sqrt{[\text{EV}(X - \text{mean})^2]} \qquad (10\text{-}6)$$

Squaring the differences between individual outcomes and the EV ensures that positive and negative deviations receive positive weights. Consequently, negative values for the standard deviation are impossible, and they instantly indicate arithmetic mistakes. The standard deviation equals 0 if only one outcome is possible. Otherwise, the standard deviation is positive.

This is not the standard deviation formula built into most calculators, just as the weighted average is not the simple average built into most calculators. The calculator formulas are for N equally likely data points from a randomly drawn sample, so that each probability is $1/N$. In economic analysis we will use a weighted average for the squared deviations since the outcomes are not equally likely.

The second difference is that for calculations (by hand or the calculator), it is easier to use Equation 10-7, which is shown to be equivalent to Equation 10-6 in introductory probability and statistics texts.

$$\text{Standard deviation } (\sigma) = \sqrt{\{\text{EV}(X^2) - [\text{EV}(X)]^2\}} \qquad (10\text{-}7)$$

$$= \sqrt{\{\text{Outcome}_A^2 \times P(A) + \text{Outcome}_B^2 \times P(B) + \cdots - \text{expected value}^2\}} \qquad (10\text{-}7')$$

This equation is the square root of the difference between the average of the squares and the square of the average. The standard deviation is used instead of the *variance* because the standard deviation is measured in the same units as the expected value. The variance is measured in "squared dollars"—whatever they are.

The calculation of a standard deviation by itself is only a descriptive statistic of limited value. However, as shown in the next section on risk/return trade-offs, it is useful when the standard deviation of each alternative is calculated and these results are compared. But first, some examples of calculating the standard deviation.

EXAMPLE 10-12   (Example 10-11 continued)

Consider the economic evaluation of collision and comprehensive (fire, theft, etc.) insurance for an auto. One example was described in Figure 10-6. The probabilities and outcomes are summarized in the calculation of the expected values, which was done using Equation 10–5.

$$\text{EV}_{\text{accident w/ins.}} = (0.9 \times 0) + (0.07 \times 300) + (0.03 \times 500) = \$36$$
$$\text{EV}_{\text{accident w/o ins.}} = (0.9 \times 0) + (0.07 \times 300) + (0.03 \times 13{,}000) = \$411$$

Calculate the standard deviations for insuring and not insuring.

SOLUTION

The first step is to calculate the EV(outcome^2) for each.

$$EV^2_{\text{accident w/ins.}} = (0.9 \times 0^2) + (0.07 \times 300^2) + (0.03 \times 500^2) = 13,800$$

$$EV^2_{\text{accident w/ins.}} = (0.9 \times 0^2) + (0.07 \times 300^2) + (0.03 \times 13,000^2) = 5,076,300$$

Then the standard deviations can be calculated.

$$\sigma_{\text{w/ins.}} = \sqrt{EV^2_{\text{w/ins.}} - (EV_{\text{w/ins.}})^2}$$

$$= \sqrt{(13,800 - 36^2)} = \sqrt{12,504} = \$112$$

$$\sigma_{\text{w/o ins.}} = \sqrt{EV^2_{\text{w/o ins.}} - (EV_{\text{w/o ins.}})^2}$$

$$= \sqrt{(5,076,300 - 411^2)} = \sqrt{4,907,379} = \$2215$$

As described in Example 10-11, the expected value cost of insuring is \$836 ($= \$36+\$800$) and the expected value cost of self-insuring is \$411. Thus the expected cost of not insuring is about half the cost of insuring. But the standard deviation of self-insuring is 20 times larger. It is clearly riskier.

Which choice is preferred depends on how much risk one is comfortable with.

As stated before, this is an example of *expected values alone not determining the decision.* Buying insurance has an expected cost that is \$425 per year higher, but that insurance limits the maximum loss to \$500 rather than \$13,000. The \$425 may be worth spending to avoid that risk.

EXAMPLE 10-13 (Example 10-6 continued)

Using the probability distribution for the PW from Example 10–6, calculate the PW's standard deviation.

SOLUTION

The following table adds a column for (PW^2) (probability) to calculate the EV(PW^2).

Annual Benefit	Probability	Life (years)	Probability	Joint Probability	PW	PW × Probability	PW^2 × Probability
\$ 5,000	30%	6	66.7%	20.0%	−\$ 3,224	−\$ 645	\$ 2,079,480
8,000	60	6	66.7	40.0	+9,842	+3,937	38,747,954
10,000	10	6	66.7	6.7	18,553	1,237	22,950,061
5,000	30	9	33.3	10.0	3,795	380	1,442,100
8,000	60	9	33.3	20.0	21,072	4,214	88,797,408
10,000	10	9	33.3	3.3	32,590	1,086	35,392,740
					EV	\$10,209	\$189,409,745

$$\text{Standard deviation} = \sqrt{\{EV(X^2) - [EV(X)]^2\}}$$

$$\sigma = \sqrt{\{189,405,745 - [10,209]^2\}} = \sqrt{85,182,064} = \$9229$$

For those with stronger backgrounds in probability than this chapter assumes, let us consider how the standard deviation in Example 10-13 depends on the assumption of independence between the variables. While exceptions exist, a positive statistical dependence between variables often increases the PW's standard deviation. Similarly, a negative statistical dependence between variables often decreases the PW's standard deviation.

RISK VERSUS RETURN

A graph of risk versus return is one way to consider these items together. Figure 10-7 in Example 10-14 illustrates the most common format. Risk measured by standard deviation is placed on the x axis, and return measured by expected value is placed on the y axis. This is usually done with the internal rates of return of the alternatives or projects.

EXAMPLE 10-14

A large firm is discontinuing an older product, so some facilities are becoming available for other uses. The following table summarizes eight new projects that would use the facilities. Considering expected return and risk, which projects are good candidates? The firm believes it can earn 4% on a risk-free investment in government securities (labeled as Project F).

Project	IRR	Standard Deviation
1	13.1%	6.5%
2	12.0	3.9
3	7.5	1.5
4	6.5	3.5
5	9.4	8.0
6	16.3	10.0
7	15.1	7.0
8	15.3	9.4
F	4.0	0.0

SOLUTION

Answering the question is far easier if we use Figure 10-7. Since a larger expected return is better, we want to select projects that are as "high up" as possible. Since a lower risk is better, we want to select projects that are as "far left" as possible. The graph lets us examine the trade-off of accepting more risk for a higher return.

FIGURE 10-7 Risk-versus-return graph.

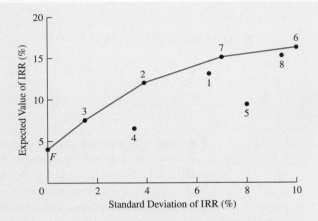

First, we can eliminate Projects 4 and 5. They are **dominated projects.** Dominated alternatives are no better than another alternative on all measures and inferior on at least one measure. Project 4 is dominated by Project 3, which has a higher expected return and a lower risk. Project 5 is dominated by Projects 1, 2, and 7. All three have a higher expected return and a lower risk.

Second, we look at the **efficient frontier.** This is the blue line in Figure 10-7 that connects Projects F, 3, 2, 7, and 6. Depending on the trade-off that we want to make between risk and return, any of these could be the best choice.

Project 1 appears to be inferior to Projects 2 and 7. Project 8 appears to be inferior to Projects 7 and 6. Projects 1 and 8 are inside and not on the efficient frontier.

There are models of risk and return that can allow us to choose between Projects F, 3, 2, 7, and 6; but those models are beyond what is covered here.

There is a simple rule of thumb for comparing a project's risk and return for which we would like to thank Joe Hartman. If the expected present worth is at least double the standard deviation of the present worth, then the project is relatively *safe*. For comparison, remember that a normal distribution has about 2.5% of its values less than 2 standard deviations below the mean.

SIMULATION

Simulation is a more advanced approach to considering risk in engineering economy problems. As such, the following discussion focuses on what it is. As the examples show, spreadsheet functions and add-in packages make simulation easier to use for economic analysis.

Economic **simulation** uses random sampling from the probability distributions of one or more variables to analyze an economic model for many iterations. For each iteration, all variables with a probability distribution are randomly sampled. These values are used to calculate the PW, IRR, or EUAC. Then the results of all iterations are combined to create a probability distribution for the PW, IRR, or EUAC.

Simulation can be done by hand, using a table of random numbers—if there are only a few random variables and iterations. However, results are more reliable as the number of iterations increases, so in practice this is usually computerized. This can be done in Excel using the RAND() function to generate random numbers, as shown in Example 10–15.

Because we were analyzing each possible outcome, the probability distributions earlier in this chapter (and in the end-of-chapter problems) used two or three discrete outcomes.

This limited the number of combinations that we needed to consider. Simulation makes it easy to use continuous probability distributions like the uniform, normal, exponential, log normal, binomial, and triangular. Examples 10-15 and 10-16 use the normal and the discrete uniform distributions.

EXAMPLE 10-15

ShipM4U is considering installing a new, more accurate scale, which will reduce the error in computing postage charges and save $250 a year. The scale's useful life is believed to be uniformly distributed over 12, 13, 14, 15, and 16 years. The initial cost of the scale is estimated to be normally distributed with a mean of $1500 and a standard deviation of $150.

FIGURE 10-8 Excel spreadsheet for simulation ($N = 25$).

	A	B	C	D
1	250	Annual Savings		
2		Life	First Cost	
3	Min	12	1500	Mean
4	Max	16	150	Std dev
5				
6	Iteration			IRR
7	1	12	1277	16.4%
8	2	15	1546	13.9%
9	3	12	1523	12.4%
10	4	16	1628	13.3%
11	5	14	1401	15.5%
12	6	12	1341	15.2%
13	7	12	1683	10.2%
14	8	14	1193	19.2%
15	9	15	1728	11.7%
16	10	12	1500	12.7%
17	11	16	1415	16.0%
18	12	12	1610	11.2%
19	13	15	1434	15.4%
20	14	12	1335	15.4%
21	15	14	1468	14.5%
22	16	13	1469	13.9%
23	17	14	1409	15.3%
24	18	15	1484	14.7%
25	19	14	1594	12.8%
26	20	15	1342	16.8%
27	21	14	1309	17.0%
28	22	12	1541	12.1%
29	23	16	1564	14.0%
30	24	13	1590	12.2%
31	25	16	1311	17.7%
32				
33	Mean	14	1468	14.4%
34	Std dev	2	135	2.2%

Use Excel to simulate 25 random samples of the problem and compute the rate of return for each sample. Construct a graph of rate of return versus frequency of occurrence.

SOLUTION

This problem is simple enough to construct a table with each iteration's values of the life and the first cost. From these values and the annual savings of $250, the IRR for each iteration can be calculated using the RATE function. These are shown in Figure 10-8. The IRR values are summarized in a relative frequency diagram in Figure 10-9.

Note: Each time Excel recalculates the spreadsheet, different values for all the random numbers are generated. Thus the results depend on the set of random numbers, and your results will be different if you create this spreadsheet.

Note for students who have had a course in probability and statistics: Creating the random values for life and first cost is done as follows. Select a random number in [0, 1) using Excel's RAND function. This is the value of the cumulative distribution function for the variable. Convert this to the variable's value by using an inverse function from Excel, or build the inverse function. For the discrete uniform life, the function is = min life + INT(range∗ RAND()). For the normally distributed first cost, the functions is = NORMINV(RAND(), mean, standard deviation).

FIGURE 10-9 Graph of IRR values.

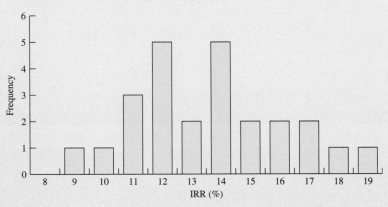

Stand-alone simulation programs and commercial spreadsheet add-in packages such as @Risk and Crystal Ball provide probability distribution functions to use for each input variable. In Example 10-16 the functions RiskUniform and RiskNormal are used. The packages also collect values for the output variables, such as the IRR for Example 10-16. In other problems the PW or EUAC could be collected. These values form a probability distribution for the PW, IRR, or EUAC. From this distribution the simulation package can calculate the expected return, P(loss), and the standard deviation of the return.

Example 10-16 uses @Risk to simulate 1000 iterations of PW for the data in Example 10-15. A simulation package makes it easy to do more iterations. More important still, since it is much easier to use different probability distributions and parameters, more accurate models can be built. Because the models are easier to build, they are less likely to contain errors.

EXAMPLE 10-16

Consider the scale described in Example 10-15. Generate 1000 iterations and construct a frequency distribution for the scale's rate of return.

SOLUTION

The first IRR (cell A8) of 14.01% that is computed in Figure 10-10 is based on the average life and the average first cost. The second IRR (cell A11) of 14.01% is computed by @Risk using the average of each distribution. The cell *content* is the RATE formula with its RiskUniform and RiskNormal function, however spreadsheets with @Risk functions *display* by default the results of using average values.

The RATE function contains two @Risk functions: RiskUniform and RiskNormal. The uniform distribution has the minimum and maximum values as parameters. The normal distribution has the average and standard deviation as parameters.

The third IRR (cell A13) is the average for 1000 iterations. It will change each time the simulation is done. The graph in Figure 10-10 with 1000 iterations is much smoother than Figure 10-9, the graph from Example 10-15, where 25 iterations were done.

FIGURE 10-10 Simulation spreadsheet for Examples 10-15 and 10-16.

	A	B	C	D	E	F
1	−1500	average first cost				
2	150	standard deviation of first cost				
3	12	minimum value of life				
4	16	maximum value of life				
5	250	annual benefit				
6						
7		IRR computed using averages				
8	14.01%	= RATE ((A3+A4)/2,A5,A1)				
9						
10		IRR for each simulation iteration				
11	14.01%	= RATE(RiskUniform(A3,A4),A5,RiskNormal(A1,A2))				
12						
13	14.15%	= average value of A11 (IRR) from 1000 iterations				
14	2.17%	= standard deviation of A11 (IRR) from 1000 iterations				
15						
16		Distribution for IRR (cell A11)				
17						
18						
19						
20						
21						
22						
23						
24						
25						
26						
27						
28						

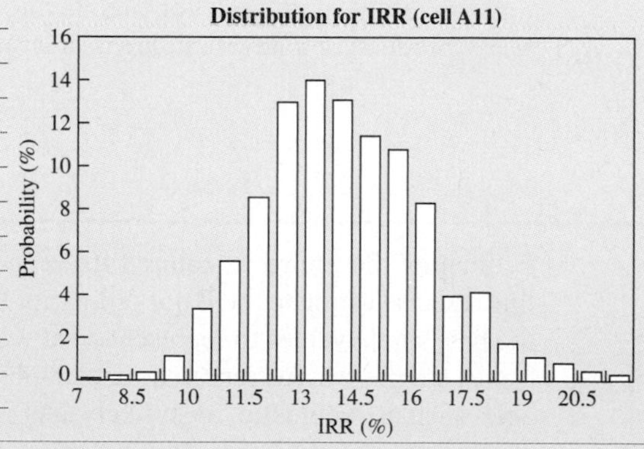

REAL OPTIONS

Real-options analysis is another approach for evaluating projects with significant future uncertainties. It is derived from the theory underlying financial options in the stock market. Rather than buying or selling a share of stock, the option gives you the right to buy or sell the stock at a given price for a given period. Since the option costs less than the stock, a $1000 investment in options has more potential gain than the same investment in the stock. It also has more risk. For engineering projects, a real option is an alternative that keeps open the opportunity by delaying the decision.

For an intuitive understanding of why real options make sense, consider the example of a lease for gas and oil development rights. Many such projects are never developed because they are not "economic." Nevertheless, it may make sense to renew the lease for a noneconomic project because of the possibility that the project will *become* economically profitable. For example, prices may rise, or pipelines and facilities may be built to support a nearby tract.

The models and assumptions of real options are beyond the scope of this text. Interested readers are referred to the third edition of the advanced text *Economic Analysis of Industrial Projects* by Eschenbach, Hartman, and Bussey (2009); the chapter on real options is coauthored by Neal Lewis.

These options may include delaying, abandoning, expanding, shrinking, changing, and replicating a project—with different models for different actions. However, some suggested guidelines may help you decide when you should consider applying real options to your project analysis.

- The value of exercising an option cannot be negative, since you can choose not to exercise it.
- An option to delay a project is valuable when the project is not currently economic but may become economic. Thus, if a project's present worth is positive or very negative, a delay option is unlikely to be worthwhile.
- Unlike projects, options become more valuable as risk (described as volatility) increases. For projects, we want more return but less risk. However, options are more likely to be worthwhile and to result in action if the risk increases.
- Many published examples of real-options analysis ignore the cost of waiting, but this is rarely appropriate for engineering projects. For example, if an R&D project, a new product, or a new building is delayed, then it will later face more competition.

SUMMARY

Estimating the future is required for economic analysis, and there are several ways to do this. Precise estimates will not ordinarily be exactly correct, but they are considered to be the best single values to represent what we think will happen.

A simple way to represent uncertainty is through a range of estimates for each variable, such as optimistic, most likely, and pessimistic. The full range of prospective results may be examined by using the optimistic values to solve the problem and then using the pessimistic values. Solving the problem with the most likely values is a good single value estimate. However, the extremes with all optimistic values or all pessimistic values are less likely—it is more likely that a mix of optimistic, most likely, and pessimistic values will occur.

One approach uses weighted values instead of a range of estimates. One set of weights suggested is:

Estimate	Relative Weight
Optimistic	1
Most likely	4
Pessimistic	1

The most commonly used approach for decision making relies on **expected values.** Here, known or estimated probabilities for future events are used as weights for the corresponding outcomes.

$$\text{Expected value} = \text{Outcome}_A \times \text{Probability}_A + \text{Outcome}_B \times \text{Probability}_B + \cdots$$

Expected value is the most useful and the most frequently used technique for estimating a project's attractiveness.

However, risk as measured by standard deviation and the probability of a loss is also important in evaluating projects. Since projects with higher expected returns also frequently have higher risk, evaluating the trade-offs between risk and return is useful in decision making.

More complicated problems can be summarized and analyzed by using decision trees, which allow logical evaluation of problems with sequential chance, decision, and outcome nodes.

Where the elements of an economic analysis are stated in terms of probability distributions, a repetitive analysis of a random sample is often done. This simulation-based approach relies on the premise that a random sampling of increasing size becomes a better and better estimate of the possible outcomes. The large number of computations means that simulation is usually computerized.

PROBLEMS

Range of Estimates

10-1 Telephone poles exemplify items that have varying useful lives. Telephone poles, once installed in a location, remain in useful service until one of a variety of events occur.

  (*a*) Name three reasons why a telephone pole might be removed from useful service at a particular location.

  (*b*) You are to estimate the total useful life of telephone poles. If the pole is removed from an original location while it is still serviceable, it will be installed elsewhere. Estimate the optimistic life, most likely life, and pessimistic life for telephone poles. What percentage of all telephone poles would you expect to have a total useful life greater than your estimated optimistic life?

10-2 The purchase of a used pickup for $9000 is being considered. Records for other vehicles show that costs for oil, tires, and repairs about equal the cost for fuel.

Fuel costs are $990 per year if the truck is driven 10,000 miles. The salvage value after 5 years of use drops about 8¢ per mile. Find the equivalent uniform annual cost if the interest rate is 8%. How much does this change if the annual mileage is 15,000? 5000?

10-3 A heat exchanger is being installed as part of a plant modernization program. It costs $80,000, including installation, and is expected to reduce the overall plant fuel cost by $20,000 per year. Estimates of the useful life of the heat exchanger range from an optimistic 12 years to a pessimistic 4 years. The most likely value is 5 years. Assume the heat exchanger has no salvage value at the end of its useful life.

(a) Determine the pessimistic, most likely, and optimistic rates of return.

(b) Using the range of estimates to compute the mean life, determine the estimated before-tax rate of return.

10-4 For the data in Problem 10-2 assume that the 5000, 10,000, and 15,000 mileage values are, respectively, pessimistic, most likely, and optimistic estimates. Use a weighted estimate to calculate the equivalent annual cost.

10-5 A new engineer is evaluating whether to use a higher-voltage transmission line. It will cost $250,000 more initially, but it will reduce transmission losses. The optimistic, most likely, and pessimistic projections for annual savings are $20,000, $15,000, and $8,000. The interest rate is 6%, and the transmission line should have a life of 30 years.

(a) What is the present worth for each estimated value?

(b) Use the range of estimates to compute the mean annual savings, and then determine the present worth.

(c) Does the answer to (b) match the present worth for the most likely value? Why or why not?

Probabilities

10-6 When a pair of dice are tossed, the results may be any whole number from 2 through 12. In the game of craps one can win by tossing either a 7 or an 11 on the first roll. What is the probability of doing this? (*Hint:* There are 36 ways that a pair of six-sided dice can be tossed. What portion of them result in either a 7 or an 11?)   (*Answer:* $8/36$)

10-7 Over the last 10 years, the hurdle or discount rate for projects from the firm's research and development division has been 10% twice, 15% three times, and 20% the rest of the time. There is no recognizable pattern. Calculate the probability distribution for next year's discount rate.

10-8 The construction time for a bridge depends on the weather. The project is expected to take 250 days if the weather is dry and hot. If the weather is damp and cool, the project is expected to take 350 days. Otherwise, it is expected to take 300 days. Historical data suggest that the probability of cool, damp weather is 30% and that of dry, hot weather is 20%. Find the project's probability distribution.

10-9 You recently had an auto accident that was your fault. If you have another accident or receive a another moving violation within the next 3 years, you will become part of the "assigned risk" pool, and you will pay an extra $600 per year for insurance. If the probability of an accident or moving violation is 20% per year, what is the probability distribution of your "extra" insurance payments over the next 4 years? Assume that insurance is purchased annually and that violations register at the end of the year—just in time to affect next year's insurance premium.

10-10 Al took a midterm examination in physics and received a score of 65. The mean was 60 and the standard deviation was 20. Bill received a score of 14 in mathematics, where the exam mean was 12 and the standard deviation was 4. Which student ranked higher in his class? Explain.

Joint Probabilities

10-11 A project has a life of 10 years, and no salvage value. The firm uses an interest rate of 12% to evaluate engineering projects. The project has an uncertain first cost and net revenue.

First Cost	P	Net Revenue	P
$300,000	0.2	$70,000	0.3
400,000	0.5	90,000	0.5
600,000	0.3	100,000	0.2

(a) What is the joint probability distribution for first cost and net revenue?

(b) Define optimistic, most likely, and pessimistic scenarios by using both optimistic, both most likely, and both pessimistic estimates. What is the present worth for each scenario?

P
0.3
0.5
0.2

10-12 A robot has just been installed at a cost of $81,000. It will have no salvage value at the end of its useful life.

Savings per Year	Probability	Useful Life (years)	Probability
$18,000	0.2	12	1/6
20,000	0.7	5	2/3
22,000	0.1	4	1/6

(a) What is the joint probability distribution for savings per year and useful life?

(b) Define optimistic, most likely, and pessimistic scenarios by using both optimistic, both most likely, and both pessimistic estimates. What is the rate of return for each scenario?

10-13 Modifying an assembly line has a first cost of $80,000, and its salvage value is $0. The firm's interest rate is 9%. The savings shown in the table depend on whether the assembly line runs one, two, or three shifts, and on whether the product is made for 3 or 5 years.

Shifts/ day	Savings/ year	Probability	Useful Life (years)	Probability
1	$15,000	0.3	3	0.6
2	30,000	0.5	5	0.4
3	45,000	0.2		

(a) Give the joint probability distribution for savings per year and the useful life.

(b) Define optimistic, most likely, and pessimistic scenarios by using both optimistic, both most likely, and both pessimistic estimates. Use a life of 4 years as the most likely value. What is the present worth for each scenario?

Expected Value

10-14 Annual savings due to an energy efficiency project have a most likely value of $30,000. The high estimate of $40,000 has a probability of 0.2, and the low estimate of $20,000 has a probability of 0.30. What is the expected value for the annual savings?   (*Answer:* $29,000)

10-15 Two instructors announced that they "grade on the curve," that is, give a fixed percentage of each of the various letter grades to each of their classes. Their curves are as follows:

Grade	Instructor A	Instructor B
A	10%	15%
B	15	15
C	45	30
D	15	20
F	15	20

If a random student came to you and said that his object was to enroll in the class in which he could expect the higher grade point average, which instructor would you recommend?   (*Answer:* $GPA_B = 1.95$, Instructor A)

10-16 For the data in Problem 10-7, compute the expected value for the next year's discount rate.

10-17 For the data in Problem 10-8, compute the project's expected completion time.

10-18 A man wants to decide whether to invest $1000 in a friend's speculative venture. He will do so if he thinks he can get his money back in one year. He believes the probabilities of the various outcomes at the end of one year are as follows:

Result	Probability
$2000 (double his money)	0.3
1500	0.1
1000	0.2
500	0.3
0 (lose everything)	0.1

What would be his expected outcome if he invests the $1000?

10-19 The MSU football team has 10 games scheduled for next season. The business manager wishes to estimate how much money the team can be expected to have left over after paying the season's expenses, including any postseason "bowl game" expenses. From records for the past season and estimates by informed people, the business manager has assembled the following data:

Situation	Prob-ability	Situation	Net Income
Regular season		Regular season	
Win 3 games	0.10	Win 5 or	$250,000
Win 4 games	0.15	fewer	
Win 5 games	0.20	games	
Win 6 games	0.15	Win 6 to 8	400,000
Win 7 games	0.15	games	
Win 8 games	0.10		
Win 9 games	0.07	Win 9 or 10	600,000
Win 10 games	0.03	games	
Postseason		Postseason	Additional
Bowl game	0.10	Bowl game	income of $100,000

What is the expected net income for the team next season? (*Answer:* $355,000)

10-20 In the New Jersey and Nevada gaming casinos, craps is a popular gambling game. One of the many bets available is the "Hard-way 8." A $1 bet in this fashion will win the player $4 if in the game the pair of dice come up 4 and 4 before one of the other ways of totaling 8. For a $1 bet, what is the expected result? (*Answer:* 80¢)

10-21 A man went to Atlantic City with $500 and placed 100 bets of $5 each, one after another, on the same number on the roulette wheel. There are 38 numbers on the wheel and the gaming casino pays 35 times the amount bet if the ball drops into the bettor's numbered slot in the roulette wheel. In addition, the bettor receives back the original $5 bet. Estimate how much money the man is expected to win or lose in Atlantic City.

10-22 For the data in Problems 10-2 and 10-4, assume that the optimistic probability is 20%, the most likely is 50%, and the pessimistic is 30%.

(*a*) What is the expected value of the equivalent uniform annual cost?

(*b*) Compute the expected value for the number of miles, and the corresponding equivalent uniform annual cost.

(*c*) Do the answers to (*a*) and (*b*) match? Why or why not?

10-23 For the data in Problem 10-3, assume that the optimistic probability is 15%, the most likely is 80%, and the pessimistic is 5%.

(*a*) What is the expected value of the rate of return?

(*b*) Compute the expected value for the life, and the corresponding rate of return.

(*c*) Do the answers to (*a*) and (*b*) match? Why or why not?

10-24 For the data in Problem 10-5, assume that the optimistic probability is 20%, the most likely is 50%, and the pessimistic is 30%.

(*a*) What is the expected value of the present worth?

(*b*) Compute the expected value for annual savings, and the corresponding present worth.

(*c*) Do the answers to (*a*) and (*b*) match? Why or why not?

10-25 The energy efficiency project described in Problem 10-14 has a first cost of $150,000, a life of 10 years, and no salvage value. Assume that the interest rate is 8%.

(*a*) What is the equivalent uniform annual worth for the expected annual savings?

(*b*) Compute the equivalent uniform annual worth for the pessimistic, most likely, and optimistic estimates of the annual savings. What is the expected value of the equivalent uniform annual worth?

(*c*) Do the answers to (*a*) and (*b*) match? Why or why not?

10-26 An industrial park is being planned for a tract of land near the river. To prevent flood damage to the industrial buildings that will be built on this low-lying land, an earthen embankment can be constructed. The height of the embankment will be determined by an economic analysis of the costs and benefits. The following data have been gathered.

Embankment Height Above Roadway (m)	Initial Cost
2.0	$100,000
2.5	165,000
3.0	300,000
3.5	400,000
4.0	550,000

Flood Level Above Roadway (m)	Average Frequency That Flood Level Will Exceed Height in Col. 1
2.0	Once in 3 years
2.5	Once in 8 years
3.0	Once in 25 years
3.5	Once in 50 years
4.0	Once in 100 years

The embankment can be expected to last 50 years and will require no maintenance. Whenever the flood water flows over the embankment, $300,000 of damage occurs. Should the embankment be built? If so, to which of the five heights above the roadway? A 12% rate of return is required.

10-27 If your interest rate is 8%, what is the expected value of the present worth of the "extra" insurance payments in Problem 10-9? (*Answer:* $528.7)

10-28 Should the project in problem 10-11 be undertaken, if the firm uses an expected value of present worth to evaluate engineering projects?

(*a*) Compute the PW for each combination of first cost and revenue and the corresponding expected worth.

(*b*) What are the expected first cost, expected net revenue, and corresponding present worth of the expected values?

(*c*) Do the answers for (*a*) and (*b*) match? Why or why not?

(*Answer:* (*b*) $45,900, do project).

10-29 For the data in Problem 10–12.

(*a*) What are the expected savings per year, life, and corresponding rate of return for the expected values?

(*b*) Compute the rate of return for each combination of savings per year and life. What is the expected rate of return?

(*c*) Do the answers for (*a*) and (*b*) match? Why or why not?

10-30 For the data in Problem 10-13.

(*a*) What are the expected savings per year, life, and corresponding present worth for the expected values?

(*b*) Compute the present worth for each combination of savings per year and life. What is the expected present worth?

(*c*) Do the answers for (*a*) and (*b*) match? Why or why not?

Decision Trees

10-31 A decision has been made to perform certain repairs on the outlet works of a small dam. For a particular 36-inch gate valve, there are three available alternatives:

A. Leave the valve as it is.

B. Repair the valve.

C. Replace the valve.

If the valve is left as it is, the probability of a failure of the valve seats, over the life of the project, is 60%; the probability of failure of the valve stem is 50%; and of failure of the valve body is 40%.

If the valve is repaired, the probability of a failure of the seats, over the life of the project, is 40%; of failure of the stem is 30%; and of failure of the body is 20%. If the valve is replaced, the probability of a failure of the seats, over the life of the project, is 30%; of failure of the stem is 20%; and of failure of the body is 10%.

The present worth of cost of future repairs and service disruption of a failure of the seats is $10,000; the present worth of cost of a failure of the stem is $20,000; the present worth of cost of a failure of the body is $30,000. The cost of repairing the valve now is $10,000; and of replacing it is $20,000. If the criterion is to minimize expected costs, which alternative is best?

10-32 A factory building is located in an area subject to occasional flooding by a nearby river. You have been brought in as a consultant to determine whether flood-proofing of the building is economically justified. The alternatives are as follows:

A. Do nothing. Damage in a moderate flood is $10,000 and in a severe flood, $25,000.

B. Alter the factory building at a cost of $15,000 to withstand moderate flooding without damage and to withstand severe flooding with $10,000 damages.

C. Alter the factory building at a cost of $20,000 to withstand a severe flood without damage.

In any year the probability of flooding is as follows: 0.70, no flooding of the river; 0.20, moderate flooding; and 0.10, severe flooding. If interest is 15%

and a 15-year analysis period is used, what do you recommend?

10-33 Five years ago a dam was constructed to impound irrigation water and to provide flood protection for the area below the dam. Last winter a 100-year flood caused extensive damage both to the dam and to the surrounding area. This was not surprising, since the dam was designed for a 50-year flood.

The cost to repair the dam now will be $250,000. Damage in the valley below amounts to $750,000. If the spillway is redesigned at a cost of $250,000 and the dam is repaired for another $250,000, the dam may be expected to withstand a 100-year flood without sustaining damage. However, the storage capacity of the dam will not be increased and the probability of damage to the surrounding area below the dam will be unchanged. A second dam can be constructed up the river from the existing dam for $1 million. The capacity of the second dam would be more than adequate to provide the desired flood protection. If the second dam is built, redesign of the existing dam spillway will not be necessary, but the $250,000 of repairs must be done.

The development in the area below the dam is expected to be complete in 10 years. A new 100-year flood in the meantime would cause a $1 million loss. After 10 years the loss would be $2 million. In addition, there would be $250,000 of spillway damage if the spillway is not redesigned. A 50-year flood is also likely to cause about $200,000 of damage, but the spillway would be adequate. Similarly, a 25-year flood would cause about $50,000 of damage.

There are three alternatives: (1) repair the existing dam for $250,000 but make no other alterations, (2) repair the existing dam ($250,000) and redesign the spillway to take a 100-year flood ($250,000), and (3) repair the existing dam ($250,000) and build the second dam ($1 million). Based on an expected annual cash flow analysis, and a 7% interest rate, which alternative should be selected? Draw a decision tree to clearly describe the problem.

10-34 In Problems 10-13 and 10-30, how much is it worth to the firm to be able to extend the product's life by 3 years, at a cost of $50,000, at the end of the product's initial useful life?

Risk

10-35 An engineer decided to make a careful analysis of the cost of fire insurance for his $200,000 home. From a fire rating bureau he found the following risk of fire loss in any year.

Outcome	Probability
No fire loss	0.986
$ 10,000 fire loss	0.010
40,000 fire loss	0.003
200,000 fire loss	0.001

(a) Compute his expected fire loss in any year.

(b) He finds that the expected fire loss in any year is less than the $550 annual cost of fire insurance. In fact, an insurance agent explains that this is always true. Nevertheless, the engineer buys fire insurance. Explain why this is or is not a logical decision.

10-36 The Graham Telephone Company may invest in new switching equipment. There are three possible outcomes, having net present worth of $6570, $8590, and $9730. The outcomes have probabilities of 0.3, 0.5, and 0.2, respectively. Calculate the expected return and risk associated with this proposal. (*Answer:* $E_{PW} = \$8212, \sigma_{PW} = \$1158$)

10-37 A new machine will cost $25,000. The machine is expected to last 4 years and have no salvage value. If the interest rate is 12%, determine the return and the risk associated with the purchase.

P	0.3	0.4	0.3
Annual savings	$7000	$8500	$9500

10-38 A new product's chief uncertainty is its annual net revenue. So far, $35,000 has been spent on development, but an additional $30,000 is required to finish development. The firm's interest rate is 10%.

(a) What is the expected PW for deciding whether to proceed?

(b) Find the P(loss) and the standard deviation for proceeding.

	State		
	Bad	**OK**	**Great**
Probability	0.3	0.5	0.2
Net revenue	−$15,000	$15,000	$20,000
Life, in years	5	5	10

10-39 (a) In Problem 10-38 how much is it worth to the firm to terminate the product after 1 year if the net revenues are negative?

(b) How much does the ability to terminate early change the P(loss) and the standard deviation?

10-40 What is your risk associated with Problem 10-27?

10-41 Measure the risk for Problems 10-13 and 10-28 using the P(loss), range of PW values, and standard deviation of the PWs. (*Answer:* $\sigma_{PW} = \$127,900$)

10-42 (a) In Problems 10-13 and 10-30, describe the risk using the P(loss) and standard deviation of the PWs.

(b) How much do the answers change if the possible life extension in Problem 10-34 is allowed?

Risk versus Return

10-43 A firm wants to select one new research and development project. The following table summarizes six possibilities. Considering expected return and risk, which projects are good candidates? The firm believes it can earn 5% on a risk-free investment in government securities (labeled as Project F).

Project	IRR	Standard Deviation
1	15.8%	6.5%
2	12.0	4.1
3	10.4	6.3
4	12.1	5.1
5	14.2	8.0
6	18.5	10.0
F	5.0	0.0

10-44 A firm is choosing a new product. The following table summarizes six new potential products. Considering expected return and risk, which products are good candidates? The firm believes it can earn 4% on a risk-free investment in government securities (labeled as Product F).

Product	IRR	Standard Deviation
1	10.4%	3.2%
2	9.8	2.3
3	6.0	1.6
4	12.1	3.6
5	12.2	8.0
6	13.8	6.5
F	4.0	0.0

Simulation

10-45 A project's first cost is $25,000, and it has no salvage value. The interest rate for evaluation is 7%. The project's life is from a discrete uniform distribution that takes on the values 7, 8, 9, and 10. The annual benefit is normally distributed with a mean of $4400 and a standard deviation of $1000. Using Excel's RAND function simulate 25 iterations. What are the expected value and standard deviation of the present worth?

10-46 A factory's power bill is $55,000 a year. The first cost of a small geothermal power plant is normally distributed with a mean of $150,000 and a standard deviation of $50,000. The power plant has no salvage value. The interest rate for evaluation is 8%. The project's life is from a discrete uniform distribution that takes on the values 3, 4, 5, 6, and 7. (The life is relatively short due to corrosion.) The annual operating cost is expected to be about $10,000 per year. Using Excel's RAND function, simulate 25 iterations. What are the expected value and standard deviation of the present worth?

CHAPTER 11

DEPRECIATION

Depreciation and Competitiveness

Depreciation sounds like a boring accounting term, but it is part of determining how much every firm pays in taxes. Reducing taxes is a common goal. Depreciation apportions an asset's capital cost into annual deductions from taxable income. Thus, factories, bulldozers, computers, sports stadiums, and even professional baseball players are depreciated. In each case what can be depreciated, and how fast, is linked to a firm's after-tax profits and competitiveness. More and faster depreciation improves competitiveness.

Why is depreciation important to governments? The answer is that depreciation rules and regulations can be crafted to promote specific outcomes such as total revenue received, increased capital investment, and improved employment rates. In 1981 and 1986, the federal government responded to concerns about U.S. firms building plants overseas by changing the tax code to allow depreciation over shorter recovery periods. The tax code was also changed to give capital investments full depreciation, using a salvage value of $0. In 2002 the tax code was again changed by the Job Creation and Worker Assistance Act to allow an additional first year's depreciation of 30% for a 3-year period. This was later amended to increase the added depreciation in the first year to 50% for tax year 2003. This was intended to stimulate the level of investment by firms in capital assets, and thus to create new jobs in the U.S.

Courtesy of the Internal Revenue Service.

Form 4562 — Department of the Treasury, Internal Revenue Service

Depreciation and Amortization (Including Information on Listed Property)

▶ See separate instructions. ▶ Attach to your tax return.

OMB No. 1545-0172

2006

Attachment Sequence No. 67

Name(s) shown on return | Business or activity to which this form relates | Identifying number

Part I Election To Expense Certain Property Under Section 179
Note: *If you have any listed property, complete Part V before you complete Part I.*

1 Maximum amount. See the instructions for a higher limit for certain businesses	1	$108,000
2 Total cost of section 179 property placed in service (see instructions)	2	
3 Threshold cost of section 179 property before reduction in limitation	3	$430,000
4 Reduction in limitation. Subtract line 3 from line 2. If zero or less, enter -0-	4	
5 Dollar limitation for tax year. Subtract line 4 from line 1. If zero or less, enter -0-. If married filing separately, see instructions	5	

(a) Description of property	(b) Cost (business use only)	(c) Elected cost
6		

7 Listed property. Enter the amount from line 29	7	
8 Total elected cost of section 179 property. Add amounts in column (c), lines 6 and 7	8	
9 Tentative deduction. Enter the **smaller** of line 5 or line 8	9	
10 Carryover of disallowed deduction from line 13 of your 2005 Form 4562	10	
11 Business income limitation. Enter the smaller of business income (not less than zero) or line 5 (see instructions)	11	
12 Section 179 expense deduction. Add lines 9 and 10, but do not enter more than line 11	12	
13 Carryover of disallowed deduction to 2007. Add lines 9 and 10, less line 12 ▶	13	

Note: *Do not use Part II or Part III below for listed property. Instead, use Part V.*

QUESTIONS TO CONSIDER

1. How can the importance of depreciation be measured?
2. What are some of the similarities and differences between depreciating a bulldozer and a ballplayer?
3. Is the added first-year depreciation still part of the tax code? How has it changed since it was legislated in 2002?
4. At times legislative bodies have been criticized for using potential changes in the depreciation and tax codes to stimulate lobbying and campaign donations. Is this truly a practical or ethical concern?

After Completing This Chapter...

The student should be able to:

- Describe depreciation, deterioration, and obsolescence.
- Distinguish various types of depreciable property and differentiate between depreciation expenses and other business expenses.
- Use *historical* depreciation methods to calculate the *annual depreciation charge* and *book value* over the asset's life.
- Explain the differences between the historical depreciation methods and the modified accelerated cost recovery system (MACRS).
- Use MACRS to calculate allowable *annual depreciation charge* and *book value* over the asset's life for various cost bases, property classes, and recovery periods.
- Fully account for *capital gains/losses, ordinary losses,* and *depreciation recapture* due to the disposal of a depreciated business asset.
- Use the *units of production* and *depletion* depreciation methods as needed in engineering economic analysis problems.
- Use spreadsheets to calculate depreciation.

Wₑ have so far dealt with a variety of economic analysis problems and many techniques for their solution. In the process we have avoided income taxes, which are an important element in many private sector economic analyses. Now, we move to more realistic—and more complex—situations.

Our government taxes individuals and businesses to support its processes—lawmaking, domestic and foreign economic policy making, even the making and issuing of money itself. The omnipresence of taxes requires that they be included in economic analyses, which means that we must understand the *way* taxes are imposed. For capital equipment, knowledge about depreciation is required to compute income taxes. Chapter 11 examines depreciation, and Chapter 12 illustrates how depreciation and other effects are incorporated in income tax computations. The goal is to support decision making on engineering projects, not to support final tax calculations.

BASIC ASPECTS OF DEPRECIATION

The word **depreciation** is defined as a "decrease in value." This is somewhat ambiguous because *value* has several meanings. In economic analysis, value may refer to either *market value* or *value to the owner*. For example, an assembly line is far more valuable to the manufacturing firm that it was designed for, than it is to a used equipment market. Thus, we now have two definitions of depreciation: a decrease in value to the market or a decrease to the owner.

Deterioration and Obsolescence

A machine may depreciate because it is **deteriorating** or wearing out and no longer performing its function as well as when it was new. Many kinds of machinery require increased maintenance as they age, reflecting a slow but continuing failure of individual parts. In other types of equipment, the quality of output may decline due to wear on components and resulting poorer mating of parts. Anyone who has worked to maintain a car has observed deterioration due to failure of individual parts (such as fan belts, mufflers, and batteries) and the wear on components (such as bearings, piston rings, and alternator brushes).

Depreciation is also caused by **obsolescence.** A machine that is in excellent working condition, and serving a needed purpose, may still be obsolete. In the 1970s, mechanical business calculators with hundreds of gears and levers became obsolete. The advance of integrated circuits resulted in a completely different and far superior approach to calculator design. Thus, mechanical calculators rapidly declined or depreciated in value. Generations of computers have followed this pattern. The continuing stream of newer models makes older ones obsolete.

The accounting profession defines depreciation in yet another way, as allocating an asset's cost over its **useful** or **depreciable life.** Thus, we now have *three distinct definitions of depreciation:*

1. Decline in market value of an asset.
2. Decline in value of an asset to its owner.
3. Systematic allocation of an asset's cost over its depreciable life.

Depreciation and Expenses

It is the third (accountant's) definition that is used to compute depreciation for business assets. Business costs are generally either **expensed** or **depreciated. Expensed** items, such as labor, utilities, materials, and insurance, are part of regular business operations and are "consumed" over short periods of time (sometimes recurring). These costs do not lose value gradually over time. For tax purposes they are subtracted from business revenues when they occur. Expensed costs reduce income taxes because businesses are able to *write off* their full amount when they occur.

In contrast, business costs due to capital assets (buildings, forklifts, chemical plants, etc.) are not fully written off when they occur. Capital assets lose value gradually and must be written off or **depreciated** over an extended period. For instance, consider a injection-molding machine used to produce the plastic beverage cups found at sporting events. The plastic pellets melted into the cup shape lose their value as raw material directly after manufacturing. The raw material cost for production material (plastic pellets) is expensed immediately. On the other hand, the injection-molding machine itself will lose value over time, and thus its cost (purchase price and installation expenses) are written off (or depreciated) over its **depreciable life** or **recovery period.** This is often different from the asset's useful or most economic life. Depreciable life is determined by the depreciation method used to spread out the cost—depreciated assets of many types operate well beyond their depreciable life.

Depreciation is a **noncash** cost that requires no exchange of dollars. Companies do not write a check to someone to **pay** their depreciation expenses. Rather, these are business expenses that are allowed by the government to offset the loss in value of business assets. Remember, the company has paid for assets up front; depreciation is simply a way to claim these "business expenses" over time. Depreciation deductions reduce the taxable income of businesses and thus reduce the amount of taxes paid. Since taxes are cash flows, depreciation must be considered in after-tax economic analysis.

In general, business assets can be depreciated only if they meet the following basic requirements:

1. The property must be used for business purposes to produce income.
2. The property must have a useful life that can be determined, and this life must be longer than one year.
3. The property must be an asset that decays, gets used up, wears out, becomes obsolete, or loses value to the owner from natural causes.

EXAMPLE 11-1

Consider the costs that are incurred by a local pizza business. Identify each cost as either *expensed* or *depreciated* and describe why that term applies.

- Cost for pizza dough and toppings
- Cost to pay wages for janitor
- Cost of a new baking oven
- Cost of new delivery van
- Cost of furnishings in dining room
- Utility costs for soda refrigerator

Cost Item	Type of Cost	Why
Pizza dough and toppings	Expensed	Life < 1 year; lose value immediately
New delivery van	Depreciated	Meets 3 requirements for depreciation
Wages for janitor	Expensed	Life < 1 year; lose value immediately
Furnishings in dining room	Depreciated	Meet 3 requirements for depreciation
New baking oven	Depreciated	Meets 3 requirements for depreciation
Utilities for soda refrigerator	Expensed	Life < 1 year; lose value immediately

Types of Property

The rules for depreciation are linked to the classification of business property as either tangible or intangible. Tangible property is further classified as either real or personal.

Tangible property can be seen, touched, and felt.

Real property includes land, buildings, and all things growing on, built upon, constructed on, or attached to the land.

Personal property includes equipment, furnishings, vehicles, office machinery, and anything that is tangible excluding assets defined as *real property*.

Intangible property is all property that has value to the owner but cannot be directly seen or touched. Examples include patents, copyrights, trademarks, trade names, and franchises.

Many different types of property that wear out, decay, or lose value can be depreciated as business assets. This wide range includes copy machines, helicopters, buildings, interior furnishings, production equipment, and computer networks. Almost all tangible property can be depreciated.

One important and notable exception is land, which is *never* depreciated. Land does not wear out, lose value, or have a determinable useful life and thus does not qualify as a depreciable property. Rather than decreasing in value, most land becomes more valuable as time passes. In addition to the land itself, expenses for clearing, grading, preparing, planting, and landscaping are not generally depreciated because they have no fixed useful life. Other tangible property that *cannot* be depreciated includes factory inventory, containers considered as inventory, and leased property. The leased property exception highlights the fact that only the owner of property may claim depreciation expenses.

Tangible properties used in *both* business and personal activities, such as a vehicle used in a consulting engineering firm that is also used to take one's kids to school, can be depreciated, but only in proportion to the use for business purposes.

Depreciation Calculation Fundamentals

To understand the complexities of depreciation, the first step is to examine the fundamentals of depreciation calculations. Figure 11-1 illustrates the general depreciation problem of allocating the total depreciation charges over the asset's depreciable life. The vertical axis is labeled **book value.** At time zero the curve of book value starts at the cost basis (= the

FIGURE 11-1 General
depreciation.

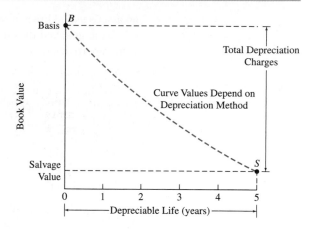

first cost plus installation cost). Over time, the book value declines to the salvage value. Thus, at any point in time:

Book value = Asset cost − Depreciation charges made to date

Looked at another way, book value is the asset's remaining unallocated cost.

In Figure 11-1, *book value* goes from a value of B at time zero in the recovery period to a value of S at the end of Year 5. Thus, book value is a *dynamic* variable that changes over an asset's recovery period. The equation used to calculate an asset's book value over time is:

$$BV_t = \text{Cost basis} - \sum_{j=1}^{t} d_j \qquad (11\text{-}1)$$

where BV_t = book value of the depreciated asset at the end of time t

Cost basis = B = dollar amount that is being depreciated; this includes the asset's purchase price as well as any other costs necessary to make the asset "ready for use"

$\displaystyle\sum_{j=1}^{t} d_j$ = sum of depreciation deductions taken from time 0 to time t, where d_j is the

depreciation deduction in Year j

Equation 11-1 shows that year-to-year depreciation charges reduce an asset's book value over its life. The following section describes methods that are or have been allowed under federal tax law for quantifying these yearly depreciation deductions.

HISTORICAL DEPRECIATION METHODS

Allowing businesses to depreciate capital assets has long been part of the tax code. However, over time various depreciation methods have been used to calculate these deductions. In general, accounting depreciation methods can be categorized as follows.

Pre-1981 historical methods: These methods include the *straight-line, sum-of-the-years'-digits,* and *declining balance* methods. Each method required estimates of an asset's useful life and salvage value. Firms could elect to use any of these methods for assets, and thus there was little uniformity in how depreciation expenses were reported.

1981–1986 method: With the Economic Recovery Tax Act (ERTA) of 1981, Congress created the accelerated cost recovery system (ACRS). The ACRS method had three key features: (1) property class lives were created, and all depreciated assets were assigned to one particular category; (2) the need to estimate salvage values was eliminated because all assets were *fully* depreciated over their recovery period; and (3) shorter recovery periods were used to calculate annual depreciation, which *accelerated* the write-off of capital costs more quickly than did the historical methods—thus the name.

1986 to present: The modified accelerated cost recovery system (MACRS) has been in effect since the Tax Reform Act of 1986 (TRA-86). The MACRS method is similar to the ACRS system except that (1) the number of property classes was expanded and (2) the annual depreciation percentages were modified to include a half-year convention for the first and final years.

In this chapter, our primary focus is to describe the MACRS depreciation method. However, it is useful to first describe three historical depreciation methods. These methods are used in some countries, and MACRS is based on two of them.

Straight-Line Depreciation

The simplest and best known depreciation method is **straight-line depreciation.** To calculate the constant **annual depreciation charge,** the total amount to be depreciated, $B - S$, is divided by the depreciable life, in years, N:[1]

$$\text{Annual depreciation charge} = d_t = \frac{B - S}{N} \qquad (11\text{-}2)$$

EXAMPLE 11-2

Consider the following:

Cost of the asset, B	$900
Depreciable life, in years, N	5
Salvage value, S	$70

Compute the straight-line depreciation schedule.

SOLUTION

$$\text{Annual depreciation charge} = d_t = \frac{B - S}{N} = \frac{900 - 70}{5} = \$166$$

---

[1] N is used for the depreciation period because it may be shorter than n, the horizon (or project life).

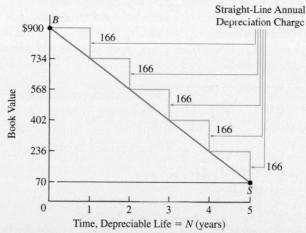

FIGURE 11-2 Straight-line depreciation.

Year, t	Depreciation for Year t, d_t	Sum of Depreciation Charges Up to Year t, $\sum_{j=1}^{t} d_j$	Book Value at the End of Year t, $BV_t = B - \sum_{j=1}^{t} d_j$
1	\$166	\$166	$900 - 166 = 734$
2	166	332	$900 - 332 = 568$
3	166	498	$900 - 498 = 402$
4	166	664	$900 - 664 = 236$
5	166	830	$900 - 830 = 70 = S$

This situation is illustrated in Figure 11-2. Notice the constant \$166 d_t each year for 5 years, and that the asset has been depreciated down to a book value of \$70, which was the estimated salvage value.

The straight-line (SL) method is often used for intangible property. For example, Veronica's firm bought a patent in April that was not acquired as part of acquiring a business. She paid \$6800 for this patent and must use the straight-line method to depreciate it over 17 years with no salvage value. The annual depreciation is \$400 (=\$6800/17). Since the patent was purchased in April, the first year's deduction must be prorated over the 9 months of ownership. This year the deduction is \$300 (=\$400 × 9/12), and then next year she can begin taking the full \$400 per year.

Sum-of-Years'-Digits Depreciation

Another method for allocating an asset's cost *minus* salvage value *over* its depreciable life is called **sum-of-years'-digits (SOYD) depreciation.** This method results in larger-than-straight-line depreciation charges during an asset's early years and smaller charges as the asset nears the end of its depreciable life. Each year, the depreciation charge equals a fraction of the total amount to be depreciated $(B - S)$. The denominator of the fraction is the sum of the years' digits. For example, if the depreciable life is 5 years, $1 + 2 + 3 + 4 + 5 = 15 =$ SOYD. Then $5/15, 4/15, 3/15, 2/15$, and $1/15$ are the fractions from Year 1 to Year 5. Each year the depreciation charge shrinks by $1/15$ of $B - S$. Because this change is the same every year, SOYD depreciation can be modeled as an arithmetic gradient, G. The equations can also be written as

$$\begin{pmatrix} \text{Sum-of-years'-digits} \\ \text{depreciation charge for} \\ \text{any year} \end{pmatrix} = \dfrac{\begin{pmatrix} \text{Remaining depreciable life} \\ \text{at beginning of year} \end{pmatrix}}{\begin{pmatrix} \text{Sum of years' digits} \\ \text{for total depreciable life} \end{pmatrix}} (\text{Total amount depreciated})$$

$$d_t = \frac{N - t + 1}{\text{SOYD}} (B - S) \tag{11-3}$$

where $d_t =$ depreciation charge in any year t

$N =$ number of years in depreciable life

SOYD $=$ sum of years' digits, calculated as $N(N + 1)/2 = \text{SOYD}$

$B =$ cost of the asset made ready for use

$S =$ estimated salvage value after depreciable life

EXAMPLE 11-3

Compute the SOYD depreciation schedule for the situation in Example 11.2.

Cost of the asset, B	$900
Depreciable life, in years, N	5
Salvage value, S	$70

SOLUTION

$$\text{SOYD} = \frac{5 \times 6}{2} = 15$$

Thus,

$$d_1 = \frac{5 - 1 + 1}{15} (900 - 70) = 277$$

$$d_2 = \frac{5 - 2 + 1}{15} (900 - 70) = 221$$

$$d_3 = \frac{5 - 3 + 1}{15} (900 - 70) = 166$$

$$d_4 = \frac{5 - 4 + 1}{15} (900 - 70) = 111$$

$$d_5 = \frac{5 - 5 + 1}{15} (900 - 70) = 55$$

FIGURE 11-3 Sum-of-years'-digits depreciation.

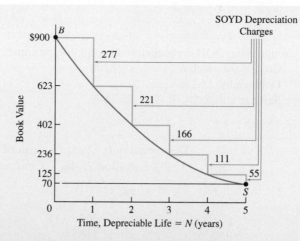

Year,	Depreciation for Year t,	Sum of Depreciation Charges Up to Year t,	Book Value at End of Year t,
t	d_t	$\sum_{j=1}^{t} d_j$	$BV_t = B - \sum_{j=1}^{t} d_j$
1	$277	$277	$900 - 277 = 623$
2	221	498	$900 - 498 = 402$
3	166	664	$900 - 664 = 236$
4	111	775	$900 - 775 = 125$
5	55	830	$900 - 830 = \ 70 = S$

These data are plotted in Figure 11-3.

Declining Balance Depreciation

Declining balance depreciation applies a *constant depreciation rate* to the property's declining book value. Two rates were commonly used before the 1981 and 1986 tax revisions, and they are used to compute MACRS depreciation percentages. These are 150 and 200% of the straight-line rate. Since 200% is twice the straight-line rate, it is called **double declining balance,** or DDB; the general equation is

$$\text{Double declining balance} \quad d_t = \frac{2}{N}(\text{Book value}_{t-1}) \tag{11-4a}$$

Since book value equals cost *minus* depreciation charges to date,

$$\text{DDB} \quad d_t = \frac{2}{N}(\text{Cost} - \text{Depreciation charges to date})$$

or

$$d_t = \frac{2}{N}\left(B - \sum_{j=1}^{t-1} d_j \right) \tag{11-4b}$$

EXAMPLE 11-4

Compute the DDB depreciation schedule for the situations in Examples 11-2 and 11-3.

Cost of the asset, B	$900
Depreciable life, in years, N	5
Salvage value, S	$70

SOLUTION

Year, t	Depreciation for Year t Using Equation 11-4a, d_t	Sum of Depreciation Charges Up to Year t, $\sum_{j=1}^{t} d_j$	Book Value at End of Year t, $BV_t = B - \sum_{j=1}^{t} d_j$
1	$(2/5)900 = 360$	$360	$900 - 360 = 540$
2	$(2/5)540 = 216$	576	$900 - 576 = 324$
3	$(2/5)324 = 130$	706	$900 - 706 = 194$
4	$(2/5)194 = 78$	784	$900 - 784 = 116$
5	$(2/5)116 = 46$	830	$900 - 830 = 70 = S$

Figure 11-4 illustrates the situation.

FIGURE 11-4 Declining balance depreciation.

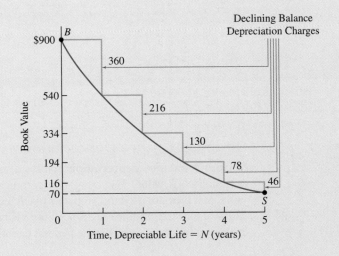

The final salvage value of $70 for Examples 11-2, 11-3, and 11-4 was chosen to match the ending value for the double declining balance method. This does not normally happen. If the final salvage value of Example 11-4 had not been $70, the double declining balance method would have had to be modified. One modification stops further depreciation once the book value has come to equal the salvage value—this prevents taking too much depreciation. The other modification would switch from declining balance depreciation to straight line—this ensures taking enough depreciation.

These modifications are not detailed here because (1) MACRS has been the legally appropriate system since 1986 and (2) MACRS incorporates the shift from declining balance to straight-line depreciation.

MODIFIED ACCELERATED COST RECOVERY SYSTEM (MACRS)

The modified accelerated cost recovery system (MACRS) depreciation method, introduced by the Tax Reform Act of 1986, was continued with the Taxpayer Relief Act of 1997. Three major advantages of MACRS are that (1) the "property class lives" are less than the "actual useful lives," (2) salvage values are assumed to be zero, and (3) tables of annual percentages simplify computations.

The definition of the MACRS classes of depreciable property is based on work by the U.S. Treasury Department. In 1971 the Treasury published guidelines for about 100 broad asset classes. For each class there was a lower limit, a midpoint, and an upper limit of useful life, called the **asset depreciation range (ADR).** The ADR midpoint lives were somewhat shorter than the actual average useful lives. These guidelines have been incorporated into MACRS so that the property class lives are again shorter than the ADR midpoint lives.

Use of MACRS focuses on the general depreciation system (GDS), which is based on declining balance with switch to straight-line depreciation. The alternative depreciation system (ADS) provides for a longer period of recovery and uses straight-line depreciation. Thus it is much less economically attractive. Under law, ADS must be used for (1) any tangible property used primarily outside the U.S., (2) any property that is tax exempt or financed by tax-exempt bonds, and (3) farming property placed in service when uniform capitalization rules are not applied. The ADS may also be *elected* for property that can be depreciated using the GDS system. However, once ADS has been elected for an asset, it is not possible to switch back to the GDS system. Because the ADS makes the depreciation deductions less valuable, unless ADS is specifically mentioned, subsequent discussion assumes the GDS system when reference is made to MACRS.

Once a property has been determined to be eligible for depreciation, the next step is to calculate its depreciation deductions over its life. The following information is required to calculate these deductions:

- The asset's cost basis.
- The asset's *property class* and *recovery period.*
- The asset's placed-in-service date.

Cost Basis and Placed-in-Service Date

The cost basis, B, is the cost to obtain and place the asset in service fit for use. However, for real property the basis may also include certain fees and charges that the buyer pays as part of the purchase. Examples of such fees include legal and recording fees, abstract fees, survey charges, transfer taxes, title insurance, and amounts that the seller owes that you pay (back taxes, interest, sales commissions, etc.).

Depreciation for a business asset begins when the asset is *placed in service* for a business purpose. If an asset is purchased and used in a personal context, depreciation may not be taken. If that asset is later used in business for income-producing activity, depreciation may begin with the change in usage.

Property Class and Recovery Period

Each depreciated asset is placed in a *MACRS property class,* which defines the **recovery period** and the depreciation percentage for each year. Historically the IRS assigned each type of depreciable asset a *class life* or an *asset depreciation range*. With MACRS, asset class lives have been pooled together in the *property classes*. Table 11-1 lists the class lives and GDS and ADS property classes for several example depreciable assets. Table 11-2 lists the MACRS GDS property classes.

The MACRS GDS property classes are described in more detail in Table 11-2. The proper MACRS property class can be found several different ways. Of the five approaches listed, the first one that works should be used.

1. Property class given in problem.
2. Asset is named in Table 11-2.
3. IRS tables or Table 11-1.
4. Class life.
5. Seven-year property for "all other property not assigned to another class."

Once the MACRS property class is known, as well as the placed-in-service date and the cost basis, the year-to-year depreciation deductions can be calculated for GDS assets

TABLE 11-1 Example Class Lives and MACRS Property Classes

IRS Asset Class	Asset Description	Class Life (years) ADR	MACRS Property Class (years) GDS	ADS
00.11	Office furniture, fixtures, and equipment	10	7	10
00.12	Information systems: computers/peripheral	6	5	6
00.22	Automobiles, taxis	3	5	5
00.241	Light general-purpose trucks	4	5	5
00.25	Railroad cars and locomotives	15	7	15
00.40	Industrial steam and electric distribution	22	15	22
01.11	Cotton gin assets	12	7	12
01.21	Cattle, breeding or dairy	7	5	7
13.00	Offshore drilling assets	7.5	5	7.5
13.30	Petroleum refining assets	16	10	16
15.00	Construction assets	6	5	6
20.10	Manufacture of grain and grain mill products	17	10	17
22.2	Manufacture of yarn, thread, and woven fabric	11	7	11
24.10	Cutting of timber	6	5	6
32.20	Manufacture of cement	20	15	20
37.11	Manufacture of motor vehicles	12	7	12
48.11	Telephone communications assets and buildings	24	15	24
48.2	Radio and television broadcasting equipment	6	5	6
49.12	Electric utility nuclear production plant	20	15	20
49.13	Electric utility steam production plant	28	20	28
49.23	Natural gas production plant	14	7	14
50.00	Municipal wastewater treatment plant	24	15	24
80.00	Theme and amusement park assets	12.5	7	12.5

TABLE 11-2 MACRS GDS Property Classes

Property Class	Personal Property (all property except real estate)
3-year property	Special handling devices for food and beverage manufacture Special tools for the manufacture of finished plastic products, fabricated metal products, and motor vehicles Property with ADR class life of 4 years or less
5-year property	Automobiles* and trucks Aircraft (of non-air-transport companies) Equipment used in research and experimentation Computers Petroleum drilling equipment Property with ADR class life of more than 4 years and less than 10 years
7-year property	All other property not assigned to another class Office furniture, fixtures, and equipment Property with ADR class life of 10 years or more and less than 16 years
10-year property	Assets used in petroleum refining and certain food products Vessels and water transportation equipment Property with ADR class life of 16 years or more and less than 20 years
15-year property	Telephone distribution plants Municipal sewage treatment plants Property with ADR class life of 20 years or more and less than 25 years
20-year property	Municipal sewers Property with ADR class life of 25 years or more

Property Class	Real Property (real estate)
27.5 years	Residential rental property (does not include hotels and motels)
39 years	Nonresidential real property

*The depreciation deduction for most automobiles placed in service after 2005 is limited to $2960 (maximum) the first tax year, $4700 the second year, $2850 the third year, and $1675 per year in subsequent years.
Source: U.S. Department of the Treasury, Internal Revenue Service Publication 946, *How to Depreciate Property.* Washington, DC: Government Printing Office.

over their depreciable life using

$$d_t = B \times r_t \tag{11-5}$$

where d_t = depreciation deduction in year t
B = cost basis being depreciated
r_t = appropriate MACRS percentage rate

Percentage Tables

The IRS has prepared tables to assist in calculating depreciation charges when MACRS GDS depreciation is used. Table 11-3 gives the yearly depreciation percentages (r_t) that are used for the six personal property classes (3-, 5-, 7-, 10-, 15-, and 20-year property classes), and Table 11-4 gives the percentages for nonresidential real property. Notice that the values are given in *percentages*—thus, for example, the value of 33.33% (given in Table 11-3 for Year 1 for a 3-year MACRS GDS property) is 0.3333.

TABLE 11-3 MACRS Depreciation for Personal Property: Half-Year Convention

Recovery Year	Applicable Percentage for Property Class					
	3-Year Property	5-Year Property	7-Year Property	10-Year Property	15-Year Property	20-Year Property
1	33.33	20.00	14.29	10.00	5.00	3.750
2	44.45	32.00	24.49	18.00	9.50	7.219
3	14.81*	19.20	17.49	14.40	8.55	6.677
4	7.41	11.52*	12.49	11.52	7.70	6.177
5		11.52	8.93*	9.22	6.93	5.713
6		5.76	8.92	7.37	6.23	5.285
7			8.93	6.55*	5.90*	4.888
8			4.46	6.55	5.90	4.522
9				6.56	5.91	4.462*
10				6.55	5.90	4.461
11				3.28	5.91	4.462
12					5.90	4.461
13					5.91	4.462
14					5.90	4.461
15					5.91	4.462
16					2.95	4.461
17						4.462
18						4.461
19						4.462
20						4.461
21						2.231

Computation method

- The 3-, 5-, 7-, and 10-year classes use 200% and the 15- and 20-year classes use 150% declining balance depreciation.
- All classes convert to straight-line depreciation in the optimal year, shown with asterisk (*).
- A half-year of depreciation is allowed in the first and last recovery years.
- If more than 40% of the year's MACRS property is placed in service in the last 3 months, then a midquarter convention must be used with depreciation tables that are not shown here.

Notice in Table 11-3 that the depreciation percentages continue for *one year beyond* the property class life. For example, a MACRS 10-year property has an r_t value of 3.28% in Year 11. This is due to the *half-year convention* that also halves the percentage for the first year. The half-year convention assumes that all assets are placed in service at the midpoint of the first year.

Another characteristic of the MACRS percentage tables is that the r_t values in any column sum to 100%. This means that assets depreciated using MACRS are *fully depreciated* at the end of the recovery period. This assumes a salvage value of zero. This is a departure from the pre-1981 historical methods, where an estimated salvage value was considered.

TABLE 11-4 MACRS Depreciation for Real Property (real estate)*

Recovery Year	Recovery Percentages for Nonresidential Real Property (month placed in service)											
	1	2	3	4	5	6	7	8	9	10	11	12
1	2.461	2.247	2.033	1.819	1.605	1.391	1.177	0.963	0.749	0.535	0.321	0.107
2–39	2.564	2.564	2.564	2.564	2.564	2.564	2.564	2.564	2.564	2.564	2.564	2.564
40	0.107	0.321	0.535	0.749	0.963	1.177	1.391	1.605	1.819	2.033	2.247	2.461

*The useful life is 39 years for nonresidential real property. Depreciation is straight line using the midmonth convention. Thus a property placed in service in January would be allowed $11\frac{1}{2}$ months depreciation for recovery Year 1.

Where MACRS Percentage Rates (r_t) Come From

This section describes the connection between historical depreciation methods and the MACRS percentages that are shown in Table 11-3. Before ACRS and MACRS, the most common depreciation method was declining balance with a switch to straight line. That combined method is used for MACRS with three further assumptions.

1. Salvage values are assumed to be zero for all assets.
2. The first and last years of the recovery period are each assumed to be *half-year*.
3. The declining balance rate is 200% for 3-, 5-, 7-, and 10-year property and 150% for 15- and 20-year property.

As shown in Example 11-5, the MACRS percentage rates can be derived from these rules and the declining balance and straight-line methods. However, it is obviously much easier to simply use the r_t values from Tables 11-3 and 11-4.

EXAMPLE 11-5

Consider a 5-year MACRS property asset with an installed and "made ready for use" cost basis of $100. Develop the MACRS percentage rates (r_t) for the asset based on the underlying depreciation methods.

SOLUTION

To develop the 5-year MACRS property percentage rates, we use the 200% declining balance method, switching over to straight line at the optimal point. Since the assumed salvage value is zero, the entire cost basis of $100 is depreciated. Also the $100 basis mimics the 100% that is used in Table 11-3.

Let's explain the accompanying table year by year. In Year 1 the basis is $100 − 0, and the d_t values are halved for the initial half-year assumption. Double declining balance has a rate of 40% for 5 years ($= 2/5$). This is larger than straight-line for Year 1. So one-half of the 40% is used for Year 1. The rest of the declining balance computations are simply 40%×(basis minus the cumulative depreciation).

In Year 2 there are 4.5 years remaining for straight line, so 4.5 is the denominator for dividing the remaining $80 in book value. Similarly in Year 3 there are 3.5 years remaining. In Year 4 the DDB and SL calculations happen to be identical, so the switch from DDB to SL can be done in either Year 4 or Year 5. Once we know that the SL depreciation is 11.52 at the switch point, then the only further calculation is to halve that for the last year.

Notice that the DDB calculations get smaller every year, so that at some point the straight-line calculations lead to faster depreciation. This point is the optimal switch point, and it is built into Table 11-3 for MACRs.

Year	DDB Calculation	SL Calculation	MACRS r_t (%) Rates	Cumulative Depreciation (%)
1	$1/(2/5)(100 - 0) = \mathbf{20.00}$	$^1/_2(100 - 0)/5 = 10.00$	20.00 (DDB)	20.00
2	$(2/5)(100 - 20.00) = \mathbf{32.00}$	$(100 - 20)/4.5 = 17.78$	32.00 (DDB)	52.00
3	$(2/5)(100 - 52.00) = \mathbf{19.20}$	$(100 - 52)/3.5 = 13.71$	19.20 (DDB)	71.20
4	$(2/5)(100 - 71.20) = \mathbf{11.52}$	$(100 - 71.20)/2.5 = \mathbf{11.52}$	11.52 (either)	82.72
5		$\mathbf{11.52}$	11.52 (SL)	94.24
6		$(^1/_2)(11.52) = \mathbf{5.76}$	5.76 (SL)	100.00

The values given in this example match the r_t percentage rates given in Table 11-3 for a 5-year MACRS property.

MACRS Method Examples

Remember the key points in using MACRS: (1) what type of asset you have, and whether it qualifies as depreciable property, (2) the amount you are depreciating [cost basis], and (3) when you are placing the asset in service. Let's look at several examples of using MACRS to calculate both depreciation deductions and book values.

EXAMPLE 11-6

Use the MACRS GDS method to calculate the yearly depreciation allowances and book values for a firm that has purchased $150,000 worth of office equipment that qualifies as depreciable property. The equipment is estimated to have a salvage (market) value of $30,000 (20% of the original cost) after the end of its depreciable life.

SOLUTION

1. The assets qualify as depreciable property.
2. The cost basis is given as $150,000.
3. The assets are being placed in service in Year 1 of our analysis.
4. MACRS GDS applies.
5. The salvage value is not used with MACRS to calculate depreciation or book value.

Office equipment is listed in Table 11-2 as a 7-year property. We now use the MACRS GDS 7-year property percentages from Table 11-3 and Equation 11-5 to calculate the year-to-year depreciation allowances. We use Equation 11-1 to calculate the book value of the asset.

Year, t	MACRS, r_t		Cost Basis	d_t	Cumulative d_t	$BV_t = B - \text{Cum.}\, d_t$
1	14.29%	×	$150,000	$ 21,435	$ 21,435	$128,565
2	24.49		150,000	36,735	58,170	91,830
3	17.49		150,000	26,235	84,405	65,595
4	12.49		150,000	18,735	103,140	46,860
5	8.93		150,000	13,395	116,535	33,465
6	8.92		150,000	13,380	129,915	20,085
7	8.93		150,000	13,395	143,310	6,690
8	4.46		150,000	6,690	150,000	0
	100.00%			$150,000		

Notice in this example several aspects of the MACRS depreciation method: (1) the sum of the r_t values is 100.00%, (2) this 7-year MACRS GDS property is depreciated over 8 years (= property class life + 1), and (3) the book value after 8 years is $0.

EXAMPLE 11-7

Investors in the JMJ Group purchased a hotel resort in April. The group paid $2.0 million for the hotel resort and $500,000 for the grounds surrounding the resort. The group sold the resort 5 years later in August. Calculate the depreciation deductions for Years 1 through 6. What was the book value at the time the resort was sold?

SOLUTION

Hotels are nonresidential real property and are depreciated over a 39-year life. Table 11-4 lists the percentages for each year. In this case the cost basis is $2.0 million, and the $500,000 paid for the land is not depreciated. JMJ's depreciation is calculated as follows:

Year 1 (obtained in April)	$d_1 = 2,000,000(1.819\%) = \$36,380$
Year 2	$d_2 = 2,000,000(2.564\%) = 51,280$
Year 3	$d_3 = 2,000,000(2.564\%) = 51,280$
Year 4	$d_4 = 2,000,000(2.564\%) = 51,280$
Year 5	$d_5 = 2,000,000(2.564\%) = 51,280$
Year 6 (disposed of in August)	$d_6 = 2,000,000(1.605\%) = 32,100$

Thus the hotel's book value when it was sold was:

$$BV_6 = B - (d_1 + d_2 + d_3 + d_4 + d_5 + d_6)$$
$$= 2,000,000 - (273,600) = \$1,726,400$$

The value of the land has not changed in terms of book value.

Comparing MACRS and Historical Methods

In Examples 11-2 through 11-4 we used the *straight-line, sum-of-the-years'-digits,* and *declining balance* depreciation methods to illustrate how the book value of an asset that cost $900 and had a salvage value of $70 changed over its 5-year depreciation life. Figures 11-2 through 11-4 provided a graphical view of book value over the 5-year depreciation period using these methods. Example 11-8 compares the MACRS GDS depreciation method directly against the historical methods.

EXAMPLE 11-8

Consider the equipment that was purchased in Example 11-6. Calculate the asset's depreciation deductions and book values over its depreciable life for MACRS and the historical methods.

SOLUTION

Table 11-5 and Figure 11-5 compare MACRS and historical depreciation methods. MACRS depreciation is the most *accelerated* or fastest depreciation method—remember its name is the modified *accelerated* cost recovery system. The book value drops fastest and furthest with MACRS, thus the present worth is the largest for the MACRS depreciation deductions.

Depreciation deductions *benefit* a firm after taxes because they reduce taxable income and taxes. The time value of money ensures that it is better to take these deductions as soon as possible. In general, MACRS, which allocates larger deductions earlier in the depreciable life, provides more economic benefits than historical methods.

TABLE 11-5 Comparison of MACRS and Historical Methods for Asset in Example 11-6

Year, t	MACRS d_t	MACRS BV_t	Straight Line d_t	Straight Line BV_t	Double Declining d_t	Double Declining BV_t	Sum-of-Years' Digits d_t	Sum-of-Years' Digits BV_t
1	21,435	128,565	12,000	138,000	30,000	120,000	21,818	128,182
2	36,735	91,830	12,000	126,000	24,000	96,000	19,636	108,545
3	26,235	65,595	12,000	114,000	19,200	76,800	17,455	91,091
4	18,735	46,860	12,000	102,000	15,360	61,440	15,273	75,818
5	13,395	33,465	12,000	90,000	12,288	49,152	13,091	62,727
6	13,380	20,085	12,000	78,000	9,830	39,322	10,909	51,818
7	13,395	6,690	12,000	66,000	7,864	31,457	8,727	43,091
8	6,690	0	12,000	54,000	1,457	30,000	6,545	36,545
9	0	0	12,000	42,000	0	30,000	4,364	32,182
10	0	0	12,000	30,000	0	30,000	2,182	30,000
PW (10%)	$108,217		$73,734		$89,918		$84,118	

DEPRECIATION AND ASSET DISPOSAL

When a depreciated asset is disposed of, the key question is, Which is larger, the asset's *book value, BV*, or the asset's *market value, MV*? If the book value is lower than the market

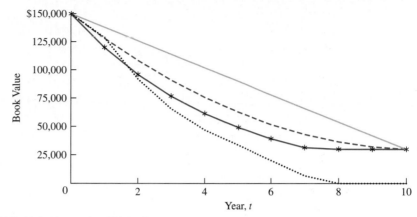

FIGURE 11-5 Comparing MACRS (·············) and historical depreciation methods: double declining balance (———*———), sum of the years' digits (– – – –), and straight line (—————).

value, then excess depreciation will be recaptured and taxed. On the other hand, if the book value is higher than the market value, there is a *loss* on the disposal. In either case, the level of taxes owed changes.

Depreciation recapture (ordinary gains): Depreciation recapture, also called ordinary gains, is necessary when an asset is sold for more than an asset's current book value. If more than the original cost basis is received, only the amount up to the original cost basis is recaptured depreciation. Since MACRS assumes $S = 0$ for its annual calculations, MACRS often has recaptured depreciation at disposal.

Losses: A *loss* occurs when less than book value is received for a depreciated asset. In the accounting records we've disposed of an asset for a dollar amount less than its book value, which is a loss.

Capital gains: Capital gains occur when more than the asset's original cost basis is received for it. The excess over the original cost basis is the *capital gain*. As described in Chapter 12, the tax rate on such gains is sometimes lower than the rate on ordinary income, but this depends on how long the investment has been held ("short," ≤ 1 year; "long," ≥ 1 year). In most engineering economic analyses capital gains are very uncommon because business and production equipment and facilities almost always *lose* value over time. Capital gains are much more likely to occur for nondepreciated assets like stocks, bonds, real estate, jewelry, art, and collectibles.

The relationship between depreciation recapture, loss, and capital gain is illustrated in Figure 11-6. Each case given is at a point in time in the life of the depreciated asset, where the original cost basis is $10,000 and the book value is $5000. Case (a) represents depreciation recapture (ordinary gain), Case (b) represents a loss, and in Case (c) both recaptured depreciation and a capital gain are present.

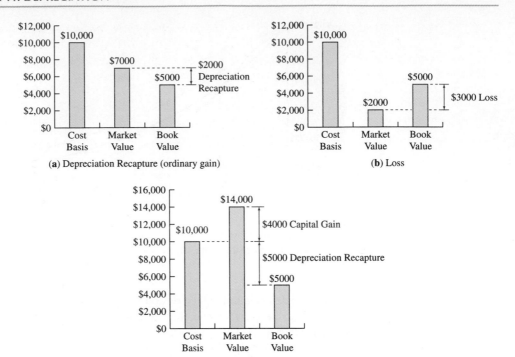

FIGURE 11-6 Recaptured depreciation, loss on sale, and capital gain.

EXAMPLE 11-9

Consider an asset with a cost basis of $10,000 that has been depreciated using the MACRS method. This asset is a 3-year MACRS property. What is the gain or loss if the asset is disposed of after 5 years of operation for **(a)** $7000, **(b)** $0, and **(c)** a cost of $2000.

SOLUTION

To find *gain* or *loss* at disposal we compare *market* and *book value*. Since MACRS depreciates to a salvage value of 0, and 5 years is greater than the recovery period, the book value equals $0.

(a) Recaptured depreciation = $7000, since the book value is $7000 higher.

(b) Since market value equals salvage value, there is no recaptured depreciation or loss.

(c) Since the money is paid for disposal, this is less than the book value, and there is a loss.

This general method for calculating recaptured depreciation or loss applies to all of the depreciation methods described in this chapter.

If the asset is in the middle of its depreciable life, then recaptured depreciation and losses are calculated in a similar manner—compare the *market* and *book values* at the time of disposal. However, in computing the book value with MACRS depreciation, a special rule must be applied for assets disposed of before the end of the recovery period. The rule is to *take one half of the allowable depreciation deduction for that year.* This rule assumes that disposals take place on average halfway through the year. Thus for a 5-year asset disposed of in the middle of Year 4, the rate allowed for MACRS depreciation is half of 11.52% or 5.76%. If the asset is disposed of in Year 6, it is already past the recovery period, and a half-year assumption has already been built into the MACRS schedule. Thus, the full r_6 is taken. Example 11-10 illustrates several cases of disposal before the asset is fully depreciated.

EXAMPLE 11-10

Consider the $10,000 asset in Example 11-9. Do the following:

1. Calculate the effect of disposal if this asset is sold during Year 2 for $2500.
2. Calculate the effect of disposal if the asset is sold during Year 3 for $2500.
3. Calculate the effect of disposal if the asset is sold after Year 3 for $4000 if straight-line depreciation is used over the asset's 5-year life, and a salvage value of $5000 is assumed.

SOLUTION

1. Market value$_2$ = $2500

$$\text{Book value}_2 = 10,000 - 10,000[r_1 + r_2/2] = 10,000 - 10,000[0.3333 + 0.4445/2]$$
$$= \$4444.5$$
$$\text{Loss} = \$1944.5 \, (= 4444.5 - 2500) \quad \text{(Loss, since BV greater than MV)}$$

Note: If the full rather than half-deduction were taken in Year 2, then the book value would be $2222.50 less and the loss would become a gain of $278. Since depreciation would increase by that $2222.50, the total of (depreciation) + (loss or gain) would be $4167 in either calculation.

2. $$\text{Market value}_3 = \$2500$$
$$\text{Book value}_3 = 10,000 - 10,000[r_1 + r_2 + r_3/2]$$
$$= 10,000 - 10,000[0.3333 + 0.4445 + 0.1481/2] = \$1481.50$$

$$\text{Recaptured depreciation} = 2500 - 1481.50 = \$1018.50 \quad \text{(Recaptured, since MV > BV)}$$

3. $$\text{Straight-line rate} = (B - S)/N = (10,000 - 5000)/5 = \$1000$$
$$\text{Market value}_4 = \$4000$$
$$\text{Book value}_4 = 10,000 - (4)(1000) = \$6000$$
$$\text{Loss} = 6000 - 4000 = \$2000$$

If the asset were disposed of *during the year rather than at year's end,* then the straight-line depreciation deduction would have to be prorated for the number of months during the year that the asset was in service. There is no half-year convention with the historical depreciation methods. For example, if disposal occurred on September 30 of Year 4, $d_4 = (9/12)(1000) = \$750$.

Note: In Case 1 it was shown that the required half-year convention did not affect the total deductions from taxable income. This is true for the other cases as well, since both recaptured depreciation and losses are treated as ordinary income.

UNIT-OF-PRODUCTION DEPRECIATION

At times the recovery of depreciation on a particular asset is more closely related to use than to time. In these few situations (and they are rare), the **unit-of-production (UOP) depreciation** in any year is

$$\text{UOP depreciation in any year} = \frac{\text{Production for year}}{\text{Total lifetime production for asset}}(B - S) \qquad (11\text{-}6)$$

This method might be useful for machinery that processes natural resources if the resources will be exhausted before the machinery wears out. Historically, this method was sometimes used for construction equipment that had very heavy use in some years and very light use in others. It is not considered an acceptable method for general use in depreciating industrial equipment.

EXAMPLE 11-11

For numerical similarity with previous examples, assume that equipment costing $900 has been purchased for use in a sand and gravel pit. The pit will operate for 5 years, while a nearby airport is being reconstructed and paved. Then the pit will be shut down, and the equipment removed and sold for $70. Compute the unit-of-production (UOP) depreciation schedule if the airport reconstruction schedule calls for 40,000 m³ of sand and gravel as follows:

Year	Required Sand and Gravel (m³)
1	4,000
2	8,000
3	16,000
4	8,000
5	4,000

SOLUTION

The cost basis, B, is $900. The salvage value, S, is $70. The total lifetime production for the asset is 40,000 m³ of sand and gravel. From the airport reconstruction schedule, the first-year UOP depreciation would be:

$$\text{First-year UOP depreciation} = \frac{4000 \text{ m}^3}{40,000 \text{ m}^3}(\$900 - \$70) = \$83$$

Similar calculations for the subsequent 4 years give the complete depreciation schedule:

Year	UOP Depreciation
1	$ 83
2	166
3	332
4	166
5	83
	$830

It should be noted that the actual unit-of-production depreciation charge in any year is based on the actual production for the year rather than the scheduled production.

DEPLETION

Depletion is the exhaustion of natural resources as a result of their removal. Since depletion covers such things as mineral properties, oil and gas wells, and standing timber, removal may take the form of digging up metallic or nonmetallic minerals, producing petroleum or natural gas from wells, or cutting down trees.

Depletion is recognized for income tax purposes for the same reason depreciation is—capital investment is being consumed or used up. Thus a portion of the gross income should be considered to be a return of the capital investment. The calculation of the depletion allowance is different from depreciation because there are two distinct methods of calculating depletion: *cost depletion* and *percentage depletion*. Except for standing timber and most oil and gas wells, depletion is calculated by both methods and the larger value is taken as depletion for the year. For standing timber and most oil and gas wells, only cost depletion is permissible.

Cost Depletion

Depreciation relied on an asset's cost, depreciable life, and salvage value to apportion the cost *minus* salvage value *over* the depreciable life. In some cases, where the asset is used at fluctuating rates, we might use the unit-of-production (UOP) method of depreciation. For mines, oil wells, and standing timber, fluctuating production rates are the usual situation. Thus, *cost depletion* is computed like unit-of-production depreciation using:

1. Property cost, less cost for land.
2. Estimated number of recoverable units (tons of ore, cubic meters of gravel, barrels of oil, million cubic feet of natural gas, thousand board-feet of timber, etc.).
3. Salvage value, if any, of the property.

EXAMPLE 11-12

A small lumber company bought a tract of timber for $35,000, of which $5000 was the land's value and $30,000 was the value of the estimated 1.5 million board-feet of standing timber. The first year, the company cut 100,000 board-feet of standing timber. What was the year's depletion allowance?

SOLUTION

$$\text{Depletion allowance per 1000 board-ft} = \frac{\$35,000 - \$5000}{1500 \times 1000 \text{ board-ft}}$$

$$= \$20 \text{ per 1000 board-ft}$$

The depletion allowance for the year would be

$$100,000 \text{ board-ft} \times \$20 \text{ per 1000 board-ft} = \$2000$$

Percentage Depletion

Percentage depletion is an alternate method for mineral property. The allowance is a certain percentage of the property's gross income during the year. This is an entirely different concept from depreciation. Unlike depreciation, which allocates cost *over* useful life, the **percentage depletion** allowance (see Table 11-6) is based on the property's gross income.

Since percentage depletion is computed on the *income* rather than the property's cost, the total depletion *may exceed the cost of the property*. In computing the *allowable*

TABLE 11-6 Percentage Depletion Allowances for Selected Deposits

Deposits	Rate
Sulfur, uranium, and, if from deposits in the U.S., asbestos, lead ore, zinc ore, nickel ore, and mica	22%
Gold, silver, copper, iron ore, and certain oil shale, if from deposits in the U.S.	15%
Borax, granite, limestone, marble, mollusk shells, potash, slate, soapstone, and carbon dioxide produced from a well	14%
Coal, lignite, and sodium chloride	10%
Clay and shale used or sold for use in making sewer pipe or bricks or used or sold for use as sintered or burned lightweight aggregates	$7\frac{1}{2}\%$
Clay used or sold for use in making drainage and roofing tile, flower pots, and kindred products, and gravel, sand, and stone (other than stone used or sold for use by a mine owner or operator as dimension or ornamental stone)	5%

Source: Internal Revenue Service, Publication 535, Chapter 9. Section 613(b) of the Internal Revenue Code gives a complete list of minerals and their percentage depletion rates.

percentage depletion on a property in any year, the *percentage depletion allowance* cannot exceed 50% of the property's taxable income computed without the depletion deduction. The percentage depletion calculations are illustrated by Example 11-13.

EXAMPLE 11-13

A coal mine has a gross income of $250,000. Mining expenses equal $210,000. Compute the allowable percentage depletion deduction.

SOLUTION

From Table 11-6, coal has a 10% depletion allowance based on gross mining income. The allowable percentage depletion deduction is also limited to a maximum of 50% of taxable income.

Computed Percentage Depletion

Gross income from mine	$250,000
Depletion percentage	× 10%
Computed percentage depletion	$ 25,000

Taxable Income Limitation

Gross income from mine	$250,000
Less: Expenses other than depletion	−210,000
Taxable income from mine	40,000
Deduction limitation	× 50%
Taxable income limitation	$20,000

Since the taxable income limitation ($20,000) is less than the computed percentage depletion ($25,000), the allowable percentage depletion deduction is $20,000.

As previously stated, on mineral property the depletion deduction can be based on either cost or percentage depletion. Each year, depletion is computed by both methods, and the allowable depletion deduction is the larger of the two amounts.

SPREADSHEETS AND DEPRECIATION

The spreadsheet functions for straight-line, double declining balance, and sum-of-years'-digits depreciation are listed in Table 11-7. Because these techniques are simple and were

TABLE 11-7 Spreadsheet Functions for Depreciation

Depreciation Technique	Excel
Straight line	SLN(cost, salvage, life)
Declining balance	DDB(cost, salvage, life, period, factor)
Sum of years' digits	SYD(cost, salvage, life, period)
MACRS	VDB(cost, salvage, life, start_period, end_period, factor, no_switch)

replaced by MACRS in 1986, they are not covered in detail here. All three functions include parameters for *cost* (initial book value), *salvage* (final salvage value), and *life* (depreciation period). Both DDB and SYD change depreciation amounts every year, so they include a parameter to pick the *period* (year). Finally, DDB includes a *factor*. The default value is 2 for 200% or double declining balance, but another commonly used value is 1.5 for 150%.

Using VDB for MACRS

The Excel function VDB is a flexible or variable declining balance method. It includes the ability to specify the starting and ending periods, rather than simply a year. It also includes an optional no_switch for problems where a switch from declining balance to straight-line depreciation is NOT desired.

To use VDB to calculate MACRS depreciation, the following are true.

1. Salvage $= 0$, since MACRS assumes no salvage value.
2. Life $=$ recovery period of 3, 5, 7, 10, 15, or 20 years.
3. First period runs from 0 to 0.5, 2nd period from 0.5 to 1.5, 3rd from 1.5 to 2.5, t^{th} from $t - 1.5$ to $t - 0.5$, and last period from $life - 0.5$ to $life$.
4. Factor $= 2$ for recovery periods of 3, 5, 7, or 10 years and $= 1.5$ for recovery periods of 15 or 20 years.
5. Since MACRS includes a switch to straight line, no_switch can be omitted.

The start_period and end_period arguments are from $t - 1.5$ to $t - 0.5$, because MACRS uses a half-year convention for the first year. Thus the first year has 0 to 0.5 year of depreciation, and the second year starts where the first year stops. When one is writing the Excel function, either the first and last periods must be edited individually, or start_period must be defined with a minimum of 0 and end_period with a maximum of life. This prevents the calculation of depreciation from -0.5 to 0 and from $life$ to $life +0.5$.

The results of using the VDB function match Table 11-4, except that the VDB function has more significant digits rather than being rounded to 2 decimals. Example 11-14 illustrates the use of the VDB function.

EXAMPLE 11-14

Return to the data of Example 11-5 which had $150,000 of office equipment, which is 7-year MACRS property. Use VDB to compute the depreciation amounts.

SOLUTION

The spreadsheet in Figure 11-7 defines the start_period with a minimum of 0 and the end_period with a maximum of life. Remember that this is the *depreciable* life or recovery period. Thus this formula could be used for any year of any recovery schedule. Notice that the VDB formula uses the value 0 for the salvage value, rather than referring to the data cell for the salvage value. MACRS assumes a salvage value of zero—no matter what the value truly is.

	A	B	C	D	E	F	G
1	150,000	First Cost					
2	0	Salvage					
3	7	Life = recovery period					
4	200%	Factor					
5							
6	Period	Depreciation					
7	1	$21,428.57	=VDB($A$1,0,$A$3,MAX(0,A7−1.5),MIN($A$3,A7−0.5),$A$4)				
8	2	$36,734.69	or (cost, salvage, life, max(0, t−1.5), min (life, t−.5), factor)				
9	3	$26,239.07					
10	4	$18,742.19					
11	5	$13,387.28					
12	6	$13,387.28					
13	7	$13,387.28					
14	8	$6,693.64					
15		$150,000	= Sum				

FIGURE 11-7 Using VDB to calculate MACRS depreciation.

SUMMARY

Depreciation is part of computing income taxes in economic analysis. There are three distinct definitions of depreciation:

1. Decline in asset's market value.
2. Decline in asset's value to its owner.
3. Allocating the asset's cost *less* its salvage value *over* its recovery period or depreciable life.

While the first two definitions are used in everyday discussions, it is the third, or accountant's, definition that is used in tax computations and in this chapter. Book value is the remaining unallocated cost of an asset, or

$$\text{Book value} = \text{Asset cost} - \text{Depreciation charges made to date}$$

This chapter describes how depreciable assets are *written off* (or claimed as a business expense) over a period of years instead of *expensed* in a single period (like wages, material costs, etc.). The depreciation methods described include the historical pre-1981 methods: *straight line, sum of the years' digits,* and *declining balance.* These methods required estimating the asset's salvage value and depreciable life.

The current tax law specifies use of the modified accelerated capital recovery system (MACRS). This chapter has focused on the general depreciation system (GDS) with limited discussion of the less attractive alternative depreciation system (ADS). MACRS (GDS) specifies faster *recovery periods* and a salvage value of zero, so it is generally economically more attractive than the historical methods.

The MACRS system is the current tax law, and it assumes a salvage value of zero. This is in contrast with historical methods, which ensured the final book value would equal the predicted salvage value. Thus, when one is using MACRS it is often necessary to consider recaptured depreciation. This is the excess of salvage value over book value, and it is taxed as ordinary income. Similarly, losses on sale or disposal are taxed as ordinary income.

Unit-of-production (UOP) depreciation relies on usage to quantify the loss in value. UOP is appropriate for assets that lose value based on the number of units produced, the tons of gravel moved, and so on (vs. number of years in service). However, this method is not considered to be acceptable for most business assets.

Depletion is the exhaustion of natural resources like minerals, oil and gas, and standing timber. The owners of the natural resources are consuming their investments as the natural resources are removed and sold. Cost depletion is computed based on the fraction of the resource that is removed or sold. For minerals and some oil and gas wells, an alternate calculation called percentage depletion is allowed. Percentage depletion is based on income, so the total allowable depletion deductions may *exceed* the invested cost.

Integrating depreciation schedules with cash flows often involves a lot of arithmetic. Thus, the tool of spreadsheets can be quite helpful. The functions for the historical methods, straight line, sum of the years' digits, and declining balance, are straightforward. Rather than individually entering MACRS percentages into the spreadsheet, the function VDB can be used to calculate MACRS depreciation percentages.

PROBLEMS

Depreciation Schedules

11-1 A depreciable asset costs $10,000 and has an estimated salvage value of $1600 at the end of its 6-year depreciable life. Compute the depreciation schedule for this asset by both SOYD depreciation and DDB depreciation.

11-2 A million-dollar oil drilling rig has a 6-year depreciable life and a $75,000 salvage value at the end of that time. Determine which one of the following methods provides the preferred depreciation schedule: DDB or SOYD. Show the depreciation schedule for the preferred method.

11-3 A new machine tool is being purchased for $16,000 and is expected to have a zero salvage value at the end of its 5-year useful life. Compute the DDB depreciation schedule for this capital asset. Assume any remaining depreciation is claimed in the last year.

11-4 Some special handling devices can be obtained for $12,000. At the end of 4 years, they can be sold for $600. Compute the depreciation schedule for the devices using the following methods:

(a) Straight-line depreciation
(b) Sum-of-years'-digits depreciation
(c) Double declining balance depreciation
(d) MACRS depreciation

11-5 The company treasurer is uncertain which of four depreciation methods the firm should use for office furniture that costs $50,000, and has a zero salvage value at the end of a 10-year depreciable life. Compute the depreciation schedule for the office furniture using the methods listed:

(a) Straight line
(b) Double declining balance
(c) Sum-of-years'-digits
(d) Modified accelerated cost recovery system

11-6 The RX Drug Company has just purchased a capsulating machine for $76,000. The plant engineer estimates the machine has a useful life of 5 years and little or no salvage value. He will use zero salvage value in the computations.

Compute the depreciation schedule for the machine using:

(a) Straight-line depreciation

(b) Sum-of-years'-digits depreciation

(c) Double declining balance depreciation (assume any remaining depreciation is claimed in the last year)

11-7 The Acme Chemical Company paid $45,000 for research equipment, which it believes will have zero salvage value at the end of its 5-year life. Compute the depreciation schedule for the equipment by each of the following methods:

(a) Straight line

(b) Sum-of-years'-digits

(c) Double declining balance

(d) Modified accelerated cost recovery system

11-8 Consider a $6500 piece of machinery, with a 5-year depreciable life and an estimated $1200 salvage value. The projected utilization of the machinery when it was purchased, and its actual production to date, are as follows:

Year	Projected Production (tons)	Actual Production (tons)
1	3500	3000
2	4000	5000
3	4500	[Not
4	5000	yet
5	5500	known]

Compute the machinery depreciation schedule by each of the following methods:

(a) Straight line

(b) Sum-of-years'-digits

(c) Double declining balance

(d) Unit of production (for first 2 years only)

(e) Modified accelerated cost recovery system

11-9 A large profitable corporation bought a small jet plane for use by the firm's executives in January. The plane cost $1.5 million and, for depreciation purposes, is assumed to have a zero salvage value at the end of 5 years. Compute the MACRS depreciation schedule.

11-10 For an asset that fits into the MACRS "all property not assigned to another class" designation, show in a table the depreciation and book value over the asset's 10-year life of use. The cost basis of the asset is $10,000.

11-11 A company that manufactures food and beverages in the vending industry has purchased some handling equipment that cost $75,000 and will be depreciated using MACRS GDS; the class life of the asset is 4 years. Show in a table the yearly depreciation amount and book value of the asset over its depreciation life.

11-12 Consider five depreciation schedules:

Year	A	B	C	D	E
1	$45.00	$35.00	$29.00	$58.00	$43.50
2	36.00	20.00	46.40	34.80	30.45
3	27.00	30.00	27.84	20.88	21.32
4	18.00	30.00	16.70	12.53	14.92
5	9.00	20.00	16.70	7.52	10.44
6			8.36		

They are based on the same initial cost, useful life, and salvage value. Identify each schedule as one of the following

- Straight-line depreciation
- Sum-of-years'-digits depreciation
- 150% declining balance depreciation
- Double declining balance depreciation
- Unit-of-production depreciation
- Modified accelerated cost recovery system

11-13 The depreciation schedule for an asset, with a salvage value of $90 at the end of the recovery period, has been computed by several methods. Identify the depreciation method used for each schedule.

Year	A	B	C	D	E
1	$323.3	$212.0	$424.0	$194.0	$107.0
2	258.7	339.2	254.4	194.0	216.0
3	194.0	203.5	152.6	194.0	324.0
4	129.3	122.1	91.6	194.0	216.0
5	64.7	122.1	47.4	194.0	107.0
6		61.1			
	970.0	1060.0	970.0	970.0	970.0

11-14 The depreciation schedule for a microcomputer has been arrived at by several methods. The estimated salvage value of the equipment at the end of its 6-year useful life is $600. Identify the resulting depreciation schedules.

Year	A	B	C	D
1	$2114	$2000	$1600	$1233
2	1762	1500	2560	1233
3	1410	1125	1536	1233
4	1057	844	922	1233
5	705	633	922	1233
6	352	475	460	1233

11-15 A heavy construction firm has been awarded a contract to build a large concrete dam. It is expected that a total of 8 years will be required to complete the work. The firm will buy $600,000 worth of special equipment for the job. During the preparation of the job cost estimate, the following utilization schedule was computed for the special equipment:

Year	Utilization (hr/yr)	Year	Utilization (hr/yr)
1	6000	5	800
2	4000	6	800
3	4000	7	2200
4	1600	8	2200

At the end of the job, it is estimated that the equipment can be sold at auction for $60,000.

(a) Compute the sum-of-years-digits' depreciation schedule.

(b) Compute the unit-of-production depreciation schedule.

11-16 Given the data in Problem 11-22, use a spreadsheet function to compute the MACRS depreciation schedule. Show the total depreciation taken (=sum( )) as well as the PW of the depreciation charges discounted at the MARR%.

11-17 Office equipment whose initial cost is $100,000 has an estimated actual life of 6 years, with an estimated salvage value of $10,000. Prepare tables listing the annual costs of depreciation and the book value at the end of each 6 years, based on the straight-line, sum-of-years'-digits, and MACRS depreciation. Use spreadsheet functions for the depreciation methods.

11-18 You are equipping an office. The total office equipment will have a first cost of $1,750,000 and a salvage value of $200,000. You expect the equipment will last 10 years. Use a spreadsheet function to compute the MACRS depreciation schedule.

11-19 Units-of-production depreciation is being used for a machine that, based on usage, has an allowable

depreciation charge of $6500 the first year and increasing by $1000 each year until complete depreciation. If the machine's cost basis is $110,000, set up a depreciation schedule that shows depreciation charge and book value over the machine's 10-year useful life.

11-20 A custom-built production machine is being depreciated using the units-of-production method. The machine costs $65,000 and is expected to produce 1.5 million units, after which it will have a $5000 salvage value. In the first 2 years of operation the machine was used to produce 140,000 units each year. In the 3rd and 4th years, production went up to 400,000 units. After that time annual production returned to 135,000 units. Use a spreadsheet to develop a depreciation schedule showing the machine's depreciation allowance and book value over its depreciable life.

Comparing Depreciation Methods

11-21 TELCO Corp. has leased some industrial land near its plant. It is building a small warehouse on the site at a cost of $250,000. The building will be ready for use January 1. The lease will expire 15 years after the building is occupied. The warehouse will belong at that time to the landowner, with the result that there will be no salvage value to TELCO. The warehouse is to be depreciated either by MACRS or SOYD depreciation. If 10% interest is appropriate, which depreciation method should be selected?

11-22 A profitable company making earthmoving equipment is considering an investment of $100,000 on equipment that will have a 5-year useful life and a $20,000 salvage value. If money is worth 10%, which one of the following three methods of depreciation would be preferable?

(a) Straight-line method

(b) Double declining balance method

(c) MACRS method

11-23 The White Swan Talc Company paid $120,000 for mining equipment for a small talc mine. The mining engineer's report indicates the mine contains 40,000 cubic meters of commercial quality talc. The company plans to mine all the talc in the next 5 years as follows:

Year	Talc Production (m³)
1	15,000
2	11,000
3	4,000
4	6,000
5	4,000

At the end of 5 years, the mine will be exhausted and the mining equipment will be worthless. The company accountant must now decide whether to use sum-of-years'-digits depreciation or unit-of-production depreciation. The company considers 8% to be an appropriate time value of money. Compute the depreciation schedule for each of the two methods. Which method would you recommend that the company adopt? Show the computations to justify your decision.

11-24 Some equipment that costs $1000 has a 5-year depreciable life and an estimated $50 salvage value at the end of that time. You have been assigned to determine whether to use straight-line or SOYD depreciation. If a 10% interest rate is appropriate, which is the preferred depreciation method for this profitable corporation? Use a spreadsheet to show your computations of the difference in present worths.

11-25 The FOURX Corp. has purchased $12,000 of experimental equipment. The anticipated salvage value is $400 at the end of its 5-year depreciable life. This profitable corporation is considering two methods of depreciation: sum of years' digits and double declining balance. If it uses 7% interest in its comparison, which method do you recommend? Show computations to support your recommendation. Use a spreadsheet to develop your solution.

Depreciation and Book Value

11-26 For its fabricated metal products, the Able Corp. is paying $10,000 for special tools that have a 4-year useful life and no salvage value. Compute the depreciation charge for the *second* year by each of the following methods:

(a) DDB

(b) Sum-of-years'-digits

(c) Modified accelerated cost recovery system

11-27 The MACRS depreciation percentages for 7-year personal property are given in Table 11-3. Make the necessary computations to determine if the percentages shown are correct.

11-28 The MACRS depreciation percentages for 10-year personal property are given in Table 11-3. Make the necessary computations to determine if the percentages shown are correct.

11-29 Use MACRS GDS depreciation for each of the assets, 1–3, to calculate the following items, (a)–(c).

1. A light general-purpose truck used by a delivery business, cost = $17,000

2. Production equipment used by a Detroit automaker to produce vehicles, cost = $30,000

3. Cement production facilities used by a construction firm, cost $130,000

(a) The MACRS GDS property class

(b) The depreciation deduction for Year 3

(c) The book value of the asset after 6 years

11-30 On July 1, Nancy paid $600,000 for a commercial building and an additional $150,000 for the land on which it stands. Four years later, also on July 1, she sold the property for $850,000. Compute the modified accelerated cost recovery system depreciation for each of the *five* calendar years during which she had the property.

11-31 A group of investors has formed Trump Corporation to purchase a small hotel. The asking price is $150,000 for the land and $850,000 for the hotel building. If the purchase takes place in June, compute the MACRS depreciation for the first three calendar years. Then assume the hotel is sold in June of the fourth year, and compute the MACRS depreciation in that year also.

11-32 Mr. Donald Spade bought a computer in January to keep records on all the property he owns. The computer cost $70,000 and is to be depreciated using MACRS. Donald's accountant pointed out that under a special tax rule (the rule applies when the value of property placed in service in the last 3 months of the tax year exceeds 40% of all the property placed in service during the tax year), the computer and all property that year would be subject to the midquarter convention. The midquarter convention assumes that all property placed in service in any quarter-year is placed in service at the midpoint of the quarter. Use the midquarter convention to compute Donald's MACRS depreciation for the first year.

11-33 A company is considering buying a new piece of machinery. A 10% interest rate will be used in the computations. Two models of the machine are available.

	Machine I	Machine II
Initial cost	$80,000	$100,000
End-of-useful-life salvage value, S	20,000	25,000
Annual operating cost	18,000	15,000 first 10 years 20,000 thereafter
Useful life, in years	20	25

(a) Determine which machine should be purchased, based on equivalent uniform annual cost.

(b) What is the capitalized cost of Machine I?

(c) Machine I is purchased and a fund is set up to replace Machine I at the end of 20 years. Compute the required uniform annual deposit.

(d) Machine I will produce an annual saving of material of $28,000. What is the rate of return if Machine I is installed?

(e) What will be the book value of Machine I after 2 years, based on sum-of-years'-digits depreciation?

(f) What will be the book value of Machine II after 3 years, based on double declining balance depreciation?

(g) Assuming that Machine II is in the 7-year property class, what would be the MACRS depreciation in the third year?

11-34 Explain in your own words the difference between capital gains and ordinary gains. In addition, explain why it is important to our analysis as engineering economists. Do we see capital gains much in industry-based economic analyses or in our personal lives?

Gain/Loss on Disposal

11-35 Equipment costing $20,000 that is a MACRS 3-year property is disposed of during the second year for $14,000. Calculate any depreciation recapture, ordinary losses, or capital gains associated with disposal of the equipment.

11-36 An asset with an 8-year ADR class life costs $50,000 and was purchased on January 1, 2001. Calculate any depreciation recapture, ordinary losses, or capital gains associated with selling the equipment on December 31, 2003, for $15,000, $25,000, and $60,000. Consider two cases of depreciation for the problem: if MACRS GDS is used, and if straight-line depreciation over the ADR class life is used with a $10,000 salvage value.

11-37 A $150,000 asset has been depreciated with the straight-line method over an 8-year life. The estimated salvage value was $30,000. At the end of the 5th year the asset was sold for $90,000. From a tax perspective, what is happening at the time of disposal, and what is the dollar amount?

11-38 O'Leary Engineering Corp. has been depreciating a $50,000 machine for the last 3 years. The asset was just sold for 60% of its first cost. What is the size of the recaptured depreciation or loss at disposal using the following depreciation methods?

(a) Sum of years, digits with $N = 8$ and $S = 2000$

(b) Straight-line depreciation with $N = 8$ and $S = 2000$

(c) MACRS GDS depreciation, classified as a 7-year property

Depletion

11-39 When a major highway was to be constructed nearby, a farmer realized that a dry streambed running through his property might be a valuable source of sand and gravel. He shipped samples to a testing laboratory and learned that the material met the requirements for certain low-grade fill material. The farmer contacted the highway construction contractor, who offered 65¢ per cubic meter for 45,000 cubic meters of sand and gravel. The contractor would build a haul road and would use his own equipment. All activity would take place during a single summer.

The farmer hired an engineering student for $2500 to count the truckloads of material hauled away. The farmer estimated that 2 acres of streambed had been stripped of the sand and gravel. The 640-acre farm had cost him $300 per acre, and the farmer felt the property had not changed in value. He knew that there had been no use for the sand and gravel prior to the construction of the highway, and he could foresee no future use for any of the remaining 50,000 cubic meters of sand and gravel.

Determine the farmer's depletion allowance. (*Answer:* $1462.50)

11-40 Mr. H. Salt purchased an $1/8$ interest in a producing oil well for $45,000. Recoverable oil reserves for the well were estimated at that time at 15,000 barrels, $1/8$ of which represented Mr. Salt's share of the reserves. During the subsequent year, Mr. Salt received $12,000 as his $1/8$ share of the gross income from the sale of 1000 barrels of oil. From this amount, he had to pay $3000 as his share of the expense of producing the oil. Compute Mr. Salt's depletion allowance for the year.   (*Answer:* $3000)

11-41 American Pulp Corp. (APC) has entered into a contract to harvest timber for $450,000. The total estimated available harvest is 150 million board-feet.

(a) What is the depletion allowance for Years 1 to 3, if 42, 45, and 35 million board-feet are harvested by APC in those years?

(b) After 3 years, the total available harvest for the original tract was reestimated at 180 million board-feet. Compute the depletion allowances for Years 4 and beyond.

11-42 A piece of machinery has a cost basis of $45,000. Its salvage value will be $5000 after 10,000 hours of operation. With units-of-production depreciation, what is the allowable depreciation rate per hour? What is the book value after 4000 hours of operation?

11-43 Mining recently began on a new deposit of 10 million metric tons of ore (2% nickel and 4% copper). Annual production of 350,000 metric tons begins this year. The market price of nickel is $3.75 per pound and $0.65 for copper. Mining operation costs are expected to be $0.50 per pound. XYZ Mining Company paid $600 million for the deposits. What is the maximum depletion allowance each year for the mine?

CHAPTER 12

INCOME TAXES

On with the Wind

For over three decades, environmental activists and community leaders have bemoaned American dependence on foreign oil. The war in Iraq again highlighted the precariousness of relying on an unstable region of the world for a major part of our energy requirements.

One solution to this dilemma is to rely more on renewable sources of energy, such as solar power and wind. The technology for such alternative energy sources has been around for many years, and most voters seem to favor more reliance on renewables. If good intentions were all it took, we'd be getting much of our electricity from windmills.

But, of course, making the transition to greater use of wind power requires a significant investment in infrastructure, especially costly wind turbines. And few investors are willing to plunk down their money unless they have a solid expectation of earning a competitive return.

Until fairly recently, cost factors kept wind energy from becoming an attractive investment. As recently as the late 1980s, wind-generated power cost roughly twice as much to produce as energy from conventional sources.

More recently, however, wind energy has decreased dramatically in price. The American Wind Energy Association (AWEA) reports that many modern, state-of-the-art wind plants can now produce power for less than 5¢ per kilowatt-hour, making them competitive with conventional sources. Unsurprisingly, investment in wind power production has also increased substantially.

How did this happen? In part, the trend was driven by advances in wind turbine technology. But it was helped significantly by a provision of the federal tax law contained in the Energy Policy Act of 1992. This statute allowed utilities and other electricity suppliers a "production tax credit" of 1.5¢ per kilowatt-hour (later adjusted to 1.9¢ to account for the effects of inflation).

This was a key incentive to the wind power industry which, like all energy producers, must expend large sums on capital assets. The tax credit had a dramatic effect on the wind energy market. During 2001, for instance, energy producers added almost 1700 megawatts of wind-generating capacity—enough to power nearly half a million homes.

But what the government giveth, the government can take away. When the production tax credit (PTC) briefly expired at the end of 2001, an estimated $3 billion worth of wind projects were suspended, and hundreds of workers were laid off. Fortunately for the industry, the credit was subsequently extended to the end of 2003. Later, the PTC was extended again, until 2008.

QUESTIONS TO CONSIDER

1. The wind energy tax credit has both advocates and opponents. One camp argues that the government should not support wind energy through tax credits—instead, let the market determine the future of alternative energy sources. Others argue that government has historically supported the oil and coal industries through subsidies, and in addition we as a society "pay" for the negative effects that these forms of energy have on health and the environment. Which side has the stronger argument in your view? Why?

2. Clearly wind energy has both costs and benefits as an alternate energy technology, and government may have some role in enabling its development and promoting investments in it. Develop a table of the costs and benefits of wind energy technology. In doing so, think about the producers and consumers of the technology as well as society in general. Think about who the winners and losers would be if we as a nation could dramatically increase the fraction of our energy from wind sources. Add a column that identifies the ethical issues that arise? Can we solve any of the ethical issues that you've identified?

3. Developing a wind power project takes many years and requires the commitment of large sums of investment capital before the project begins to return a profit. What is the effect on investment when the wind power production tax credit is allowed to expire, or is extended for periods of only a few years?

4. The federal income tax was introduced in the U.S. in 1913. Using the Internet, can you determine how tax rates changed throughout the course of the 20th century? How has this affected the value of tax credits to industry?

After Completing This Chapter...

The student should be able to:

- Calculate *taxes due* or *taxes owed* for both individuals and corporations.
- Understand the incremental nature of the individual and corporate tax rates used for calculating taxes on income.
- Calculate a combined income tax rate for state and federal income taxes and select an appropriate tax rate for engineering economic analyses.
- Utilize an *after-tax tax table* to find the after-tax cash flows for a prospective investment project.
- Calculate after-tax measures of merit, such as present worth, annual worth, payback period, internal rate of return, and benefit–cost ratio, from developed after-tax cash flows.
- Evaluate investment alternatives on an after-tax basis including asset disposal.
- Use spreadsheets for solving after-tax economic analysis problems.

As Benjamin Franklin said, two things are inevitable: death and taxes. In this chapter we will examine the structure of taxes in the U.S. These include sales taxes, gasoline taxes, property taxes, and state and federal income taxes. Here we will concentrate our attention on federal income taxes. Income taxes are part of most real problems and often have a substantial impact that must be considered.

First, we must understand the way in which taxes are imposed. Chapter 11 concerning depreciation is an integral part of this analysis, so the principles covered there must be well understood. Then, having understood the mechanism, we will see how federal income taxes affect our economic analysis.

A PARTNER IN THE BUSINESS

Probably the most straightforward way to understand the role of federal income taxes is to consider the U.S. government as a partner in every business activity. As a partner, the government shares in the profits from every successful venture. In a somewhat more complex way, the government shares in the losses of unprofitable ventures too. The tax laws are complex, and it is not our purpose to fully explain them.[1] Instead, we will examine the fundamental concepts of the federal income tax laws—and we emphasize at the start that there are exceptions and variations to almost every statement we shall make!

CALCULATION OF TAXABLE INCOME

At the mention of income taxes, one can visualize dozens of elaborate and complex calculations. There is some truth to that vision, for there can be complexities in computing income taxes. Yet incomes taxes are just another type of disbursement. Our economic analysis calculations in prior chapters have dealt with all sorts of disbursements: operating costs, maintenance, labor and materials, and so forth. Now we simply add one more prospective disbursement to the list—income taxes.

Taxable Income of Individuals

The amount of federal income taxes to be paid depends on taxable income and the income tax rates. Therefore, our first concern is the definition of **taxable income.** To begin, one must compute his or her **gross income:**

$$\text{Gross income} = \text{Wages, salary, etc.} + \text{Interest income} + \text{Dividends} + \text{Capital gains} + \text{Unemployment compensation} + \text{Other income}$$

From gross income, we subtract any allowable retirement plan contributions and other **adjustments.** The result is **adjusted gross income (AGI).** From adjusted gross income, individuals may deduct the following items:[2]

---

[1] Many government and private sources exist that describe detailed taxation information. These include the Internal Revenue Service (www.irs.gov), U.S. Treasury (www.treasury.gov), Commerce Clearing House (www.cch.com), and Research Institute of America (www.riahome.com).

[2] The 2007 returns itemized deductions are limited if adjusted gross income is more than $156,400 ($78,200 if married filing separately). Data given here is for 2007 returns, filed in 2008.)

1. **Personal Exemptions.** One exemption ($3400 for 2007 returns) is provided for each person who depends on the gross income for his or her living.[3]

2. **Itemized Deductions.** Some of these are:

   (*a*) Excessive medical and dental expenses (exceeding $7\frac{1}{2}\%$ of adjusted gross income)

   (*b*) State and local income, real estate and personal property tax

   (*c*) Home mortgage interest

   (*d*) Charitable contributions

   (*e*) Casualty and theft losses (exceeding $100 + 10\%$ of adjusted gross income)

   (*f*) Job expenses and certain miscellaneous deductions (some categories must exceed 2% of adjusted gross income)

3. **Standard Deduction**. Each taxpayer may either itemize his or her deductions, or instead take a standard deduction as follows:

   (*a*) Single taxpayers, $5350 (for 2007 returns)

   (*b*) Married taxpayers filing a joint return, $10,700 (for 2007 returns)

The result is **taxable income.**

For individuals, taxable income is computed as follows:

$$\text{Adjusted gross income} = \text{Gross income} - \text{Adjustments}$$

$$\textbf{Taxable income} = \text{Adjusted gross income}$$
$$- \text{ Personal exemption(s)}$$
$$- \text{ Itemized deductions or Standard deduction} \qquad (12\text{-}1)$$

Classification of Business Expenditures

When an individual or a firm operates a business, there are three distinct types of business expenditure:

1. For depreciable assets.
2. For nondepreciable assets.
3. All other business expenditures.

Expenditures for depreciable assets: When facilities or productive equipment with useful lives in excess of one year are acquired, the taxpayer will recover the investment through depreciation charges.[4] Chapter 11 detailed how to allocate an asset's cost over its useful life.

---

[3]For 2007 returns if adjusted gross income is over $117,300 and depending upon filing status, the personal exemption deduction may be reduced.

[4]There is an exception. In 2007, businesses may immediately deduct (*expense*) up to $112,000 of business equipment (via the Section 179 Business Deduction) in a year, provided their total equipment expenditure for the year does not exceed $450,000. Between $450,000 and $562,000 in qualifying Section 179 property, the dollar limit is reduced by the amount over $450,000. Above $562,000 a Section 179 deduction cannot be taken.

Expenditures for nondepreciable assets: Land is considered to be a nondepreciable asset, for there is no finite life associated with it. Other nondepreciable assets are properties *not* used either in a trade, in a business, or for the production of income. An individual's home and car are generally nondepreciable assets. The final category of nondepreciable assets comprises those subject to *depletion,* rather than *depreciation.* Since business firms generally acquire assets for use in the business, their only nondepreciable assets normally are land and assets subject to depletion.

All other business expenditures: This category is probably the largest of all, for it includes all the ordinary and necessary expenditures of operating a business. Labor costs, materials, all direct and indirect costs, and facilities and productive equipment with a useful life of one year or less are part of routine expenditures. They are charged as a business expense—*expensed*—when they occur.

Business expenditures in the first two categories—that is, for either depreciable or nondepreciable assets—are called **capital expenditures.** In the accounting records of the firm, they are **capitalized;** all ordinary and necessary expenditures in the third category are **expensed.**

Taxable Income of Business Firms

The starting point in computing a firm's taxable income is *gross income.* All ordinary and necessary expenses to conduct the business—*except* capital expenditures—are deducted from gross income. Capital expenditures may *not* be deducted from gross income. Except for land, business capital expenditures are allowed on a period by period basis through depreciation or depletion charges.

For business firms, taxable income is computed as follows:

$$\textbf{Taxable income} = \text{Gross income}$$
$$- \text{All expenditures except capital expenditures}$$
$$- \text{Depreciation and depletion charges} \qquad (12\text{-}2)$$

Because of the treatment of capital expenditures for tax purposes, the taxable income of a firm may be quite different from the actual cash flows.

EXAMPLE 12-1

During a 3-year period, a firm had the following cash flows (in millions of dollars):

	Year 1	Year 2	Year 3
Gross income from sales	$200	$200	$200
Purchase of special tooling (useful life: 3 years)	−60	0	0
All other expenditures	−140	−140	−140
Cash flows for the year	$ 0	$ 60	$ 60

Compute the taxable income for each of the 3 years.

SOLUTION

The cash flows for each year would suggest that Year 1 was a poor one, while Years 2 and 3 were very profitable. A closer look reveals that the firm's cash flows were adversely affected in Year 1 by the purchase of special tooling. Since the special tooling has a 3-year useful life, it is a capital expenditure with its cost allocated over the useful life. If we assume that straight-line depreciation applies with no salvage value, we use Equation 11-2 to find the annual charge:

$$\text{Annual depreciation charge} = \frac{B - S}{N} = \frac{60 - 0}{3} = \$20 \text{ million}$$

Applying Equation 12-2, we write

$$\text{Taxable income} = 200 - 140 - 20 = \$40 \text{ million}$$

In each of the 3 years, the taxable income is $40 million.

An examination of the cash flows and the taxable income in Example 12-1 indicates that taxable income is a better indicator of the firm's annual performance.

INCOME TAX RATES

Income tax rates for individuals changed many times between 1960 and 1995, as illustrated in Figure 12-1. Recent movements have been less dramatic. From 1995 to 2000 the maximum rate was 39.6%; it fell to 39.1% in 2001 and to 38.6% in 2002. From 2003 through 2007 it was 35%.

Individual Tax Rates

There are four schedules of federal income tax rates for individuals. Single taxpayers use the Table 12-1a schedule. Married taxpayers filing a joint return use the Table 12-1b

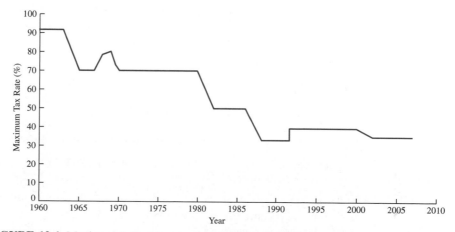

FIGURE 12-1 Maximum federal income tax rates for individuals.

TABLE 12-1a 2007 Tax Rates for Single Taxpayers

Taxable Income		Tax		
Over	But Not Over	Base Tax	Plus	On Income Over
$ 0	$ 7,825	$ 0.00	10%	$ 0
7,825	31,850	782.50	15%	7,825
31,850	77,100	4,386.50	25%	31,850
77,100	160,850	15,698.75	28%	77,100
160,850	349,700	39,148.75	33%	160,850
Over 349,700		101,469.25	35%	349,700

TABLE 12-1b 2007 Tax Rates for Married Individuals Filing Jointly

Taxable Income		Tax		
Over	But Not Over	Base Tax	Plus	On Income Over
$ 0	$ 15,650	$ 0.00	10%	$ 0
15,650	63,700	1,565.00	15%	15,650
63,700	128,500	8,772.50	25%	63,700
128,500	195,850	24,972.50	28%	128,500
195,850	349,700	43,830.50	33%	195,850
Over 349,700		94,601.00	35%	349,700

schedule. Two other schedules (not shown here) are applicable to unmarried individuals with dependent relatives ("head of household"), and married taxpayers filing separately.

EXAMPLE 12-2

An unmarried student earned $10,000 in the summer plus another $6000 during the rest of the year. When he files an income tax return, he will be allowed one exemption (for himself). He estimates that he spent $1000 on allowable itemized deductions. How much income tax will he pay?

$$\textbf{Adjusted gross income} = \$10,000 + \$6000 = \$16,000$$

$$\textbf{Taxable income} = \text{Adjusted gross income}$$

$$- \text{Deduction for one exemption (\$3400)}$$

$$- \text{Standard deduction (\$5350)}$$

$$= 16,000 - 3400 - 5350 = \$7250$$

$$\textbf{Federal income tax} = 10\% \ (7250) = \$725$$

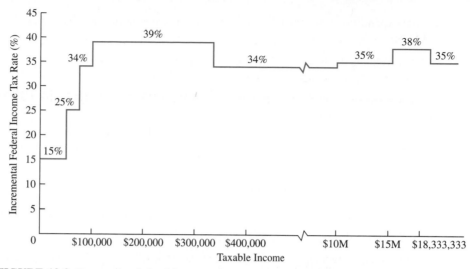

FIGURE 12-2 Corporation federal income tax rates (2006) rates. Federal corporate tax rates have been constant over the past several years.

TABLE 12-2 Corporate Income Tax Rates

Taxable Income	Tax Rate	Corporate Income Tax
Not over $50,000	15%	15% over $0
$50,000–75,000	25%	$7,500 + 25% over $50,000
$75,000–100,000	34%	$13,750 + 34% over $75,000
$100,000–335,000	39%*	$22,250 + 39% over $100,000
$335,000–10 million	34%	$113,900 + 34% over $335,000
$10 million–15 million	35%	$3,400,000 + 35% over $10 million
$15 million–18,333,333	38%	$5,159,000 + 38% over $15 million
≥$18,333,333	35%	$6,425,667 + 35% over $18,333,333

*The extra 5% from $100,000 to $335,000 was chosen so that firms in the $335,000 to $10 million bracket pay a flat 34% tax rate. [(0.39 − 0.34)](335,000 − 100,000) = (0.34 − 0.15)(50,000) + (0.34 − 0.25)(75,000 − 50,000)] so tax = 0.34 (tax income) in $335,000 to $10 million bracket. Similarly, for corporate incomes over $18,333,333 the tax rate schedule is equivalent to a flat tax rate of 35%.

Corporate Tax Rates

Income tax for corporations is computed in a manner similar to that for individuals. The data in Figure 12-2 can be presented in schedule form (Table 12-2).

EXAMPLE 12-3

The French Chemical Corporation was formed to produce household bleach. The firm bought land for $220,000, had a $900,000 factory building erected, and installed $650,000 worth of chemical and packaging equipment. The plant was completed and operations begun on April 1. The gross income for the calendar year was $450,000. Supplies and all operating expenses, excluding the capital expenditures, were \$100,000. The firm will use modified accelerated cost recovery system (MACRS) depreciation.

  (*a*) What is the first-year depreciation charge?

  (*b*) What is the first-year taxable income?

  (*c*) How much will the corporation pay in federal income taxes for the year?

SOLUTION TO PART a

MACRS depreciation: Chemical equipment is personal property. From Table 11-2, it is probably in the "7-year, all other property" class.

$$\text{First-year depreciation(equipment)} = \$650,000 \times 14.29\% = \$92,885$$

The building is in the 39-year real property class. Since it was placed in service on April 1, the first-year depreciation is:

$$\text{First-year depreciation(building)} = \$900,000 \times 1.819\% = \$16,371 \quad \text{(see Table 11-4)}$$

The land is a nondepreciable asset. Thus we have

$$\text{Total first year MACRS depreciation} = \$92,885 + \$16,371 = \$109,256$$

SOLUTION TO PART b

$$
\begin{aligned}
\text{Taxable income} = {}& \text{Gross income} \\
& - \text{All expenditures except capital expenditures} \\
& - \text{Depreciation and depletion charges} \\
= {}& \$450,000 - 100,000 - 109,256 = \$240,744
\end{aligned}
$$

SOLUTION TO PART c

$$
\begin{aligned}
\text{Federal income tax} &= \$22,250 + 39\%(240,744 - 100,000) \\
&= \$77,140
\end{aligned}
$$

Combined Federal and State Income Taxes

In addition to federal income taxes, most individuals and corporations pay state income taxes. It would be convenient if we could derive a single tax rate to represent both the state and federal incremental tax rates. In the computation of taxable income for federal taxes, the amount of state taxes paid is one of the allowable itemized deductions. Federal income taxes are not, however, generally deductible in the computation of state taxable income. Therefore, the state income tax is applied to a *larger* taxable income than is the federal income tax rate. As a result, the combined incremental tax rate will not be the sum of two tax rates.

For an increment of income (ΔIncome) and tax rate on incremental income (ΔTax rate):

$$\text{State income taxes} = (\Delta\text{State tax rate})(\Delta\text{Income})$$

$$\text{Federal taxable income} = (\Delta\text{Income})(1 - \Delta\text{State tax rate})$$

$$\text{Federal income taxes} = (\Delta\text{Federal tax rate})(\Delta\text{Income}) \times (1 - \Delta\text{State tax rate})$$

The total of state and federal income taxes is

$$[\Delta\text{State tax rate} + (\Delta\text{Federal tax rate})(1 - \Delta\text{State tax rate})](\Delta\text{Income})$$

The term in the brackets gives the combined incremental tax rate.

Combined incremental tax rate

$$= \Delta\text{State tax rate} + (\Delta\text{Federal tax rate})(1 - \Delta\text{State tax rate}) \quad (12\text{-}3)$$

EXAMPLE 12-4

An engineer has an income that puts her in the 25% federal income tax bracket and at the 10% state incremental tax rate. She has an opportunity to earn an extra $500 by doing a small consulting job. What will be her combined state and federal income tax rate on the additional income?

SOLUTION

Use Equation 12-3 to find the combined incremental tax rate: $0.10 + 0.25(1 - 0.10) = 32.5\%$.

Selecting an Income Tax Rate for Economy Studies

Since income tax rates vary with the level of taxable income for both individuals and corporations, one must decide which tax rate to use in a particular situation. The simple answer is that the tax rate to use is the incremental tax rate that applies to the change in taxable income projected in the economic analysis. If a married couple filing jointly has a taxable income of $40,000 and can increase their income by $2000, what tax rate should be used for the $2000 of incremental income? From Table 12-1b, we see the $2000 falls within the 10% tax bracket.

Now suppose this couple could increase their $40,000 income by $10,000. In this situation, Table 12-1b shows that the 10% incremental tax rate should be applied to the first $7825 and a 25% incremental tax rate to the last $2175 of extra income. The appropriate incremental tax rate for corporations is equally easy to determine. For larger corporations, the federal incremental tax rate is 35%. In addition, there may be up to a 12 to 15% state tax.

ECONOMIC ANALYSIS TAKING INCOME TAXES INTO ACCOUNT

An important step in economic analysis has been to resolve the consequences of alternatives into a cash flow. Because income taxes have been ignored, the result has been a *before-tax cash flow*. This before-tax cash flow is an essential component in economic analysis that also considers the consequences of income tax. The principal elements in an *after-tax analysis* are as follows:

- Before-tax cash flow
- Depreciation
- Taxable income (Before-tax cash flow − Depreciation)
- Income taxes (Taxable income × Incremental tax rate)
- After-tax cash flow (Before-tax cash flow − Income taxes)

These elements are usually arranged to form an after-tax *cash flow table*. This is illustrated by Example 12–5.

EXAMPLE 12-5

A medium-sized profitable corporation may buy a $3000 used pickup truck for use by the shipping and receiving department. During the truck's 5-year useful life, it is estimated the firm will save $800 per year after all the costs of owning and operating the truck have been paid. Truck salvage value is estimated at $750.

(a) What is the before-tax rate of return?

(b) What is the after-tax rate of return on this capital expenditure? Assume straight-line depreciation.

SOLUTION TO PART a

For a before-tax rate of return, we must first compute the before-tax cash flow.

Year	Before-Tax Cash Flow
0	−$3000
1	+800
2	+800
3	+800
4	+800
5	$\begin{cases} +800 \\ +750 \end{cases}$

Solve for the before-tax rate of return, IRR_{BT}:

$$3000 = 800(P/A, i, 5) + 750(P/F, i, 5)$$

Try $i = 15\%$:

$$3000 \neq 800(3.352) + 750(0.4972) \neq 2682 + 373 = 3055$$

Since i is slightly low, try $i = 18\%$:

$$3000 \neq 800(3.127) + 750(0.4371) \neq 2502 + 328 = 2830$$

$$IRR_{BT} = 15\% + 3\% \left(\frac{3055 - 3000}{3055 - 2830} \right) = 15\% + 3\%(0.15) = 15.7\%$$

SOLUTION TO PART b

For an after-tax rate of return, we must set up an after-tax cash flow table (Table 12-3). The starting point is the before-tax cash flow. Then we will need the depreciation schedule for the truck:

$$\text{Straight-line depreciation} = \frac{B - S}{N} = \frac{3000 - 750}{5} = \$450 \text{ per year}$$

TABLE 12-3 After-Tax Cash Flow Table for Example 12-5

Year	(a) Before-Tax Cash Flow	(b) Straight-Line Depreciation	(c) Δ(Taxable Income) (a)−(b)	(d) 34% Income Taxes −0.34 (c)	(e) After-Tax Cash Flow (a) + (d)†
0	−$3000				−$3000
1	800	$450	$350	−$119	681
2	800	450	350	−119	681
3	800	450	350	−119	681
4	800	450	350	−119	681
5	{800 {750	450	350	−119	{681 {750

(handwritten: Ch P 11; Δ(Taxable Income) 350 − 450; −119 +153; 681 153; no tax on salvage because you already pay)

*Sign convention for income taxes: a minus (−) represents a disbursement of money to pay income taxes; a plus (+) represents the receipt of money by a decrease in the tax liability.

†The after-tax cash flow is the before-tax cash flow minus income taxes. Based on the income tax sign convention, this is accomplished by *adding* columns (a) and (d).

Taxable income is the before-tax cash flow *minus* depreciation. For this medium-sized profitable corporation, let's assume an incremental federal income tax rate of 34%. Therefore income taxes are 34% of taxable income. Finally, the after-tax cash flow equals the before-tax cash flow *minus* income taxes. These data are used to compute Table 12-3.

The after-tax cash flow may be solved to find the after-tax rate of return, IRR_{AT}. Try $i = 10\%$:

$$3000 \neq 681(P/A, 10\%, 5) + 750(P/F, 10\%, 5)$$
$$\neq 681(3.791) + 750(0.6209) = 3047$$

Since i is slightly low, try $i = 12\%$:

$$3000 \neq 681(3.605) + 750(0.5674) = 2881$$
$$\text{IRR}_{AT} = 10\% + 2\% \left(\frac{3047 - 3000}{3047 - 2881} \right) = 10.6\%$$

The calculations required to compute the after-tax rate of return in Example 12-5 were certainly more elaborate than those for the before-tax rate of return. It must be emphasized, however, that often the after-tax rate of return is the key value, since income taxes are a major disbursement that cannot be ignored.

EXAMPLE 12-6

An analysis of a firm's sales activities indicates that a number of profitable sales are lost each year because the firm cannot deliver some of its products quickly enough. By investing an additional $20,000 in inventory it is believed that the firm will realize $1000 more in before-tax profits in the first year. In the second year, before-tax extra profit will be $1500. Profits for subsequent years are expected to continue to increase on a $500-per-year gradient. The investment in the additional inventory may be recovered at the end of a 4-year analysis period simply by selling it and not replenishing the inventory. Compute:

(a) The before-tax rate of return.

(b) The after-tax rate of return assuming an incremental tax rate of 39%.

SOLUTION

Inventory is not considered to be a depreciable asset. Therefore, the investment in additional inventory is not depreciated. The after-tax cash flow table for the problem is presented in Table 12-4.

SOLUTION TO PART b

Use the following equation to calculate the before-tax rate of return:

$$20,000 = 1000(P/A, i, 4) + 500(P/G, i, 4) + 20,000(P/F, i, 4)$$

Try $i = 8\%$:

$$20,000 \doteq 1000(3.312) + 500(4.650) + 20,000(0.7350)$$

$$\doteq 3312 + 2325 + 14,700 = 20,337$$

TABLE 12-4 After-tax Cash Flow Table for Example 12-6

Year	(a) Before-Tax Cash Flow	(b) Depreciation	(c) Δ(Taxable Income) (a)−(b)	(d) 39% Income Taxes (c) x−0.39	(e) After-Tax Cash Flow (a) + (d)
0	−$20,000				−$20,000
1	1,000	—	$1000	−$390	610
2	1,500	—	1500	−585	915
3	2,000	—	2000	−780	1,220
4	{ 2,500 / 20,000	—	2500	−975	{ 1,525 / 20,000

Since i is too low, try $i = 10\%$:

$$20,000 \neq 1000(3.170) + 500(4.378) + 20,000(0.6830)$$
$$\neq 3170 + 2189 + 13,660 = 19,019$$

$$\text{Before-tax rate of return} = 8\% + 2\%\left(\frac{20,337 - 20,000}{20,337 - 19,019}\right) = 8.5\%$$

SOLUTION TO PART b

For a before-tax cash flow gradient of $500, the resulting after-tax cash flow gradient is $(1 - 0.39)(500) = \$305$.

$$20,000 = 610(P/A, i, 4) + 305(P/G, i, 4) + 20,000(P/F, i, 4)$$

Try $i = 5\%$:

$$20,000 \neq 610(3.546) + 305(5.103) + 20,000(0.8227) \neq 20,173$$

Since i is too low, try $i = 6\%$:

$$20,000 \neq 610(3.465) + 300(4.4945) + 20,000(0.7921) \neq 19,304$$

$$\text{After-tax rate of return} = 5\% + 1\%\left(\frac{20,173 - 20,000}{20,173 - 19,304}\right) = 5.2\%$$

CAPITAL GAINS AND LOSSES FOR NONDEPRECIATED ASSETS

When a nondepreciated capital asset is sold or exchanged, appropriate entries are made in the firm's accounting records. If the selling price of the capital asset exceeds the original cost basis, the excess is called a **capital gain.** If the selling price is less than the original cost basis, the difference is a **capital loss.** Examples of nondepreciated assets include stocks, land, art, and collectibles.

$$\text{Capital}\begin{bmatrix}\text{Gain}\\\text{Loss}\end{bmatrix} = \text{Selling price} - \text{Original cost basis}$$

Current tax law sets the highest net capital gains tax at 15% for most assets held by individuals for more than one year. This is in contrast to recaptured depreciation, which is taxed at the same rate as other (ordinary) income. The tax treatment of capital gains and losses for nondepreciated assets is shown in Table 12-5.

Investment Tax Credit

When the economy slows down and unemployment rises, the U.S. government frequently alters its tax laws to promote greater industrial activity. One technique used to stimulate capital investments has been the **investment tax credit (ITC).** Businesses were able to

TABLE 12-5 Tax Treatment of Capital Gains and Losses

For Individuals

Capital gain	For most assets held for less than 1 year, taxed as ordinary income. For most assets held for more than 1 year, taxed at 15% tax rate.*
Capital loss	Subtract capital losses from any capital gains; balance may be deducted from ordinary income, but not more than $3000 per year ($1500 if married filing separately). Excess capital losses may be carried into future taxable years indefinitely.

For Corporations

Capital gain	Taxed as ordinary income.
Capital loss	Corporations may deduct capital losses only to the extent of capital gains. Any capital loss in the current year that exceeds capital gains can be carried back 2 years, and, if not completely absorbed, is then carried forward for up to 20 years.

*Depending on tax bracket, type of asset, and duration of ownership, the capital gain tax rate can range from 5% to 28%. In addition, a special 0% capital gain tax rate applies to homeowners who treat their home as a primary residence for at least 2 years within a 5-year period before selling. In such cases, single taxpayers are allowed $250,000 in tax-free gain ($500,000 for married couples).

deduct a percentage of their new business equipment purchases as a *tax credit*. This meant that the firm's net cost of the equipment was reduced by the amount of the investment tax credit. Depending on the specific investment tax credit provisions, the credit might be subtracted from the basis for depreciation, or the basis for computing depreciation might remain the full cost of the equipment. The Tax Reform Act of 1986 eliminated the investment tax credit for most assets, although credits are allowed in some specialized cases such as historic building preservation and in the development of alternate energy sources. It is likely, however, that the general ITC will reappear at some future time.

ESTIMATING THE AFTER-TAX RATE OF RETURN

There is no shortcut method for computing the after-tax rate of return from the before-tax rate of return. One possible exception to this statement is in the situation of nondepreciable assets. In this special case, we have

$$\textbf{After-tax rate of return} = (1 - \text{Incremental tax rate}) \times (\text{Before-tax rate of return})$$

For Example 12-6, we could estimate the after-tax rate of return from the before-tax rate of return as follows:

$$\text{After-tax rate of return} = (1 - 0.39)(8.5\%) = 5.2\%$$

This value agrees with the value computed in Example 12-6(b).

This relationship may be helpful for selecting a trial after-tax rate of return when the before-tax rate of return is known. It must be emphasized, however, this relationship is only a rough approximation in almost all situations.

AFTER-TAX CASH FLOWS AND SPREADSHEETS

Realistic after-tax analysis requires spreadsheets. Even if costs and revenues are the same every year, MACRS depreciation percentages are not. The steps for calculating an after-tax internal rate of return are illustrated in Example 12-7. Because some cash flows are taxed and some are not, the spreadsheet is easier to build if these two types are separated. Spreadsheet construction is easier, as well, if recaptured depreciation or other gain/loss on disposal or sale is tabulated separately.

Taxes are considered even if only the costs of a project are known. The firm that does an engineering project must generate profits—or go out of business. Even if a firm has an unprofitable year, the tax law includes carry-forward and -backward provisions to transfer deductions to profitable years. The depreciation and revenues in Example 12-7 result in a negative taxable income for Year 2. Thus the *positive cash flow* due to taxes that is shown in Year 2 of Example 12-7 really represents tax savings for the firm.

EXAMPLE 12-7

Return to the data of Example 12-6, where the used truck had a first cost of $3000, a salvage value after 5 years of $750, and savings of $800 per year. Use MACRS depreciation and calculate the after-tax rate of return.

SOLUTION

Under MACRS, vehicles have a 5-year recovery period. Thus the MACRS depreciation can be calculated by using a VDB function (see "Spreadsheets and Depreciation" at the end of Chapter 11) or by lookup in Table 11-3. The depreciation in Year 5 has been halved, since it is the year of disposal.

	A	B	C	D	E	F	G	H
1	3000	First Cost						
2	800	Annual Benefit						
3	5	Recovery Period						
4	750	Salvage Value						
5	0.34	Tax Rate						
6								
7	Year	Untaxed BTCF	Taxed BTCF	MACRS	Recaptured Depreciation	Tax Income	Tax	ATCF
8	0	−3000						−3000.0
9	1		800	600.0		200.0	−68.0	732.0
10	2		800	960.0		(160.0)	54.4	854.4
11	3		800	576.0		224.0	−76.2	723.8
12	4		800	345.6		454.4	−154.5	645.5
13	5	750	800	172.8	404.4	1031.6	−350.7	1199.3
14		Cum. Depr.=		2654.4			IRR=	11.24%
15								
16						=Taxed BTCF − MACRS + Recapt.		
17					=Salvage − BookValue			

FIGURE 12-3 Spreadsheet for after-tax IRR calculation.

Note from Figure 12-3 that using MACRS rather than straight-line depreciation increases the after-tax IRR from 10.6% to 11.24%. This is due solely to the faster write-off that is allowed under MACRS.

SUMMARY

Since income taxes are part of most problems, no realistic economic analysis can ignore their consequences. Income taxes make the U.S. government a partner in every business venture. Thus the government benefits from all profitable ventures and shares in the losses of unprofitable ventures.

The first step in computing individual income taxes is to tabulate gross income. Any adjustments—for example, allowable taxpayer contributions to a retirement fund—are subtracted to yield adjusted gross income. Personal exemptions and either itemized deductions or the standard deduction are subtracted to find taxable income. This is used, together with a tax rate table, to compute the income tax liability for the year.

For corporations, taxable income equals gross income *minus* all ordinary and necessary expenditures (except capital expenditures) and depreciation and depletion charges. The income tax computation (whether for an individual or a corporation) is relatively simple, with rates ranging from 10 to 39%. The proper rate to use in an economic analysis is the incremental tax rate applicable to the increment of taxable income being considered.

Most individuals and corporations pay state income taxes in addition to federal income taxes. Since state income taxes are an allowable deduction in computing federal taxable income, it follows that the taxable income for the federal computation is lower than the state taxable income.

Combined state and federal incremental tax rate

$$= \Delta \text{State tax rate} + (\Delta \text{Federal tax rate})(1 - \Delta \text{State tax rate})$$

To introduce the effect of income taxes into an economic analysis, the starting point is a before-tax cash flow. Then the depreciation schedule is deducted from appropriate parts of the before-tax cash flow to obtain taxable income. Income taxes are obtained by multiplying taxable income by the proper tax rate. Before-tax cash flow less income taxes equals the after-tax cash flow. This data is all captured in an after-tax cash flow table.

Current tax law has decreased long-term capital gains on most nondepreciated assets for individuals to 15% when held for more than 1 year and provided an exclusion on the gain of the principal residence held for more than 2 years.

When dealing with nondepreciable assets, there is a nominal relationship between before-tax and after-tax rate of return. It is

$$\text{After-tax rate of return} = (1 - \Delta \text{Tax rate})(\text{Before-tax rate of return})$$

There is no simple relationship between before-tax and after-tax rate of return in the more usual case of investments involving depreciable assets.

PROBLEMS

These problems can be solved by hand, but most will be solved much more easily with a spreadsheet.

Individual/Joint Taxes

12-1 An unmarried taxpayer with no dependents expects an adjusted gross income of $70,000 in a given year. His nonbusiness deductions are expected to be $6000.

(a) What will his federal income tax be?

(b) He is considering an additional activity expected to increase his adjusted gross income. If this increase should be $16,000 and there should be no change in nonbusiness deductions or exemptions, what will be the increase in his federal income tax?

12-2 John Adams has a 65,000 adjusted gross income from Apple Corp. and allowable itemized deductions of $7200. Mary Eve has a $75,000 adjusted gross income and $2000 of allowable itemized deductions. Compute the total tax they would pay as unmarried individuals. Then compute their tax as a married couple filing a joint return. (*Answers:* $10,024 + 12,986.50 = 23,010.50; 23,597.50)

12-3 Bill Jackson had a total taxable income of $1800. Bill's employer wants him to work another month during the summer, but Bill had planned to spend the month hiking. If an additional month's work would increase Bill's taxable income by $1600, how much more money would he have after paying the income tax? (*Answer:* $1440)

12-4 A married couple filing jointly have a combined total adjusted gross income of $75,000. They have computed that their allowable itemized deductions are $4000. Compute their federal income tax. (*Answer:* $7,842.50)

12-5 Jane Shay operates a management consulting business. The business has been successful and now produces a taxable income of $100,000 per year after all "ordinary and necessary" expenses and depreciation have been deducted. At present the business is operated as a proprietorship; that is, Jane pays personal federal income tax on the entire $100,000. For tax purposes, it is as if she had a job that pays her a $100,000 salary per year.

As an alternative, Jane is considering incorporating the business. If she does, she will pay herself a salary of $40,000 a year from the corporation. The corporation will then pay taxes on the remaining $60,000 and retain the balance of the money as a corporate asset. Thus Jane's two alternatives are to operate the business as a proprietorship or as a corporation. Jane is single and has $3500 of itemized personal deductions. Which alternative will result in a smaller total payment of taxes to the government? (*Answer:* Incorporation, $14,296 versus $19,661)

12-6 Bill Alexander and his wife, Valerie, are both employed. Bill will have an adjusted gross income this year of $70,000. Valerie has an adjusted gross income of $2000 a month. Bill and Valerie have agreed that Valerie should continue working only until the federal income tax on their joint income tax return becomes $11,500. On what date should Valerie quit her job?

12-7 An unmarried individual in California with a taxable income of about $80,000 has a federal incremental tax rate of 28% and a state incremental tax rate of 9.3%. What is his combined incremental tax rate?

Corporate Taxes

12-8 A company wants to set up a new office in a country where the corporate tax rate is as follows: 15% of first $50,000 profits, 25% of next $25,000, 34% of next $25,000, and 39% of everything over $100,000. Executives estimate that they will have gross revenues of $500,000, total costs of $300,000, $30,000 in allowable tax deductions, and a one time business start-up credit of $8000. What is taxable income for the first year, and how much should the company expect to pay in taxes?

12-9 ARKO oil company purchased two large compressors for $125,000 each. One compressor was installed in the firm's Texas refinery and is being depreciated by MACRS depreciation. The other compressor was placed in the Oklahoma refinery, where it is being depreciated by sum-of-years'-digits depreciation with zero salvage value. Assume the company pays federal income taxes each year and the tax rate is constant. The corporate accounting department noted that the two compressors are being depreciated differently and wonders whether the corporation will wind up paying more income taxes over the life of the equipment as a result of this. What do you tell them?

12-10 Sole Brother Inc. is a shoe outlet to a major shoe manufacturing industry located in Chicago. Sole Brother uses accounts payable as one of its financing sources. Shoes are delivered to Sole Brother with a 3% discount if payment on the invoice is received within 10 days of delivery. By paying after the 10-day period, Sole is borrowing money and paying (giving up) the 3% discount. Although Sole Brother is not required to pay interest on delayed payments, the shoe manufacturers require that payments not be delayed beyond 45 days after the invoice date. To be sure of paying within 10 days, Sole Brothers decides to pay on the fifth day. Sole has a marginal corporate income tax of 40% (combined state and federal). By paying within the 10-day period, Sole is avoiding paying a fairly high price to retain the money owed shoe manufacturers. What would have been the effective annual after-tax interest rate?

12-11 A major industrialized state has a state corporate tax rate of 9.6% of taxable income. If a corporation has a state taxable income of $150,000, what is the total state and federal income tax it must pay? Also, compute its combined incremental state and federal income tax rate. (*Answers:* $50,534; 44.86%)

12-12 To increase its market share, Sole Brother Inc. decided to borrow $5000 from its banker for the purchase of newspaper advertising for its shoe retail line. The loan is to be paid in four equal annual payments with 15% interest. The loan is discounted 6 points. The 6 "points" is an additional interest charge of 6% of the loan, deducted immediately. This additional interest 6%(5000) = $300 means the actual amount received from the $5000 loan is $4700. The $300 additional interest may be deducted as four $75 additional annual interest payments. What is the after-tax interest rate on this loan?

12-13 The Lynch Bull investment company suggests that Steven Comstock, a wealthy New York City investor (his incremental income tax rate is 35%), consider the following investment.

Buy corporate bonds on the New York Stock Exchange with a face value (par value) of $100,000 and a 5% interest rate paid annually. These bonds can be purchased at their present market value of $75,000. Each year Steve will receive the $5000 interest, and after 5 years, when the bonds mature, he will receive $100,000 plus the last $5000 of interest.

Steve will pay for the bonds by borrowing $50,000 at 10% interest for 5 years. The $5000 interest paid on the loan each year will equal the $5000

of interest income from the bonds. As a result Steve will have no net taxable income during the five years due to this bond purchase and borrowing money scheme.

At the end of 5 years, Steve will receive $100,000 plus $5000 interest from the bonds and will repay the $50,000 loan and pay the last $5000 interest. The net result is that he will have a $25,000 capital gain; that is, he will receive $100,000 from a $75,000 investment. (*Note:* This situation represents an actual recommendation of a brokerage firm.)

(*a*) Compute Steve's after-tax rate of return on this dual bond-plus-loan investment package.

(*b*) What would be Steve's after-tax rate of return if he purchased the bonds for $75,000 cash and *did not* borrow the $50,000?

Historical Depreciation

12-14 Albert Chan decided to buy an old duplex as an investment. After looking for several months, he found a desirable duplex that could be bought for $93,000 cash. He decided that he would rent both sides of the duplex, and determined that the total expected income would be $800 per month. The total annual expenses for property taxes, repairs, gardening, and so forth are estimated at $600 per year. For tax purposes, Al plans to depreciate the building by the sum-of-years'-digits method, assuming that the building has a 20-year remaining life and no salvage value. Of the total $93,000 cost of the property, $84,000 represents the value of the building and $9000 is the value of the lot. Assume that Al is in the 38% incremental income tax bracket (combined state and federal taxes) throughout the 20 years.

In this analysis Al estimates that the income and expenses will remain constant at their present levels. If he buys and holds the property for 20 years, what after-tax rate of return can he expect to receive on his investment, using the following assumptions?

A. Al believes the building and the lot can be sold at the end of 20 years for the $9000 estimated value of the lot.

B. A more optimistic estimate of the future value of the building and the lot is that the property can be sold for $100,000 at the end of 20 years.

12-15 Zeon, a large, profitable corporation, is considering adding some automatic equipment to its production facilities. An investment of $120,000 will produce an initial annual benefit of $29,000, but the benefits are expected to decline $3000 per year, making second-year benefits $26,000, third-year benefits $23,000, and so forth. If the firm uses sum-of-years'-digits depreciation, an 8-year useful life, and $12,000 salvage value, will it obtain the desired 6% after-tax rate of return? Assume that the equipment can be sold for its $12,000 salvage value at the end of the 8 years. Also assume a 46% income tax rate for state and federal taxes combined.

12-16 A group of businessmen formed a corporation to lease for 5 years a piece of land at the intersection of two busy streets. The corporation has invested $50,000 in car-washing equipment. They will depreciate the equipment by sum-of-years'-digits depreciation, assuming a $5000 salvage value at the end of the 5 year useful life. The corporation is expected to have a before-tax cash flow, after meeting all expenses of operation (except depreciation), of $20,000 the first year, declining $3000 per year in future years (second year = $17,000, third year = $14,000, etc.). The corporation has other income, so it is taxed at a combined corporate tax rate of 20%. If the projected income is correct, and the equipment can be sold for $5000 at the end of 5 years, what after-tax rate of return would the corporation receive from this venture? (*Answer:* 14%)

12-17 The effective combined tax rate in an owner-managed corporation is 40%. An outlay of $20,000 for certain new assets is under consideration. It is estimated that for the next 8 years, these assets will be responsible for annual receipts of $9000 and annual disbursements (other than for income taxes) of $4000. After this time, they will be used only for stand-by purposes, and no future excess of receipts over disbursements is estimated.

(*a*) What is the prospective rate of return before income taxes?

(*b*) What is the prospective rate of return after taxes if straight-line depreciation can be used to write off these assets for tax purposes in 8 years?

(*c*) What is the prospective rate of return after taxes if it is assumed that these assets must be written off for tax purposes over the next 20 years, using straight-line depreciation?

12-18 A firm is considering the following investment project:

Year	Before-Tax Cash Flow (thousands)
0	−$1000
1	+500
2	+340
3	+244
4	+100
5	+100 / +125 Salvage value

The project has a 5-year useful life with a $125,000 salvage value, as shown. Double declining balance depreciation will be used, assuming the $125,000 salvage value. The combined income tax rate is 34%. If the firm requires a 10% after-tax rate of return, should the project be undertaken?

12-19 The Shellout Corp. owns a piece of petroleum drilling equipment that costs $100,000 and will be depreciated in 10 years by double declining balance depreciation, with conversion to straight-line depreciation at the optimal point. Assume no salvage value in the depreciation computation and a combined 34% tax rate. Shellout will lease the equipment to others and each year receive $30,000 in rent. At the end of 5 years, the firm will sell the equipment for $35,000. (Note that this is different from the zero-salvage-value assumption used in computing the depreciation.) What is the after-tax rate of return Shellout will receive from this equipment investment?

12-20 A mining corporation purchased $120,000 of production machinery and depreciated it using SOYD depreciation, a 5-year depreciable life, and zero salvage value. The corporation is a profitable one that has a 34% combined incremental tax rate.

At the end of 5 years the mining company changed its method of operation and sold the production machinery for $40,000. During the 5 years the machinery was used, it reduced mine operating costs by $32,000 a year, before taxes. If the company MARR is 12% after taxes, was the investment in the machinery a satisfactory one?

12-21 An automaker is buying some special tools for $100,000. The tools are being depreciated by double declining balance depreciation using a 4-year depreciable life and a $6250 salvage value. It is expected the tools will actually be kept in service for 6 years and then sold for $6250. The before-tax benefit of owning the tools is as follows:

Year	Before-Tax Cash Flow
1	$30,000
2	30,000
3	35,000
4	40,000
5	10,000
6	10,000
	6,250 Selling price

Compute the after-tax rate of return for this investment situation, assuming a 46% incremental tax rate. (*Answer:* 11.6%)

12-22 This is the continuation of Problem 12-21. Instead of paying $100,000 cash for the tools, the corporation will pay $20,000 now and borrow the remaining $80,000. The depreciation schedule will remain unchanged. The loan will be repaid by 4 equal end-of-year payments of $25,240.

Prepare an expanded cash flow table that takes into account both the special tools and the loan.

(*a*) Compute the after-tax rate of return for the tools, taking into account the $80,000 loan.

(*b*) Explain why the rate of return obtained in part (*a*) is different from the rate of return obtained in Problem 12-21.

Hints: **1.** Interest on the loan is 10%, $25,240 = 80,000 (*A/P*, 10%, 4). Each payment is made up of part interest and part principal. Interest portion for any year is 10% of balance due at the beginning of the year.

2. Interest payments are tax deductible (i.e., they reduce taxable income and thus taxes paid). Principal payments are not. Separate each $25,240 payment into interest and principal portions.

3. The Year-0 cash flow is −$20,000.

4. After-tax cash flow will be before-tax cash flow − interest payment − principal payment − taxes.

12-23 A project will require the investment of $108,000 in equipment (sum-of-years'-digits depreciation with a depreciable life of 8 years and zero salvage value) and $25,000 in raw materials (not depreciable). The annual project income after all expenses except depreciation have been paid is projected to be $24,000. At the end of 8 years the project will be discontinued and the $25,000 investment in raw materials will be recovered.

Assume a 34% combined income tax rate for this corporation. The corporation wants a 15% after-tax rate of return on its investments. Determine by present worth analysis whether this project should be undertaken.

12-24 A profitable incorporated business is considering an investment in equipment having the following before-tax cash flow. The equipment will be depreciated by double declining balance depreciation with conversion, if appropriate, to straight-line depreciation at the preferred time. For depreciation purposes a $700 salvage value at the end of 6 years is assumed. But the actual value is thought to be $1000 and it is this sum that is shown in the before-tax cash flow.

Year	Before-Tax Cash Flow
0	$12,000
1	1,727
2	2,414
3	2,872
4	3,177
5	3,358
6	1,997
	1,000 Salvage value

If the firm wants a 9% after-tax rate of return and its combined incremental income tax rate is 34%, determine by annual cash flow analysis whether the investment is desirable.

12-25 A salad oil bottling plant can either buy caps for the glass bottles at 5¢ each or install $500,000 worth of plastic molding equipment and manufacture the caps at the plant. The manufacturing engineer estimates the material, labor, and other costs would be 3¢ per cap.

(*a*) If 12 million caps per year are needed and the molding equipment is installed, what is the payback period?

(*b*) The plastic molding equipment would be depreciated by straight-line depreciation using a 5-year useful life and no salvage value. Assuming a combined 40% income tax rate, what is the after-tax payback period, and what is the after-tax rate of return?

12-26 A firm has invested $14,000 in machinery with a 7-year useful life. The machinery will have no salvage value, as the cost to remove it will equal its scrap value. The uniform annual benefits from the machinery are $3600. For a combined 47% income tax rate, and sum-of-years'-digits depreciation, compute the after-tax rate of return.

12-27 A firm manufactures padded shipping bags. A cardboard carton should contain 100 bags, but machine operators fill the cardboard cartons by eye, so a carton may contain anywhere from 98 to 123 bags (average = 105.5 bags).

  Management realizes that they are giving away $5^1/2$% of their output by overfilling the cartons. The solution would be to weigh each filled shipping carton. Underweight cartons would have additional shipping bags added, and overweight cartons would have some shipping bags removed. If the weighing is done, it is believed that the average quantity of bags per carton could be reduced to 102, with almost no cartons containing fewer than 100 bags.

  The weighing equipment would cost $18,600. The equipment would be depreciated by straight-line depreciation using a 10-year depreciable life and a $3600 salvage value at the end of 10 years. Assume the $18,600 worth of equipment qualifies for a 10% investment tax credit. One person, hired at a cost of $16,000 per year, would be required to operate the weighing equipment and to add or remove padded bags from the cardboard cartons. 20 0,000 cartons will be checked on the weighing equipment each year, with an average removal of 3.5 padded bags per carton with a manufacturing cost of 3¢ per bag. This large profitable corporation has a 50% combined federal-plus-state incremental tax rate. Assume a 10-year study period for the analysis and an after-tax MARR of 20%. Compute:

(a) The after-tax present worth.

(b) The after-tax internal rate of return.

(c) The after-tax simple payback period.

MACRS Depreciation

12-28 Mr. Sam K. Jones, a successful businessman, is considering erecting a small building on a commercial lot. A local furniture company is willing to lease the building for $9000 per year, paid at the end of each year. It is a net lease, which means the furniture company must also pay the property taxes, fire insurance, and all other annual costs. The furniture company will require a 5-year lease with an option to buy the building and land on which it stands for $125,000 after 5 years. Mr. Jones could have the building constructed for $82,000. He could sell the commercial lot now for $30,000, the same price he paid for it. Mr. Jones files a joint return and has an annual taxable income from other sources of $63,900. He would depreciate the commercial building by modified accelerated cost recovery system (MACRS) depreciation. Mr. Jones believes that at the end of the 5-year lease he could easily sell the property for $125,000. What is the after-tax present worth of this 5-year venture if Mr. Jones uses a 10% after-tax MARR?

12-29 One January Gerald Adair bought a small house and lot for $99,700. He estimated that $9700 of this amount represented the land's value. He rented the house for $6500 a year during the 4 years he owned the house. Expenses for property taxes, maintenance, and so forth were $500 per year. For tax purposes the house was depreciated by MACRS depreciation (27.5-year straight-line depreciation with a midmonth convention is used for rental property). At the end of 4 years the property was sold for $105,000. Gerald is married and works as an engineer. He estimates that his incremental state and federal combined tax rate is 24%. What after-tax rate of return did Gerald obtain on his investment?

12-30 A corporation with a 34% combined income tax rate is considering the following investment in research equipment, and has projected the benefits as follows:

Year	Before-Tax Cash Flow
0	−$50,000
1	+2,000
2	8,000
3	17,600
4	13,760
5	5,760
6	2,880

Prepare an after-tax cash flow table assuming MACRS depreciation.

(a) What is the after-tax rate of return?

(b) What is the before-tax rate of return?

12-31 An engineer is working on the layout of a new research and experimentation facility. Two plant operators will be required. If, however, an additional

$100,000 of instrumentation and remote controls were added, the plant could be run by a single operator. The total before-tax cost of each plant operator is projected to be $35,000 per year. The instrumentation and controls will be depreciated by means of the modified accelerated cost recovery system (MACRS).

If this corporation (34% combined corporate tax rate) invests in the additional instrumentation and controls, how long will it take for the after-tax benefits to equal the $100,000 cost? In other words, what is the after-tax payback period? (*Answer:* 3.24 years).

12-32 A special power tool for plastic products costs $400, has a 4-year useful life, no salvage value, and a 2-year before-tax payback period. Assume uniform annual end-of-year benefits.

(*a*) Compute the before-tax rate of return.

(*b*) Compute the after-tax rate of return, based on MACRS depreciation and a 34% combined corporate income tax rate.

12-33 The Ogi Corporation, a construction company, purchased a pickup truck for $14,000 and used MACRS depreciation in the income tax return. During the time the company had the truck, they estimated that it saved $5000 a year. At the end of 4 years, Ogi sold the truck for $3000. The combined federal and state income tax rate for Ogi is 45%. Compute the after-tax rate of return for the truck. (*Answer:* 12.5%)

12-34 A profitable wood products corporation is considering buying a parcel of land for $50,000, building a small factory building at a cost of $200,000, and equipping it with $150,000 of MACRS 5-year class machinery.

If the project is undertaken, MACRS depreciation will be used. Assume the plant is put in service October 1. The before-tax net annual benefit from the project is estimated at $70,000 per year. The analysis period is to be 5 years, and planners assume the sale of the total property (land, building, and machinery) at the end of 5 years, also on October 1, for $328,000. Compute the after-tax cash flow based on a 34% combined income tax rate. If the corporation's criterion is a 15% after-tax rate of return, should it proceed with the project?

12-35 A chemical company bought a small vessel for $55,000; it is to be depreciated by MACRS depreciation. When requirements changed suddenly, the chemical company leased the vessel to an oil company for 6 years at $10,000 per year. The lease also provided that the oil company could buy the vessel at the end of 6 years for $35,000. At the end of the 6 years, the oil company exercised its option and bought the vessel. The chemical company has a 34% combined incremental tax rate. Compute its after-tax rate of return on the vessel. (*Answer:* 9.86%)

12-36 Xon, a small oil company, purchased a new petroleum drilling rig for $1,800,000. Xon will depreciate the drilling rig using MACRS depreciation. The drilling rig has been leased to a drilling company, which will pay Xon $450,000 per year for 8 years. At the end of 8 years the drilling rig will belong to the drilling company. If Xon has a 34% combined incremental tax rate and a 10% after-tax MARR, does the investment appear to be satisfactory?

12-37 The profitable Palmer Golf Cart Corp. is considering investing $300,000 in special tools for some of the plastic golf cart components. Executives of the company believe the present golf cart model will continue to be manufactured and sold for 5 years, after which a new cart design will be needed, together with a different set of special tools.

The saving in manufacturing costs, owing to the special tools, is estimated to be $150,000 per year for 5 years. Assume MACRS depreciation for the special tools and a 39% combined income tax rate.

(*a*) What is the after-tax payback period for this investment?

(*b*) If the company wants a 12% after-tax rate of return, is this a desirable investment?

12-38 Uncle Elmo is contemplating a $10,000 investment in a methane gas generator. He estimates his gross income would be $2000 the first year and increase by $200 each year over the next 10 years. His expenses of $200 the first year would increase by $200 each year over the next 10 years. He would depreciate the generator by MACRS depreciation, assuming a 7-year property class. A 10-year-old methane generator has no market value. The income tax rate is 40%. (Remember that recaptured depreciation is taxed at the same 40% rate).

(*a*) Construct the after-tax cash flow for the 10-year project life.

(b) Determine the after-tax rate of return on this investment. Uncle Elmo thinks it should be at least 8%.

(c) If Uncle Elmo could sell the generator for $7000 at the end of the fifth year, would his rate of return be better than if he kept the generator for 10 years? You don't have to actually find the rate of return, just do enough calculations to see whether it is higher than that of part (b).

12-39 Granny's Butter and Egg Business is such that she pays an effective tax rate of 40%. Granny is considering the purchase of a new Turbo Churn for $25,000. This churn is a special handling device for food manufacture and has an estimated life of 4 years and a salvage value of $5000. The new churn is expected to increase net income by $8000 per year for each of the 4 years of use. If Granny works with an after-tax MARR of 10% and uses MACRS depreciation, should she buy the churn?

12-40 Eric has a house and lot for sale for $70,000. It is estimated that $10,000 is the value of the land and $60,000 is the value of the house. Bonnie is purchasing the house on January 1 to rent and plans to own the house for 5 years. After 5 years, it is expected that the house and land can be sold on December 31 for $80,000. Total annual expenses (maintenance, property taxes, insurance, etc.) are expected to be $3000 per year. The house would be depreciated by MACRS depreciation using a 27.5-year straight-line rate with midmonth convention for rental property. For depreciation, a salvage value of zero was used. Bonnie wants a 15% after-tax rate of return on her investment. You may assume that Bonnie has an incremental income tax rate of 28%

in each of the 5 years. Capital gains are taxed at 15%. Determine the following:

(a) The annual depreciation

(b) The capital gain (loss) resulting from the sale of the house

(c) The annual rent Bonnie must charge to produce an after-tax rate of return of 15%. (*Hint:* Write an algebraic equation to solve for rent.)

12-41 Bill owns a data processing company. He plans to buy an additional computer for $20,000, use the computer for 3 years, and sell it for $10,000. He expects that use of the computer will produce a net income of $8000 per year. The combined federal and state incremental tax rate is 45%. Using MACRS depreciation, complete Table P12-41 to determine the net present worth of the after-tax cash flow using an interest rate of 12%.

12-42 Refer to Problem 12-33. To help pay for the pickup truck the Ogi Corp. obtained a $10,000 loan from the truck dealer, payable in four end-of-year payments of $2500 plus 10% interest on the loan balance each year.

(a) Compute the after-tax rate of return for the truck together with the loan. Note that the interest on the loan is tax deductible, but the $2500 principal payments are not.

(b) Why is the after-tax rate of return computed in part (a) so much different from the 12.5% obtained in Problem 12-33?

Solving for Unknowns

12-43 A store owner, Joe Lang, believes his business has suffered from the lack of adequate customer parking

TABLE P12-41 Worksheet for Problem 12-41

Year	Before-Tax Cash Flow	MACRS Depreciation	Taxable Income	Income Tax (45%)	After-Tax Cash Flow	Present Worth (12%)
0	−$20,000					
1	+8,000					
2	+8,000					
3	+8,000					
	+10,000					
				Net Present worth =		

space. Thus, when he was offered an opportunity to buy an old building and lot next to his store, he was interested. He would demolish the old building and make off-street parking for 20 customers' cars. Joe estimates that the new parking would increase his business and produce an additional before-income-tax profit of $7000 per year. It would cost $2500 to demolish the old building. Mr. Lang's accountant advised that both costs (the property and demolishing the old building) would be considered to comprise the total value of the land for tax purposes, and it would not be depreciable. Mr. Lang would spend an additional $3000 right away to put a light gravel surface on the lot. This expenditure, he believes, may be charged as an operating expense immediately and need not be capitalized. To compute the tax consequences of adding the parking lot, Joe estimates that his combined state and federal incremental income tax rate will average 40%. If Joe wants a 15% after-tax rate of return from this project, how much could he pay to purchase the adjoining land with the old building? Assume that the analysis period is 10 years and that the parking lot could always be sold to recover the costs of buying the property and demolishing the old building. (*Answer:* $23,100)

12-44 The management of a private hospital is considering the installation of an automatic telephone switchboard, which would replace a manual switchboard and eliminate the attendant operator's position. The class of service provided by the new equipment is estimated to be at least equal to the present method of operation. To provide telephone service, five operators will work three shifts per day, 365 days per year. Each operator earns $14,000 per year. Company-paid benefits and overhead are 25% of wages. Money costs 8% after income taxes. Combined federal and state income taxes are 40%. Annual property taxes and maintenance are $2\frac{1}{2}$ and 4% of investment, respectively. Depreciation is 15-year straight line. Disregarding inflation, how large an investment in the new equipment can be economically justified by savings obtained by eliminating the present equipment and labor costs? The existing equipment has zero salvage value.

12-45 A contractor has to choose one of the following alternatives in performing earthmoving contracts:

  A. Buy a heavy-duty truck for $13,000. Salvage value is expected to be $3000 at the end of the vehicle's 7-year depreciable life. Maintenance

is $1100 per year. Daily operating expenses are $35.

  B. Hire a similar unit for $83 per day.

Based on a 10% after-tax rate of return, how many days per year must the truck be used to justify its purchase? Base your calculations on straight-line depreciation and a 50% income tax rate. (*Answer:* $91\frac{1}{2}$ days)

12-46 The Able Corporation is considering the installation of a small electronic testing device for use in conjunction with a government contract the firm has just won. The testing device will cost $20,000, and will have an estimated salvage value of $5000 in 5 years when the government contract is finished. The firm will depreciate the instrument by the sum-of-years'-digits method, using 5 years as the useful life and a $5000 salvage value. Assume that Able pays 50% federal and state corporate income taxes and uses 8% *after-tax* in economic analysis. What minimum equal annual benefit must Able obtain *before taxes* in each of the 5 years to justify purchasing the electronic testing device? (*Answer:* $5150)

12-47 A house and lot are for sale for $155,000. It is estimated that $45,000 is the value of the land and $110,000 is the value of the house. If purchased, the house can be rented to provide a net income of $12,000 per year after taking all expenses, except depreciation, into account. The house would be depreciated by straight-line depreciation using a 27.5-year depreciable life and zero salvage value.

  Mary Silva, the prospective purchaser, wants a 10% after-tax rate of return on her investment after considering both annual income taxes and a capital gain when she sells the house and lot. At what price would she have to sell the house at the end of 10 years to achieve her objective? You may assume that Mary has an incremental income tax rate of 28% in each of the 10 years.

12-48 A corporation is considering buying a medium-sized computer that will eliminate a task that must be performed three shifts per day, 7 days per week, except for one 8-hour shift per week when the operation is shut down for maintenance. At present four people are needed to perform the day and night tasks. Thus the computer will replace four employees. Each employee costs the company $32,000 per year ($24,000 in direct wages plus $8000 per year in other company employee costs). It will cost $18,000 per year to maintain and operate the computer.

The computer will be depreciated by sum-of-years'-digits depreciation using a 6-year depreciable life, at which time it will be assumed to have zero salvage value.

The corporation has a combined federal and state incremental tax rate of 50%. If the firm wants a 15% rate of return, after considering both state and federal income taxes, how much can it afford to pay for the computer?

12-49 A sales engineer has the following alternatives to consider in touring his sales territory.

A. Buy a new car for $14,500. Salvage value is expected to be about $5000 after 3 years. Maintenance and insurance cost is $1000 in the first year and increases at the rate of $500/year in subsequent years. Daily operating expenses are $50/day.

B. Rent a similar car for $80/day.

Based on a 12% after-tax rate of return, how many days per year must he use the car to justify its purchase? You may assume that this sales engineer is in the 28% incremental tax bracket. Use MACRS depreciation.

12-50 A large profitable company, in the 40% combined federal/state tax bracket, is considering the purchase of a new piece of equipment that will yield benefits of $10,000 in Year 1, $15,000 in Year 2, $20,000 in Year 3, and $20,000 in Year 4. The equipment is to be depreciated using 5-year MACRS depreciation starting in the year of purchase (Year 0). It is expected that the equipment will be sold at the end of Year 4 at 20% of its purchase price. What is the maximum equipment purchase price the company can pay if its after-tax MARR is 10%?

12-51 A prosperous businessman is considering two alternative investments in bonds. In both cases the first interest payment would be received at the end of the first year. If his personal taxable income is fixed at $40,000 and he is single, which investment produces the greater after-tax rate of return? Compute the after-tax rate of return for each bond to within $1/4$ of 1 percent.

Ann Arbor Municipal Bonds: A bond with a face value of $1000 pays $60 per annum. At the end of 15 years, the bond becomes due ("matures"), at which time the owner of the bond will receive $1000 plus the final $60 annual payment. The bond may be purchased for $800. Since it is a municipal bond, the annual interest is *not* subject to federal income tax. The difference between what the businessman would pay for

the bond ($800) and the $1000 face value he would receive at the end of 15 years must be included in taxable income when the $1000 is received.

Southern Coal Corporation Bonds: A thousand-dollar bond yields $100 per year in annual interest payments. When the bonds mature at the end of 20 years, the bondholder will receive $1000 plus the final $100 interest. The bonds may be purchased now for $1000. The income from corporation bonds must be included in federal taxable income.

Multiple Alternatives

12-52 Use the after-tax IRR method to evaluate the following three alternatives for MACRS 3-year property, and offer a recommendation. The after-tax MARR is 25%, the project life is 5 years, and the firm has a combined incremental tax rate of 45%.

Alt.	First Cost	Annual Costs	Salvage Value
A	$14,000	$2500	$ 5,000
B	18,000	1000	10,000
C	10,000	5000	0

12-53 A small-business corporation is considering whether to replace some equipment in the plant. An analysis indicates there are five alternatives in addition to the do-nothing option, Alt. A. The alternatives have a 5-year useful life with no salvage value. Straight-line depreciation would be used.

Alternatives	Cost (thousands)	Before-Tax Uniform Annual Benefits (thousands)
A	$ 0	$ 0
B	25	7.5
C	10	3
D	5	1.7
E	15	5
F	30	8.7

The corporation has a combined federal and state income tax rate of 20%. If the corporation expects a 10% after-tax rate of return for any new investments, which alternative should be selected?

12-54 A corporation with $7 million in annual taxable income is considering two alternatives:

Before-Tax Cash Flow

Year	Alt. 1	Alt. 2
0	−$10,000	−$20,000
1–10	4,500	4,500
11–20	0	4,500

Both alternatives will be depreciated by straight-line depreciation assuming a 10-year depreciable life and no salvage value. Neither alternative is to be replaced at the end of its useful life. If the corporation has a minimum attractive rate of return of 10% *after taxes,* which alternative should it choose? Solve the problem by:

(a) Present worth analysis

(b) Annual cash flow analysis

(c) Rate of return analysis

(d) Future worth analysis

(e) Benefit–cost ratio analysis

12-55 Two mutually exclusive alternatives are being considered by a profitable corporation with an annual taxable income between $5 million and $10 million.

Before-Tax Cash Flow

Year	Alt. *A*	Alt. *B*
0	−$3000	−$5000
1	1000	1000
2	1000	1200
3	1000	1400
4	1000	2600
5	1000	2800

Both alternatives have a 5-year useful and depreciable life and no salvage value. Alternative *A* would be depreciated by sum-of-years'-digits depreciation, and Alt. *B* by straight-line depreciation. If the MARR is 10% after taxes, which alternative should be selected?   (*Answer:* Alt. *B*)

12-56 A large profitable corporation is considering two mutually exclusive capital investments:

	Alt. *A*	Alt. *B*
Initial cost	$11,000	$33,000
Uniform annual benefit	3,000	9,000
End-of-depreciable-life salvage value	2,000	3,000
Depreciation method	SL	SOYD
End-of-useful-life salvage value obtained	2,000	5,000
Depreciable life, in years	3	4
Useful life, in years	5	5

If the firm's after-tax minimum attractive rate of return is 12% and its combined incremental income tax rate is 34%, which project should be selected?

12-57 Assume that you are deciding whether to lease a car or buy one to use exclusively in business. The estimated mileage in both cases will be 12,000/year. Insurance costs will be the same in either case at $600 per year. You want to examine the tax impacts of the two options to see which is preferred. Use MACRS depreciation.

Assume that you will have leased or purchased the car on June 30, 2006, and that you intend to keep the car for 36 months. List any assumptions that you make in your analysis. Use a 12% MARR, a 40% tax rate, and assume end of year payments.

Lease option: $2250 down and $369 a month for 36 months.

Purchase option: The car that you are interested in lists for $29,188 and the dealer finance plan calls for 30 monthly payments of $973 with no money down and 0% interest. Estimated value after 36 months is $15,200. Remember that the tax law limits the depreciation on cars according to the following rule:

Depreciation Limitation on Automobiles Purchased After 2005

Tax Year	Amount
First	$2960
Second	4700
Third	2850
Fourth and later	1675

12-58 A plant can be purchased for $1,000,000 or it can be leased for $200,000 per year. The annual income is expected to be $800,000 with the annual operating cost of $200,000. The resale value of the plant is estimated to be $400,000 at the end of its 10-year life. The company's combined federal and state income tax rate is 40%. A straight-line depreciation can be used over the 10 years with the full first-year depreciation rate.

(a) If the company uses the after-tax minimum attractive rate of return of 10%, should it lease or purchase the plant?

(b) What is the breakeven rate of return of purchase versus lease?

12-59 VML Industries has need of specialized yarn manufacturing equipment for operations over the next 3 years. The firm could buy the machinery for $95,000 and depreciate it using MACRS. Annual maintenance would be $7500, and it would have a salvage value of $25,000 after 3 years. Another alternative would be to lease the same machine for $45,000 per year on an "all costs" inclusive lease (maintenance costs included in lease payment). These lease payments are due at the beginning of each year. VML Industries uses an after-tax MARR of 18%, and a combined tax rate of 40%. Do an after-tax present worth analysis to determine which option is preferred.

12-60 Padre Pio owns a small business and has taxable income of $150,000. He is considering four mutually exclusive alternative models of machinery. Which machine should be selected on an after-tax basis? The after-tax MARR is 15%. Assume that each machine is MACRS 5-year property and can be sold for a market value that is 25% of the purchase cost, and the project life is 10 years.

Model	I	II	III	IV
First cost	$9000	$8000	$7500	$6200
Annual costs	25	200	300	600

CHAPTER 13

REPLACEMENT ANALYSIS

The $2 Billion Upgrade

In February 2003, Intel Corporation announced it was spending $2 billion to modernize and update its silicon wafer manufacturing plant in Chandler, Arizona. The upgrade

allowed Intel to manufacture 300-millimeter chips rather than the 200-millimeter size it had been producing at the plant, known as Fab 12. The project was estimated to double manufacturing capacity while at the same time reducing costs. Upgrade work included remodeling the plant's interior and the purchase of new wafer fabrication tools.

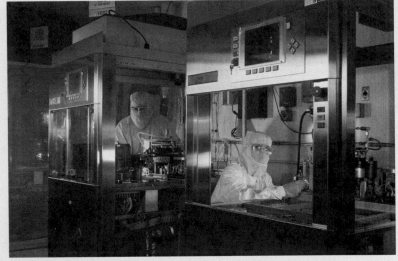

Despite the price tag for the project, Intel planned on saving money overall by upgrading the existing plant instead of designing and building a new one. The upgraded plant used newer processes that allowed chips to hold smaller and faster transistors, and the new larger wafer accommodates more chips. These all lead to lower production costs.

The company also noted that by deciding to remain in its current location it was able to retain its highly skilled workforce.

Questions to Consider

1. Intel made the decision to upgrade existing plant facilities versus building new ones. In their analysis they claimed to have saved money by doing so. Use the Internet and search for articles that discuss what new manufacturing plants cost in different type of industries. Does Intel's $2 billion upgrade decision seem in line? What issues are you not considering if you think the cost seems very high?
2. Companies that upgrade existing production assets may or may not scrap their current assets. Give a realistic scenario for each case.
3. The upgrade vs. replace decision often includes economic as well as noneconomic factors. List three economic and three noneconomic factors that may have been at the top of Intel's list in making the decision.
4. When companies build new manufacturing plants they are often sited outside the U.S. Discuss the ethical impacts that are or should be part of a firm's decision process.

After Completing This Chapter...

The student should be able to:

- Recast an equipment reinvestment decision as a *challenger vs. defender* analysis.
- Use the *replacement analysis decision map* to select the correct economic analysis technique to apply.
- Calculate the *minimum cost life* of economic challengers.
- Incorporate concepts such as *repeatability assumption for replacement analysis* and *marginal cost data for the defender* to select the correct economic analysis techniques.
- Perform replacement problems on an after-tax basis, utilizing the *defender sign change procedure* when appropriate.
- Use spreadsheets for solving before-tax and after-tax replacement analysis problems.

U p to this point in our economic analysis we have considered the evaluation and selection of *new* alternatives. Which new car or production machine should we buy? What new material handling system or ceramic grinder should we install? More frequently, however, economic analysis weighs *existing* versus *new* facilities. For most engineers the problem is less likely to be one of building a new plant; rather, the goal is more often keeping a present plant operating economically. We are not choosing between new ways to perform the desired task. Instead, we have equipment performing the task, and the question is: Should the existing equipment be retained or replaced? This adversarial situation has given rise to the terms **defender** and **challenger.** The defender is the existing equipment; the challenger is the best available replacement equipment. Economically evaluating the existing defender and its challengers is the domain of **replacement analysis.**

THE REPLACEMENT PROBLEM

Replacement of an existing asset may be appropriate due to obsolescence, depletion, and deterioration due to aging. In each of these cases, the ability of a previously implemented business asset to produce a desired output is challenged. For cases of obsolescence, depletion, and aging, it may be economical to replace the existing asset. We define each of these situations.

Obsolescence: occurs when an asset's technology is surpassed by newer and/or different technologies. As an example, today's personal computers (PCs) with more RAM, faster clock speeds, larger hard drives, and more powerful central processors have made older, less powerful PCs obsolete. Thus, obsolete assets may need to be replaced with newer, more technologically advanced ones.

Depletion: the gradual loss of market value of an asset as it is being consumed or exhausted. Oil wells and timber stands are examples of such assets. In most cases the asset will be used until it is depleted, at which time a replacement asset will be obtained. Depletion was treated in Chapter 11.

Deterioration due to aging: the general condition of loss in value of some asset due to the aging process. Production machinery and other business assets that were once new eventually become aged. To compensate for a loss in functionality due to the aging process, additional operating and maintenance expenses are usually incurred to maintain the asset at its operating efficiency.

Aging equipment often has a greater risk of breakdowns. Planned replacements can be scheduled to minimize the time and cost of disruptions. Unplanned replacements can be very costly or even, as with an airplane engine, potentially catastrophic.

There are variations of the replacement problem: an existing asset may be abandoned or retired, augmented by a new asset but kept in service, or overhauled to reduce its operating and maintenance costs. These variations are most easily considered as potential new challengers.

Replacement problems are normally analyzed by looking only at the *costs* of the existing and replacement assets. Since the assets typically perform the same function, the value of using the vehicle, machine, or other equipment can be ignored. If the new asset has new features or better performance, these can be included as a cost savings.

Alternatives in a replacement problem almost always have *different lives*, and the problem includes picking the best one. This is because an existing asset will often be kept for at most a few years longer, while the potential replacements may have lives of any length. Thus replacement problems are focused on annual marginal costs and on EUAC values. We can calculate present costs, but only as a step in calculating EUAC values.

In industry, as in government, expenditures are normally monitored by means of *annual budgets.* One important facet of a budget is the allocation of money for new capital expenditures, either new facilities or replacement and upgrading of existing facilities.

Replacement analysis may recommend that certain equipment be replaced, with the cost included in the capital expenditures budget. Even if no recommendation to replace is made, such a recommendation may be made the following year or subsequently. At *some* point, existing equipment will be replaced, either when it is no longer necessary or when better equipment is available. Thus, the question is not *if* the defender will be replaced, but *when* it will be replaced. This leads us to the first aspect of the defender–challenger comparison:

Shall we replace the defender now, or shall we keep it for one or more additional years?

If we do decide to keep the asset for another year, we will often reanalyze the problem next year. The operating environment and costs may have changed, or new challengers with lower costs or better performance may have emerged.

REPLACEMENT ANALYSIS DECISION MAP

Figure 13-1 is a basic decision map for conducting a replacement analysis.

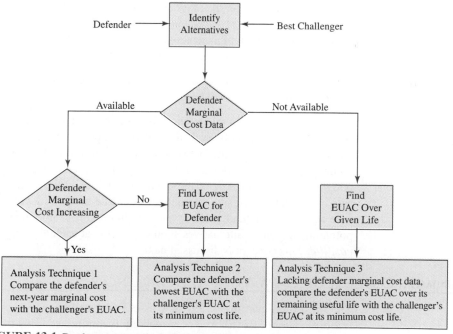

FIGURE 13-1 Replacement analysis decision map.

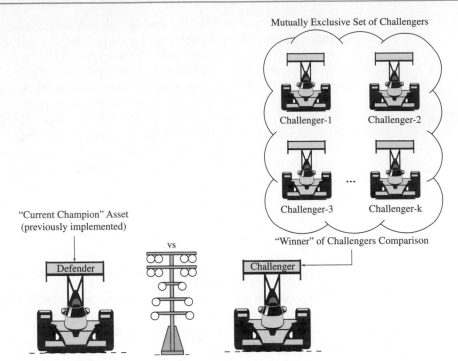

FIGURE 13-2 Defender–challenger comparison.

Looking at the map, we can see there are three *replacement analysis techniques* that are correct under different circumstances. The correct replacement analysis technique depends on the data available for the alternatives and how the data behave over time.

By looking at the replacement analysis map, we see that the first step is to identify the alternatives. Again, in replacement analysis we are interested in comparing our previously implemented asset (the *defender*) against the best current available *challenger*.

If the defender proves more economical, it will be retained. If the challenger proves more economical, it will be installed.

In this comparison the defender is being evaluated against a challenger that has been selected from a mutually exclusive set of competing challengers. Figure 13-2 illustrates this concept as a drag race between the defender and a challenger. The challenger that is competing against the defender has emerged from an earlier competition among a set of potential challengers. Any of the methods previously discussed in this text for evaluating sets of mutually exclusive alternatives could be used to identify the "best" challenger to race against the defender. However, it is important to note that the comparison of these potential challenger alternatives should be made at each alternative's respective *minimum cost life*. This concept is discussed next.

MINIMUM COST LIFE OF A NEW ASSET—THE CHALLENGER

The **minimum cost life** of any new asset is the number of years at which the equivalent uniform annual cost (EUAC) of ownership is minimized. This minimum cost life is often shorter than either the asset's physical or useful life, because of increasing operating and maintenance costs in the later years of asset ownership. The challenger asset selected to "race" against the defender (in Figure 13-2) is the one having the lowest minimum cost of all the competing mutually exclusive challengers.

To calculate the minimum cost life of an asset, we determine the EUAC for each possible life less than or equal to the useful life. As illustrated in Example 13-1, the EUAC tends to be high if the asset is kept only a few years; then it decreases to some minimum EUAC, and then increases again as the asset ages. By identifying the number of years at which the EUAC is a minimum and then keeping the asset for that number of years, we are minimizing the yearly cost of ownership.

EXAMPLE 13-1

A piece of machinery costs \$7500 and has no salvage value after it is installed. The manufacturer's warranty will pay the first year's maintenance and repair costs. In the second year, maintenance costs will be \$900, and this item will increase on a \$900 arithmetic gradient in subsequent years. Also, operating expenses for the machinery will be \$500 the first year and will increase on a \$400 arithmetic gradient in the following years. If interest is 8%, compute the machinery's economic life that minimizes the EUAC. That is, find its minimum cost life.

SOLUTION

	If Retired at the End of Year n			
Year, n	EUAC of Capital Recovery Costs: \$7500 $(A/P, 8\%, n)$	EUAC of Maintenance and Repair Costs: \$900 $(A/G, 8\%, n)$	EUAC of Operating Costs: \$500 + \$400$(A/G, 8\%, n)$	EUAC Total
1	\$8100	\$ 0	\$ 500	\$8600
2	4206	433	692	5331
3	2910	854	880	4644
4	2264	1264	1062	4589 ←
5	1878	1661	1238	4779
6	1622	2048	1410	5081
7	1440	2425	1578	5443
8	1305	2789	1740	5834
9	1200	3142	1896	6239
10	1117	3484	2048	6650
11	1050	3816	2196	7063
12	995	4136	2338	7470
13	948	4446	2476	7871
14	909	4746	2609	8265
15	876	5035	2738	8648

The total EUAC data are plotted in Figure 13-3. From either the tabulation or the figure, we see that the machinery's minimum cost life is 4 years, with a minimum EUAC of \$4589 for each of those 4 years.

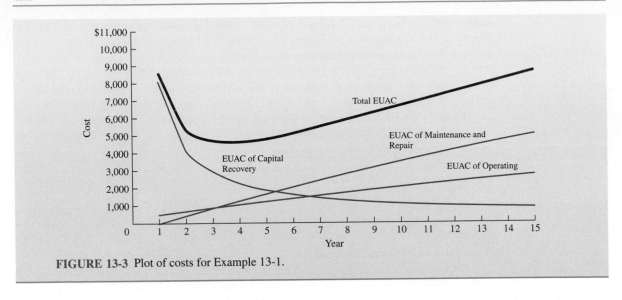

FIGURE 13-3 Plot of costs for Example 13-1.

Looking at Figure 13-3 a bit more closely, we see the effects of each of the individual cost components on total EUAC (capital recovery, maintenance/repair, and operating expense EUACs) and how they behave over time. The total EUAC curve of most assets tends to follow this concave shape—high at the beginning due to capital recovery costs, and high at the end due to increased maintenance/repair and operating expenses. The minimum EUAC occurs somewhere between these high points.

Like many pieces of installed equipment, the item considered in Example 13-1 had no salvage value. However, some assets, like your car, can easily be sold for a value than depends on the car's age and condition. Another possible complication is that repair costs may be reduced in early years by a warranty. The resulting cost curves will look like Figure 13-3, but the calculations are more work and a spreadsheet is very helpful. Example 13-11 in the last section of this chapter illustrates finding the minimum cost life for a new vehicle.

DEFENDER'S MARGINAL COST DATA

Once the basic participants in the defender–challenger comparison have been identified (see Figure 13-1), two specific questions regarding marginal costs must be answered: *Do we have marginal cost data for the defender?* and *Are the defender's marginal costs increasing on a year-to-year basis?* Let us first define marginal cost and then discuss why it is important to answer these two questions.

Marginal costs, as opposed to an EUAC, are the year-by-year costs of keeping an asset. Therefore, the "period" of any yearly marginal cost of ownership is always *1 year.* The marginal cost is compared with the EUAC, which is an end-of-year cash flow. Therefore, the marginal cost is calculated as an equivalent end-of-year cash flow.

On the other hand, an EUAC can apply to any number of consecutive years. Thus, the marginal cost of ownership for any year in an asset's life is the cost for *that year only.* In replacement problems, the total marginal cost for any year can include the capital recovery cost (loss in market value and lost interest for the year), yearly operating and maintenance costs, yearly taxes and insurance, and any other expense that occurs during that year. To calculate an asset's yearly marginal cost of ownership, it is necessary to have estimates of an asset's market value on a year-to-year basis over its useful life, as well as ordinary yearly expenses. Example 13-2 illustrates how total marginal cost can be calculated for an asset.

EXAMPLE 13-2

A new piece of production machinery has the following costs.

$$\text{Investment cost} = \$25,000$$

$$\text{Annual operating and maintenance cost} = \$2000 \text{ in Year 1 and then increasing by}$$
$$\$500 \text{ per year}$$

$$\text{Annual cost for risk of breakdown} = \$5000 \text{ per year for 3 years, then}$$
$$\text{increasing by } \$1500 \text{ per year}$$

$$\text{Useful life} = 7 \text{ years}$$

$$\text{MARR} = 15\%$$

Calculate the marginal cost of keeping this asset over its useful life.

SOLUTION

From the problem data we can easily find the marginal costs for O&M and risk of breakdowns. However, to calculate the marginal capital recovery cost, we need estimates of each year's market value.

Year	Market Value
1	$18,000
2	13,000
3	9,000
4	6,000
5	4,000
6	3,000
7	2,500

We can now calculate the machinery's *marginal cost* (year-to-year cost of ownership) over its 7-year useful life.

Year, n	Loss in Market Value in Year n	Interest in Year n	O&M Cost in Year n	Cost of Breakdown Risk in Year n	Total Marginal Cost in Year n
1	$25,000 - 18,000 = \$7000$	$25,000(0.15) = \$3750$	$2000	$5,000	$17,750
2	$18,000 - 13,000 = 5000$	$18,000(0.15) = 2700$	2500	5,000	15,200
3	$13,000 - 9,000 = 4000$	$13,000(0.15) = 1950$	3000	5,000	13,950
4	$9,000 - 6,000 = 3000$	$9,000(0.15) = 1350$	3500	6,500	14,350
5	$6,000 - 4,000 = 2000$	$6,000(0.15) = 900$	4000	8,000	14,900
6	$4,000 - 3,000 = 1000$	$4,000(0.15) = 600$	4500	9,500	15,600
7	$3,000 - 2,500 = 500$	$3,000(0.15) = 450$	5000	11,000	16,950

Notice that each year's total marginal cost includes loss in market value, interest, O&M cost, and cost for risk of breakdowns. For example, the Year-5 marginal cost of $14,900 is calculated as $2000 + 900 + 4000 + 8000$.

Do We Have Marginal Cost Data for the Defender?

Our decision map indicates that it is necessary to know whether marginal cost data are available for the defender asset to determine the appropriate replacement technique to use. Usually in engineering economic problems, annual savings and expenses are given for all alternatives. However, as in Example 13-2, it is also necessary to have year-to-year salvage value estimates to calculate total marginal costs. If the total marginal costs for the defender can be calculated, and if the data are increasing from year to year, then *replacement analysis technique 1* should be used to compare the defender to the challenger.

Are These Marginal Costs Increasing?

We have seen that it is important to know whether the marginal cost for the defender is increasing from year to year. This is determined by inspecting the total marginal cost of ownership of the defender over its remaining life. In most replacement analyses the defender is nearing the end of its economic life. The question usually is, Should we replace it now, next year, or perhaps the year after? In the early years of an asset's life we rarely analyze whether it is time for replacement. Thus, the defender's marginal costs are usually increasing, as shown in Example 13-3.

EXAMPLE 13-3

An asset purchased 5 years ago for $75,000 can be sold today for $15,000. Operating expenses will be $10,000 this year, but these will increase by $1500 per year. It is estimated that the asset's market value will decrease by $1000 per year over the next 5 years. If the MARR used by the company is 15%, calculate the total marginal cost of ownership of this old asset (that is, the defender) for each of the next 5 years.

SOLUTION

We calculate the total marginal cost of maintaining the old asset for the next 5-year period as follows:

Year, n	Loss in Market Value in Year n	Interest in Year n	Operating Cost in Year n	Marginal Cost in Year n
1	$15,000 - 14,000 = \$1000$	$15,000(0.15) = \$2250$	$10,000	$13,250
2	$14,000 - 13,000 = 1000$	$14,000(0.15) = 2100$	11,500	14,600
3	$13,000 - 12,000 = 1000$	$13,000(0.15) = 1950$	13,000	15,950
4	$12,000 - 11,000 = 1000$	$12,000(0.15) = 1800$	14,500	17,300
5	$11,000 - 10,000 = 1000$	$11,000(0.15) = 1650$	16,000	18,650

We can see that the marginal costs increase in each subsequent year of ownership. When the condition of increasing marginal costs for the defender has been met, then the defender–challenger comparison should be made with *replacement analysis technique 1*.

REPLACEMENT ANALYSIS TECHNIQUE 1: DEFENDER MARGINAL COSTS CAN BE COMPUTED AND ARE INCREASING

When our first method of analyzing the defender asset against the best available challenger is used, the basic comparison involves *the defender's marginal cost data and the challenger's minimum cost life data.*

When the defender's marginal cost is increasing from year to year, we will maintain that defender as long as the marginal cost of keeping it one more year is less than the challenger's minimum EUAC. Thus our decision rule is

Maintain the defender as long as the marginal cost of ownership for one more year is less than the challenger's minimum EUAC. When the defender's marginal cost becomes greater than the challenger's minimum EUAC, then replace the defender with the challenger.

One can see that this technique assumes that the current best challenger, with its minimum EUAC, will be available and unchanged in the future. However, it is easy to update a replacement analysis when marginal costs for the defender change or when there is a change in the cost and/or performance of available challengers. Example 13-4 illustrates the use of this technique for comparing defender and challenger assets.

EXAMPLE 13-4

Taking the machinery in Example 13-2 as the *challenger* and the machinery in Example 13-3 as the *defender,* use *replacement analysis technique 1* to determine when, if at all, a replacement decision should be made.

SOLUTION

Replacement analysis technique 1 should be used only in the condition of increasing marginal costs for the defender. Since these marginal costs are increasing for the defender (from Example 13-3), we can proceed by comparing defender marginal costs against the minimum EUAC of the challenger asset. In Example 13-2 we calculated only the marginal costs of the challenger; thus it is necessary to calculate the challenger's minimum EUAC. The EUAC of keeping this asset for each year of its useful life is worked out as follows.

Year, n	Challenger Total Marginal Cost in Year n	Present Cost If Kept Through Year n (PC_n)	EUAC If Kept Through Year n
1	$17,750	$[17,750(P/F,15\%,1)]$	$\times (A/P,15\%,1) = \$17,750$
2	15,200	$PC_1 + 15,200(P/F,15\%,2)$	$\times (A/P,15\%,2) = \ 16,560$
3	13,950	$PC_2 + 13,950(P/F,15\%,3)$	$\times (A/P,15\%,3) = \ 15,810$
4	14,350	$PC_3 + 14,350(P/F,15\%,4)$	$\times (A/P,15\%,4) = \ 15,520$
5	14,900	$PC_4 + 14,900(P/F,15\%,5)$	$\times (A/P,15\%,5) = \ 15,430$
6	15,600	$PC_5 + 15,600(P/F,15\%,6)$	$\times (A/P,15\%,6) = \ 15,450$
7	16,950	$PC_6 + 16,950(P/F,15\%,7)$	$\times (A/P,15\%,7) = \ 15,580$

A minimum EUAC of $15,430 is attained for the challenger at Year 5, which is the challenger's *minimum cost life*. We proceed by comparing this value against the *marginal* costs of the defender from Example 13-3:

Year, n	Defender Total Marginal Cost in Year n	Challenger Minimum EUAC	Comparison Result and Recommendation
1	$13,250	$15,430	Since $13,250 is *less than* $15,430, keep defender.
2	14,600	15,430	Since $14,600 is *less than* $15,430, keep defender.
3	15,950	15,430	Since $15,950 is *greater than* $15,430, replace defender.
4	17,300		
5	18,650		

Based on the data given for the challenger and for the defender, we would keep the defender for 2 more years and then replace it with the challenger because at that point the defender's marginal cost of another year of ownership would be greater than the challenger's minimum EUAC.

REPLACEMENT REPEATABILITY ASSUMPTIONS

The decision to use the challenger's minimum EUAC reflects two assumptions: the best challenger will be available "with the same minimum EUAC" in the future; and the period of needed service is indefinitely long. In other words, we assume that once the decision has been made to replace, there will be an indefinite cycle of replacement with the current best challenger asset. These assumptions must be satisfied for our calculations to be correct into an indefinite future. However, because the near future is economically more important than the distant future, and because our analysis is done with the *best data currently available*, our results and recommendations are robust or stable for reasonable changes in the estimated data.

The repeatability assumptions together are much like the repeatability assumptions that allowed us to use the annual cost method to compare competing alternatives with different useful lives. Taken together, we call these the **replacement repeatability assumptions.** They allow us to greatly simplify comparing the defender and the challenger.

Stated formally, these two assumptions are:

1. The currently available best challenger will continue to be available in subsequent years and will be unchanged in its economic costs. When the defender is ultimately replaced, it will be replaced with this challenger. Any challengers put into service will also be replaced with the same currently available challenger.

2. The period of needed service of the asset is indefinitely long. Thus the challenger asset, once put into service, will continuously replace itself in repeating cycles.

If these two assumptions are satisfied completely, then our calculations are exact. Often, however, future challengers represent further improvements so that Assumption 1 is not satisfied. While the calculations are no longer exact, the repeatability assumptions allow us to make the best decision we can with the data we have.

If the defender's marginal cost is increasing, once it rises above the challenger's minimum EUAC, it will continue to be greater. Under the repeatability assumptions, we would never want to incur a defender's marginal cost that was greater than the challenger's minimum EUAC. Thus, we use replacement analysis technique 1 when the defender's marginal costs are increasing.

REPLACEMENT ANALYSIS TECHNIQUE 2: DEFENDER MARGINAL COSTS CAN BE COMPUTED AND ARE NOT INCREASING

If the defender's marginal costs do not increase, we have no guarantee that *replacement analysis technique 1* will produce the alternative that is of the greatest economic advantage. Consider the new asset in Example 13-2, which has marginal costs that begin at a high of $17,750, then *decrease* over the next years to a low of $13,950, and then *increase* thereafter to $16,950 in Year 7. If evaluated *one year after implementation*, the asset would not have increasing marginal costs. Defenders in their early stages typically do not fit the requirements of *replacement analysis technique 1*. In the situation graphed in Figure 13-3, such defender assets would be in the downward slope of a concave marginal cost curve.

Example 13-5 details why *replacement analysis technique 1* cannot be applied when defenders do not have consistently increasing marginal cost curves. Instead we apply *replacement analysis technique 2*. That is, we calculate the defender's minimum EUAC to see whether the replacement should occur immediately. If not, as shown in Example 13-5, the replacement occurs after the defender's minimum cost life when the marginal costs are increasing. Then *replacement analysis technique 1* applies again.

EXAMPLE 13-5

Let us look again at the defender and challenger assets in Example 13-4. This time let us arbitrarily change the defender's marginal costs for its 5-year useful life. Now when, if at all, should the defender be replaced with the challenger?

Year, n	Defender Total Marginal Cost in Year n
1	$16,000
2	14,000
3	13,500
4	15,300
5	17,500

SOLUTION

In this case the defender's total marginal costs are *not* consistently increasing from year to year. However, if we ignore this fact and apply *replacement analysis technique 1*, the recommendation would be to replace the defender now, because the defender's marginal cost for the first year ($16,000) is greater than the minimum EUAC of the challenger ($15,430). This would be the wrong choice.

Since the defender's marginal costs are below the challenger's minimum EUAC in the second through fourth years, we must calculate the EUAC of keeping the defender asset in each of its remaining 5 years, at $i = 15\%$.

Year, n	Present Cost If Kept n Years (PC_n)	EUAC If Kept n Years
1	$16,000(P/F, 15\%, 1)$	$\times (A/P, 15\%, 1) = \$16,000$
2	$PC_1 + 14,000(P/F, 15\%, 2)$	$\times (A/P, 15\%, 2) = \ \ 15,070$
3	$PC_2 + 13,500(P/F, 15\%, 3)$	$\times (A/P, 15\%, 3) = \ \ 14,618$
4	$PC_3 + 15,300(P/F, 15\%, 4)$	$\times (A/P, 15\%, 4) = \ \ 14,754$
5	$PC_4 + 17,500(P/F, 15\%, 5)$	$\times (A/P, 15\%, 5) = \ \ 15,162$

The minimum EUAC of the defender for 3 years is $14,618, which is less than that of the challenger's minimum EUAC of $15,430. Thus, under the replacement repeatability assumptions we will keep the defender for at least 3 years. We must still decide how much longer.

The defender's EUAC begins to rise in Year 4, because the marginal costs are increasing, and because they are above the defender's minimum EUAC. Thus, we can use *replacement analysis technique 1* for Year 4 and later. The defender's marginal cost in Year 4 is $130 below the challenger's minimum EUAC of $15,430. Since the defender's marginal cost is higher in Year 5, we replace it with the new challenger at the end of Year 5. Notice that we did *not* keep the defender for its minimum cost life of 3 years, we kept it for 5 years.

If the challenger's minimum EUAC were less than the defender's minimum EUAC of $14,618, then the defender would be immediately replaced.

Example 13-5 illustrates several potentially confusing points about replacement analysis.

- If the defender's marginal cost data is not increasing, the defender's minimum EUAC must be calculated.
- If the defender's minimum EUAC exceeds the challenger's minimum EUAC, then replace immediately. If the defender's minimum EUAC is lower than the challenger's minimum EUAC, then under the replacement repeatability assumptions the defender will be kept *at least* the number of years for its minimum EUAC.
- After this number of years, then replace when the defender's increasing marginal cost exceeds the challenger's minimum EUAC.

The problem statement for Example 13-6 illustrates a second approach to calculating the defender's marginal costs for its capital costs. The value at the year's beginning is multiplied by $(1 + i)$ and the salvage value at the year's end is subtracted. Each year's total marginal cost also includes the operations and maintenance costs.

Then the solution to Example 13-6 details the calculation of the minimum EUAC when the defender's data is presented as costs and salvage values in each year rather than as marginal costs. Notice that this is calculated the same way as the minimum cost life was calculated for new assets—the challengers.

EXAMPLE 13-6

A 5-year-old machine, whose current market value is $5000, is being analyzed to determine its minimum EUAC at a 10% interest rate. Salvage value and maintenance estimates and the corresponding marginal costs are given in the following table.

	Data		Calculating Marginal Costs		
Year	Salvage Value	O&M Cost	$S_{t-1}(1+i)$	$-S_t$	Marginal Cost
0	$5000				
1	4000	$0	$5500	−4000	$1500
2	3500	100	4400	−3500	1000
3	3000	200	3850	−3000	1050
4	2500	300	3300	−2500	1100
5	2000	400	2750	−2000	1150
6	2000	500	2200	−2000	700
7	2000	600	2200	−2000	800
8	2000	700	2200	−2000	900
9	2000	800	2200	−2000	1000
10	2000	900	2200	−2000	1100
11	2000	1000	2200	−2000	1200

SOLUTION

Because the marginal costs have a complex, nonincreasing pattern, we must calculate the defender's minimum EUAC.

			If Retired at End of Year n		
Years Kept, n	Salvage Value (S) at End of Year n	Maintenance Cost for Year	EUAC of Capital Recovery $(P-S) \times (A/P, 10\%,n) + Si$	EUAC of Maintenance $100(A/G, 10\%,n)$	Total EUAC
0	$P = $5000				
1	4000	$ 0	$1100 + 400	$ 0	$1500
2	3500	100	864 + 350	48	1262
3	3000	200	804 + 300	94	1198
4	2500	300	789 + 250	138	1177
5	2000	400	791 + 200	181	1172
6	2000	500	689 + 200	222	1111
7	2000	600	616 + 200	262	1078
8	2000	700	562 + 200	300	1062
9	2000	800	521 + 200	337	1058 ←
10	2000	900	488 + 200	372	1060
11	2000	1000	462 + 200	406	1068

A minimum EUAC of $1058 is computed at Year 9 for the existing machine. Notice that the EUAC begins to increase with n when the marginal cost in Year 10 exceeds the EUAC for 9 years.

Now to apply *replacement analysis technique 2* to Example 13-6, we ask: Is the challenger's minimum EUAC higher or lower than the defender's minimum EUAC of $1058? If the challenger's minimum EUAC is lower, then we replace the defender now.

Under the repeatability assumptions, if the challenger's minimum EUAC is higher, we would keep the defender at least 9 years. Replacement would occur in Year 10 or later when the defender's marginal costs exceed the challenger's minimum EUAC. Relaxing the repeatability assumptions to allow for better challengers, we may replace the defender whenever a new challenger has an EUAC that is lower than $1058.

Example 13-7 illustrates the common situation of a current defender that may be kept if overhauled. This can also be analyzed as a potential new challenger.

EXAMPLE 13-7

We must decide whether existing (defender) equipment in an industrial plant should be replaced. A $4000 overhaul must be done now if the equipment is to be retained in service. Maintenance is $1800 in each of the next 2 years, after which it increases by $1000 each year. The defender has no present or future salvage value. The equipment described in Example 13-1 is the challenger (EUAC = $4589). Should the defender be kept or replaced if the interest rate is 8%?

FIGURE 13-4 Overhaul and maintenance costs for the defender in Example 13-7.

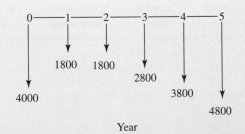

SOLUTION

The first step is to determine the defender's lowest EUAC. The pattern of overhaul and maintenance costs (Figure 13-4) suggests that if the overhaul is done, the equipment should be kept for several years. The computation is as follows:

		If Retired at End of Year n	
Year, n	EUAC of Overhaul $\$4000(A/P, 8\%, n)$	EUAC of Maintenance $\$1800 + \$1000$ Gradient from Year 3 on	Total EUAC
1	$4320	$1800	$6120
2	2243	1800	4043
3	1552	1800 + 308*	3660 ←
4	1208	1800 + 683†	3691
5	1002	1800 + 1079	3881

*For the first 3 years, the maintenance is $1800, $1800, and $2800. Thus, EUAC $=$ $1800 + 1000(A/F, 8\%, 3) = 1800 + 308$.

†EUAC $= 1800 + 1000(P/G, 8\%, 3)(P/F, 8\%, 1)(A/P, 8\%, 4) = 1800 + 683$.

The lowest EUAC of the overhauled defender is $3660. In Example 13-1, the challenger's minimum cost life was 4 years with an EUAC of $4589. If we assume the equipment is needed for at least 4 years, the overhauled defender's EUAC ($3660) is less than the challenger's EUAC ($4589). Overhaul the defender.

If the defender's and challenger's cost data do not change, we can use *replacement analysis technique 1* to determine when the overhauled defender should be replaced. We know from the minimum EUAC calculation that the defender should be kept at least 3 years. Is this the best life? The following table computes the marginal cost to answer this question.

Year, n	Overhaul Cost	Maintenance Cost	Marginal Cost to Extend Service
0	$4000	$ 0	
1	0	1800	$6120 = 4000(1.08) + 1800
2	0	1800	1800
3	0	2800	2800
4	0	3800	3800
5	0	4800	4800

Year 5 is the first year after Year 3, which has the overhauled defender's lowest EUAC (in which the $4800 marginal cost exceeds the challenger's $4589 minimum EUAC). Thus, the overhauled defender should be kept 4 more years if costs do not change. (Note that if the defender can be overhauled again after 3 or 4 years, that might be an even better choice.)

REPLACEMENT ANALYSIS TECHNIQUE 3: WHEN DEFENDER MARGINAL COST DATA ARE NOT AVAILABLE

In this case, we simply compare the defender's *EUAC over its stated useful life, and the challenger's minimum EUAC*. Pick the *EUAC* that is lower.

If the defender's marginal cost data is not known and cannot be estimated, it is impossible to apply *replacement analysis techniques 1 or 2* to decide *when* the defender should be replaced. Instead we must assume that the defender's stated useful life is the only one to consider. From a student problem-solving perspective, the defender in Example 13-7 might be described as follows.

The defender can be overhauled for $4800 to extend its life for 5 years. Maintenance costs will average $3000 per year and there will be no salvage value. In this case the only possibility is to compare the defender's EUAC for a 5-year life with the best challenger.

In the real world the most likely scenario for this approach involves a facility-wide overhaul every 3, 5, 10, etc. years. Pipelines and many process plants, such as refineries, chemical plants, and steel mills, must shut down to do major maintenance. All equipment is overhauled or replaced with a new challenger as needed, and the facility is expected to operate until the next maintenance shutdown.

The defender's EUAC over its remaining useful life is compared with the challenger's EUAC at its minimum cost life, and the lower cost is chosen. However, in making this basic comparison an often complicating factor is deciding what first cost to assign to the challenger and the defender.

COMPLICATIONS IN REPLACEMENT ANALYSIS

Defining Defender and Challenger First Costs

Because the defender is already in service, analysts often misunderstand what first cost to assign it. Example 13-8 demonstrates this problem.

EXAMPLE 13-8

A model SK-30 was purchased 2 years ago for $1600; it has been depreciated by straight-line depreciation using a 4-year life and zero salvage value. Because of recent innovations, the current price of the SK-30 is $995. An equipment firm has offered a trade-in allowance of $350 for the SK-30 on a new $1200 model EL-40. Some discussion revealed that without a trade-in, the EL-40 can be purchased for $1050. Thus, the originally quoted price of the EL-40 was overstated to allow a larger trade-in allowance. The true current market value of the SK-30 is probably only $200. In a replacement analysis, what value should be assigned to the SK-30?

SOLUTION

In the example, five different dollar amounts relating to the SK-30 have been outlined:

1. *Original cost:* It cost $1600 2 years ago.
2. *Present cost:* It now sells for $995.
3. *Book value:* The original cost less 2 years of depreciation is $1600 - \frac{2}{4}(1600 - 0) = \$800$.
4. *Trade-in value:* The offer was $350.
5. *Market value:* The estimate was $200.

We know that an economic analysis is based on the current situation, not on the past. We refer to past costs as *sunk* costs to emphasize this. These costs cannot be altered, and they are not relevant. (There is one exception: past costs may affect present or future income taxes.)

We want to use actual cash flows for each alternative. Here the question is, What value should be used in an economic analysis for the SK-30? The relevant cost is the equipment's present market value of $200. Neither the original cost, the present cost, the book value, nor the trade-in value is relevant.

At first glance, an asset's trade-in value would appear to be a suitable present value for the equipment. Often the trade-in price is inflated *along with* the price for the new item. (This practice is so common in new-car showrooms that the term *overtrade* is used to describe the excessive portion of the trade-in allowance. The buyer is also quoted a higher price for the new car.) Distorting the defender's present value, or the challenger's price can be serious because these distortions do not cancel out in an economic analysis.

Example 13-8 illustrated that of the several different values that can be assigned to the defender, the correct one is the present market value. If a trade-in value is obtained, care should be taken to ensure that it actually represents a fair market value.

Determining the value for the challenger's installed cost should be less difficult. In such cases the first cost is usually made up of purchase price, sales tax, installation costs, and other items that occur initially on a one-time basis if the challenger is selected. These values are usually rather straightforward to obtain. One aspect to consider in assigning a first cost to the challenger is the defender's potential market or salvage value. One must not arbitrarily subtract the defender's disposition value from the challenger's first cost, for this practice can lead to an incorrect analysis.

As described in Example 13-8, the correct first cost to assign to the defender SK-30 is its $200 current market value. This value represents the present economic benefit that we would be *forgoing* to keep the defender. This can be called our *opportunity first cost*. If, instead of assuming that this is the defender's *opportunity cost*, we assume it is a *cash benefit* to the challenger, a potential error arises.

The error lies in incorrect use of a *cash flow* perspective when the lives of the challenger and the defender are not equal, which is usually the case. The approach of subtracting the defender's salvage value from the challenger's first cost is called the *cash flow* perspective. From this perspective, keeping the defender in place causes $0 in cash to flow, but selecting the challenger causes the cash flow now to equal the challenger's first cost minus the defender's salvage value.

If the lives of the defender and the challenger are the same, then the cash flow perspective will lead to the correct answer. However, it is normally the case that the defender is an aging asset with a relatively short horizon of possible lives and the challenger is a new asset with a longer life. The *opportunity cost* perspective will always lead to the correct answer, so it is the one that should be used.

For example, consider the SK-30 and EL-40 from Example 13-8. It is reasonable to assume that the 2-year-old SK-30 has 3 years of life left and that the new EL-40 would have a 5-year life. Assume that neither will have any salvage value at the end of its life. Compare the difference in their annual capital costs with the correct *opportunity cost* perspective and the incorrect *cash flow* perspective.

SK-30		EL-40	
Market value	$200	First cost	$1050
Remaining life	3 years	Useful life	5 years

Looking at this from an *opportunity cost* perspective, the annual cost comparison of the first costs is:

$$\text{Annualized first cost}_{\text{SK-30}} = \$200(A/P, 10\%, 3) = \$80$$

$$\text{Annualized first cost}_{\text{EL-40}} = \$1050(A/P, 10\%, 5) = \$277$$

The *difference* in annualized first cost between the SK-30 and EL-40 is:

$$\text{AFC}_{\text{EL-40}} - \text{AFC}_{\text{SK-30}} = \$277 - \$80 = \$197$$

Now using an incorrect *cash flow* perspective to look at the first costs, we can calculate the *difference* due to first cost between the SK-30 and EL-40.

$$\text{Annualized first cost}_{\text{SK-30}} = \$0(A/P, 10\%, 3) = \$0$$

$$\text{Annualized first cost}_{\text{EL-40}} = (\$1050 - 200)(A/P, 10\%, 5) = \$224$$

$$\text{AFC}_{\text{EL-40}} - \text{AFC}_{\text{SK-30}} = \$224 - \$0 = \$224$$

When the defender's remaining life (3 years) differs from the challenger's useful life (5 years), the two perspectives yield different results. The correct difference of $197 is shown by using the *opportunity cost* approach, and an inaccurate difference of $224 is obtained if the *cash flow* perspective is used. From the opportunity cost perspective, the $200 is spread out over 3 years as a cost to the defender, and in the cash flow perspective, the opportunity cost is spread out over 5 years as a benefit to the challenger. Spreading the $200 over 3 years in one case and 5 years in the other case does not produce equivalent annualized amounts. Because of the difference in the lives of the assets, the annualized $200 opportunity cost for the defender cannot be called an equivalent benefit to the challenger.

In the case of unequal lives, the correct method is to assign the defender's current market value as its Time-0 opportunity costs, rather than subtracting this amount from the challenger's first cost. Because the cash flow approach yields an incorrect value when challenger and defender have unequal lives, the *opportunity cost* approach for assigning a first cost to the challenger and defender assets should *always* be used.

REPEATABILITY ASSUMPTIONS NOT ACCEPTABLE

Under certain circumstances, the repeatability assumptions described earlier may not apply. Then replacement analysis techniques 1, 2, and 3 may not be valid. For instance, there may be a specific study period instead of an indefinite need for the asset. For example, consider the case of phasing out production after a certain number of years—perhaps a person who is about to retire is closing down a business and selling all the assets. Another example is production equipment such as molds and dies that are no longer needed when a new model with new shapes is introduced. Yet another is a construction camp that may be needed for only a year or two or three.

This specific study period could potentially be any number of years relative to the lives of the defender and the challenger, such as equal to the defender's life, equal to the challenger's life, less than the defender's life, greater than the challenger's life, or somewhere between the lives of the defender and challenger. The analyst must be explicit about the challenger's and defender's economic costs and benefits, as well as residual or salvage values at the end of the specific study period. In this case the repeatability replacement assumptions do not apply, costs must be analyzed over the study period. The analysis techniques in the decision map also may not apply when future challengers are not assumed to be identical to the current best challenger. This concept is discussed in the next section.

A Closer Look at Future Challengers

We defined the challenger as the best available alternative to replace the defender. But in time, the best available alternative can change. And given the trend in our technological society, it seems likely that future challengers will be better than the present challenger. If so, the prospect of improved future challengers may affect the present decision between the defender and the challenger.

Figure 13-5 illustrates two possible estimates of future challengers. In many technological areas it seems likely that the equivalent uniform annual costs associated with future challengers will decrease by a constant amount each year. In other fields, however, a rapidly changing technology will produce a sudden and substantially improved challenger—with decreased costs or increased benefits. The uniform decline curve of Figure 13-5 reflects the assumption that each future challenger has a minimum EUAC that is a fixed amount less than the previous year's challenger. This assumption, of course, is only one of many possible assumptions that could be made regarding future challengers.

If future challengers will be better than the present challenger, what impact will this have on an analysis now? The prospect of better future challengers may make it more desirable to retain the defender and to reject the present challenger. By keeping the defender for now, we may be able to replace it later by a better future challenger. Or, to state it another way, the present challenger may be made less desirable by the prospect of improved future challengers.

As engineering economic analysts, we must familiarize ourselves with potential technological advances in assets targeted for replacement. This part of the decision process is much like the search for all available alternatives, from which we select the best. Upon finding out more about what alternatives and technologies are emerging, we will be better able

FIGURE 13-5 Two possible ways the EUAC of future challengers may decline.

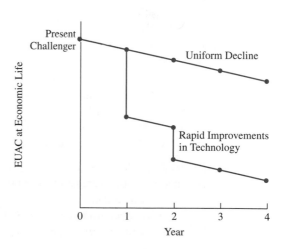

to understand the repercussions of investing in the current best available challenger. Selecting the current best challenger asset can be particularly risky when (1) the costs are very high and/or (2) the challenger's useful minimum cost life is relatively long (5–10 years or more). When one or both of these conditions exist, it may be better to keep or even augment our defender asset until better future challengers emerge.

There are, of course, many assumptions that *could* be made regarding future challengers. However, if the replacement repeatability assumptions do not hold, the analysis needed becomes more complicated.

AFTER-TAX REPLACEMENT ANALYSIS

As described in Chapter 12, an after-tax analysis provides greater realism and insight. Tax effects can alter before-tax recommendations. After-tax effects may influence calculations in the defender–challenger comparisons discussed earlier. Consequently, one should always perform or check these analyses on an after-tax basis.

Marginal Costs on an After-Tax Basis

Marginal costs on an after-tax basis represent the cost that would be incurred through ownership of the defender *in each year*. On an after-tax basis we must consider the effects of ordinary taxes as well as gains and losses due to asset disposal. Consider Example 13-9.

EXAMPLE 13-9

Refer to Example 13-2, where we calculated the before-tax marginal costs for a new piece of production machinery. Calculate the asset's after-tax marginal costs considering this additional information.

- Depreciation is by the straight-line method, with $S = \$0$ and $n = 5$ years, so $d_t = (\$25,000 - \$0)/5 = \$5000$.
- Ordinary income, recaptured depreciation, and losses on sales are taxed at a rate of 40%.
- The after-tax MARR is 10%.

Some classes skip or have not yet covered Chapter 10's explanation of expected value. Thus, the expected cost for risk of breakdowns is described here as an insurance cost.

SOLUTION

The after-tax marginal cost of ownership will involve the following elements: incurred or forgone loss or recaptured depreciation, interest on invested capital, tax savings due to depreciation, and annual after-tax operating/maintenance and insurance. Figure 13-6 shows example cash flows for the marginal cost detailed in Table 13-1.

As a refresher of the recaptured depreciation calculations in Chapter 12:

The market value in Year 0 = 25,000.
The market value decreases to \$18,000 at Year 1.
The book value at Year 1 = 25,000 − 5000 = \$20,000.
So loss on depreciation = 20,000 − 18,000 = \$2000.
This results in a tax savings of (2000)(0.4) = \$800.

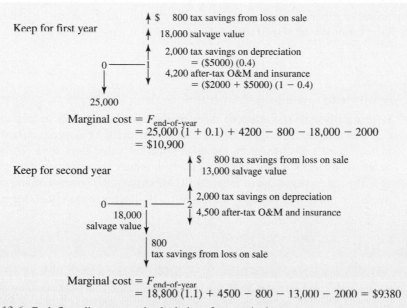

FIGURE 13-6 Cash flow diagrams and calculations for marginal cost.

TABLE 13-1 Marginal Costs of Ownership

Year	Market Value	Book Value	Recaptured Depr. or Loss	Taxes or Tax Savings	After-Tax Market Value
0	$25,000	$25,000			$25,000
1	18,000	20,000	−$2000	−$800	18,800
2	13,000	15,000	−2000	−800	13,800
3	9,000	10,000	−1000	−400	9,400
4	6,000	5,000	1000	400	5,600
5	4,000	0	4000	1600	2,400
6	3,000		3000	1200	1,800
7	2,500		2500	1000	1,500

	Col. B After-Tax	Col. C Beg. Yr	Col. D Tax Savings	O&M and	Col. F After-Tax	= C + D + F − B Marginal
Year	Market Value	Value × (1+i)	from Depr. Deduct.	Insurance Cost	Annual Expense	Cost
0	$25,000					
1	18,800	$27,500	−$2000	$7,000	$4200	$10,900
2	13,800	20,680	−2000	7,500	4500	9,380
3	9,400	15,180	−2000	8,000	4800	8,580
4	5,600	10,340	−2000	10,000	6000	8,740
5	2,400	6,160	−2000	12,000	7200	8,960
6	1,800	2,640	0	14,000	8400	9,240
7	1,500	1,980	0	16,000	9600	10,080

The marginal cost in each year is much lower after taxes than the pretax numbers shown in Example 13-2. This is true because depreciation and expenses can be subtracted from taxable income. However, the pattern of declining and then increasing marginal costs is the same, and Year 3 is still the year of lowest marginal costs.

Minimum Cost Life Problems

Here we illustrate the effect of taxes on the calculation of the minimum cost life of the defender and the challenger. The after-tax minimum EUAC depends on both the depreciation method used and changes in the asset's market value over time. Using an accelerated depreciation method (like MACRS) tends to reduce the after-tax costs early in the asset's life. This alters the shape of the total EUAC curve—the concave shape can be shifted and the minimum EUAC changed. Example 13-10 illustrates the effect that taxes can have.

EXAMPLE 13-10

Some new production machinery has a first cost of $100,000 and a useful life of 10 years. Its estimated operating and maintenance (O&M) costs are $10,000 the first year, which will increase annually by $4000. The asset's before-tax market value will be $50,000 at the end of the first year and then will decrease by $5000 annually. This property is a 7-year MACRS property. The company uses a 6% after tax MARR and is subject to a combined federal/state tax rate of 40%.

Calculate the after-tax cash flows.

SOLUTION

To find this new production machinery's minimum cost life, we first find the after-tax cash flow (ATCF) effect of the O&M costs and depreciation (Table 13-2). Then, we find the ATCFs of disposal if the equipment is sold in each of the 10 years (Table 13-3). Finally in the closing section on spreadsheets, we combine these two ATCFs (in Figure 13-7) and choose the minimum cost life.

In Table 13-2, the O&M expense simply starts at $10,000 and increases at $4000 per year. The depreciation entries equal the 7-year r_t MACRS depreciation values given in Table 11-3 multiplied by the $100,000 first cost. The taxable income, which is simply the O&M costs minus the depreciation values, is then multiplied by minus the tax rate to determine the tax savings. The O&M expense plus the tax savings is the Table 13-2 portion of the total ATCF.

Regarding the market value data in this problem, it should be pointed out that the initial decrease of $50,000 in Year 1 is not uncommon. This is especially true for custom-built equipment for a particular and unique application at a specific plant. Such equipment would not be valuable to others in the marketplace. Also, the $100,000 first cost (cost basis) could have included costs due to installation, facility modifications, or removal of old equipment. The $50,000 is realistic for the market value of one-year-old equipment.

The next step is to determine the ATCFs that would occur in each possible year of disposal. (The ATCF for Year 0 is easy; it is −$100,000.) For example, as shown in Table 13-3, in Year 1 there is a $35,710 loss as the book value exceeds the market value. The tax savings from this loss are added to the salvage (market) value to determine the ATCF (*If the asset is disposed during Year 1*).

TABLE 13-2 ATCF for O&M and Depreciation for Example 13-10

Year, t	O&M Expense	MACRS Depreciation, d_t	Taxable Income	Tax Savings (at 40%)	O&M Depreciation ATCF
1	−$10,000	$14,290	−$24,290	$ 9,716	−$284
2	−14,000	24,490	−38,490	15,396	1,396
3	−18,000	17,490	−35,490	14,196	−3,804
4	−22,000	12,490	−34,490	13,796	−8,204
5	−26,000	8,930	−34,930	13,972	−12,028
6	−30,000	8,920	−38,920	15,568	−14,432
7	−34,000	8,930	−42,930	17,172	−16,828
8	−38,000	4,460	−42,460	16,984	−21,016
9	−42,000	0	−42,000	16,800	−25,200
10	−46,000	0	−46,000	18,400	−27,600

TABLE 13-3 ATCF in Year of Disposal for Example 13-10

Year, t	Market Value	Book Value	Gain or Loss	Gain/Loss Tax (at 40%)	ATCF if Disposed of
1	$50,000	$85,710	−$35,710	$14,284	$64,284
2	45,000	61,220	−16,220	6,488	51,488
3	40,000	43,730	−3,730	1,492	41,492
4	35,000	31,240	3,760	−1,504	33,496
5	30,000	22,310	7,690	−3,076	26,924
6	25,000	13,390	11,610	−4,644	20,356
7	20,000	4,460	15,540	−6,216	13,784
8	15,000	0	15,000	−6,000	9,000
9	10,000	0	10,000	−4,000	6,000
10	5,000	0	5,000	−2,000	3,000

These tables assume that depreciation is taken during the year of disposal and then calculates the recaptured depreciation (gain) or loss on the year-end book value.

SPREADSHEETS AND REPLACEMENT ANALYSIS

Spreadsheets are obviously useful in nearly all after-tax calculations. However, they are absolutely required for optimal life calculations in after-tax situations. Because MACRS is the tax law, the after-tax cash flows are different in every year. Thus, the NPV function and the PMT function are both needed to find the minimum EUAC after taxes. Figure 13-7 illustrates the calculation of the minimum cost life for Example 13-10.

In Figure 13-7, the NPV finds the present worth of the irregular cash flows from Period 1 through Period t for $t = 1$ to life. The PV function is used to find the PW of the salvage value. Then PMT can be used to find the EUAC over each potential life. Before-tax replacement analysis can also be done this way. The spreadsheet block function NPV is used to find the PW of cash flows from Period 1 to Period t. Note that the cell for Period 1 is an absolute address and the cell for period t is a relative address. This allows the formula to be copied.

Example 13-11 illustrates the before-tax optimal economic life of a new car.

	A	B	C	D	E	F
1		**Table 13-2**	**Table 13-3**	6%	Interest Rate	
2		O&M & Depr.	if disposed of			
3	Year	ATCF	ATCF	PW	EUAC	
4	0			−$100,000		
5	1	−$284	$64,284	−39,623	$42,000	=PMT($D$1,A5,D5)
6	2	1,396	51,488	−53,201	29,018	
7	3	−3,804	41,492	−67,382	25,208	
8	4	−8,204	33,496	−82,186	23,718	
9	5	−12,028	26,924	−97,587	23,167	
10	6	−14,432	20,356	−113,530	23,088	optimal life
11	7	−16,828	13,784	−129,904	23,270	
12	8	−21,016	9,000	−146,610	23,610	
13	9	−25,200	6,000	−163,621	24,056	
14	10	−27,600	3,000	−180,909	24,580	=PMT($D$1,A14,D14)
15						
16			=NPV($D$1,$B$5:B14)+$D$4+PV($D$1,A14,0,−C14)			
17			=NPV(i, B column) + year 0 + present value of a future salvage			

FIGURE 13-7 Spreadsheet for life with minimum after-tax cost.

EXAMPLE 13-11

A new vehicle costs $19,999 plus $400 in fees. Its value drops 30% the first year, 20% per year for Years 2 through 4, and 15% each additional year. When the car is sold, detailing and advertising will cost $250. Repairs on similar vehicles have averaged $50 annually in lost time (driving to/from the dealer's shop) during the 3-year warranty period. After the warranty period, the cost of repairs and the associated inconvenience climbs at $400 annually. If the MARR is 8%, what is the optimal economic life for the vehicle?

SOLUTION

Figure 13-8 is a spreadsheet that uses the same functions used in Figure 13-7. The present worth for each life is found by using the NPV function for the irregular costs from Years 1 to n, and the PV function for salvage value minus the cost to sell, and adding in the Time 0 costs to buy the vehicle. Then a PMT function is used to find the annual cost.

The optimal economic life is 10 years, but it is not much more expensive for lives that are 3 years shorter or 5 years longer. However, keeping a vehicle for less than 5 years is significantly more expensive.

	A	B	C	D	E	F
1	$19,999	first cost				
2	$400	cost to buy				
3	$250	cost to sell				
4	30%	salvage value drop yr 1				
5	20%	salvage value drop yr 2–4				
6	15%	salvage value drop yr 5+				
7	$50	repair during 3 yr warranty				
8	$400	gradient for repair after 3 yr warranty				
9	8%	interest rate				
10	year	cost	salvage value	PW	EUAC	
11	0	$20,399				
12	1	50	$13,999	−$7,714		=PMT($A$9,A12,D12
13	2	50	11,199	−11,101	6,225	
14	3	50	8,960	−13,614	5,283	
15	4	450	7,168	−15,774	4,762	
16	5	850	6,092	−17,461	4,373	
17	6	1250	5,179	−19,119	4,136	
18	7	1650	4,402	−20,765	3,988	
19	8	2050	3,742	−22,409	3,899	
20	9	2450	3,180	−24,055	3,851	
21	10	2850	2,703	−25,705	3,831	economic life & min cost
22	11	3250	2,298	−27,356	3,832	
23	12	3650	1,953	−29,008	3,849	
24	13	4050	1,660	−30,655	3,879	
25	14	4450	1,411	−32,293	3,917	
26	15	4850	1,199	−33,918	3,963	
27						
28	=−$B$11−NPV($A$9,$B$12:B26)+PV($A$9,A26,,−C26+$A$3)					
29	= year 0 + NPV(i, B column) + PV(salvage value − sale cost)					

FIGURE 13-8 Spreadsheet for vehicle's optimal economic life.

Example 13-11 shows that even with complicated cost structures, it is still relatively easy to find the minimum cost EUAC with the power of spreadsheets.

SUMMARY

The question in selecting new equipment is, Which machine will be more economical? But when there is an existing machine (called the **defender**), the question is, Shall we replace it now, or shall we keep it for one or more years? When replacement is indicated, the best available replacement equipment (called the **challenger**), will be acquired. When we already have equipment, there is a tendency to use past or sunk costs in the replacement analysis. But only present and future costs are relevant.

This chapter has presented three distinctly different **replacement analysis techniques.** All use the simplifying **replacement repeatability assumptions.** These state that the defender will ultimately be replaced by the current best challenger (as will any challengers implemented in the future), and that we have an indefinite need for the asset's service.

In the usual case, marginal cost data are both available and increasing on a year-to-year basis, and **replacement analysis technique 1** compares *the defender's marginal cost data* *against the challenger's minimum EUAC.* We keep the defender as long as its marginal cost is less than the challenger's minimum EUAC.

When marginal cost data are available for the defender but are not increasing from year to year, **replacement analysis technique 2** compares *the defender's lowest EUAC against the challenger's minimum EUAC.* If the challenger's EUAC is less, select this asset in place of the defender today. If the defender's lowest EUAC is smaller, we do not replace it yet. If the cost data for the challenger and the defender do not change, replace the defender after the life that minimizes its EUAC when its marginal cost data exceed the challenger's minimum EUAC.

In the case of no marginal cost data being available for the defender, **replacement analysis technique 3** compares *the defender's EUAC over its stated life, against the challenger's minimum EUAC.* As in the case of replacement analysis technique 2, select the alternative that has the smallest EUAC. An important concept when calculating the EUAC of both defender and challenger is the first cost to be assigned to each alternative for calculation purposes. When the lives of the two alternatives match, either an **opportunity cost** or a **cash flow approach** may be used. However, in the more common case of different useful lives, only the opportunity cost approach accurately assigns first costs to the defender and challenger.

It is important when performing engineering economic analyses to include the effects of taxes. This is much easier to do with spreadsheets. Spreadsheets also make it easy to compute the optimal economic life of vehicles and equipment—even when this includes complex patterns of declining salvage values, warranty periods, and increasing repair costs.

Replacement analyses are vastly important, yet often ignored by companies as they invest in equipment and facilities. Investments in business and personal assets should not be forgotten once an initial economic evaluation has produced a "buy" recommendation. It is important to continue to evaluate assets over their respective life cycles to ensure that invested monies continue to yield the greatest benefit. Replacement analyses help us to ensure this.

PROBLEMS

Replacement Problems

13-1 Typically there are two alternatives in a replacement analysis. One alternative is to replace the defender now. The other alternative is which one of the following?

A. Keep the defender for its remaining useful life.

B. Keep the defender for another year and then reexamine the situation.

C. Keep the defender until there is an improved challenger that is better than the present challenger.

D. The answer to this question depends on the data available for the defender and challenger as well as the assumptions made regarding the period of needed service and future challengers.

13-2 The defender's economic life can be found if certain estimates about the defender can be made. Assuming those estimates prove to be exactly correct, one can accurately predict the year when the defender should be replaced, even if nothing is known about the challenger. True or false? Explain.

13-3 A proposal has been made to replace a large heat exchanger (3 years ago, the initial cost was $85,000) with a new, more efficient unit at a cost of $120,000. The existing heat exchanger is being depreciated by the MACRS method. Its present book value is $20,400, but its scrap value just equals the cost to remove it from the plant. In preparing the before-tax economic analysis, should the $20,400 book value of the old heat exchanger be

A. *added* to the cost of the new exchanger?

B. *subtracted* from the cost of the new exchanger?

C. *ignored* in this before-tax economic analysis?

13-4 Which one of the following is the proper dollar value of defender equipment to use in replacement analysis?

A. Original cost

B. Present market value

C. Present trade-in value

D. Present book value

E. Present replacement cost, if different from original cost

13-5 Consider the following data for a defender asset. What is the correct replacement analysis technique to compare this asset against a competing challenger? How is this method used? That is, what comparison is made, and how do we choose?

Year, n	BTCF in Year n (marginal costs)
1	−$2000
2	−1750
3	−1500
4	−1250
5	−1000
6	−1000
7	−1000
8	−1500
9	−2000
10	−3000

Challenger Min EUAC

13-6 A machine has a first cost of $50,000. Its market value declines by 20% annually. The operating and maintenance costs start at $3500 per year and climb by $2000 per year. The firm's MARR is 9%. Find the minimum EUAC for this machine and its economic life.   (*Answer:* $17,240, 4 years)

13-7 A machine has a first cost of $10,000. Its market value declines by 20% annually. The repair costs are covered by the warranty in Year 1, and then they increase $600 per year. The firm's MARR is 15%. Find the minimum EUAC for this machine and its economic life.

13-8 A vehicle has a first cost of $20,000. Its market value declines by 15% annually. It is used by a firm that estimates the effect of older vehicles on the firm's image. A new car has no "image cost." But the image cost of older vehicles climbs by $700 per year. The firm's MARR is 10%. Find is the minimum EUAC for this vehicle and its economic life.

13-9 The Clap Chemical Company needs a large insulated stainless steel tank to expand its plant. Clap has located such a tank at a recently closed brewery. The brewery has offered to sell the tank for $15,000 delivered. The price is so low that Clap believes it can sell the tank at any future time and recover its $15,000 investment.

The outside of the tank is covered with heavy insulation that requires considerable maintenance.

Year	Insulation Maintenance Cost
0	$2000
1	500
2	1000
3	1500
4	2000
5	2500

(*a*) Based on a 15% before-tax MARR, what life of the insulated tank has the lowest EUAC?

(*b*) Is it likely that the insulated tank will be replaced by another tank at the end of the period with the lowest EUAC? Explain.

13-10 The plant manager may purchase a piece of unusual machinery for $10,000. Its resale value after 1 year is estimated to be $3000. Because the device is sought by antique collectors, resale value is rising $500 per year.

The maintenance cost is $300 per year for each of the first 3 years, and then it is expected to double each year. Thus the fourth-year maintenance will be $600; the fifth-year maintenance, $1200, and so on. Based on a 15% before-tax MARR, what life of this machinery has the lowest EUAC?

Replacement Technique 1

13-11 A machine tool, which has been used in a plant for 10 years, is being considered for replacement. It cost

$9500 and was depreciated by MACRS depreciation using a 5-year recovery period. An equipment dealer indicates that the machine has no resale value. Maintenance on the machine tool has been a problem, with an $800 cost this year. Future annual maintenance costs are expected to be higher. What is the economic life of this machine tool if it is kept in service?

13-12 In a replacement analysis problem, the following facts are known:

Initial cost	$12,000
Annual maintenance	Year
None	1–3
$2000	4–5
$4500	6
$7000	7
+$2500/year	

Salvage value in any year is zero. Assume a 10% interest rate and ignore income taxes. Compute the life for this challenger having the lowest EUAC. (*Answer:* 5 years)

13-13 An injection-molding machine has a first cost of $1,050,000 and a salvage value of $225,000 in any year. The maintenance and operating cost is $235,000 with an annual gradient of $75,000. The MARR is 10%. What is the most economic life?

13-14 Five years ago, Thomas Martin installed production machinery that had a first cost of $25,000. At that time initial yearly costs were estimated at $1250, increasing by $500 each year. The market value of this machinery each year would be 90% of the previous year's value. There is a new machine available now that has a first cost of $27,900 and no yearly costs over its 5 year-minimum cost life. If Thomas Martin uses an 8% before-tax MARR, when, if at all, should he replace the existing machinery with the new unit?

13-15 Consider Problem 13-14 involving Thomas Martin. When, if at all, should the old machinery be replaced with the new, given the following changes in the data. The old machine retains only 70% of its value in the market from year-to-year. The yearly costs of the old machine were $3000 in Year 1 and increase at 10% thereafter.

13-16 Mary O'Leary's company ships fine wool garments from County Cork, Ireland. Five years ago she purchased some new automated packing equipment having a first cost of $125,000 and a MACRS class life of 7 years. The annual costs for operating, maintenance, and insurance, as well as market value data for each year of the equipment's 10-year useful life are as follows.

Year	Annual Costs in Year *n* for			Market Value
n	Operating	Maintenance	Insurance	in Year *n*
1	$16,000	$ 5,000	$17,000	$80,000
2	20,000	10,000	16,000	78,000
3	24,000	15,000	15,000	76,000
4	28,000	20,000	14,000	74,000
5	32,000	25,000	12,000	72,000
6	36,000	30,000	11,000	70,000
7	40,000	35,000	10,000	68,000
8	44,000	40,000	10,000	66,000
9	48,000	45,000	10,000	64,000
10	52,000	50,000	10,000	62,000

Now Mary is looking at the remaining 5 years of her investment in this equipment, which she had initially evaluated on the basis of an after-tax MARR of 25% and a tax rate of 35%. Assume that the replacement repeatability assumptions are valid.

(a) What is the before-tax marginal cost for the remaining 5 years?

(b) When, if at all, should Mary replace this packing equipment if a new challenger, with a minimum EUAC of $110,000, has been identified?

13-17 SHOJ Enterprises has asked you to look at the following data. The interest rate is 10%.

Year, *n*	Marginal Cost Data Defender	EUAC if Kept *n* Years Challenger
1	$3000	$4500
2	3150	4000
3	3400	3300
4	3800	4100
5	4250	4400
6	4950	6000

(a) What is the *defender's* lowest EUAC?

(b) What is the *challenger's* economic life?

(c) When, if at all, should we replace the *defender* with the *challenger*?

13-18 Bill's father read that each year a car's value declines by 25%. After a car is 3 years old, the rate of decline falls to 15%. Maintenance and operating costs increase as the car ages. Because of the manufacturer's warranty, first-year maintenance is very low.

Age of Car (years)	Maintenance Expense
1	$ 50
2	150
3	180
4	200
5	300
6	390
7	500

Bill's dad wants to keep his annual cost of car ownership low. The car Bill's dad prefers costs $11,200 new. Should he buy a new or a used car and, if used, when would you suggest he buy it, and how long should it be kept? Give a practical, rather than a theoretical, solution. *(Answer:* Buy a 3-year-old car and keep it 3 years.)

Replacement Technique 2

13-19 A new $40,000 bottling machine has just been installed in a plant. It will have no salvage value when it is removed. The plant manager has asked you to estimate the machine's economic service life, ignoring income taxes. He estimates that the annual maintenance cost will be constant at $2500 per year. What service life will result in the lowest equivalent uniform annual cost?

13-20 Big-J Construction Company, Inc. (Big-J CC) is conducting a routine periodic review of existing field equipment. They use a MAAR of 20%. This includes a replacement evaluation of a paving machine now in use. The machine was purchased 3 years ago for $200,000, The paver's current market value is $120,000, and yearly operating and maintenance costs are as follows.

Year, n	Operating Cost in Year n	Maintenance Cost in Year n	Market Value if Sold in Year n
1	$15,000	$ 9,000	$85,000
2	15,000	10,000	65,000
3	17,000	12,000	50,000
4	20,000	18,000	40,000
5	25,000	20,000	35,000
6	30,000	25,000	30,000
7	35,000	30,000	25,000

Data for a new paving machine have been analyzed. Its most economic life is at 8 years with a minimum EUAC of $62,000. When should the existing paving machine be replaced?

13-21 VMIC Corp. has asked you to look at the following data. The interest rate is 10%.

Year, n	Marginal Cost Data Defender	EUAC If Kept n Years Challenger
1	$2500	$4500
2	2400	3600
3	2300	3000
4	2550	2600
5	2900	2700
6	3400	3500
7	4000	4000

(a) What is the *defender's* lowest EUAC?

(b) What is the *challenger's* minimum cost life?

(c) When, if at all, should we replace the *defender* with the *challenger*?

Replacement Technique 3

13-22 You are considering the purchase of a new high-efficiency machine to replace older machines now. The new machine can replace four of the older machines, each with a current market value of $600. The new machine will cost $5000 and will save the equivalent of 10,000 kWh of electricity per year. After a period of 10 years, neither option (new or old) will have any market value. If you use a before-tax MARR of 25% and pay $0.075 per kilowatt-hour, would you replace the old machines today with the new one?

13-23 The Quick Manufacturing Company, a large profitable corporation, may replace a production machine tool. A new machine would cost $3700, have a 4-year useful and depreciable life, and have no salvage value. For tax purposes, sum-of-years'-digits depreciation would be used. The existing machine tool cost $4000 4 years ago and has been depreciated by straight-line depreciation assuming an 8-year life and no salvage value. The tool could be sold now to a used equipment dealer for $1000 or be kept in service for another 4 years. It would then have no salvage value. The new machine tool would save about $900 per year in operating costs compared to the existing machine. Assume a 40% combined state and federal tax rate.

Compute the before-tax rate of return on the replacement proposal of installing the new machine rather than keeping the existing machine. (*Answer:* 12.6%)

13-24 A professor of engineering economics owns a 1996 car. In the past 12 months, he has paid $2000 to replace the transmission, bought two new tires for $160, and installed a CD player for $110. He wants to keep the car for 2 more years because he invested money 3 years ago in a 5-year certificate of deposit, which is earmarked to pay for his dream machine, a red European sports car. Today the old car's engine failed. The professor has two alternatives. He can have the engine overhauled at a cost of $1800 and then most likely have to pay another $800 per year for the next 2 years for maintenance. The car will have no salvage value at that time. Alternatively, a colleague offered to make the professor a $5000 loan to buy another used car. He must pay the loan back in two equal installments of $2500 due at the end of Year 1 and Year 2, and at the end of the second year he must give the colleague the car. The "new" used car has an expected annual maintenance cost of $300. If the professor selects this alternative, he can sell his current vehicle to a junkyard for $1500. Interest is 5%. Using present worth analysis, which alternative should he select and why?

13-25 The Plant Department of the local telephone company purchased four special pole hole diggers 8 years ago for $14,000 each. They have been in constant use to the present. Owing to an increased workload, additional machines will soon be required.

Recently an improved model of the digger was announced. The new machines have a higher production rate and lower maintenance expense than the old machines but will cost $32,000 each with a service life of 8 years and salvage value of $750 each. The four original diggers have an immediate salvage of $2000 each and an estimated salvage value of $500 each 8 years hence. The average annual maintenance expense of the old machines is about $1500 each, compared with $600 each for the new machines.

A field study and trial show that the workload would require three additional new machines if the old machines continue in service. However, if the old machines were all retired from service, the workload could be carried by six new machines with an annual savings of $12,000 in operation costs. A training program to prepare employees to run the machines will be necessary at an estimated cost of $700 per new machine. If the MARR is 9% before taxes, what should the company do?

After-Tax

13-26 The Ajax Corporation purchased a railroad tank car 8 years ago for $60,000. It is being depreciated by SOYD depreciation, assuming a 10-year depreciable life and a $7000 salvage value. The tank car needs to be reconditioned now at a cost of $35,000. If this is done, it is estimated the equipment will last for 10 more years and have a $10,000 salvage value at the end of the 10 years.

On the other hand, the existing tank car could be sold now for $10,000 and a new tank car purchased for $85,000. The new tank car would be depreciated by MACRS depreciation. Its estimated actual salvage value after 10 years would be $15,000. In addition, the new tank car would save $7000 per year in maintenance costs, compared to the reconditioned tank car.

Based on a 15% before-tax rate of return, determine whether the existing tank car should be reconditioned or a new one purchased. *Note:* The problem statement provides more data than are needed, which is typical of real situations. (*Answer:* Recondition the old tank car.)

13-27 State the advantages and disadvantages with respect to after-tax benefits of the following options for a major equipment unit:

A. Buy new.

B. Trade in and buy a similar, rebuilt equipment from the manufacturer.

C. Have the manufacturer rebuild your equipment with all new available options.

D. Have the manufacturer rebuild your equipment to the original specifications.

E. Buy used equipment.

13-28 Fifteen years ago the Acme Manufacturing Company bought a propane-powered forklift truck for $4800. The company depreciated the forklift using straight-line depreciation, a 12-year life, and zero salvage value. Over the years, the forklift has been a good piece of equipment, but lately the maintenance cost has risen sharply. Estimated end-of-year maintenance costs for the next 10 years are as follows:

Year	Maintenance Cost
1	$400
2	600
3	800
4	1000
5–10	1400/year

The old forklift has no present or future net salvage value, since its scrap metal value just equals the cost to haul it away. A replacement is now being considered for the old forklift. A modern unit can be purchased for $6500. It has an economic life equal to its 10-year depreciable life. Straight-line depreciation will be employed, with zero salvage value at the end of the 10-year depreciable life. At any time the new forklift can be sold for its book value. Maintenance on the new forklift is estimated to be a constant $50 per year for the next 10 years, after which maintenance is expected to increase sharply. Should Acme Manufacturing keep its old forklift truck for the present, or replace it now with a new one? The firm expects an 8% after-tax rate of return on its investments. Assume a 40% combined state-and-federal tax rate.   (*Answer:* Keep the old forklift truck.)

13-29 Machine *A* has been completely overhauled for $9000 and is expected to last another 12 years. The $9000 was treated as an expense for tax purposes last year. Machine *A* can be sold now for $30,000 net after selling expenses, but will have no salvage value 12 years hence. It was bought new 9 years ago for $54,000 and has been depreciated since then by straight-line depreciation using a 12-year depreciable life.

Because less output is now required, Machine *A* can be replaced with a smaller machine: Machine *B* costs $42,000, has an anticipated life of 12 years, and would reduce operating costs $2500 per year. It would be depreciated by straight-line depreciation with a 12-year depreciable life and no salvage value.

The income tax rate is 40%. Compare the after-tax annual costs and decide whether Machine *A* should be retained or replaced by Machine *B*. Use a 10% after-tax rate of return.

13-30 (*a*) A new employee at CLL Engineering Consulting Inc., you are asked to join a team performing an economic analysis for a client. Your task is to find the Time-0 ATCFs. CLL Inc. has a combined federal/state tax rate of 45% on ordinary income, depreciation recapture, and losses.

Defender: This asset was placed in service 7 years ago. At that time the $50,000 cost basis was set up on a straight-line depreciation schedule with an estimated salvage value of $15,000 over its 10-year ADR life. This asset has a present market value of $30,000.

Challenger: The new asset has a first cost of $85,000 and will be depreciated by MACRS depreciation over its 10-year class life. This asset qualifies for a 10% investment tax credit.

(*b*) How would your calculations change if the present market value of the *defender* is $25,500?

(*c*) How would your calculations change if the present market value of the *defender* is $18,000?

13-31 Foghorn Leghorn may replace an old egg-sorting machine used by his business, Foggy's Farm Fresh Eggs. The old egg machine is not quite running eggs-actly the way it was originally designed and will require an additional investment now of $2500 (expensed at Time 0) to get it back in working shape. This old machine was purchased 6 years ago for $5000 and has been depreciated by the straight-line method at $500 per year. Six years ago the estimated salvage value for tax purposes was $1000. Operating expenses for the old machine are projected at $600 this next year and are increasing by $150 per year each year thereafter. Foggy projects that with refurbishing, the machine will last another 3 years. Foggy believes that he could sell the old machine as-is today for $1000 to his friend Fido to sort bones. He also believes he could sell it 3 years from now at the barnyard flea market for $500.

The new egg-sorting machine, a deluxe model, has a purchase price of $10,000 and will last 6 years, at which time it will have a salvage value of $1000. The new machine qualifies as a MACRS 7-year property and will have operating expenses of $100 the first year, increasing by $50 per year thereafter. Foghorn uses an after-tax MARR of 18% and a tax rate of 35% on original income.

(a) What was the depreciation life used with the defender asset (the old egg sorter)?

(b) Calculate the after-tax cash flows for both the defender and challenger assets.

(c) Use the annual cash flow method to offer a recommendation to Foggy. What assumptions did you make in this analysis?

13-32 A firm is concerned about the condition of some of its plant machinery. Bill James, a newly hired engineer, reviewed the situation and identified five feasible, mutually exclusive alternatives.

Alternative A: Spend $44,000 now repairing various items. The $44,000 can be charged as a current operating expense (rather than capitalized) and deducted from other taxable income immediately. These repairs will keep the plant functioning for 7 years with current operating costs.

Alternative B: Spend $49,000 to buy general-purpose equipment. Depreciation would be straight line over the equipment's 7-year useful life. The equipment has no salvage value. The new equipment will reduce annual operating costs by $6000.

Alternative C: Spend $56,000 to buy new specialized equipment. This equipment would be depreciated by sum-of-years'-digits depreciation over its 7-year useful life. This equipment would reduce annual operating costs by $12,000. It will have no salvage value.

Alternative D: This alternative is the same as Alternative B, except it reduces annual operating costs by $7000.

Alternative E: This is the "do-nothing" alternative, with annual operating costs $8000 above the present level.

This profitable firm pays 40% corporate income taxes and uses a 10% after-tax rate of return. Which alternative should the firm adopt?

13-33 Fred's Rodent Control Corporation has been using a low-frequency sonar device to locate subterranean pests. This device was purchased 5 years ago for $18,000. The device has been depreciated using SOYD depreciation with an 8-year depreciable life and a salvage value of $3600. Presently, it could be sold for $7000. If it is kept for the next 3 years, its market value is expected to drop to $1600.

A new lightweight subsurface heat-sensing searcher (SHSS) that is available for $10,000 would improve the annual net income by $500 for each of the next 3 years. The SHSS would be depreciated as a 5-year class property, using MACRS. At the end of 3 years, the SHSS should have a market value of $4000. Fred's Rodent Control is a profitable enterprise subject to a 40% tax rate.

(a) Construct the after-tax cash flow for the old sonar unit for the next 3 years.

(b) Construct the after-tax cash flow for the SHSS unit for the next 3 years.

(c) Construct the after-tax cash flow for the difference between the SHSS unit and the old sonar unit for the next 3 years.

(d) Should Fred buy the new SHSS unit if his MARR is 20%? You do not have to calculate the incremental rate of return; just show how you reach your decision.

13-34 Compute the after-tax rate of return on the replacement proposal for Problem 13-23.

13-35 BC Junction purchased some embroidering equipment for their Denver facility 3 years ago for $15,000. This equipment qualified as MACRS 5-year property. Maintenance costs are estimated to be $1000 this next year and will increase by $1000 per year thereafter. The market (salvage) value for the equipment is $10,000 at the end of this year and declines by $1000 per year in the future. If BC Junction has an after-tax MARR of 30%, a marginal tax rate of 45% on ordinary income, depreciation recapture, and losses, what after-tax life of this previously purchased equipment has the lowest EUAC? Use a spreadsheet to develop your solution.

13-36 Reconsider the acquisition of packing equipment for Mary O'Leary's business, as described in Problem 13-16. Given the data tabulated there, and again using an after-tax MARR of 25% and a tax rate of 35% on ordinary income to evaluate the investment, determine the after-tax lowest EUAC of the equipment. Use a spreadsheet to develop your solution.

CHAPTER 14

INFLATION AND PRICE CHANGE

The Athabasca Oil Sands

For centuries, people have known about the sticky bitumen that lines the banks of the Athabasca River in northern Alberta. Even before the coming of the European fur traders, Native people used it to seal canoes. Over the years, many people dreamed of producing usable oil from the bitumen, but the sand and oil are not easily separated, and the recovery was not viewed as economically viable. In the 1950s, the world price of oil was around $3 a barrel and the estimated cost of mining and separating oil from the sands was over $30 a barrel.

In 1964, the Sun Oil Company, with government support, formed the Great Canadian Oil Sands Company, and in 1967 it started to mine and process shallow deposits of oil sands. The target production was 31,000 barrels per day, and the initial production cost would be about $25 a barrel. The world price was then about $3.50, but the planners were predicting that production costs would decline and market prices would increase.

Forty years later, in 2004, two major oil firms together supply one-third of Canada's oil from these bitumen deposits. Other firms extract bitumen from deeper deposits by steam heating, and $40 billion worth of new oil sands projects are on the books. The successor to Great Canadian Oil Sands, Suncor Energy, has 5,500 employees and is producing 260,000 barrels of oil per day. The production costs are in the range of $22 a barrel, and the world price of oil is much higher.

Contributed by Peter Flynn, University of Alberta

QUESTIONS TO CONSIDER

1. In the 1960s, market analysts estimated that at a world price of $25 per barrel, oil sands projects could provide about a 10% rate of return. How might inflation have affected these estimates?
2. The prices of some items—for example, gas at the pumps, houses—increase over time while others, such as calculators and computers, decline in price. Given these variations, how can we know if inflation is occurring, and how could we measure it?
3. In 1967, the Canadian consumer price index was 21.5; in 2004, the CPI was 123.2. What was the 2004 production cost in 1967 dollars?
4. Use the Internet and other resources to learn more about the mining of Athabasca oil sands. Are there any environmental and ethical dimensions to this story?

After Completing This Chapter...
The student should be able to:

- Describe inflation, explain how it happens, and list its effects on purchasing power.
- Define real and actual dollars and interest rates.
- Conduct constant dollar and nominal dollar analyses.
- Define and use composite and commodity-specific price indexes.
- Develop and use cash flows that inflate at different interest rates and cash flows subject to different interest rates per period.
- Incorporate the effects of inflation in before-tax and after-tax calculations.
- Develop spreadsheets to incorporate the effects of inflation and price change.

Thus far we have assumed that the dollars in our analyses were unaffected by inflation or price change. However, this assumption is not always valid or realistic. If inflation occurs in the general economy, or if there are price changes in economic costs and benefits, the impact can be substantial on both before- and after-tax analyses. In this chapter we develop several key concepts and illustrate how inflation and price changes may be incorporated.

Meaning and Effect of Inflation

Inflation is an important concept because the purchasing power of money used in most world economies rarely stays constant. Rather, over time the amount of goods and services that can be bought with a fixed amount of money tends to change. Inflation causes money to lose **purchasing power.** That is, when prices inflate we can buy less with the same amount of money. **Inflation makes future dollars less valuable than present dollars.** Think about examples in your own life, or for an even starker comparison, ask your grandparents how much a loaf of bread or a new car cost 50 years ago. Then compare these prices with what you would pay today for the same items. This exercise will reveal the effect of inflation: as time passes, goods and services cost more, and more monetary units are needed to buy the same goods and services.

Because of inflation, dollars in one period of time are not equivalent to dollars in another. We know that engineering economic analysis requires that comparisons be made on an **equivalent** basis. So, it is important for us to be able to incorporate the effects of inflation.

When the purchasing power of a monetary unit *increases* rather than decreases as time passes, the result is **deflation.** Deflation, very rare in the modern world, nonetheless can exist. Deflation has the opposite effect of inflation—one can buy **more** with money in future years than can be bought today. Thus, deflation makes future dollars more valuable than current dollars.

How Does Inflation Happen?

Economists generally believe that inflation depends on the following, either in isolation or in combination.

Money supply: The amount of money in our national economy has an effect on its purchasing power. If there is too much money in the system (the Federal Reserve controls the flow of money) versus goods and services to purchase with that money, the value of dollars tends to decrease. When there are fewer dollars in the system, they become more valuable. The Federal Reserve seeks to increase the volume of money in the system at the same rate that the economy is growing.

Exchange rates: The strength of the dollar in world markets affects the profitability of international companies. Prices may be adjusted to compensate for the dollar's relative strength or weakness in the world market. As corporations' profits are weakened or eliminated in some markets owing to fluctuations in exchange rates, prices may be raised in other markets to compensate.

Cost–push: This cause of inflation develops as producers of goods and services "push" their increasing operating costs along to the customer through higher prices. These operating costs include fabrication/manufacturing, marketing, and sales.

Demand–pull: This cause is realized when consumers spend money freely on goods and services. As more and more people demand certain goods and services, the prices of those goods and services will rise (demand exceeding supply).

A further consideration in analyzing how inflation works is the usually different rates at which prices and wages rise. Do workers benefit if, as their wages increase, the prices of goods and services increase? To determine the net effect of differing rates of inflation, we must be able to make comparisons and understand costs and benefits from an equivalent perspective. In this chapter we will learn how to make such comparisons.

Definitions for Considering Inflation in Engineering Economy

The following definitions are used throughout this chapter to illustrate how inflation and price change affect two quantities: interest rates and cash flows.

Inflation rate (f): The inflation rate captures the effect of goods and services costing more—a decrease in the purchasing power of dollars. More money is required to buy a good or service whose price has inflated. The inflation rate is measured as the annual rate of increase in the number of dollars needed to pay for the same amount of goods and services.

Real interest rate (i'): This interest rate measures the "real" growth of our money excluding the effect of inflation. Because it does not include inflation, it is sometimes called the *inflation-free interest rate.*

Market interest rate (i): This is the rate of interest that one obtains in the general marketplace. For instance, the interest rates on passbook savings, checking plus, and certificates of deposit quoted by banks are all market rates. The lending interest rate for autos and boats is also a market rate. This rate is sometimes called the *combined interest* rate because it incorporates the effect of both real money growth **and** inflation. We can view i as follows:

Market interest rate	**has in it**	**"Real" growth of money**	**and**	**Effect of inflation**

The mathematical relationship between the inflation, real and market interest rates is given as:

$$i = i' + f + (i')(f) \qquad (14\text{-}1)$$

This is the first point where we have defined a real interest rate and a market or combined interest rate. Which naturally leads to the question of what meaning should be attached to the interest rate i, which is found throughout the text. In fact, both meanings have been used. In problems about savings accounts and loans, the interest rate is usually a market rate. In problems about engineering projects where costs and benefits are often estimated as $x per year, the interest rate is a real rate.

EXAMPLE 14-1

Suppose Tiger Woods wants to invest some recent golf winnings in his hometown bank for one year. Currently, the bank is paying a rate of 5.5% *compounded annually*. Assume inflation is expected to be 2% per year. Identify i, f, and i'. Repeat for inflation of 8% per year.

SOLUTION

If Inflation Is 2% per Year

The bank is paying a *market rate* (i). The *inflation rate* (f) is given. What then is the *real interest rate* (i')?

$$i = 5.5\% \qquad f = 2\% \qquad i' = ?$$

Solving for i' in Equation 14-1, we have

$$i = i' + f + (i')(f)$$
$$i - f = i'(1 + f)$$
$$i' = (i - f)/(1 + f)$$
$$= (0.055 - 0.02)/(1 + 0.02) = 0.034 \quad \text{or} \quad \textbf{3.4\% per year}$$

This means that Tiger Woods will have 3.4% **more** purchasing power than he had a year ago. At the end of the year he can buy 3.4% more goods and services than he could have at the beginning of the year. For example, assume he was buying golf balls that cost $5 each and that he had invested $1000.

At the *beginning* of the year he could buy:

$$\text{Number of balls purchased today} = \frac{\text{Dollars today available to buy balls}}{\text{Cost of balls today}}$$

$$= 1000/\$5 = 200 \text{ golf balls}$$

At the *end* of the year he could buy:

$$\text{Number of balls bought at end of year} = \frac{\text{Dollars available for purchase at end of year}}{\text{Cost per ball at end of year}}$$

$$= \frac{(\$1000)(F/P, 5.5\%, 1)}{(\$5)(1 + 0.02)^1} = \frac{\$1055}{\$5.10} = 207 \text{ golf balls}$$

Tiger Woods can, after one year, buy 3.4% more golf balls than he could before. With rounding, this is 207 balls.

If Inflation Is 8%

As for the lower inflation rate, we would solve for i':

$$i' = (i - f)/(1 + f)$$

$$= (0.055 - 0.08)/(1 + 0.08)$$

$$= -0.023 \quad \text{or} \quad \textbf{-2.3\% per year}$$

In this case the real growth in money has *decreased* by 2.3%, so that Tiger can now buy 2.3% fewer balls with the money he had invested. Even though he has more money year-end, it is worth less, so he can purchase less.

Regardless of how inflation behaves over the year, the bank will pay Tiger $1055 at the end of the year. However, as we have seen, inflation can greatly affect the "real" growth of dollars over time. In a presidential speech, inflation has been called "that thief" because it steals real purchasing power from our dollars.

Let us continue the discussion of inflation by focusing on cash flows. We define dollars of two types:

Actual dollars (A$): These are the dollars that we ordinarily think of when we think of money. These dollars circulate in our economy and are used for investments and payments. We can touch these dollars and often keep them in our purses and wallets—they are "actual" and exist physically. Sometimes they are called *inflated dollars* because they carry any inflation that has reduced their worth. These are also the dollars shown on paychecks, credit card receipts, and normal financial transactions.

Real dollars (R$): These dollars are a bit harder to define. They are always expressed in terms of some constant purchasing power "base" year, for example, 2008-based dollars. Real dollars are sometimes called *constant dollars* or *constant purchasing power dollars,* and because they do not carry the effects of inflation, they are also known as *inflation-free dollars.*

Having defined *market, inflation,* and *real interest rates* as well as *actual* and *real dollars,* let us describe how these quantities relate. Figure 14-1 illustrates the relationship between these quantities.

Figure 14-1 illustrates the following principles:

When dealing with actual dollars (A$), use a market interest rate (i), and when discounting A$ over time, also use i.

When dealing with real dollars (R$), use a real interest rate (i'), and when discounting R$ over time, also use i'.

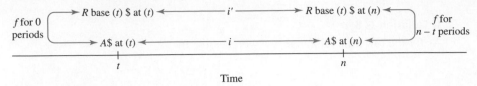

FIGURE 14-1 Relationship between i, f, i', $A\$$, and $R\$$.

Figure 14-1 shows the relationships between $A\$$ and $R\$$ that occur **at the same period of time.** Actual and real dollars are related by the *inflation rate,* in this case, over the period of years defined by $-t$. To translate between dollars of one type to dollars of the other ($A\$$ to $R\$$ or $R\$$ to $A\$$), use the inflation rate for the right number of periods. The following example illustrates many of these relationships.

EXAMPLE 14-2

When the university's stadium was completed in 1955, the total cost was $1.2 million. At that time a wealthy alumnus gifted the university with $1.2 million to be used for a future replacement. University administrators are now considering building the new facility in the year 2010. Assume that:

- Inflation is 6.0% per year from 1955 to 2010.
- In 1955 the university invested the gift at a market interest rate of 8.0% per year.

(a) Define i, i', f, $A\$$, and $R\$$.
(b) How many actual dollars in the year 2010 will the gift be worth?
(c) How much would the actual dollars in 2010 be in terms of 1955 *purchasing power?*
(d) How much better or worse should the new stadium be?

SOLUTION TO PART a

Since 6.0% is the inflation rate (f) and 8.0% is the market interest rate (i), we can write

$$i' = (0.08 - 0.06)/(1 + 0.06) = 0.01887, \quad \text{or} \quad 1.887\%$$

The building's cost in 1955 was $1,200,000, which were the actual dollars ($A\$$) spent in 1955.

SOLUTION TO PART b

From Figure 14-1 we are going from *actual dollars at t, in 1955,* to *actual dollars at n, in 2010.* To do so, we use the *market interest rate* and compound this amount forward 55 years, as illustrated in Figure 14-2.

$$\text{Actual dollars in 2010} = \text{Actual dollars in 1955} \ (F/P, i, 55 \text{ years})$$
$$= \$1,200,000(F/P, 8\%, 55)$$
$$= \$82,701,600$$

FIGURE 14-2 Compounding A\$ in 1955 to A\$ in 2010.

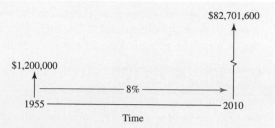

SOLUTION TO PART c

Now we want to determine the amount of *real 1955 dollars that occur in the year 2010,* which are equivalent to the \$82.7 million from the solution to part **b**. Let us solve this problem two ways.

1. Translate *actual dollars in the year 2010 to real 1955 dollars in the year 2010.* From Figure 14-1 we can use the inflation rate to **strip 55 years of inflation** from the actual dollars. We do this by using the P/F factor for 55 years at the inflation rate. We are not physically moving the dollars in time; rather, we are simply removing inflation from these dollars one year at a time—the P/F factor does that for us. This is illustrated in the following equation and Figure 14-3.

$$\text{Real 1955 dollars in 2010} = (\text{Actual dollars in 2010})(P/F, f, 55)$$
$$= (\$82{,}701{,}600)(P/F, 6\%, 55)$$
$$= \$3{,}357{,}000$$

FIGURE 14-3 Translation of A\$ in 2010 to R 1955-based dollars in 2010.

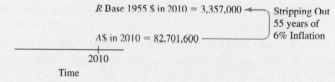

2. Translate real 1950 dollars in 1950 to real 1955 dollars in 2010. The \$1.2 million can also be said to be *real 1955 dollars that circulated in 1955.* So, let us translate those real dollars from 1955 to the year 2010 (Figure 14-4). Since they are *real dollars,* we use the *real interest rate.*

$$\text{Real 1955 dollars in 2010} = (\text{Real 1955 dollars in 1955})(P/F, i', 55)$$
$$= (\$1{,}200{,}000)(F/P, 1.887\%, 55)$$
$$= \$3{,}355{,}000$$

FIGURE 14-4 Translation of R 1955 dollars in 1955 to R 1955 dollars in 2010.

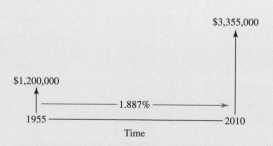

Note: The answers differ due to rounding the market interest rate off to 1.887% versus carrying it out to more significant digits. The difference due to this rounding is less than 0.1%. If i' and the factors have enough digits, the answers to the two parts would be identical.

SOLUTION TO PART d

Assuming that construction costs increased at the rate of 6% per year, then the amount available for the project *in terms of 1955 dollars* is almost $3.4 million. This means that the new stadium will be about 3.4/1.2 or approximately 2.8 times "better" than the original one using *real dollars*.

EXAMPLE 14-3

In 1924 Mr. O'Leary buried $1000 worth of quarters in his backyard. Over the years he had always thought that the money would be a nice nest egg to give to his first grandchild. His first granddaughter, Gabrielle, arrived in 1994. From 1924 to 1994, inflation averaged 4.5%, the stock market increased an average of 15% per year, and investments in guaranteed government bonds averaged 6.5% return per year. What was the relative purchasing power of the jar of quarters that Mr. O'Leary gave to his granddaughter Gabrielle at birth? What might have been a better choice for his "backyard investment"?

SOLUTION

Mr. O'Leary's $1000 are *actual dollars* in both 1924 **and** in 1994.

To obtain the *real 1924 dollar equivalent* of the $1000 that Gabrielle received in 1994, we would **strip 70 years of inflation out of those dollars.** As it turned out, Gabrielle's grandfather gave her $45.90 worth of 1924 purchasing power. Because inflation has "stolen" purchasing power from his stash of quarters during the 70-year period, Mr. O'Leary gave his grandaughter much less than the amount he first spaded underground. This loss of purchasing power caused by inflation can be calculated as follows:

$$\text{Real 1924 dollars in 1994} = (\text{Actual dollars in 1994})(P/F, f, 1994 - 1924)$$

$$= \$1000(P/F, 4.5\%, 70) = \$45.90$$

On the other hand, if Mr. O'Leary had put his $1000 in the stock market in 1924, he would have made baby Gabrielle an instant multimillionaire by giving her $17,735,000. We calculate this as follows:

$$\text{Actual dollars in 1994} = (\text{Actual dollars in 1924})(F/P, i, 1994 - 1924)$$

$$= \$1000(F/P, 15\%, 70) = \$17,735,000$$

At the time of Gabrielle's birth that $17.7 million translates to $814,069 in 1924 purchasing power. This is quite a bit different from the $45.90 in 1924 purchasing power calculated for the unearthed jar of quarters.

Real 1924 dollars in 1994 = $17,735,000(P/F, 4.5\%, 70) = \$814,069$

Mr. O'Leary was never a risk taker, so it is doubtful he would have chosen the stock market for his future grandchild's nest egg. If he had chosen guaranteed government bonds instead of his backyard, by 1994 the investment would have grown to $59,076 (actual dollars)—the equivalent of $2712 in 1924 purchasing power.

$$\text{Actual dollars in 1994} = (\text{Actual dollars in 1924})(F/P, i, 1994 - 1924)$$
$$= \$1000(F/P, 6\%, 70) = \$59,076$$
$$\text{Real 1924 dollars in 1994} = \$59,076(P/F, 4.5\%, 70) = \$2712$$

Obviously, either option would have been better than the choice Mr. O'Leary made. This example illustrates the effects of inflation and purchasing power, as well as the power of compound interest. However, in Mr. O'Leary's defense, if the country had experienced 70 years of *deflation* instead of *inflation,* he might have had the last laugh!

There are in general two ways to approach an economic analysis problem after the effects of inflation have been recognized. The first is to ignore these effects in conducting the analysis, as we've done so far in the text.

> *Ignoring inflation in the analysis:* Use **real dollars** and a **real interest rate** that does not reflect inflation.

The second approach is to systematically include the effects of inflation, as we study in this chapter.

> *Incorporating inflation in the analysis:* Use a **market interest rate** and **actual dollars** that include inflation.

Since inflation is so common, why do many economic analyses of engineering projects and most of this text choose *not* to explicitly address inflation? This question is best answered by referring to the many examples and problems that contained statements like "Operations and maintenance costs are expected to be $30,000 annually for the equipment's 20-year life."

Does such a statement mean that accounting records for the next 20 years will show constant costs? Obviously not. Instead, it means that in constant-dollar terms the O&M costs are not expected to increase. In real dollars, O&M costs are uniform. In actual or inflated dollars, we will pay more each year, but each of those dollars will be worth less.

Most costs and benefits in the real world and in this text have prices that increase at about the same rate of inflation as the economy as a whole. In most analyses these inflation increases are addressed by simply stating everything in real dollar terms and using a real interest rate.

There are specific items, such as computers and depreciation deductions, where inflation is clearly expected to differ from the general rate of inflation. It is for these cases, that this chapter is included in this text.

ANALYSIS IN CONSTANT DOLLARS VERSUS THEN-CURRENT DOLLARS

Performing an analysis requires that we distinguish cash flows as being either constant dollars (real dollars, expressed in terms of some purchasing power base) or then-current dollars (actual dollars that are then-current when they occur). As previously stated, constant (real) dollars require the use of a *real interest rate* for discounting, then-current dollars require a *market (or combined) interest rate*. We must not mix these two dollar types when performing an analysis. If both types are stated in the problem, one must be converted to the other, so that a consistent comparison can be made.

EXAMPLE 14-4

The Waygate Corporation is interested in evaluating a major new video display technology (VDT). Two competing computer innovation companies have approached Waygate to develop the technology. Waygate believes that both companies will be able to deliver equivalent products at the end of a 5-year period. From the yearly development costs of the VDT for each firm, determine which one Waygate should choose if the corporate MARR (investment market rate) is 25% and price inflation is assumed to be 3.5% per year over the next 5 years.

Company Alpha costs: Development costs will be $150,000 the first year and will increase at a rate of 5% over the 5-year period.

Company Beta costs: Development costs will be a constant $150,000 per year in terms of today's dollars over the 5-year period.

SOLUTION

The costs for each of the two alternatives are as follows:

Year	Then-Current Costs Stated by Alpha	Constant Dollar Costs Stated by Beta
1	$150,000 \times (1.05)^0 = \$150,000$	$150,000
2	$150,000 \times (1.05)^1 = \ 157,500$	150,000
3	$150,000 \times (1.05)^2 = \ 165,375$	150,000
4	$150,000 \times (1.05)^3 = \ 173,644$	150,000
5	$150,000 \times (1.05)^4 = \ 182,326$	150,000

We inflate (or escalate) the stated yearly cost given by Company Alpha by 5% per year to obtain the then-current (actual) dollars each year. Company Beta's costs are given in terms of today-based constant dollars.

Using a Constant Dollar Analysis

Here we must convert the then-current costs given by Company Alpha to constant today-based dollars. We do this by stripping the number of years of general inflation from each year's cost using $(P/F, f, n)$ or $(1 + f)^{-n}$.

Year	Constant Dollar Costs Stated by Alpha	Constant Dollar Costs Stated by Beta
1	$150,000 \times (1.035)^{-1} = \$144,928$	$150,000
2	$157,500 \times (1.035)^{-2} = 147,028$	150,000
3	$165,375 \times (1.035)^{-3} = 149,159$	150,000
4	$173,644 \times (1.035)^{-4} = 151,321$	150,000
5	$182,326 \times (1.035)^{-5} = 153,514$	150,000

We use the *real interest rate* (i') calculated from Equation 14-1 to calculate the present worth of costs for each alternative:

$$i' = (i - f)/(1 + f) = (0.25 - 0.035)/(1 + 0.035) = 0.208$$

PW of cost (Alpha) $= \$144,928(P/F, 20.8\%, 1) + \$147,028(P/F, 20.8\%, 2)$
$$+ \$149,159(P/F, 20.8\%, 3) + \$151,321(P/F, 20.8\%, 4)$$
$$+ \$153,514(P/F, 20.8\%, 5) = \$436,000$$

PW of cost (Beta) $= \$150,000(P/A, 20.8\%, 5) = \$150,000(2.9387) = \$441,000$

Using a Then-Current Dollar Analysis

Here we must convert the constant dollar costs of Company Beta to then-current dollars. We do this by using $(F/P, f, n)$ or $(1 + f)^n$ to "add in" the appropriate number of years of general inflation to each year's cost.

Year	Then-Current Costs Stated by Alpha	Then-Current Costs Stated by Beta
1	$150,000 \times (1.05)^0 = \$150,000$	$150,000 \times (1.035)^1 = \$155,250$
2	$150,000 \times (1.05)^1 = 157,500$	$150,000 \times (1.035)^2 = 160,684$
3	$150,000 \times (1.05)^2 = 165,375$	$150,000 \times (1.035)^3 = 166,308$
4	$150,000 \times (1.05)^3 = 173,644$	$150,000 \times (1.035)^4 = 172,128$
5	$150,000 \times (1.05)^4 = 182,326$	$150,000 \times (1.035)^5 = 178,153$

Calculate the present worth of costs for each alternative using the *market interest rate* (i).

PW of cost (Alpha) $= \$150,000(P/F, 25\%, 1) + \$157,500(P/F, 25\%, 2)$
$$+ \$165,375(P/F, 25\%, 3) + \$173,644(P/F, 25\%, 4)$$
$$+ \$182,326(P/F, 25\%, 5) = \$436,000$$

PW of cost (Beta) $= \$155,250(P/F, 25\%, 1) + \$160,684(P/F, 25\%, 2)$
$$+ \$166,308(P/F, 25\%, 3) + \$172,128(P/F, 25\%, 4)$$
$$+ \$178,153(P/F, 25\%, 5) = \$441,000$$

Using either a constant dollar or then-current dollar analysis, Waygate should chose Company Alpha's offer, which has the lower present worth of costs. There may, of course, be intangible elements in the decision that are more important than a 1% difference in the costs.

PRICE CHANGE WITH INDEXES

We have already described the effects that inflation can have on money over time. Also, several definitions and relationships regarding dollars and interest rates have been given. We have seen that it is not correct to compare the benefits of an investment in 2008-based dollars with costs in 2010-based dollars. This is like comparing apples and oranges. Such comparisons of benefits and costs can be meaningful only if a standard purchasing power base of money is used. Thus we ask, How do I know what inflation rate to use in my studies? and, How can we measure price changes over time?

What Is a Price Index?

Price indexes (introduced in Chapter 2) describe the relative price fluctuation of goods and services. They provide a *historical* record of prices over time. Price indexes are tracked for *specific commodities* as well as for **bundles (composites) of commodities.** As such, price indexes can be used to measure historical price changes for individual cost items (like labor and material costs) as well as general costs (like consumer products). We use **past** price fluctuations to predict **future** prices.

Table 14-1 lists the historic prices of sending a first-class letter in the U.S. from 1970 to 2007. The cost is given both in terms of dollars (cents) and as measured by a fictitious price index that we could call the letter cost index (LCI).

Notice two important aspects of the LCI. First, as with all cost or price indexes, there is a **base year,** which is assigned a value of 100. Our LCI has a base year of 1970—thus for 1970, LCI = 100. Values for subsequent years are stated in relation to the 1970 value. Second, the LCI changes only when the cost of first-class postage changes. In years when this quantity does not change, the LCI is not affected. These general observations apply to all price indexes.

In general, engineering economists are the "users" of cost indexes such as our hypothetical LCI. That is, cost indexes are calculated or tabulated by some other party, and our interest is in assessing what the index tells us about the historical prices and how these may affect our estimate of future costs. However, we should understand how the LCI in Table 14-1 was calculated.

In Table 14-1, the LCI is assigned a value of 100 because 1970 serves as our base year. In the following years the LCI is calculated on a year-to-year basis based on the annual percentage increase in first-class mail. Equation 14-2 illustrates the arithmetic used.

$$\text{LCI year, } n = \frac{\text{cost }(n)}{\text{cost 1970}} \times 100 \tag{14-2}$$

For example, consider the LCI for the year 1980. We calculate the LCI as follows.

$$\text{LCI year 1980} = \frac{0.15}{0.06} \times 100 = 250$$

As mentioned, engineering economists often use cost indexes to project future cash flows. As such, our first job is to use a cost index to **calculate** the *year-to-year* percentage increase (or *inflation*) of prices tracked by an index. We can use Equation 14-3:

$$\text{Annual percentage increase, } n = \frac{\text{Index}(n) - \text{Index}(n-1)}{\text{Index}(n-1)} \times 100\% \tag{14-3}$$

TABLE 14-1 Historic Prices of First-Class Mail, 1970–2003, and Letter Cost Index

Year, n	Cost of First-Class Mail	LCI	Annual Increase for n	Year, n	Cost of First-Class Mail	LCI	Annual Increase for n
1970	$0.06	100	0.00%	1989	0.25	417	0.00%
1971	0.08	133	33.33	1990	0.25	417	0.00
1972	0.08	133	0.00	1991	0.29	483	16.00
1973	0.08	133	0.00	1992	0.29	483	0.00
1974	0.10	166	25.00	1993	0.29	483	0.00
1975	0.13	216	30.00	1994	0.29	483	0.00
1976	0.13	216	0.00	1995	0.32	533	10.34
1977	0.13	216	0.00	1996	0.32	533	0.00
1978	0.15	250	15.74	1997	0.32	533	0.00
1979	0.15	250	0.00	1998	0.33	550	3.13
1980	0.15	250	0.00	1999	0.33	550	0.00
1981	0.20	333	33.33	2000	0.33	550	0.00
1982	0.20	333	0.00	2001	0.34	567	3.03
1983	0.20	333	0.00	2002	0.37	617	8.82
1984	0.20	333	0.00	2003	0.37	617	0.00
1985	0.22	367	10.00	2004	0.37	617	0.00
1986	0.22	367	0.00	2005	0.37	617	0.00
1987	0.22	367	0.00	2006	0.39	650	5.41
1988	0.25	417	13.64	2007	0.41	683	5.13

To illustrate, let us look at the percent change from 1977 to 1978 for the LCI.

$$\text{Annual percentage increase (1978)} = \frac{250 - 216}{216} \times 100\% = 15.74\%$$

For 1978 the price of mailing a first-class letter increased by 15.74% over the previous year. This is tabulated in Table 14-1.

An engineering economist often wants to know how a particular cost quantity changes over time. Often we are interested in calculating the *average* rate of price increases over a period of time. For instance, we might want to know the average yearly increase in postal prices from 1987 to 2007. If we generalize Equation 14-3 to calculate the percent change from 1987 to 2007, we obtain:

$$\% \text{ Increase (1987 to 2007)} = \frac{683 - 367}{367} \times 100\% = 866\%$$

How do we use this to obtain the **average** rate of increase over those 20 years? Should we divide 86% by 20 years (86/20 = 4.31%)? Of course not! Inflation, like interest, compounds. Such a simple division treats inflation like simple interest—without compounding. So the question remains: How do we calculate an *equivalent average rate of increase* in postage rates over a period of time? If we think of the index numbers as cash flows, we have:

$$P = 367 \qquad F = 683 \qquad n = 20 \text{ years} \qquad i = ?$$

Using $F = P(1 + i)^n$ $683 = 367(1 + i)^{20}$ $i = (683/367)^{1/20} - 1$ $i = 0.0315 = 3.2\%$

We can use a cost index to calculate the average rate of increase over any period of years, which should provide insight into how prices may behave in the future.

Composite Versus Commodity Indexes

Cost indexes come in two types: commodity-specific indexes and composite indexes. Commodity-specific indexes measure the historical change in price for specific items—such as construction labor or iron ore. Commodity indexes, like our letter cost index, are useful when an economic analysis includes individual cost items that are tracked by such indexes. For example, if we need to estimate the direct-labor cost portion of a construction project, we could use an index that tracks the inflation, or escalation, of labor costs. The U.S. Departments of Commerce and Labor track many cost quantities through the Department of Economic Analysis and Bureau of Labor Statistics. Example 14-5 uses data from a California government website (www.resd.dgs.ca.gov/CaliforniaConstructionCostIndexPage.htm) to demonstrate using a commodity index. This data is compiled from the *Engineering News-Record*.

EXAMPLE 14-5

In January 1997 bids were opened for a new building in Los Angeles. The low bid and the final construction cost were $1.30 million. Another building of the same size, quality, and purpose is planned with a bid opening in January 2010. Estimate the new building's low bid and cost.

SOLUTION

In January 2007 the California Construction Cost Index (CCCI) had a value of 4869 and in January 1997 the value was 3473. If we wanted a cost estimate for January 2007, we could simply use the ratio of these values and Equation 14-2. But we want a value for January 2010, which is outside our data set. (This is true for all future estimates.)

The solution is to estimate the average annual rate of increase, and then to apply this for the longer period.

$$F = 4869, P = 3473, n = 10, \text{ find } f$$
$$F = P(1 + f)^n$$
$$f = (4869/3473)^{1/10} - 1 = 3.44\% \text{ per year}$$

Now we can apply the inflation rate for $n = 13$ years to the building's cost in 1997.

$$F = \$1.3 \text{ million} \times (1.0344)^{13} = \$2.02 \text{ million}$$

TABLE 14-2 CPI Index Values and Yearly Percentage Increases, 1973–2006

Year	CPI Value*	CPI Increase	Year	CPI Value*	CPI Increase
1973	44.4	6.2%	1990	130.7	5.4%
1974	49.3	11.0	1991	136.2	4.2
1975	53.8	9.1	1992	140.3	3.0
1976	56.9	5.8	1993	144.5	3.0
1977	60.6	6.5	1994	148.2	2.6
1978	65.2	7.6	1995	152.4	2.8
1979	72.6	11.3	1996	156.9	2.9
1980	82.4	13.5	1997	160.5	2.3
1981	90.9	10.3	1998	163.0	1.6
1982	96.5	6.2	1999	166.6	2.2
1983	99.6	3.2	2000	172.2	3.4
1984	103.9	4.3	2001	177.1	2.8
1985	107.6	3.6	2002	179.9	1.1
1986	109.6	1.9	2003	184.0	2.7
1987	113.6	3.6	2004	188.9	2.7
1988	118.3	4.1	2005	195.3	3.4
1989	124.0	4.8	2006	201.6	3.2

*Reference base: 1982–1984 = 100.

Composite cost indexes do not track historical prices for individual classes of items. Instead, they measure the historical prices of *bundles* or *market baskets* of assets. Examples of composite indexes include the *Consumer Price Index* (CPI) and the *Producer Price Index* (PPI). The CPI measures prices for consumers in the U.S. marketplace, and each PPI measures prices for categories of producers in the U.S. economy.

The CPI, an index calculated by the Bureau of Labor Statistics, tracks the cost of a standard *bundle of consumer goods* from year to year. This "consumer bundle" or "market basket" includes housing, clothing, food, transportation, and entertainment. The CPI enjoys popular identification as the "inflation" indicator. Table 14-2 gives yearly index values and annual percent increases in the CPI. Figure 14-5 charts the CPI inflation rate for the same period.

Composite indexes are used the same way as commodity-specific indexes. That is, we can pick a single value from the table if we are interested in measuring the historic price for a single year, or we can calculate an *average inflation rate* or *average rate of price increase* as measured by the index over a time period extending several years.

How to Use Price Indexes in Engineering Economic Analysis

One may question the usefulness of *historical* data (as provided by price indexes) when engineering economic analysis deals with economic effects projected to occur in the *future*.

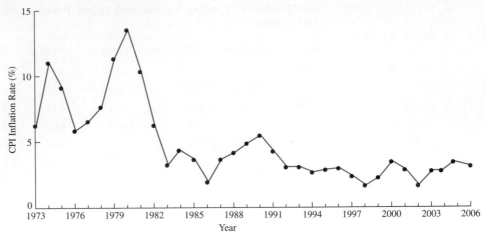

FIGURE 14-5 CPI inflation rate: 1973–2006.

However, historical index data are often better predictors of future prices than official government predictions, which may be influenced by political considerations. The engineering economist can use *average historical percentage increases (or decreases)* from commodity-specific and composite indexes, along with data from market analyses and other sources, to estimate future costs and benefits.

> When the estimated quantities are items that are tracked by commodity specific indexes, then those indexes should be used to calculate *average historical percentage increases (or decreases)*. If no commodity-specific indexes are kept, use an appropriate composite index to make this calculation.

For example, to estimate electric usage costs for a turret lathe over a 5-year period, you would first want to refer to a commodity-specific index for electric power in your area. If such an index does not exist, you might use a specific index for a very closely related commodity—perhaps an index of electric power costs nationally. In the absence of such substitute or related commodity indexes, you could use a composite index for national energy prices. The key point is that you should try to identify and use the price index that most closely relates to the quantity being estimated.

CASH FLOWS THAT INFLATE AT DIFFERENT RATES

Engineering economic analysis requires the estimation of various parameters. Over time, it is not uncommon for these parameters to *inflate* or *increase* (or even decrease) at different rates. For instance, one parameter might *increase* 5% per year and another 15% per year, and a third might *decrease* 3.5% per year. Since we are looking at the behavior of cash flows over time, we must have a way of handling this effect.

EXAMPLE 14-6

On your first assignment as an engineer, your boss asks you to develop the utility cost estimate for a new manufacturing facility. After some research you define the problem as finding the present worth of utility costs given the following data:

- Your company uses a minimum attractive rate of return (MARR) = 35% as i (not adjusted for inflation.)
- The project has a useful life of 25 years.
- Utilities to be estimated include electricity, water, and natural gas.
- The 35-year historical data reveal:

    Electricity costs increase at 8.5% per year
    Water costs increase at 5.5% per year
    Natural gas costs increase at 6.5% per year

- First-year estimates of the utility costs (in today's dollars) are as follows:

    Electricity will cost $55,000
    Water will cost $18,000
    Natural gas will cost $38,000

SOLUTION

For this problem we will take each of the utilities used in our manufacturing facility and inflate them independently at their respective historical annual rates. Once we have these actual dollar amounts ($A\$$), we can total them and then discount each year's total at 35% back to the present.

Year	Electricity		Water		Natural Gas		Total
1	$55,000(1.085)^0$ =	\$ 55,000	$18,000(1.055)^0$ =	\$18,000	$38,000(1.065)^0$ =	\$ 38,000	\$111,000
2	$55,000(1.085)^1$ =	59,675	$18,000(1.055)^1$ =	18,990	$38,000(1.065)^1$ =	40,470	119,135
3	$55,000(1.085)^2$ =	64,747	$18,000(1.055)^2$ =	20,034	$38,000(1.065)^2$ =	43,101	127,882
4	$55,000(1.085)^3$ =	70,251	$18,000(1.055)^3$ =	21,136	$38,000(1.065)^3$ =	45,902	137,289
5	$55,000(1.085)^4$ =	76,222	$18,000(1.055)^4$ =	22,299	$38,000(1.065)^4$ =	48,886	147,407
6	$55,000(1.085)^5$ =	82,701	$18,000(1.055)^5$ =	23,525	$38,000(1.065)^5$ =	52,063	158,290
7	$55,000(1.085)^6$ =	89,731	$18,000(1.055)^6$ =	24,819	$38,000(1.065)^6$ =	55,447	169,997
8	$55,000(1.085)^7$ =	97,358	$18,000(1.055)^7$ =	26,184	$38,000(1.065)^7$ =	59,051	182,594
⋮	⋮		⋮		⋮		⋮
24	$55,000(1.085)^{23}$ =	359,126	$18,000(1.055)^{23}$ =	61,671	$38,000(1.065)^{23}$ =	161,743	582,539
25	$55,000(1.085)^{24}$ =	389,652	$18,000(1.055)^{24}$ =	65,063	$38,000(1.065)^{24}$ =	172,256	626,970

The present worth of the total yearly utility costs is:

$$PW = \$111,000(P/F, 35\%, 1) + \$119,135(P/F, 35\%, 2) + \cdots + \$626,970(P/F, 35\%, 25)$$

$$= \$5,540,000$$

In Example 14-6 several commodity prices changed at different rates. By using the respective individual inflation rates, the **actual dollar** amounts for each commodity were obtained in each year. Then, we used a market interest rate to discount these actual dollar amounts.

DIFFERENT INFLATION RATES PER PERIOD

In this section we address the situation of inflation rates that are changing over the study period. Rather than different inflation rates for different cash flows, in Example 14-7 the *interest rate* for the same cash flow is changing over time. A method for handling this situation is much like that of the preceding section. We can simply apply the inflation rates in the years in which they are projected to occur. We would do this for each cash flow. Once we have all these actual dollar amounts, we can use the market interest rate to apply PW, EUAC, or other measures of merit.

EXAMPLE 14-7

While working as a clerk at Piggly Wiggly, Elvis has learned much about the cost of different vegetables. The kitchen manager at Heartbreak Hotel called recently, requesting Elvis to estimate the raw material cost over the next 5 years to introduce succotash (lima beans and corn) to the buffet line. To develop his estimate, Elvis has used his advanced knowledge of soil growing conditions, world demand, and government subsidy programs for these two crops. He has estimated the following data:

- Costs for lima beans will inflate at 3% per year for the next 3 years and then at 4% for the following 2 years.
- Costs for corn will inflate at 8% per year for the next 2 years and then will decrease 2% in the following 3 years.

The kitchen manager wants to know the equivalent annual cost of providing succotash on the buffet line over the 5-year period. His before-tax MARR is 20%. An average of 50 pounds each of beans and corn will be needed each day. The hotel kitchen operates 6 days per week, 52 weeks per year. Current costs are $0.35/lb for lima beans and $0.80/lb for corn.

SOLUTION

Today's cost for one year's supply of vegetables is:

Lima beans   0.35 $/lb × 50 lb/day × 6 day/wk × 52 wk/yr = $ 5460/yr
Corn   0.80 $/lb × 50 lb/day × 6 day/wk × 52 wk/yr = 12,480/yr

Year	Lima Beans	Corn	Total
0	$5,460	$12,480	
1	$5,460(1.03) = 5,624$	$12,480(1.08) = 13,478$	$19,102
2	$5,624(1.03) = 5,793$	$13,478(1.08) = 14,556$	20,349
3	$5,793(1.03) = 5,967$	$14,556(1.02)^{-1} = 14,271$	20,238
4	$5,967(1.04) = 6,206$	$14,271(1.02)^{-1} = 13,991$	20,197
5	$6,206(1.04) = 6,454$	$13,991(1.02)^{-1} = 13,717$	20,171

$$\text{EUAC} = [19,102(P/F, 20\%, 1) + 20,349(P/F, 20\%, 2) + 20,238(P/F, 20\%, 3)$$
$$+ 20,197(P/F, 20\%, 4) + 20,171(P/F, 20\%, 5)](A/P, 20\%, 5)$$
$$= \$19,900 \text{ per year}$$

In Example 14-7, both today's cost of each vegetable and the respective inflation rates were used to calculate the yearly costs of purchasing the desired quantities over the 5-year period. As in Example 14-6, we obtained a total marginal cost (in terms of actual dollars) by combining the two individual yearly costs. We then calculated the equivalent uniform annual cost (EUAC) using the given market interest rate.

Example 14-8 provides another example of how the effect of changes in inflation rates over time can affect an analysis.

EXAMPLE 14-8

If general price inflation is estimated to be 5% for the next 5 years, 7.5% for the 3 years after that, and 3% the following 5 years, at what market interest rate (i) would you have to invest your money to maintain a real purchasing power growth rate (i') of 10% during those years?

SOLUTION

In Years 1–5 you must invest at $0.10 + 0.050 + (0.10)(0.050) = 0.1150 = 11.50\%$ per year.

In Years 6–8 you must invest at $0.10 + 0.075 + (0.10)(0.075) = 0.1825 = 18.25\%$ per year.

In Years 9–13 you must invest at $0.10 + 0.030 + (0.10)(0.030) = 0.1330 = 13.30\%$ per year.

Note: Most interest-bearing investments have fixed, up-front rates that the investor well understands going in. On the other hand, inflation is not quantified, and its effect on real return is not measured until the end of the year. Therefore the real investment return (i') may not turn out to be what was originally required.

INFLATION EFFECT ON AFTER-TAX CALCULATIONS

Earlier we noted the impact of inflation on before-tax calculations. If the future benefits keep up with the rate of inflation, the rate of return will not be adversely affected by the inflation. Unfortunately, we are not so lucky when we consider a situation with income taxes, as illustrated by Example 14-9. The value of the depreciation deduction is diminished by inflation.

EXAMPLE 14-9

A \$12,000 investment with no salvage value will return annual benefits for 6 years. Assume straight-line depreciation and a 46% income tax rate. Solve for both before- and after-tax rates of return for two situations:

1. *No inflation:* the annual benefits are constant at \$2918 per year.
2. *Inflation equal to 5%:* the benefits from the investment increase at this same rate, so that they continue to be the equivalent of \$2918 in Year-0 dollars.

The benefit schedules are as follows:

Year	Annual Benefit for Both Situations (Year-0 dollars)	No Inflation, Actual Dollars Received	5% Inflation Factor*	5% Inflation, Actual Dollars Received
1	\$2918	\$2918	$(1.05)^1$	\$3064
2	2918	2918	$(1.05)^2$	3217
3	2918	2918	$(1.05)^3$	3378
4	2918	2918	$(1.05)^4$	3547
5	2918	2918	$(1.05)^5$	3724
6	2918	2918	$(1.05)^6$	3910

*May be read from the 5% compound interest table as $(F/P, 5\%, n)$.

SOLUTIONS

Before-Tax Rate of Return

Since both situations (no inflation and 5% inflation) have an annual benefit, stated in Year-0 dollars of \$2918, they have the same before-tax rate of return.

$$PW \text{ of cost} = PW \text{ of benefit}$$

$$12,000 = 2918(P/A, i, 6) \qquad (P/A, i, 6) = \frac{12,000}{2918} = 4.11$$

From compound interest tables: before-tax rate of return equals 12%.

After-Tax Rate of Return, No Inflation

Year	Before-Tax Cash Flow	Straight-Line Depreciation	Taxable Income	46% Income Taxes	Actual Dollars, and Year-0 Dollars, After-Tax Cash Flow
0	−\$12,000				−\$12,000
1–6	+2,918	\$2000	\$918	−\$422	+2,496

PW of cost = PW of benefit

$$12{,}000 = 2496(P/A, i, 6) \qquad (P/A, i, 6) = \frac{12{,}000}{2496} = 4.81$$

From compound interest tables: after-tax rate of return equals 6.7%.

After-Tax Rate of Return, 5% Inflation

Year	Before-Tax Cash Flow	Straight-Line Depreciation	Taxable Income	46% Income Taxes	Actual Dollars, After-Tax Cash Flow
0	−$12,000				−$12,000
1	+3,064	$2000	$1064	−$489	+2,575
2	+3,217	2000	1217	−560	+2,657
3	+3,378	2000	1378	−634	+2,744
4	+3,547	2000	1547	−712	+2,835
5	+3,724	2000	1724	−793	+2,931
6	+3,910	2000	1910	−879	+3,031

Converting to Year-0 Dollars and Solving for Rate of Return

Year	Actual Dollars, After-Tax Cash Flow	Conversion Factor	Year-0 Dollars, After-Tax Cash Flow	Present Worth at 4%	Present Worth at 5%
0	−$12,000		−$12,000	−$12,000	−$12,000
1	+2,575	×(1.05)⁻¹ =	+2,452	+2,358	+2,335
2	+2,657	×(1.05)⁻² =	+2,410	+2,228	+2,186
3	+2,744	×(1.05)⁻³ =	+2,370	+2,107	+2,047
4	+2,835	×(1.05)⁻⁴ =	+2,332	+1,993	+1,919
5	+2,931	×(1.05)⁻⁵ =	+2,297	+1,888	+1,800
6	+3,031	×(1.05)⁻⁶ =	+2,262	+1,788	+1,688
				+362	−25

Linear interpolation between 4 and 5%:

After-tax rate of return = 4% + 1%[362/(362 + 25)] = 4.9%

From Example 14-9, we see that the before-tax rate of return for both situations (no inflation and 5% inflation) is the same. Equal before-tax rates of return are expected because the benefits in the inflation situation increased in proportion to the inflation. No special calculations are needed in before-tax calculations when future benefits are expected to respond to inflation or deflation rates.

The after-tax calculations illustrate that equal before-tax rates of return do not produce equal after-tax rates of return considering inflation.

	Rate of Return (%)	
Situation	Before Taxes	After Taxes
No inflation	12	6.7
5% inflation	12	4.9

Inflation reduces the after-tax rate of return, even though the benefits increase at the same rate as the inflation. A review of the cash flow table reveals that while benefits increase, the depreciation schedule does not. Thus, the inflation results in increased taxable income and, hence, larger income tax payments.

The result is that while the after-tax cash flow in actual dollars increases, the augmented amount is not high enough to offset *both* inflation and increased income taxes. The Year-0-dollar after-tax cash flow is smaller with inflation than the Year-0-dollar after-tax cash flow without inflation.

USING SPREADSHEETS FOR INFLATION CALCULATIONS

Spreadsheets are the perfect tool for incorporating consideration of inflation into analyses of economic problems. For example, next year's labor costs are likely to be estimated as equal to this year's costs times $(1 + f)$, where f is the inflation rate. Thus each year's value is different, so we can't use factors for uniform flows, A. Also the formulas that link different years are easy to write. The result is problems that are very tedious to do by hand, but easy by spreadsheet.

Example 14-10 illustrates two different ways to write the equation for inflating costs. Example 14-11 illustrates that inflation reduces the after-tax rate of return because inflation makes the depreciation deduction less valuable.

EXAMPLE 14-10

Two costs for construction of a small, remote mine are for labor and transportation. Labor costs are expected to be $350,000 the first year, with inflation of 6% annually. Unit transportation costs are expected to inflate at 5% annually, but the volume of material being moved changes each year. In Time-0 dollars, the transportation costs are estimated to be $40,000, $60,000, $50,000, and $30,000 in Years 1 through 4. The inflation rate for the value of the dollar is 3%. If the firm uses an i' of 7%, what is the equivalent annual cost for this 4-year project?

SOLUTION

The data for labor costs can be stated so that no inflation needs to be applied in Year 1: the cost is $350,000. In contrast, the transportation costs for Year 1 are determined by multiplying $40,000 by 1.05 $(= 1 + f)$.

Also in later years the labor cost$_t$ = labor cost$_{t-1}(1 + f)$, while each transportation cost must be computed as the time-0 value times $(1 + f)^t$. In Figure 14-6, the numbers in the Year 0 (or real) dollar column equal the values in the actual dollars column divided by $(1.03)^t$.

	A	B	C	D	E	F	G	H
1						7%	Inflation-Free Interest	
2	Inflation Rate	6%		5%		3%		
3			Transportation Costs		Total	Total		
4	Year	Labor Costs	Year 0 $s	Actual $s	Actual $s	Real $s		
5	1	120,000	40,000	42,000	162,000	157,282	= E5/(1+$F$2)^A5	
6	2	127,200	60,000	66,150	193,350	182,251		
7	3	134,832	50,000	57,881	192,713	176,360		
8	4	142,922	30,000	36,465	179,387	159,383		
9						$571,732	= NPV(F1,F5:F8)	
10					=B8+D8	$168,791	= −PMT(F1,4,F9)	
11	=B7*(1+$B$2)		=C8*(1+$D$2)^A8					

FIGURE 14-6 Spreadsheet for inflation.

The equivalent annual cost equals $168,791.

EXAMPLE 14-11

For the data of Example 14-9, calculate the IRR with and without inflation with MACRS depreciation. How are the results affected by inflation by comparison with the earlier results.

SOLUTION

Most of the formulas for this spreadsheet are given in rows 11 and 12 of Figure 14-6 for the data in Year 6. The benefits received are computed from the base value in cell B5. The depreciation is the MACRS percentage times the $120,000 spent in Year 0. This value is not influenced by inflation, so the depreciation deduction is less valuable as inflation increases. The tax paid equals the tax rate times the taxable income, which equals dollars received minus the depreciation charge. Then ATCF (after-tax cash flow) equals the before-tax cash flow minus the tax paid.

In Figure 14-7, notice that in Year 2 the depreciation charge is large enough to cause this project to pay "negative" tax. For a firm, this means that the deduction on this project will be used to offset income from other projects.

	A	B	C	D	E	F	G	H
1		0%	= Inflation Rate		46%	= Tax Rate		
2		Actual $s	MACRS	Actual $s	Actual $s	Actual $s	Real $s	
3	Year	Received	Deprec. %	Deprec.	Tax	ATCF	ATCF	
4	0	−12000				−12000	−12000	
5	1	2918	20.00%	2400	238	2680	2680	= F5/(1+$B$1)^A5
6	2	2918	32.00%	3840	−424	3342	3342	
7	3	2918	19.20%	2304	282	2636	2636	
8	4	2918	11.52%	1382	706	2212	2212	
9	5	2918	11.52%	1382	706	2212	2212	
10	6	2918	5.76%	691	1024	1894	1894	
11	Formulas		= −$B$4*C10		=B10−E10			
12	for Yr 6	=$B$5*(1+$B$1)^A10		=(B10−D10)*$E$1			7.29%	= IRR
13								
14		5%	= Inflation Rate		46%	= Tax Rate		
15		Actual $s	MACRS	Actual $s	Actual $s	Actual $s	Real $s	
16	Year	Received	Deprec. %	Deprec.	Tax	ATCF	ATCF	
17	0	−12000				−12000	−12000	
18	1	3064	20.00%	2400	305	2759	2627	
19	2	3217	32.00%	3840	−287	3504	3178	
20	3	3378	19.20%	2304	494	2884	2491	
21	4	3547	11.52%	1382	996	2551	2099	
22	5	3724	11.52%	1382	1077	2647	2074	
23	6	3910	5.76%	691	1481	2430	1813	
24							5.68%	= IRR

FIGURE 14-7 After-tax IRRs with MACRS and inflation.

The IRRs are higher in this example (7.29% without inflation vs. 6.7% with straight-line depreciation in Example 14-9, and 5.68% with inflation vs. 4.9%) because MACRS supports faster depreciation, so the depreciation deductions are more valuable. Also because the depreciation is faster, the results are affected somewhat less by inflation. Specifically, with MACRS 5% inflation lowers the IRR by 1.6% and with straight-line depreciation, 5% inflation lowers the IRR by 1.8%.

SUMMARY

Inflation is characterized by rising prices for goods and services, whereas deflation produces a fall in prices. An inflationary trend makes future dollars have less **purchasing power** than present dollars. Deflation has the opposite effect. If money is borrowed over a period of time in which deflation is occurring, then debt will be repaid with dollars that have **more** purchasing power than those originally borrowed. Inflation and deflation have opposite effects on the purchasing power of a monetary unit over time.

To distinguish and account for the effect of inflation in our engineering economic analysis, we define *inflation, real,* and *market* interest rates. These interest rates are related by the following expression:

$$i = i' + f + i'f$$

Each rate applies in a different circumstance, and it is important to apply the correct rate to the correct circumstance. Cash flows are expressed in terms of either *actual* or *real dollars*. The *market interest* rate should be used with *actual dollars* and the *real interest rate* should be used with *real dollars*.

The different cash flows in our analysis may inflate or change at different interest rates when we look over the life cycle of the investment. Also, a single cash flow may inflate or deflate at different rates over time. These two circumstances are handled easily by applying the correct inflation rates to each cash flow over the study period to obtain the actual dollar amounts occurring in each year. After the actual dollar quantities have been calculated, the analysis proceeds using the market interest rate to calculate the measure of merit of interest.

Historical price change for single commodities and bundles of commodities are tracked with price indexes. The Consumer Price Index (CPI) is an example of a composite index formed by a bundle of consumer goods. The CPI serves as a surrogate for general inflation in our economy. Indexes can be used to calculate the *average annual increase* (or decrease) of the costs and benefits in our analysis. The historical data provide valuable information about how economic quantities may behave in the future over the long run.

The effect of inflation on the computed rate of return for an investment depends on how future benefits respond to the inflation. Usually the costs and benefits increase at the same rate as inflation, so the before-tax rate of return will not be adversely affected by the inflation. This outcome is not found when an after-tax analysis is made because the allowable depreciation schedule does not increase. The result will be increased taxable income and income tax payments, which reduce the available after-tax benefits and, therefore, the after-tax rate of return. The important conclusion is that estimates of future inflation or deflation may be important in evaluating capital expenditure proposals.

PROBLEMS

Meaning and Effect

14-1 Define inflation in terms of the purchasing power of dollars.

14-2 Define and describe the relationships between the following: actual and real dollars, inflation, and real and market (combined) interest rates.

14-3 How does inflation happen? Describe a few circumstances that cause prices in an economy to increase.

14-4 Is it necessary to account for inflation in an engineering economy study? What are the two approaches for handling inflation in such analyses?

14-5 In Chapters 5 (Present Worth Analysis) and 6 (Annual Cash Flow Analysis) it is assumed that prices are stable and a machine purchased today for $5000 can be replaced for the same amount many years hence. In fact, prices have generally been rising, so the stable price assumption tends to be incorrect. Under what circumstances is it correct to use the "stable price" assumption when prices actually are changing?

14-6 An economist has predicted 7% inflation during the next 10 years. How much will an item that presently sells for $10 bring a decade hence? (*Answer:* $19.67)

14-7 A man bought a 5% tax-free municipal bond. It cost $1000 and will pay $50 interest each year for 20 years. At maturity the bond returns the original $1000. If there is 2% annual inflation, what real rate of return will the investor receive?

14-8 A man wishes to set aside some money for his daughter's college education. His goal is to have a bank savings account containing an amount equivalent to $20,000 in today's dollars at the girl's 18th birthday. The estimated inflation rate is 8%. If the bank pays 5% compounded annually, what lump sum should he deposit on the child's 4th birthday? (*Answer:* $29,670)

14-9 A newspaper reports that in the last 5 years, prices have increased a total of 50%. This is equivalent to what annual inflation rate, compounded annually? (*Answer:* 8.45%)

14-10 An economist has predicted that for the next 5 years, the U.S. will have an 8% annual inflation rate, followed by 5 years at a 6% inflation rate. This is equivalent to what average price change per year for the entire 10-year period?

14-11 An investor wants a real rate of return i' of 10% per year. If the expected annual inflation rate for the next several years is 6%, what interest rate i should be used in project analysis calculations?

14-12 A South American country has had a high rate of inflation. Recently, its exchange rate was 15 cruzados per dollar; that is, one dollar will buy 15 cruzados. It is likely that the country will continue to experience a 25% inflation rate and that the U.S. will continue at a 7% inflation rate. Assume that the exchange rate will vary the same as the inflation. In this situation, one dollar will buy how many cruzados 5 years from now? (*Answer:* 32.6)

14-13 An automaker has a car that gets 10 kilometers per liter of gasoline. Gas prices will increase 12% per year, compounded annually, for the next 8 years. The manufacturer believes that the fuel consumption for its new cars should decline as fuel prices increase to keep the fuel costs constant. To achieve this, what must be the fuel rating, in kilometers per liter, of the cars 8 years hence?

14-14 An economist has predicted that during the next 6 years, prices in the U.S. will increase 55%. He expects a further increase of 25% in the subsequent 4 years, so that prices at the end of 10 years will have increased to 180% of the present level. Compute the inflation rate, f, for the entire 10-year period.

14-15 Sally Johnson loaned a friend $10,000 at 15% interest, compounded annually. The loan will be paid in five equal end-of-year payments. Sally expects the inflation rate to be 12%. After taking inflation into account, what rate of return is Sally receiving on the loan? Compute your answer to the nearest 0.1%. (*Answer:* 2.7%)

14-16 You may pay $15,000 for an annuity that pays $2500 per year for the next 10 years. You want a real rate of return of 5%, and you estimate inflation will average 6% per year. Should you buy the annuity?

14-17 Inflation is a reality for the general economy of the U.S. for the foreseeable future. Given this assumption, calculate the number of years it will take for the purchasing power of today's dollars to equal *one-fifth* of their present value. Assume that inflation will average 6% per year.

14-18 A homebuilder's advertising has the caption, "Inflation to Continue for Many Years." The ad explains that if one buys a home now for $97,000, and inflation continues at 7%, the home will be worth $268,000 in 15 years. Thus, by buying a new home now, one can realize a profit of $171,000 in 15 years. Do you find this logic persuasive? Explain.

14-19 You were recently looking at the historical prices paid for homes in a neighborhood that interests you. Calculate on a year-to-year basis how home prices in this neighborhood have inflated (*a–e* in the table).

Year	Average Home Price	Inflation Rate for That Year
5 years ago	$165,000	(*a*)
4 years ago	167,000	(*b*)
3 years ago	172,000	(*c*)
2 years ago	180,000	(*d*)
Last year	183,000	(*e*)
This year	190,000	(*f*, see below)

(*f*) What is your estimate of the inflation rate for this year?

14-20 The average cost of a Model T Ford was $18,000 ten years ago. This year the average cost is $30,000.

(*a*) Calculate the average monthly inflation rate (f_m) for Model T cars?

(*b*) Given the monthly rate f_m, what is the effective annual rate, f, of inflation for Model T cars?

(*c*) Estimate what Model T cars will sell for 10 years from now, expressed in today's dollars.

Contributed by D. P. Loucks, Cornell University

14-21 Dale saw that the campus bookstore is having a special on pads of computation paper normally priced at $3 a pad, now on sale for $2.50 a pad. This sale is unusual and Dale assumes the paper will not be put on sale again. On the other hand, Dale expects that there will be no increase in the $3 regular price, even though the inflation rate is 2% every 3 months. Dale believes that competition in the paper industry will keep wholesale and retail prices constant. He uses a pad of computation paper every 3 months. Dale considers 19.25% a suitable minimum attractive rate of return. Dale will buy one pad of paper for his immediate needs. How many extra pads of computation paper should he buy?   (*Answer:* 4)

Before-Tax Cases

14-22 Calculate the future equivalent in Year 15 of:

Use a market interest rate of 15% and an inflation rate of 8%.

(a) Dollars having today's purchasing power.

(b) Then-current purchasing power dollars, of $10,000 today.

14-23 The City of Columbia is trying to attract a new manufacturing business. It has offered to install and operate a water pumping plant to provide service to the proposed plant site. This would cost $50,000 now, plus $5000 per year in operating costs for the next 10 years, all measured in Year-0 dollars.

To reimburse the city, the new business must pay a fixed uniform annual fee, A, at the end of each year for 10 years. In addition, it is to pay the city $50,000 at the end of 10 years. It has been agreed that the city should receive a 3% rate of return, after taking an inflation rate, f, of 7% into account.

Determine the amount of the uniform annual fee. (*Answer:* $12,100)

14-24 A firm is having a large piece of equipment overhauled. It expects that the machine will be needed for the next 12 years. The firm has an 8% minimum attractive rate of return. The contractor has suggested three alternatives:

A. A complete overhaul for $6000 that should permit 12 years of operation.

B. A major overhaul for $4500 that can be expected to provide 8 years of service. At the end of 8 years, a minor overhaul would be needed.

C. A minor overhaul now. At the end of 4 and 8 years, additional minor overhauls would be needed.

If minor overhauls cost $2500, which alternative should the firm select? If minor overhauls, which now cost $2500, increase in cost at +5% per year, but other costs remain unchanged, which alternative should the firm select? (*Answers:* Alt. C; Alt. A)

14-25 A group of students decided to lease and run a gasoline service station. The lease is for 10 years. Almost immediately the students were confronted with the need to alter the gasoline pumps to read in liters. The Dayton Company has a conversion kit available for $900 that may be expected to last 10 years. The firm also sells a $500 conversion kit that has a 5-year useful life. The students believe that any money not invested

in the conversion kits may be invested elsewhere at a 10% interest rate. Income tax consequences are to be ignored in this problem.

(a) Assuming that future replacement kits cost the same as today, which alternative should be selected?

(b) If one assumes a 7% inflation rate, which alternative should be selected?

14-26 Pollution control equipment must be purchased to remove the suspended organic material from liquid being discharged from a vegetable packing plant. Two alternative pieces of equipment are available that would accomplish the task. A Filterco unit costs $7000 and has a 5-year useful life. A Duro unit, on the other hand, costs $10,000 but will have a 10-year useful life.

With inflation, equipment costs are rising at 8% per year, compounded annually, so when the Filterco unit needed to be replaced, the cost would be much more than $7000. Based on a 10-year analysis period, and a 20% minimum attractive rate of return, which pollution control equipment should be purchased?

14-27 Sam bought a house for $150,000 with some creative financing. The bank, which agreed to lend Sam $120,000 for 6 years at 15% interest, took a first mortgage on the house. The Joneses, who sold Sam the house, agreed to lend Sam the remaining $30,000 for 6 years at 12% interest. They received a second mortgage on the house. Thus Sam became the owner without putting up any cash. Sam pays $1500 a month on the first mortgage and $300 a month on the second mortgage. In both cases these are "interest-only" loans, and the principal is due at the end of the loan.

Sam rented the house to Justin and Shannon, but after paying the taxes, insurance, and so on, he had only $800 left, so he was forced to put up $1000 a month to make the monthly mortgage payments. At the end of 3 years, Sam sold the house for $205,000. After paying off the two loans and the real estate broker, he had $40,365 left. After taking an 8% inflation rate into account, what was his before-tax rate of return?

14-28 Ima Luckygirl recently found out that her grandfather has passed away and left her his Rocky Mountain Gold savings account. The only deposit was 50 years ago when Ima's grandfather deposited $2500. If the account has earned an average rate of 10% and inflation has been 4%, answer the following:

(a) How much money is now in the account in *actual dollars*?

(b) Express the answer to part (a) in terms of the purchasing power of dollars from 50 years ago.

14-29 Auntie Frannie wants to provide tuition for her twin nephews to attend a private school. She intends to send a check for $2000 at the end of each of the next 8 years.

(a) If general price inflation, as well as tuition price inflation, is expected to average 5% per year for those 8 years, calculate the present worth of the gifts. Assume that the real interest rate will be 3% per year.

(b) If Auntie Frannie wants her gifts to keep pace with inflation, what would be the present worth of her gifts? Again assume inflation is 5% and the real interest rate is 3%.

14-30 As a recent graduate, you are considering employment offers from three different companies. However, in an effort to confuse you and perhaps make their offers seem better, each company has used a different *purchasing power base* for expressing your annual salary over the next 5 years. If you expect inflation to be 6% for the next 5 years and your personal (real) MARR is 8%, which plan would you choose?

Company A: A constant $50,000 per year in terms of today's purchasing power.

Company B: $45,000 the first year, with increases of $2500 per year thereafter.

Company C: A constant $65,000 per year in terms of Year-5-based purchasing power.

Indexes

14-31 What is the Consumer Price Index (CPI)? What is the difference between commodity-specific and composite price indexes? Can each be used in engineering economic analysis?

14-32 A composite price index for the cost of vegetarian foods called *eggs, artichokes, and tofu* (EAT) was 330 ten years ago and has averaged an annual increase of 12% since. Calculate the current value of the index.

14-33 From the data in Table 14-1 in the text, calculate the *overall rate change* of first-class postage as measured by the LCI for the following decades:

(a) The 1970s (1970–1979)

(b) The 1980s (1980–1989)

(c) The 1990s (1990–1999)

14-34 From the data in Table 14-1 in the text, calculate the *average annual inflation rate* of first class postage as measured by the LCI for the following years:

(a) End of 1970 to end of 1979

(b) End of 1980 to end of 1989

(c) End of 1990 to end of 1999

14-35 From the data in Table 14-2 in the text, calculate the *average annual inflation rate* as measured by the CPI for the following years:

(a) End of 1973 to end of 1982

(b) End of 1980 to end of 1989

(c) End of 1985 to end of 2002

14-36 (a) Compute the equivalent annual inflation rate, based on the consumer price index, for the period from 1981 to 1986.

(b) Using the equivalent annual inflation rate computed in part (a), estimate the consumer price index in 1996, working from the 1987 Consumer Price Index.

14-37 Redo Problem 14-17, but estimate the annual inflation rate using the period from 1996 to 2006 and the CPI index values in Table 14-2.

14-38 Here is some information about a professor salary index (PSI).

Year	PSI	Change in PSI
1991	82	3.22%
1992	89	8.50
1993	100	*a*
1994	*b*	4.00
1995	107	*c*
1996	116	*d*
1997	*e*	5.17
1998	132	7.58

(a) Calculate the unknown quantities *a, b, c, d, e* in the table. Review Equation 14-3.

(b) What is the *base year* of the PSI? How did you determine it?

(c) Given the data for the PSI, calculate the *average annual price increase* in salaries paid to professors for between 1991 and 1995 and between 1992 and 1998.

14-39 Homeowner Henry is building a fireplace for his house. The fireplace will require 800 bricks.

(a) If the cost of a chimney brick in 1978 was $2.10, calculate the material cost of Henry's project in 1998. The chimney brick index (CBI) was 442 in 1978 and is expected to be 618 in 1998.

(b) Estimate the material cost of a similar fireplace to be built in the year 2008. What assumption did you make?

Different Rates

14-40 General price inflation is estimated to be 3% for the next 5 years, 5% the 5 years after that, and 8% the following 5 years. If you invest $10,000 at 10% for those 15 years, what is the future worth of your investment in actual dollars at that time and in Year-0-dollars at that time?

14-41 Due to cost structures, trade policies, and corporate changes three big automakers have different inflation rates for the purchase prices of their vehicles. Which car should Mary Clare buy at graduation 3 years from now, assuming everything but purchase price is equivalent?

Automaker	Current Price	Price Will Inflate $x\%$ per Year
X	$27,500	4 %
Y	30,000	1.5
Z	25,000	8

14-42 Granny Viola has been saving money in the Bread & Butter mutual fund for 15 years. She has been a steady depositor over those years and has a pattern of putting $100 into the account every 3 months. If her original investment 15 years ago was $500 and interest in the account has varied as shown, what is the current value of her savings?

Years	Interest Earned in the Account
1–5	12% compounded quarterly
6–10	16 compounded quarterly
10–15	8 compounded quarterly

14-43 Andrew just bought a new boat for $15,000 to use on the river near his home. He has taken delivery of the boat and agreed to the terms of the following loan: all principal and interest is due in 3 years (balloon loan), first year annual interest (on the purchase price) is set at 5%, this is to be adjusted up 1.5% per year for each of the following years of the loan.

(a) How much does Andrew owe to pay off the loan in 3 years?

(b) If inflation is 4%, what is the payment in Year-0 dollars?

14-44 Given the following data, calculate the present worth of the investment.

First cost $= \$60,000$ Project life $= 10$ years
Salvage value $= \$15,000$ MARR $= 25\%$

General price inflation $= 4\%$ per year
Annual cost 1 $= \$4500$ in Year 1 and inflating at 2.5% per year
Annual cost 2 $= \$7000$ in Year 1 and inflating at 10.0% per year
Annual cost 3 $= \$10,000$ in Year 1 and inflating at 6.5% per year
Annual cost 4 $= \$8500$ in Year 1 and inflating at -2.5% per year

14-45 As the owner of Beanie Bob's Basement Brewery, you are interested in a construction project to increase production to offset competition from your crosstown rival, Bad Brad's Brewery and Poolhall. Construction cost percentage increases, as well as current cost estimates are given for a 3-year period. Use a market interest rate of 25%, and assume that general price inflation is 5% over the 3-year period.

Item	Cost if Incurred Today	Cost Percentage Increase		
		Year 1	Year 2	Year 3
Structural metal/concrete	$120,000	4.3%	3.2%	6.6%
Roofing materials	14,000	2.0	2.5	3.0
Heating/plumbing equipment/fixtures	35,000	1.6	2.1	3.6
Insulation material	9,000	5.8	6.0	7.5
Labor	85,000	5.0	4.5	4.5

(a) What would the costs be for labor in Years 1, 2, and 3?

(b) What is the *average percentage increase* of labor cost over the 3-year period?

(c) What is the present worth of the insulation cost of this project?

(d) Calculate the future worth of the labor and insulation material cost portion of the project.

(e) Calculate the present worth of the total construction project for Beanie Bob.

14-46 Philippe Marie wants to race in the Tour de France ten years from now. He wants to know what the cost of a custom-built racing bicycle will be ten years from today.

Item	Current Cost	Cost Will Inflate $x\%$ per Year
Frame	$800	2 %
Wheels	350	10
Gearing system	200	5
Braking system	150	3
Saddle	70	2.5
Finishes	125	8

After-Tax Cases

14-47 Sally Seashell bought a lot at the Salty Sea for $18,000 cash. She does not plan to build on the lot, but instead will hold it as an investment for 10 years. She wants a 10% after-tax rate of return after taking the 6% annual inflation rate into account. If income taxes amount to 15% of the capital gain, at what price must she sell the lot at the end of the 10 years? (*Answer:* $95,188)

14-48 The U.S. tax laws provide for the depreciation of equipment based on original cost. Yet owing to substantial inflation, the replacement cost of equipment is often much greater than the original cost. What effect, if any, does this have on a firm's ability to buy new equipment to replace old equipment?

14-49 Tom Ward put $10,000 in a 5-year certificate of deposit that pays 12% interest per year. At maturity he will receive his $10,000 back. Tom's marginal income tax rate is 42%. If the inflation rate is 7% per year, find his

(a) before-tax rate of return, ignoring inflation

(b) after-tax rate of return, ignoring inflation

(c) after-tax rate of return, after taking inflation into account

14-50 Dick DeWolf and his wife have a total taxable income of $60,000 this year and file a joint federal income tax return. If inflation continues for the next 20 years at a 7% annual rate, Dick wonders what their taxable income must be to provide the same purchasing power after taxes. Assuming the federal income tax rate table is unchanged, what must their taxable income be 20 years from now?

14-51 Assume that your private university's tuition is $28,000.
Contributed by D. P. Loucks, Cornell University

(a) If the inflation rate for tuition is 5% per year, calculate what the tuition will cost 20 years from now.

(b) If the general inflation rate for the economy is 3% per year, express that future tuition in today's dollars.

(c) Calculate the amount you would have to invest today to pay for tuition costs 20, 21, 22, and 23 years from now. Assume you can invest at 7% per year, your income tax rate is 40% per year, and the tuition has to be paid at the beginning of the year.

14-52 You must decide when to go on a vacation. One option is right after graduation. The other option is to wait and go after spending 2 years with the Peace Corps. Assume the vacation costs $2500 now, and your annual income tax rate is 20% now, and is expected to continue to be 20% during the next 2 years. Also assume the annual inflation rate for a week's trip to Hawaii (hotel included) is 4%.

(a) Calculate the additional money you could spend on your vacation, after taxes, by putting your vacation money ($2500) into a taxable investment at 6% per year (before taxes) and waiting 2 years until after you come out of the Peace Corps compared to taking your well-deserved vacation now.

(b) If your tax rate drops to 0% while you're in the Peace Corps, how much additional money will you have for your vacation?

Contributed by D. P. Loucks, Cornell University

14-53 Sam Johnson inherited $85,000 from his father. Sam is considering investing the money in a house, which he will then rent to tenants. The $85,000 cost of the property consists of $17,500 for the land, and $67,500 for the house. Sam believes he can rent the house and have $8000 a year net income left after paying the property taxes and other expenses. The house will be depreciated by straight-line depreciation using a 45-year depreciable life.

(a) If the property is sold at the end of 5 years for its book value at that time, what after-tax rate of return will Sam receive? Assume that his marginal personal income tax rate is 34% for federal and state taxes.

(b) Now assume there is 7% per year inflation, compounded annually. Sam will increase the rent 7% per year to match the inflation rate, so that after considering increased taxes and other expenses, the annual net income will go up 7% per year. Assume Sam's marginal income tax rate remains at 34% for all ordinary taxable income related to the property. The value of the property is now projected to increase from its present $85,000 at a rate of 10% per year, compounded annually.

If the property is sold after 5 years, compute the rate of return on the after-tax cash flow in actual dollars. Also compute the rate of return on the after-tax cash flow in Year-0 dollars.

14-54 A small research device is purchased for $10,000 and depreciated by MACRS depreciation. The net benefits from the device, before deducting depreciation, are $2000 at the end of the first year and increasing $1000 per year after that (second year equals $3000, third year equals $4000, etc.), until the device is hauled to the junkyard at the end of 7 years. During the 7-year period there is an inflation rate f of 7%.

This profitable corporation has a 40% combined federal and state income tax rate. If it requires a real 12% after-tax rate of return on its investment, should the device have been purchased?

14-55 A couple in Ruston, Louisiana, must decide whether it is more economical to buy a home or to continue to rent during an inflationary period. The couple rents a one-bedroom duplex for $450 a month plus $139 a month in basic utilities (heating and cooling). These costs have a projected inflation rate of 5%, so the couple's monthly costs per year over a 10-year planning horizon are:

$n =$	1	2	3	4	5	6	7	8	9	10
Rent	450	473	496	521	547	574	603	633	665	698
Utilities	139	146	153	161	169	177	186	196	205	216

The couple would like to buy a home that costs $75,000. A local mortgage company will provide a loan that requires a down payment of 5% plus estimated closing costs of 1% cash. The couple prefers a 30-year fixed-rate mortgage with an 8% interest rate. The couple falls in the 30% marginal income tax rate (federal plus state), and as such, buying a home will provide them some tax write-off. It is estimated that the basic utilities for the home inflating at 5% will cost $160 per month; insurance and maintenance also inflating at 5% will cost $50 per month. The home will appreciate in value about 6% per year. Assuming a nominal interest rate of 15.5%, which alternative should the couple select? Use a present worth analysis. (*Note:* Realtor's sales commission here is 5%.)

14-56 When there is little or no inflation, a homeowner can expect to rent an unfurnished home for 12% of its market value. About $^1/_8$ of the rental income is paid out for property taxes, insurance, and other operating

expenses. Thus the net annual income to the owner is 10.5% of the market value. Since prices are relatively stable, the future selling price of the property often equals the original price paid by the owner.

For a $150,000 property (where the land is estimated at $46,500 of the $150,000), compute the after-tax rate of return, assuming the selling price 59 months later (in December) equals the original purchase price. Use modified accelerated cost recovery system depreciation beginning January 1. Also, assume a 35% income tax rate. (*Answer:* 6.84%)

14-57 (This is a continuation of Problem 14-56.) As inflation has increased throughout the world, the rental income of homes has decreased and a net annual rental income of 8% of the market value is common. On the other hand, the market value of homes tends to rise about 2% per year more than the inflation rate. As a result, both annual net rental income, and the resale value of the property rise faster than the inflation rate. Consider the following situation.

A $150,000 property (with the house valued at $103,500 and the land at $46,500) is purchased for cash in Year 0. The market value of the property increases at a 12% annual rate. The annual rental income is 8% of the beginning-of-year market value of the property. Thus the rental income also increases each year. The general inflation rate f is 10%.

The individual who purchased the property has an average income tax rate of 35%.

(*a*) Use MACRS depreciation, beginning January 1, to compute the actual dollar after-tax rate of return for the owner, assuming he sells the property 59 months later (in December).

(*b*) Similarly, compute the after-tax rate of return for the owner, after taking the general inflation rate into account, assuming he sells the property 59 months later.

14-58 Consider two mutually exclusive alternatives stated in Year-0 dollars. Both alternatives have a 3-year life with no salvage value. Assume the annual inflation rate is 5%, an income tax rate of 25%, and straight-line depreciation. The minimum attractive rate of return (MARR) is 7%. Use rate of return analysis to determine which alternative is preferable.

Year	A	B
0	−$420	−$300
1	+200	+150
2	+200	+150
3	+200	+150

SELECTION OF A MINIMUM ATTRACTIVE RATE OF RETURN

BP Goes to Russia

In the early 1990s, western investors flocked to Russia, hoping to reap a fortune as markets opened up following the fall of the Soviet Union. Many didn't stay long because contracts could be impossible to enforce and bribery was common. In 1998, when Russia devalued its currency and defaulted on debt obligations, most of the remaining investors fled in panic.

Despite this business outlook, British Petroleum (BP) announced in early 2003 that it was planning to pay $6.75 billion for a 50% interest in Tyumen Oil Company, Russia's fourth largest producer of oil. BP's decision is particularly striking in view of the company's own history in Russia: in 1997 it bought a share in a small Russian oil company, only to lose part of its investment a few years later after a bitter court battle. Moreover, Russia's state-owned pipeline infrastructure is outdated and inadequate—and the government has been slow to allow private companies to build and operate their own pipelines.

Given all these drawbacks, BP would seem to be taking a big gamble by investing in Tyumen.

QUESTIONS TO CONSIDER

1. Many of BP's older sources of oil are now reaching the end of their useful life. How might this have affected the firm's decision to invest in Tyumen Oil?
2. Outside of Tyumen, there are few other oil companies left in Russia for competitors to buy. How might this have affected BP's decision?
3. Where else are the world's largest available oil reserves located? How does the business climate in these areas compare to Russia's?
4. Most of the largest oil-producing nations are members of OPEC (the Organization of Petroleum Exporting Countries), but Russia is not. Is this good or bad for BP?
5. What are some ethical issues that arise when one is investing in foreign firms and countries?

After Completing This Chapter...

The student should be able to:

- Define various sources of capital and the costs of those funds to the firm.
- Discuss the impact of inflation and the cost of borrowed money.
- Select a firm's MARR based on the opportunity cost approach for analyzing investments.
- Adjust the firm's MARR to account for risk and uncertainty.
- Use spreadsheets to develop cumulative investments and the opportunity cost of capital.

The preceding chapters have said very little about what interest rate or minimum attractive rate of return is suitable for use in a particular situation. This problem is complex, and no single answer is always appropriate. A discussion of what interest rate to use must inevitably begin by examining the sources of capital, followed by looking at the prospective investment opportunities and risk. Only in this way can an interest rate or minimum attractive rate of return (MARR) be chosen intelligently.

SOURCES OF CAPITAL

In broad terms there are three sources of capital available to a firm: money generated from the firm's operation, borrowed money, and money from selling stock.

Money Generated from the Firm's Operations

A major source of capital investment money is retained profits from the firm's operation. Overall, industrial firms retain about half of their profits and pay out the rest to stockholders. In addition to profit, the firm generates money equal to the annual depreciation charges on existing capital assets. In other words, a profitable firm will generate money equal to its depreciation charges *plus* its retained profits. Even a firm that earns zero profit will still generate money from operations equal to its depreciation charges. (A firm with a loss, of course, will have still less funds.)

External Sources of Money

When a firm requires money for a few weeks or months, it typically borrows from banks. Longer-term unsecured loans (of, say, 1–4 years) may also be arranged through banks. While banks undoubtedly finance a lot of capital expenditures, regular bank loans cannot be considered a source of permanent financing.

Longer-term financing is done by selling bonds to banks, insurance firms, pension funds, and the public. A wide variety of bonds exist, but most are interest-only loans, where interest is paid every 6 months or once a year and the principal is due at the bond's maturity. Common maturities are 10 to 30 years, although some extend to 100 years and a few even longer. Often interest rates are explicitly stated, but not always; Chapter 7 includes examples of how to calculate the interest rates when these are not given.

A firm can also raise funds by issuing new stock (shares of ownership in the firm). Many firms have also repurchased their own stock in the past, which is called *treasury stock*. Another way firms can raise funds is to sell this treasury stock.

One of the finance questions each firm must address is maintaining an appropriate balance between debt (loans and bonds) and equity (stock and retained earnings). The debt has a maturity date, and there are legal obligations to repay it unless the firm declares bankruptcy. On the other hand, stockholders expect a higher rate of return to compensate them for the risks of ownership. Those who are interested in the models used to calculate the cost of equity capital are referred to *The Economic Analysis of Industrial Projects*, 3rd edition, by Eschenbach, Hartman, and Bussey, published by Oxford University Press.

Choice of Source of Funds

Choosing the source of funds for capital expenditures is a decision for the firm's top executives, and it may require approval of the board of directors. When internal operations generate adequate funds for the desired capital expenditures, external sources of money are not likely to be used. But when the internal sources are inadequate, external sources must be employed or the capital expenditures will have to be deferred or canceled.

COST OF FUNDS

Cost of Borrowed Money

A first step in deciding on a minimum attractive rate of return might be to determine the interest rate at which money can be borrowed. Longer-term loans or bonds may be obtained from banks, insurance companies, or the variety of places in which substantial amounts of money accumulates (for example, the oil-producing nations).

A large, profitable corporation might be able to borrow money at the **prime rate,** that is, the interest rate that banks charge their best and most sought-after customers. All other firms are charged an interest rate that is higher by one-half to several percentage points. In addition to the firm's financial strength and ability to repay the debt, the interest rate will depend on the debt's duration and on whether the debt has collateral or is unsecured.

Cost of Capital

Another relevant interest rate is the **cost of capital.** This is also called the weighted average cost of capital (WACC). The general assumption concerning the cost of capital is that all the money the firm uses for investments is drawn from all of the firm's overall capitalization. The mechanics for computing the cost of capital or WACC are given in Example 15-1.

EXAMPLE 15-1

For a particular firm, the purchasers of common stock require an 11% rate of return, bonds are sold at a 7% interest rate, and bank loans are available at 9%. Compute the cost of capital or WACC for the following capital structure:

		Rate of Return	Annual Amount
$ 20 million	Bank loan	9%	$1.8 million
20	Bonds	7	1.4
60	Common stock and retained earnings	11	6.6
$100 million			$9.8 million

SOLUTION

Interest payments on debt, like bank loans and bonds, are tax-deductible business expenses. Thus:

$$\text{After-tax interest cost} = (\text{Before-tax interest cost}) \times (1 - \text{Tax rate})$$

If we assume that the firm pays 40% income taxes, the computations become:

Bank loan　After-tax interest cost $= 9\%(1 - 0.40) = 5.4\%$
Bonds　After-tax interest cost $= 7\%(1 - 0.40) = 4.2\%$

Dividends paid on the ownership in the firm (common stock + retained earnings) are not tax deductible. Combining the three components, the after-tax interest cost for the $100 million of capital is:

$$\$20 \text{ million } (5.4\%) + \$20 \text{ million } (4.2\%) + \$60 \text{ million } (11\%) = \$8.52 \text{ million}$$

$$\text{Cost of capital} = \frac{\$8.52 \text{ million}}{\$100 \text{ million}} = 8.52\%$$

In practical situations, the cost of capital is often difficult to compute. The fluctuation in the price of common stock, for example, makes it difficult to pick a cost, and because of the fluctuating prospects of the firm, it is even more difficult to estimate the future benefits that purchasers of the stock might expect to receive. Given the fluctuating costs and prospects of future benefits, what rate of return do stockholders require? There is no precise answer, but we can obtain an approximate answer. As described in the next section, inflation complicates the task of finding the *real* interest rate.

Inflation and the Cost of Borrowed Money

As inflation varies, what is its effect on the cost of borrowed money? A widely held view has been that interest rates on long-term borrowing, like 20-year Treasury bonds, will be about 3% more than the inflation rate. For borrowers this is the real—that is, after-inflation—cost of money, and for lenders the real return on loans. If inflation rates were to increase, it would follow that borrowing rates would also increase. All this suggests a rational and orderly situation, about as we might expect.

Unfortunately, things have not worked out this way. Figure 15-1 shows that the real interest rate has not always been about 3% and, in fact, there have been long periods during which the real interest rate was negative. Can this be possible? Would anyone invest money at an interest rate several percentage points below the inflation rate? Well, consider this: when the U.S. inflation rate was 12%, savings banks were paying 5½% on regular passbook

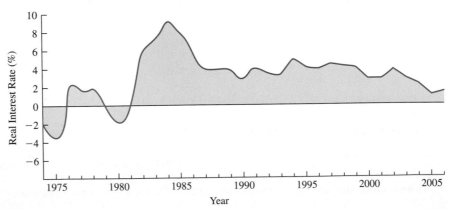

FIGURE 15-1 The real interest rate. The interest rate on 20-year Treasury bonds *minus* the inflation rate, f, as measured by changes in the Consumer Price index.

deposits. And there was a lot of money in those accounts. While there must be a relationship between interest rates and inflation, Figure 15-1 suggests that it is complex.

The relationship between the inflation rate and the rate of return on investments is quite important, because inflation reduces the real rate of return (as shown in Chapter 14). In addition, many interest rates are reported without adjusting for inflation. For example, you know the interest rate on your car loan, student loan, house loan, savings account, and so on—but all those rates are stated without adjusting for inflation.

INVESTMENT OPPORTUNITIES

An industrial firm can invest its money in many more places than are available to an individual. A firm has larger amounts of money, and this alone makes certain kinds of investment possible that are unavailable to individual investors, with their more limited investment funds. The U.S. government, for example, borrows money for short terms of 90 or 180 days by issuing certificates called Treasury bills that frequently yield a greater interest rate than savings accounts. The customary minimum purchase is $25,000.

More important, however, is the fact that a firm conducts a business, which itself offers many investment opportunities. While exceptions can be found, a good generalization is that the opportunities for investment of money within the firm are superior to the investment opportunities outside the firm. Consider the available investment opportunities for a particular firm as outlined in Table 15-1. The cumulative investment required for all projects at or above a given rate of return is given in Figure 15-2.

Figure 15-2 illustrates that a firm may have a broad range of investment opportunities available at varying rates of return and with varying lives and uncertainties. It may take some study and searching to identify the better investment projects available to a firm. Typically, the good projects will almost certainly require more money than the firm budgets for capital investment projects.

Opportunity Cost

We see that there are two aspects of investing that are basically independent. One factor is the source and quantity of money available for capital investment projects. The other aspect is the firm's investment opportunities.

These two situations are typically out of balance, with investment opportunities exceeding the available money supply. Thus some investment opportunities can be selected and others must be rejected. Obviously, we want to ensure that *all the selected projects are better than the best rejected project.* To do this, we must know something about the rate of return on the best rejected project. The best rejected project is the best opportunity forgone, and this in turn is called the **opportunity cost.**

> **Opportunity cost = Cost of the best opportunity forgone**
> **= Rate of return on the best rejected project**

If one could predict the opportunity cost for some future period (like the next 12 months), this rate of return could be one way to judge whether to accept or reject any proposed capital expenditure. Examples 15-2 and 15-3 illustrate this.

TABLE 15-1 A Firm's Available Investment Opportunities

Project Number	Project	Cost ($\times 10^3$)	Estimated Rate of Return
	Investment Related to Current Operations		
1	New equipment to reduce labor costs	$150	30%
2	Other new equipment to reduce labor costs	50	45
3	Overhaul particular machine to reduce material costs	50	38
4	New test equipment to reduce defective products produced	100	40
	New Operations		
5	Manufacture parts that previously had been purchased	200	35
6	Further processing of products previously sold in semifinished form	100	28
7	Further processing of other products	200	18
	New Production Facilities		
8	Relocate production to new plant	250	25
	External Investments		
9	Investment in a different industry	300	20
10	Other investment in a different industry	300	10
11	Overseas investment	400	15
12	Purchase of Treasury bills	Unlimited	8

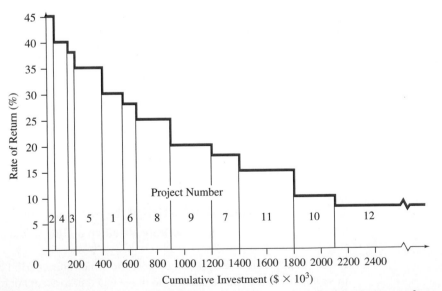

FIGURE 15-2 Cumulative investment required for all projects at or above a given rate of return.

EXAMPLE 15-2

Consider the situation represented by Table 15-1 and Figure 15-2. For a capital expenditure budget of $1.2 million ($1.2 \times 10^6$), what is the opportunity cost?

SOLUTION

From Figure 15-2 we see that the eight projects with a rate of return of 20% or more require a cumulative investment of $1.2 ($\times 10^6$). We would take on these projects and reject the other four (7, 11, 10, and 12) with rates of return of 18% or less. The best rejected project is 7, and it has an 18% rate of return. Thus the opportunity cost is 18%.

EXAMPLE 15-3

Nine independent projects are being considered. Figure 15-3 may be prepared from the following data.

Project	Cost (thousands)	Uniform Annual Benefit (thousands)	Useful Life (years)	Salvage Value (thousands)	Computed Rate of Return
1	$100	$23.85	10	$0	20%
2	200	39.85	10	0	15
3	50	34.72	2	0	25
4	100	20.00	6	100	20
5	100	20.00	10	100	20
6	100	18.00	10	100	18
7	300	94.64	4	0	10
8	300	47.40	10	100	12
9	50	7.00	10	50	14

If a capital budget of $650,000 is available, what is the opportunity cost of capital? With this model, which projects should be selected?

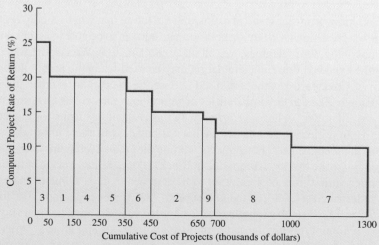

FIGURE 15-3 Cumulative cost of projects versus rate of return.

SOLUTION

Looking at the nine projects, we see that some are expected to produce a larger rate of return than others. It is natural that if we are to select from among them, we will pick those with a higher rate of return. When the projects are arrayed by rate of return, as in Figure 15-3, Project 2 is the last one funded. Thus, the opportunity cost of capital is 14% from Project 9, the highest ranked unfunded project. This model implies funding Projects 3, 1, 4, 5, 6, and 2.

SELECTING A MINIMUM ATTRACTIVE RATE OF RETURN

Focusing on the three concepts on the cost of money (the cost of borrowed money, the cost of capital, and opportunity cost), which, if any, of these values should be used as the minimum attractive rate of return (MARR) in economic analyses?

Fundamentally, we know that unless the benefits of a project exceed its cost, we cannot add to the profitability of the firm. A lower boundary for the minimum attractive rate of return must be the cost of the money invested in the project. It would be unwise, for example, to borrow money at 8% and invest it in a project yielding a 6% rate of return.

Further, we know that no firm has an unlimited ability to borrow money. Bankers— and others who evaluate the limits of a firm's ability to borrow money—look at both the profitability of the firm and the relationship between the components in the firm's capital structure. This means that continued borrowing of money will require that additional stock must be sold to maintain an acceptable ratio between **ownership** (equity) and **debt.** In other words, borrowing for a particular investment project is only a block of money from the overall capital structure of the firm. This suggests that the MARR should not be less than the cost of capital. Finally, we know that the MARR should not be less than the rate of return on the best opportunity forgone. Stated simply,

Minimum attractive rate of return should be equal to the largest of: cost of borrowed money, cost of capital, or opportunity cost.

ADJUSTING MARR TO ACCOUNT FOR RISK AND UNCERTAINTY

We know from our study of estimating the future that what actually occurs is often different from the estimate. When we are fortunate enough to be able to assign probabilities to a set of possible future outcomes, we call this a **risk** situation. We saw in Chapter 10 that techniques like expected value and simulation may be used when the probabilities are known.

Uncertainty is the term used to describe the condition when the probabilities are *not* known. Thus, if the probabilities of future outcomes are known, we have *risk*, and if they are unknown, we have *uncertainty*. In this case, adjustments for risk are more subjective.

In projects accompanied by normal business risk and uncertainty, the MARR is used without adjustment. For projects with greater than average risk or uncertainty, most firms increase the MARR. As reported in Block (2005), the percentage of firms using risk-adjusted rates varied from 66% in retail to 82% in health care. Some of the percentages for other industries are 70% for manufacturing, 73% for energy, and 78% for technology firms. Table 15-2 lists an example of risk-adjusted MARRs in manufacturing.

Some firms use the same rates for all divisions and groups. Other firms vary the rates by division for strategic reasons. There are even cases when a project-specific rate based on that project's financing may be justified. For example, a firm or joint venture may be founded to develop a specific mine, pipeline, or other resource development project.

TABLE 15-2 **Example Risk-Adjusted Interest Rates**

Rate (%)	Applies to:
6	Equipment replacement
8	New equipment
10	New product in normal market
12	New product in related market
16	New product in new market
20	New product in foreign market

However, as shown in Example 15-4, risk-adjusted rates may not work well. A preferable way deals explicitly with the probabilities using the techniques from Chapter 10. When the interest rate (MARR) used in economic analysis calculations is raised to adjust for risk or uncertainty, greater emphasis is placed on immediate or short-term results and less emphasis on longer-term results.

EXAMPLE 15-4

Consider the two following alternatives: the MARR of Alt. B has been raised from 10 to 15% to take into account the greater risk and uncertainty that Alt. B's results may not be as favorable as indicated. What is the impact of this change of MARR on the decision?

Year	Alt. A	Alt. B
0	−$80	−$80
1–10	10	13.86
11–20	20	10

SOLUTION

Year	Alt. A	NPW At 14.05%	At 10%	At 15%
0	−$80	−$80.00	−$80.00	−$80.00
1–10	10	52.05	61.45	50.19
11–20	20	27.95	47.38	24.81
		0	+28.83	−5.00

Year	Alt. B	NPW At 15.48%	At 10%	At 15%
0	−$80	−$80.00	−$80.00	−$80.00
1–10	13.86	68.31	85.14	69.56
11–20	10	11.99	23.69	12.41
		0	+28.83	+1.97

Computations at MARR of 10% Ignoring Risk and Uncertainty

Both alternatives have the same positive NPW (+$28.83) at a MARR of 10%. Also, the differences in the benefits schedules $(A - B)$ produce a 10% incremental rate of return. (The calculations are not shown here.) This must be true if NPW for the two alternatives is to remain constant at a MARR of 10%.

Considering Risk and Uncertainty with MARR of 10%

At 10%, both alternatives are equally desirable. Since Alt. B is believed to have greater risk and uncertainty, a logical conclusion is to select Alt. A rather than B.

Increase MARR to 15%

At a MARR of 15%, Alt. A has a negative NPW and Alt. B has a positive NPW. Alternative B is preferred under these circumstances.

Conclusion

Based on a business-risk MARR of 10%, the two alternatives are equivalent. Recognizing some greater risk of failure for Alt. B makes A the preferred alternative. If the MARR is increased to 15%, to add a margin of safety against risk and uncertainty, the computed decision is to select B. Since Alt. B has been shown to be less desirable than A, the decision, based on a MARR of 15%, may be an unfortunate one. The difficulty is that the same risk adjustment (increase the MARR by 5%) is applied to both alternatives even though they have different amounts of risk.

The conclusion to be drawn from Example 15-4 is that increasing the MARR to compensate for risk and uncertainty is only an approximate technique and may not always achieve the desired result. Nevertheless, it is common practice in industry to adjust the MARR upward to compensate for increased risk and uncertainty.

REPRESENTATIVE VALUES OF MARR USED IN INDUSTRY

We argued that the minimum attractive rate of return should be established at the highest one of the following: cost of borrowed money, cost of capital, or the opportunity cost.

The cost of borrowed money will vary from enterprise to enterprise, with the lowest rate being the prime interest rate. The prime rate may change several times in a year; it is widely reported in newspapers and business publications. As we pointed out, the interest rate for firms that do not qualify for the prime interest rate may be ½% to several percentage points higher.

The cost of capital of a firm is an elusive value. There is no widely accepted way to compute it; we know that as a *composite value* for the capital structure of the firm, it conventionally is higher than the cost of borrowed money. The cost of capital must consider the market valuation of the shares (common stock, etc.) of the firm, which may fluctuate widely, depending on future earnings prospects of the firm. We cannot generalize on representative costs of capital.

Somewhat related to cost of capital is the computation of the return on total capital (long-term debt, capital stock, and retained earnings) actually achieved by firms. *Fortune* magazine, among others, does an annual analysis of the rate of return on total capital. The after-tax rate of return on total capital for individual firms ranges from 0% to about 40% and averages 8%. *Business Week* does a periodic survey of corporate performance. This magazine reports an after-tax rate of return on common stock and retained earnings. We would expect the values to be higher than the rate of return on total capital, and this is the case. The after-tax return on common stock and retained earnings ranges from 0% to about 65% with an average of 14%.

Higher values for the MARR are used by firms that are short of capital, such as high-technology start-ups. They are also used in industries, such as petroleum and mining, where volatile prices increase the risk of poor returns for projects. Rates of 25 to 30% are relatively common, and even higher rates are sometimes used. For companies with more normal levels of risk, rates of 12 to 15% are more typical.

Note that the values of MARR given earlier are approximations. But the values quoted appear to be opportunity costs, rather than cost of borrowed money or cost of capital. This indicates that firms cannot or do not obtain money to fund projects whose anticipated rates of return are nearer to the cost of borrowed money or cost of capital. One reason that firms operate as they do is that they can focus limited resources of people, management, and time on a smaller number of good projects.

One cannot leave this section without noting that the MARR used by enterprises is much higher than can be obtained by individuals. (Where can you get a 30% after-tax rate of return without excessive risk?) The reason appears to be that businesses are not obliged to compete with the thousands of individuals in any region seeking a place to invest $2000 with safety, whereas the number of people who could or would want to invest $500,000 in a business is far smaller. This diminished competition, combined with a higher risk, appears to explain at least some of the difference.

CAPITAL BUDGETING OR SELECTING THE BEST PROJECTS

The opportunity cost of capital approach of ranking projects by their rate of return introduced a new type of problem. Up to that point we'd been analyzing mutually exclusive alternatives, where only one could be chosen. Engineering design problems are this type of problem, where younger engineers use engineering economy to choose the best alternative design.

At higher levels in the organization, engineering economy is applied to solve a different problem. For example, 30 projects have passed initial screening and are being proposed for funding. Every one of the 30 meets the MARR. The firm can afford to invest in only some of them. So, which ones should be chosen and how? This is called the *capital budgeting* problem.

Examples 15-2 and 15-3 applied the opportunity cost of capital approach to the capital budgeting problem. Firms often use this approach as a starting point to rank the projects from best to worst. In some cases the ranking by rate of return is used to make the decision.

More often, managers then meet and decide which projects will be funded by obtaining a consensus, or a decision by the highest-ranking manager, which will modify the rate of return ranking. At this meeting, business units argue for a larger share of the capital budget, as do plants in the same business, groups at the same plants, and individuals within the groups. Some considerations, such as strategy, necessity, and the availability and capability of particular resources and people are difficult to represent in the project's *numbers*, which are the subject of economic analysis.

Other firms rank projects using a benefit–cost ratio or present worth index. As shown in Example 15-5, the present worth index is the NPW divided by the cost's present value.

Anyone who has ever bought firecrackers probably used the practical ranking criterion of "biggest bang for the buck" in making a selection. This same criterion—stated more elegantly—is used by some firms to rank independent projects.

Rank independent projects according to their value of net present worth divided by the present worth of cost. The appropriate interest rate is MARR (as a reasonable estimate of the cutoff rate of return).

EXAMPLE 15-5

Rank the following nine independent projects in their order of desirability, based on a 14.5% minimum attractive rate of return.

Project	Cost (thousands)	Uniform Annual Benefit (thousands)	Useful Life (years)	Salvage Value (thousands)	Computed Rate of Return	Computed NPW at 14.5% (thousands)	Computed NPW/Cost (thousands)
1	$100	$23.85	10	$0	20%	$22.01	0.2201
2	200	39.85	10	0	15	3.87	0.0194
3	50	34.72	2	0	25	6.81	0.1362
4	100	20.00	6	100	20	21.10	0.2110
5	100	20.00	10	100	20	28.14	0.2814
6	100	18.00	10	100	18	17.91	0.1791
7	300	94.64	4	0	10	−27.05	−0.0902
8	300	47.40	10	100	12	−31.69	−0.1056
9	50	7.00	10	50	14	−1.28	−0.0256

SOLUTION

Ranked NPW/PW of cost, the projects are listed as follows:

Project	NPW/PW of Cost	Rate of Return
5	0.2814	20%
1	0.2201	20
4	0.2110	20
6	0.1791	18
3	0.1362	25
2	0.0194	15
9	−0.0256	14
7	−0.0902	10
8	−0.1056	12

With a 14.5% MARR, Projects 1 to 6 are recommended for funding and 7 to 9 are not. However, they are ranked in a different order by the present worth index and by the rate of return approaches. For example, Project 3 has the highest ranking for the rate of return and is fifth by the present worth index.

Some consider the present worth index to be a better measure, but this can be true only if PW is applied at the correct interest rate. It is more common for firms to simply rank on the rate of return. If independent projects can be ranked in their order of desirability, then the selection of projects to be included in a capital budget is a simple task. One may proceed down the list of ranked projects until the capital budget has been exhausted. The only difficulty with this scheme occurs, occasionally, when the capital budget is more than enough for n projects but too little for $n + 1$ projects.

In Example 15-5, suppose the capital budget is $550,000. This is more than enough for the top five projects (sum = $450,000) but not enough for the top six projects (sum = $650,000). When we have this situation, it may not be possible to say with certainty that the best use of a capital budget of $550,000 is to fund the top five projects. There may be some other set of projects that makes better use of the available $550,000. While some trial-and-error computations may indicate the proper set of projects, more elaborate techniques are needed to prove optimality.

As a practical matter, a capital budget probably has some flexibility. If in Example 15-5 the tentative capital budget is $550,000, then a careful examination of Project 2 will dictate whether to expand the capital budget to $650,000 (to be able to include Project 2) or to drop back to $450,000 (and leave Project 2 out of the capital budget). Or perhaps Project 2 can be started in this budget year and finished next year.

Spreadsheets, Cumulative Investments, and the Opportunity Cost of Capital

As shown in earlier chapters, spreadsheets make computing rates of return dramatically easier. In addition, spreadsheets can be used to sort the projects by rate of return and then calculate the cumulative first cost. This is accomplished through the following steps.

1. Enter or calculate each project's rate of return.
2. Select the data to be sorted. Do *not* include headings, but do include all information on the row that goes with each project.
3. Select the SORT tool (found in the menu under DATA), identify the rate of return column as the first key, and a sort order of descending. Also ensure that row sorting is selected. Sort.
4. Add a column for the cumulative first cost. This column is compared with the capital limit to identify the opportunity cost of capital and which projects should be funded.

Example 15-6 illustrates these steps.

EXAMPLE 15-6

A firm has a budget of $800,000 for projects this year. Which of the following projects should be accepted? What is the opportunity cost of capital?

Project	First Cost	Annual Benefit	Salvage Value	Life (years)
A	$200,000	$25,000	$50,000	15
B	250,000	47,000	−25,000	10
C	150,000	17,500	20,000	15
D	100,000	20,000	15,000	10
E	200,000	24,000	25,000	20
F	300,000	35,000	15,000	15
G	100,000	18,000	0	10
H	200,000	22,500	15,000	20
I	350,000	50,000	0	25

SOLUTION

The first step is to use the RATE function to find the rate of return for each project. The results of this step are shown in the top portion of Figure 15-4. Next the projects are sorted in descending order by their rates of return. Finally, the cumulative first cost is computed. Projects D, I, B, and G should be funded. The opportunity cost of capital is 10.6% the rate for the first project rejected.

	A	B	C	D	E	F	G	H
1	Project	First Cost	Annual Benefit	Salvage Value	Life	IRR		
2	A	200,000	25,000	50,000	15	10.2%	=RATE(E2,C2,−B2,D2)	
3	B	250,000	47,000	−25,000	10	12.8%		
4	C	150,000	17,000	20,000	15	8.6%		
5	D	100,000	20,000	15,000	10	16.0%		
6	E	200,000	24,000	25,000	20	10.6%		
7	F	300,000	35,000	15,000	15	8.2%		
8	G	100,000	18,000	0	10	12.4%		
9	H	200,000	22,500	15,000	20	9.6%		
10	I	350,000	50,000	0	25	13.7%		
11	Projects Sorted by IRR						Cumulative First Cost	
12	D	100,000	20,000	15,000	10	16.0%	100,000	
13	I	350,000	50,000	0	25	13.7%	450,000	
14	B	250,000	47,000	−25,000	10	12.8%	700,000	
15	G	100,000	18,000	0	10	12.4%	800,000	
16	E	200,000	24,000	25,000	20	10.6%	1,000,000	
17	A	200,000	25,000	50,000	15	10.2%	1,200,000	
18	H	200,000	22,500	15,000	20	9.6%	1,400,000	
19	C	150,000	17,500	20,000	15	8.6%	1,550,000	
20	F	300,000	35,000	15,000	15	8.2%	1,850,000	

FIGURE 15-4 Spreadsheet for finding opportunity cost of capital.

Example 15-7 outlines the more complicated case of independent projects with mutually exclusive alternatives.

EXAMPLE 15-7

A company is preparing its capital budget for next year. The amount has been set at $250,000 by the board of directors. Rank the following project proposals for the board's consideration and recommend which should be funded.

Project Proposals	Cost (thousands)	Uniform Annual Benefit (thousands)	Salvage Value (thousands)	Useful Life (years)
Proposal 1				
Alt. *A*	$100	$23.85	$0	10
Alt. *B*	150	32.20	0	10
Alt. *C*	200	39.85	0	10
Proposal 2	50	14.92	0	5
Proposal 3				
Alt. *A*	100	18.69	25	10
Alt. *B*	150	19.42	125	10

SOLUTION

For project proposals with two or more alternatives, incremental rate of return analysis is required.

Combination of Alternatives	Cost (thousands)	Uniform Annual Benefit (thousands)	Salvage Value (thousands)	Computed Rate of Return	Incremental Analysis Cost (thousands)	Incremental Analysis Uniform Annual Benefit (thousands)	Incremental Analysis Salvage Value (thousands)	Incremental Analysis Computed Rate of Return
Proposal 1								
A	$100	$23.85	$0	20.0%				
B − *A*					$50	8.35	$0	10.6%
B	150	32.20	0	17.0				
C − *B*					50	7.65	0	8.6
C − *A*					100	16.00	0	9.6
C	200	39.85	0	15.0				
Proposal 2	50	14.92	0	15.0				
Proposal 3								
A	100	18.69	25	15.0				
B − *A*					50	0.73	100	8.3
B	150	19.42	125	12.0				

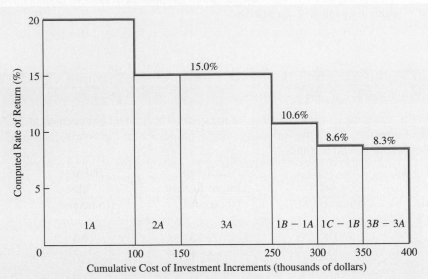

FIGURE 15-5 Cumulative cost versus incremental rate of return.

The various separable increments of investment may be ranked by the rate of return. They are plotted in a graph of cumulative cost versus rate of return in Figure 15-5. The ranking of projects by rate of return gives the following:

Project	
1A	20.0%
2	15.0%
3A	15.0%
1B in place of 1A	10.6%
1C in place of 1B	8.6%
3B in place of 3A	8.3%

For a budget of $250,000 the selected projects are 1A, 2, and 3A. Note that if a budget of $300,000 were available, 1B would replace 1A, making the proper set of projects 1B, 2, and 3A. At a budget of $400,000 1C would replace 1B; and 3B would replace 3A, making the selected projects 1C, 2, and 3B.

In Example 15-7, each of the more expensive mutually exclusive alternatives (1B, 1C, and 3B) had lower rates of return than the less expensive alternatives (1A and 3A). This is common because many projects exhibit decreasing returns to scale. As in Example 15-7, when it occurs, then it makes the ranking of increments easy to use. Sometimes more expensive mutually exclusive alternatives have higher rates of return, then several combinations or more advanced techniques must be tried.

SUMMARY

There are three general sources of capital available to a firm. The most important one is money generated from the firm's operations. This has two components: there is the portion of profit that is retained in the business; in addition, funds equal to its depreciation charges that are available for reinvestment.

The two other sources of capital are from outside the firm's operations.

Debt: borrowed as loans from banks, insurance companies, and so forth.

Longer-term borrowing: from selling bonds.

Equity: sale of equity securities like common or preferred stock.

Retained profits and cash equal to depreciation charges are the primary sources of investment capital for most firms, and the only sources for many enterprises.

In selecting a value of MARR, three values are frequently considered:

1. Cost of borrowed money.
2. Cost of capital. This is a composite cost of the components of the firm's overall capitalization.
3. Opportunity cost, which is the rate of return on the best investment project that is rejected.

The MARR should be equal to the highest one of these three values.

When there is a risk aspect to the problem (probabilities are known or reasonably estimated), this can be handled by techniques like expected value and simulation. Where there is uncertainty (probabilities of the various outcomes are not known), there are analytical techniques, but they are less satisfactory. A method commonly used to adjust for risk and uncertainty is to increase the MARR. This method can distort the time-value-of-money relationship. The effect is to discount longer-term consequences more heavily than short-term consequences, which may or may not be desirable. Prior to this chapter we had assumed that all worthwhile projects are approved and implemented. But industrial firms, like individuals and governments, are typically faced with more good projects than can be funded with the money available. The task is to select the best projects and reject, or at least delay, the rest.

Capital may be rationed among competing investment opportunities by either rate of return or present worth methods. The results may not always be the same for these two methods in many practical situations.

If projects are ranked by rate of return, a proper procedure is to go down the list until the capital budget has been exhausted. The rate of return at this point is the cutoff rate of return. This procedure gives the best group of projects, but does not necessarily have them in the proper priority order.

It has been shown in earlier chapters that the usual business objective is to maximize NPW, and this is not necessarily the same as maximizing rate of return. One suitable procedure is to use the ratio (NPW/PW of cost) to rank the projects, letting the MARR equal the cutoff rate of return (which is the opportunity cost of capital). This present worth ranking method will order the projects so that, for a limited capital budget, NPW will be maximized. The MARR must equal the cutoff rate of return for the rate of return and present worth methods to yield compatible results.

PROBLEMS

Cost of Funds

15-1 Examine the financial pages of your newspaper (or *The Wall St. Journal*) and determine the current interest rate on the following securities, and explain why the interest rates are different for these different bonds.

(*a*) U.S. Treasury bond due in 5 years.

(*b*) General obligation bond of a municipal district, city, or a state due in 20 years.

(*c*) Corporate debenture bond of a U.S. industrial firm due in 20 years.

15-2 A small engineering firm has borrowed $125,000 at 8%. The partners have invested another $75,000. If the partners require a 12% rate of return, what is the firm's cost of capital?   (*Answer:* 9.05%)

15-3 An engineering firm has borrowed $725,000 at 7%. The stockholders have invested another $600,000. The firm's retained earnings total $1.2M. The return on equity is estimated to be 11%. What is the firm's cost of capital?

15-4 A firm's stockholders expect a 15% rate of return, and there is $12M in common stock and retained earnings. The firm has $5M in loans at an average rate of 7%. The firm has raised $8M by selling bonds at an average rate of 6%. What is the firm's cost of capital?

15-5 A firm's stockholders expect an 18% rate of return, and there is $22M in common stock and retained earnings. The firm has $9M in loans at an average rate of 8%. The firm has raised $14M by selling bonds at an average rate of 4%. What is the firm's cost of capital?

15-6 Assume you have $2000 available for investment for a 5-year period. You wish to *invest* the money— not just spend it on fun things. There are obviously many alternatives available. You should be willing to assume a modest amount of risk of loss of some or all of the money if this is necessary, but not a great amount of risk (no investments in poker games or at horse races). How would you invest the money? What is your minimum attractive rate of return? Explain.

15-7 There are many venture capital syndicates that consist of a few (say, eight or ten) wealthy people who combine to make investments in small and (hopefully) growing businesses. Typically, the investors hire a young investment manager (often an engineer with an MBA) who seeks and analyzes investment opportunities for the group. Would you estimate that the MARR sought by this group is more or less than 12%? Explain.

Inflation

15-8 What is the interest rate on a 2-year certificate of deposit at a bank or credit union in your area? What is the most recent value of the Consumer Price Index (CPI)? If inflation continues at that rate, what is the real rate of return on the 2-year CD? Include references for the sources of your data.

15-9 What is the interest rate on a 4- or 5-year new-car loan at a bank or credit union in your area? What is the most recent value of the Consumer Price Index (CPI)? If inflation continues at that rate, what is the real interest rate you would pay on the car loan? Include references for the sources of your data.

15-10 Over the last 10 years, what has been the inflation rate? Compare this with the rate of return on the "Dow" average over the same period. What has been the real rate of return for investing in this mix of stocks? Include references for the sources of your data.

15-11 Over the last 10 years, what has been the inflation rate? Compare this with the rate of return on the NASDAQ average over the same period. What has been the real rate of return for investing in this mix of stocks? Include references for the sources of your data.

Opportunity Cost of Capital

15-12 A factory has a $100,000 capital budget. Determine which project(s) should be funded and the opportunity cost of capital.

Project	First Cost	Annual Benefits	Life (years)	Salvage Value
A	$50,000	$13,500	5	$5000
B	50,000	9,000	10	0
C	50,000	13,250	5	1000
D	50,000	9,575	8	6000

15-13 Chips USA is considering the following projects to improve their production process. Chips have a short life, so a 3-year horizon is used in evaluation. Which projects should be done if the budget is $70,000? What is the opportunity cost of capital?

Project	First Cost	Benefit
1	$20,000	$11,000
2	30,000	14,000
3	10,000	6,000
4	5,000	2,400
5	25,000	13,000
6	15,000	7,000
7	40,000	21,000

15-14 The National Motors Rock Creek plant is considering the following projects to improve the company's production process. Which projects should be done if the budget is $500,000? What is the opportunity cost of capital?

Project	First Cost	Annual Benefit	Life (years)
1	$200,000	$50,000	15
2	300,000	70,000	10
3	100,000	40,000	5
4	50,000	12,500	10
5	250,000	75,000	5
6	150,000	32,000	20
7	400,000	125,000	5

15-15 The WhatZit Company has decided to fund six of nine project proposals for the coming budget year. Determine the next capital budget for WhatZit. What is the MARR?

Project	First Cost	Annual Benefits	Life (years)
A	$15,000	$ 4,429	4
B	20,000	6,173	4
C	30,000	9,878	4
D	25,000	6,261	5
E	40,000	11,933	5
F	50,000	11,550	5
G	35,000	6,794	8
H	60,000	12,692	8
I	75,000	14,058	8

15-16 Which projects should be done if the budget is $100,000? What is the opportunity cost of capital?

Project	Life (years)	First Cost	Annual Benefit	Salvage Value
1	20	$20,000	$4000	
2	20	20,000	3200	$20,000
3	30	20,000	3300	10,000
4	15	20,000	4500	
5	25	20,000	4500	−20,000
6	10	20,000	5800	
7	15	20,000	4000	10,000

Risk-Adjusted MARR

15-17 Use the example risk-adjusted interest rates for manufacturing projects in Table 15-2. Assume Project B in Problem 15-12 is a new product in a new market. What is the interest rate for evaluating this project? Should it be done?

15-18 Use the example risk-adjusted interest rates for manufacturing projects in Table 15-2. Assume Project E in Problem 15-15 is a new product in an existing market. What is the interest rate for evaluating this project? Should it be done?

15-19 Use the example risk-adjusted interest rates for manufacturing projects in Table 15-2. Assume Project 1 in Problem 15-16 is a new product in a foreign market. What is the interest rate for evaluating this project? Should it be done?

Capital Budgeting

15-20 Each of the following 10 independent projects has a 10-year life and no salvage value.

Project	Cost (thousands)	Uniform Annual Benefits (thousands)	Computed Rate of Return
1	$ 5	$1.03	16%
2	15	3.22	17
3	10	1.77	12
4	30	4.88	10
5	5	1.19	20
6	20	3.83	14
7	5	1.00	15
8	20	3.69	13
9	5	1.15	19
10	10	2.23	18

The projects have been proposed by the staff of the Ace Card Company. The MARR of Ace has been 12% for several years.

(a) If there is ample money available, what projects should Ace approve?

(b) Rank-order all the acceptable projects in according to desirability.

(c) If only $55,000 is available, which projects should be approved?

15-21 Ten capital spending proposals have been made to the budget committee as the members prepare the annual budget for their firm. Each independent project has a 5-year life and no salvage value.

Project	Initial Cost (thousands)	Uniform Annual Benefit (thousands)	Computed Rate of Return
A	$10	$2.98	15%
B	15	5.58	25
C	5	1.53	16
D	20	5.55	12
E	15	4.37	14
F	30	9.81	19
G	25	7.81	17
H	10	3.49	22
I	5	1.67	20
J	10	3.20	18

(a) Based on a MARR of 14%, which projects should be approved?

(b) Rank-order all the projects according to desirability.

(c) If only $85,000 is available, which projects should be approved?

15-22 At Miami Products, four project proposals (three with mutually exclusive alternatives) are being considered. All the alternatives have a 10-year useful life and no salvage value.

Project Proposal	Cost (thousands)	Uniform Annual Benefits (thousands)	Computed Rate of Return
Project 1			
Alt. A	$25	$4.61	13%
Alt. B	50	9.96	15
Alt. C	10	2.39	20
Project 2			
Alt. A	20	4.14	16
Alt. B	35	6.71	14
Project 3			
Alt. A	25	5.56	18
Alt. B	10	2.15	17
Project 4	10	1.70	11

(a) Use rate of return methods to determine which set of projects should be undertaken if the capital budget is limited to about $100,000.

(b) For a budget of about $100,000, what interest rate should be used in rationing capital by present worth methods? (Limit your answer to a value for which there is a compound interest table available in Appendix B).

(c) Using the interest rate determined in part (b), rank-order the eight different investment opportunities by means of the present worth method.

(d) For a budget of about $100,000 and the ranking in part (c), which of the investment opportunities should be selected?

15-23 Al Dale is planning his Christmas shopping for seven people. To quantify how much his various relatives would enjoy receiving items from a list of prospective gifts, Al has assigned appropriateness units (called "ohs") for each gift for each person (Table P15-23). A rating of 5 ohs represents a gift that the recipient would really like. A rating of 4 ohs indicates the recipient would like it four-fifths as much; 3 ohs, three-fifths as much, and so forth. A zero rating indicates an inappropriate gift that cannot be given to that person. Everyone must get a gift.

The objective is to maximize total ohs that can be obtained with the selected budget.

(a) How much will it cost to buy the seven gifts the people would like best, if there is ample money for Christmas shopping?

(b) If the Christmas shopping budget is set at $112, which gifts should be purchased, and what is their total appropriateness rating in ohs?

(c) If the Christmas shopping budget must be cut to $90, which gifts should be purchased, and what is their total appropriateness rating in ohs? (*Answer:* (a), $168)

15-24 A financier has a staff of three people whose job it is to examine possible business ventures for him. Periodically they present their findings concerning business opportunities. On a particular occasion, they presented the following investment opportunities:

Project A: This is a project for the use of commercial land the financier already owns. There are three mutually exclusive alternatives.

- *A1.* Sell the land for $500,000.
- *A2.* Lease the property for a car-washing business. An annual income, after all costs

TABLE P15-23 Data

Prospective Gift	"Oh" Rating of Gift If Given to Various Family Members						
	Father	Mother	Sister	Brother	Aunt	Uncle	Cousin
1. $20 box of candy	4	4	2	1	5	2	3
2. $12 box of cigars	3	0	0	1	0	1	2
3. $16 necktie	2	0	0	3	0	3	2
4. $20 shirt or blouse	5	3	4	4	4	1	4
5. $24 sweater	3	4	5	4	3	4	2
6. $30 camera	1	5	2	5	1	2	0
7. $ 6 calendar	0	0	1	0	1	0	1
8. $16 magazine subscription	4	3	4	4	3	1	3
9. $18 book	3	4	2	3	4	0	3
10. $16 game	2	2	3	2	2	1	2

(property taxes, etc.) of $98,700 would be received at the end of each year for 20 years. At the end of the 20 years, it is believed that the property could be sold for $750,000.

• A3. Construct an office building on the land. The building will cost $4.5 million to construct and will not produce any net income for the first 2 years. The probabilities of various levels of rental income, after all expenses, for the subsequent 18 years are as follows:

Annual Rental Income	Probability
$1,000,000	0.1
1,100,000	0.3
1,200,000	0.4
1,900,000	0.2

The property (building and land) probably can be sold for $3 million at the end of 20 years.

Project B: An insurance company is seeking to borrow money for 90 days at $13\,^3/_4\%$ per annum, compounded continuously.

Project C: A financier owns a manufacturing company. The firm desires additional working capital to allow it to increase its inventories of raw materials and finished products. An investment of $2 million will allow the company to obtain sales that in the past the company had to forgo. The additional capital will increase company profits by $500,000 a year. The financier can recover this additional investment by ordering the company to reduce its inventories and to return the $2 million. For planning purposes, assume the additional investment will be returned at the end of 10 years.

Project D: The owners of *Sunrise* magazine are seeking a loan of $500,000 for 10 years at a 16% interest rate.

Project E: The Galveston Bank has indicated a willingness to accept a deposit of any sum of money over $100,000, for any desired duration, at a 14.06% interest rate, compounded monthly. It seems likely that this interest rate will be available from Galveston, or some other bank, for the next several years.

Project F: A car rental firm is seeking a loan of $2 million to expand its fleet. The firm offers to repay the loan by paying $1 million at the end of Year 1 and $1,604,800 at the end of Year 2.

• If there is $4 million available for investment now (or $4.5 million if the Project A land is sold), which projects should be selected? What is the MARR in this situation?

• If there is $9 million available for investment now (or $9.5 million if the Project A land is sold), which projects should be selected?

15-25 The Raleigh Soap Company has been offered a 5-year contract to manufacture and package a leading brand of soap for Taker Bros. It is understood that the contract will not be extended past the 5 years because Taker Bros. plans to build its own plant nearby. The contract calls for 10,000 metric tons (one metric ton equals 1000 kg) of soap a year. Raleigh normally produces 12,000 metric tons of soap a year, so production for the 5-year period would be increased to 22,000 metric tons. Raleigh must decide what changes, if any, to make to accommodate this increased production. Five projects are under consideration.

Project 1: Increase liquid storage capacity. Raleigh has been forced to buy caustic soda in tank truck quantities owing to inadequate storage capacity. If another liquid caustic soda tank is installed to hold 1000 cubic meters, the caustic soda may be purchased in railroad tank car quantities at a more favorable price. The result would be a saving of 0.1¢ per kilogram of soap. The tank, which would cost $83,400, has no net salvage value.

Project 2: Acquire another sulfonation unit. The present capacity of the plant is limited by the sulfonation unit. The additional 12,000 metric tons of soap cannot be produced without an additional sulfonation unit. Another unit can be installed for $320,000.

Project 3: Expand the packaging department. With the new contract, the packaging department must either work two 8-hour shifts or have another packaging line installed. If the two-shift operation is used, a 20% wage premium must be paid for the second shift. This premium would amount to $35,000 a year. The second packaging line could be installed for $150,000. It would have a $42,000 salvage value at the end of 5 years.

Project 4: Build a new warehouse. The existing warehouse will be inadequate for the greater production. It is estimated that 400 square meters of additional warehouse is needed. A new warehouse can be built on a lot beside the existing warehouse for $225,000, including the land. The annual taxes, insurance, and other ownership costs would be $5000 a year. It is believed the warehouse could be sold at the end of 5 years for $200,000.

Project 5: Lease a warehouse. An alternative to building an additional warehouse would be to lease warehouse space. A suitable warehouse one mile away could be leased for $15,000 per year. The $15,000 includes taxes, insurance, and so forth. The annual cost of moving materials to this more remote warehouse would be $34,000 a year.

The contract offered by Taker Bros. is a favorable one, which Raleigh Soap plans to accept. Raleigh management has set a 15% before-tax minimum attractive rate of return as the criterion for any of the projects. Which projects should be undertaken?

15-26 Mike Moore's microbrewery is considering production of a new ale called Mike's Honey Harvest Brew. To produce this new offering he is considering two independent projects. Each of these projects has two mutually exclusive alternatives and each alternative has a useful life of 10 years and no salvage value. Mike's MARR is 8%. Information regarding the projects and alternatives are given in the following table:

Project/Alternative	Cost	Annual Benefit
Project 1. Purchase new fermenting tanks		
Alt. *A*: 5000-gallon tank	$ 5000	$1192
Alt. *B*: 15,000-gallon tank	10,000	1992
Project 2. Purchase bottle filler and capper		
Alt. *A*: 2500-bottle/hour machine	15,000	3337
Alt. *B*: 5000-bottle/hour machine	25,000	4425

Use incremental rate of return analysis to complete the following worksheet.

Proj./Alt.	Cost, P	Annual Benefit, A	A/P, i, 10	IRR
1A	$ 5,000	$1192	0.2385	20%
1B–1A	5,000	800	0.1601	
2A	15,000	3337		
2B–2A	10,000			

Use this information to determine:

(a) which projects should be funded if only $15,000 is available.

(b) the cutoff rate of return if only $15,000 is available.

(c) which projects should be funded if $25,000 is available.

CHAPTER 16

ECONOMIC ANALYSIS IN THE PUBLIC SECTOR

From Waste to Power and Money

Many of the world's leading climate scientists have concluded that global warming is "unequivocal" and that human activity is the main driver. One such activity is the dumping into landfills of organic material that decomposes over time, producing a significant amount of landfill gas (LFG). About 50% of LFG is methane, which is a greenhouse gas that contributes to global warming. Methane is of particular concern because it traps heat in the atmosphere with an effect 20 times greater than that of carbon dioxide—also a large contributor to global climate change.

Landfills that do not control the escape of landfill gas are the largest single human source of methane emissions in the U.S. Of additional concern are emissions of volatile organic compounds that create smog in the form of ground-level ozone and produce respiratory problems. Landfills also generate hazardous air pollutants such as benzene and toluene, and odorous compounds such as hydrogen sulfide that negatively impact the quality of life for individuals nearby.

But humans must produce waste to live—it is one of the basic laws of thermodynamics. Bad news all around—except that using engineering principles, bringing together various "constituencies," and evaluating alternatives with engineering economy—can bring benefits out of almost every one of these issues.

If it is extracted and properly utilized, landfill gas can be an asset. This medium-BTU gas has a heating value of between 350 and 600 BTU per cubic foot, about half the heating value of natural gas. As a result, more and more owners of municipal solid waste landfills extract this gas and use it to generate electricity. Such electricity is used directly for the landfills' power needs or is sold to the general power grid.

The environment benefits from the gas extraction and combustion because the release of methane, volatile organics, hazardous air pollutants, and odorous gases to the environment is greatly reduced. The generation of electrical power by utilizing a previously wasted source (i.e., methane from LFG) reduces the use of fossil fuel and the associated emissions. The income from selling power can assist in paying landfill operating expenses.

Extracting and using a landfill gas to generate electricity is a "*6 win*" situation.

1. It benefits the environment by reducing unwanted gas emission.
2. It adds electrical power to the grid.
3. It produces cash flow for the landfill owner.
4. It lowers the use of nonrenewable fossil fuels.
5. It reduces financial costs to the local population.
6. It reduces the hazardous, noxious, and odorous gases for those nearby and downwind.

Contributed by William R. Truran, Stevens Institute of Technology, and Peter A. Cerenzio, Cerenzio & Panaro Consulting Engineers

QUESTIONS TO CONSIDER

1. How can this proposal best be summarized for a nontechnical public audience? This is an important "skill" of an engineer—conveying sometimes complex and confusing scientific issues to a less informed, less scientific, and less interested audience with a short attention span.
2. What engineering economic principles would you apply in analyzing this application? Which measures would be most effective for the public audience? The important measures, and the metrics used to describe them, may be considered "key performance indicators."
3. This waste-to-power application cannot be answered with a simple equation with a closed-ended answer that is "correct" or "wrong." This problem requires the cooperation of several (at least) agencies and communities of interest. Name some, and identify how their objectives are aligned or in conflict.
4. Like this vignette, the case of Enron, a "poster child" of what can go wrong in business, is related to electric power generation. Discuss how there may be ways for some of the stakeholders to be "shady" here. Discuss how you—as an engineer—can not only influence factual answers but project an image that is not reality. You as an engineer can be honored (or guilty) of your representations.

After Completing This Chapter...
The student should be able to:

- Distinguish the unique objective and viewpoint of public decisions.
- Explain methods for determining the interest rates for evaluating public projects.
- Use the benefit–cost ratio to analysis projects.
- Distinguish between the conventional and modified versions of the benefit–cost ratio.
- Use an incremental benefit–cost ratio to evaluate a set of mutually exclusive projects.
- Discuss the impact of financing, duration, quantifying and valuing consequences, and politics in public investment analysis.

Public organizations, such as federal, state, and local governments, port authorities, school districts, and government agencies, no less than firms and individuals in the private sector, make investment decisions. For these decision-making bodies, economic analysis is complicated by several factors that do not affect companies in the private sector. These factors include the overall purpose of investment, the viewpoint for analysis, and how to select the interest rate. Other factors include project financing sources, expected project duration, quantifying and valuing benefits and disbenefits, effects of politics, beneficiaries of investment, and the multipurpose nature of investments. The overall mission in the public sector is the same as that in the private sector—to make prudent investment decisions that promote the organization's overall objectives.

The primary economic measure used in the public sector is the benefit–cost (B/C) ratio, which was introduced in Chapter 9. This measure is calculated as a ratio of the equivalent worth of the project's benefits to the equivalent worth of the project's costs. If the B/C ratio is *greater than 1.0,* the project under evaluation is accepted; if not, it is rejected. The B/C ratio is used to evaluate both single investments and sets of mutually exclusive projects (where the incremental B/C ratio is used). The uncertainties of quantifying cash flows, long project lives, and low interest rates all tend to lessen the reliability of public sector engineering economic analysis. There are two versions of the B/C ratio: *conventional* and *modified.* Both provide consistent recommendations to decision makers for single investment decisions and for decisions involving sets of mutually exclusive alternatives. The B/C ratio is a widely used and accepted measure in government economic analysis and decision making.

INVESTMENT OBJECTIVE

Organizations exist to promote the overall goals of those they serve. In private sector firms, investment decisions are based on increasing the firm's wealth and economic stability. Beneficiaries of investments generally are clearly identified as the firm's owners and/or stockholders.

In the public sector, the purpose of investment decisions is sometimes ambiguous. For people in the United States of America, the Preamble to the Constitution establishes the overall theme or objective for why the public body exists:

> We the People of the United States, in order to form a more perfect Union, establish justice, insure domestic tranquillity, provide for the common defense, *promote the general welfare* [italics supplied], and secure the blessings of liberty to ourselves and our posterity, do ordain and establish this Constitution for the United States of America.

The catch phrase *promote the general welfare* serves as a guideline for public decision making. But what does this phrase mean? At best it is a general guideline; at worst it is a vague slogan that can be used to justify any action. Projects some citizens want may be opposed by other citizens. In government economic analysis, it is not always easy to distinguish which investments promote the "general welfare" and which do not.

Consider the case of a dam construction project to provide water, electricity, flood control, and recreational facilities. Such a project might seem to be advantageous for a region's entire population. But on closer inspection, decision makers must consider that the dam will require the loss of land upstream due to backed-up water. Farmers will lose pastures or cropland, and nature lovers will lose canyon lands. Or perhaps the land to be lost is a pivotal breeding ground for protected species, and environmentalists will oppose the project. The project may also have a negative impact on towns, cities, and states downstream. How will it affect their water supply? Thus, a project initially deemed to have many benefits, on closer inspection, reveals many conflicts. Projects' conflicting aspects are characteristic of public sector investment and decision making.

Public investment decisions are more difficult than those in the private sector owing to the many people, organizations, and political units potentially affected. Opposition to a proposal is more likely in public investment decisions than in those made by private sector companies because for every group that benefits from a particular project, there is usually an opposing group. Many conflicts in opinion arise when the project involves the use of public lands, including industrial parks, housing projects, business districts, roadways, sewage and power facilities, and landfills. Opposition may be based on the belief that development of *any* kind is bad or that the proposed development should not be near "our" homes, schools, or businesses.

Consider the decision that a small town might face when considering whether to establish a municipal rose garden, seemingly a beneficial public investment with no adverse consequences. However, an economic analysis of the project must consider *all* effects of the project, including potential unforeseen outcomes. Where will visitors park their vehicles? Will increased travel around the park necessitate new traffic lights and signage? Will traffic and visitors to the park increase noise levels for adjacent homes? Will special varieties of roses create a disease hazard for local gardens? Will the garden require high levels of fertilizers and insecticides, and where will these substances wind up after they have been applied? Clearly, many issues must be addressed. What appeared to be a simple proposal for a city rose garden, in fact, brings up many aspects to be considered.

Our simple rose garden illustrates how effects on *all parties involved* must be identified, even for projects that seem very useful. Public decision makers must reach a compromise between the positive effects enjoyed by some groups and the negative effects on other groups. The overall objective is to make prudent decisions that *promote the general welfare,* but in the public sector the decision process is not so straightforward as in the private sector.

The **Flood Control Act of 1936** specified that waterway improvements for flood control could be made as long as "the benefits *to whomsoever they accrue* [italics supplied] are in excess of the estimated costs." Perhaps the overall general objective of investment decision analysis in government should be a dual one: to promote the general welfare and to ensure that the value to those who can potentially benefit exceeds the overall costs to those who do not benefit.

VIEWPOINT FOR ANALYSIS

When governmental bodies do economic analysis, an important concern is the proper viewpoint of the analysis. A look at industry will help to explain how the viewpoint, or perspective, from which an analysis is conducted influences the final recommendation. A firm pays its costs and counts *its* benefits. Thus, both the costs and benefits are measured from the firm's perspective.

Costs and benefits that occur outside the firm are referred to as external consequences (Figure 16-1). In years past, private sector companies generally ignored the external consequences of their actions. Ask anyone who has lived near a cement plant, a slaughterhouse, or a steel mill about external consequences! More recently, governments have forced industries to reduce pollution and other undesirable external consequences, with the result that today many companies are evaluating the consequences of their action from a broader, or community-oriented, viewpoint.

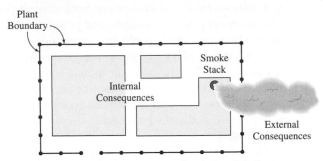

FIGURE 16-1 Internal and external consequences of an industrial plant.

The council members of a small town that levies taxes can be expected to take the "viewpoint of the town" in making decisions: unless it can be shown that the money from taxes can be used *effectively,* the town council is unlikely to spend it. But what happens when the money is contributed to the town by the federal government, as in "revenue sharing" or by means of some other federal grant? Often the federal government pays a share of project costs varying from 10 to 90%. Example 16-1 illustrates the viewpoint problem that is created.

EXAMPLE 16-1

A municipal project will cost $1 million. The federal government will pay 50% of the cost if the project is undertaken. Although the original economic analysis showed that the PW of benefits was $1.5 million, a subsequent detailed analysis by the city engineer indicates a more realistic estimate of the PW of benefits is $750,000. The city council must decide whether to proceed with the project. What would you advise?

SOLUTION

From the viewpoint of the city, the project is still a good one. If the city puts up half the cost ($500,000) it will receive all the benefits ($750,000). On the other hand, from an *overall* viewpoint, the revised estimate of $750,000 of benefits does not justify the $1 million expenditure. This illustrates the dilemma caused by varying viewpoints. For economic efficiency, one does not want to encourage the expenditure of money, regardless of the source, unless the benefits at least equal the costs.

Possible viewpoints that may be taken include those of an individual, a business firm or corporation, a town or city district, a city, a state, a nation, or a group of nations. To avoid suboptimizing, the proper approach is to *take a viewpoint at least as broad as those who pay the costs and those who receive the benefits.* When the costs and benefits are totally confined to a town, for example, the town's viewpoint seems to be an appropriate basis for the analysis. But when the costs or the benefits are spread beyond the proposed viewpoint, then the viewpoint should be enlarged to this broader population.

Other than investments in defense and social programs, most of the benefits provided by government projects are realized at a regional or local level. Projects, such as dams for electricity, flood control, and recreation, as well as transportation facilities, such as roads, bridges, and harbors, all benefit most those in the region in which they are constructed. Even smaller-scale projects, such as the municipal rose garden, although funded by public monies at a local or state level, provide most benefit to those nearby. As in the case of private decision making, it is important to adopt an appropriate and *consistent* viewpoint and to designate all the costs and benefits that arise from the prospective investment from that perspective. To shift perspective when quantifying costs and benefits could greatly skew the results of the analysis and subsequent decision, and thus such changes in perspective are inappropriate.

SELECTING AN INTEREST RATE

Several factors, not present for private sector firms, influence selecting an interest rate for economic analysis in the government sector. Recall that for private sector firms the overall objective is wealth maximization, and the interest rate is selected consistent with this goal. Most firms use *cost of capital* or *opportunity cost* concepts when setting an interest rate. The objective of public investment involves the use of public resources to *promote the general welfare* and to secure the benefits of a given project *to whomsoever they may accrue*, as long as those benefits outweigh the costs. How to set an interest rate is less clear-cut in this case. Possibilities include no interest rate, cost of capital concepts, and an opportunity cost concept.

No Time-Value-of-Money Concept

In government, monies are obtained through taxation and spent about as quickly as they are obtained. Often, there is little time delay between collecting money from taxpayers and spending it. (Remember that the federal government and many states collect taxes every paycheck in the form of withholding tax.) The collection of taxes, like their disbursement, although based on an annual budget, is actually a continuous process. Using this line of reasoning, some would argue that there is little or no time lag between collecting and spending tax dollars. Thus, they would advocate the use of a 0% interest rate for economic analysis of public projects. Not surprisingly, this viewpoint is most often expressed by people who are *not* engineering economists and who are pushing a marginal project.

Cost of Capital Concept

Another approach in determining interest rates in public investments is that most levels of government (federal, state, and local) borrow money for capital expenditures in addition to collecting taxes. Where money is borrowed for a specific project, one line of reasoning is to use an interest rate equal to the *cost of borrowed money*. This argument is less valid for state and local governments than for private firms because the federal government, through the income tax laws, subsidizes state and local bonded debt. If a state, county, city, or special assessment district raises money by selling bonds, the interest paid on these bonds is exempt from federal taxes. In this way the federal government is *subsidizing* the debt, thereby encouraging investors to purchase such bonds.

These bonds, called municipal bonds, can be either general obligation or revenue bonds. **General obligation municipal bonds** pay interest and are retired (paid off) through taxes raised by the issuing government unit. A school district may use property taxes it receives to finance bond debt for construction of new language labs. **Revenue bonds** are not supported by the taxing authority of the government unit; rather, they are supported by revenues earned by the project being funded. As an example, a city could use toll revenues from a new bridge to retire debt on revenue bonds sold for the bridge's construction.

For those who purchase municipal bonds, the instruments' tax-free status means that the expected return on this investment is somewhat less than that required of fully taxed bond investments (of similar risk). As a rough estimate, when fully taxed bonds yield an 8% interest rate, municipal bonds might make interest payments at a rate of 6%. The difference of 2% is due to the preferred treatment for federal taxation, and the federal subsidy on tax-free bonds. This can skew the *cost of capital* approach by lowering the apparent cost of long-term bonds.

Opportunity Cost Concept

Opportunity cost, which is related to the interest rate on the best opportunity forgone, may take two forms in governmental economic analysis: government opportunity cost and taxpayer opportunity cost. In public decision making, if the interest rate is based on the opportunity cost to a government agency or other governing body, this interest rate is known as **government opportunity cost.** In this case the interest rate is set at that of the best prospective project for which funding is not available. One disadvantage of the government opportunity cost concept is that different agencies and subdivisions of government will have different opportunities. Therefore, political units could use different interest rates for economic analysis, and a project that is rejected in one branch might be accepted in another. Differing interest rates lead to inconsistent evaluation and decision making across government.

Dollars used for public investments are generally gathered by taxing the citizenry. The concept of **taxpayer opportunity cost** suggests that a correct interest rate for evaluating public investments is that which the *taxpayer* could have received if the government had not collected those dollars through taxation. This philosophy holds that through taxation the government is taking away the taxpayers' opportunity to use the same dollars for investment. The interest rate that the government requires should not be less than what the taxpayer would have received.

The most widely followed standard is found in the Office of Management and Budget (OMB) A94 directive, which stipulates that a 7% interest rate be used in economic analysis for a wide range of federal projects. It is not economically desirable to take money from a taxpayer with a 12% opportunity cost, for example, and invest it in a government project yielding 4%.

Recommended Concept

The general rule of thumb in setting an interest rate for government investments has been to select the *largest* of the cost of capital, the government opportunity cost, or the taxpayer opportunity cost interest rates. However, as is the case in the private sector, there is no hard and fast rule, universally applied in all decision circumstances. Setting an interest rate for use in economic analysis is at the discretion of the government entity performing the analysis.

THE BENEFIT–COST RATIO

The benefit–cost ratio was described briefly in Chapter 9. This method is used almost exclusively in public investment analysis, and because of the magnitude of the amount of public dollars committed each year through such analysis, the benefit–cost ratio deserves our attention and understanding.

One of the primary reasons for using the benefit–cost ratio (B/C ratio) in public decision making is its simplicity. The ratio is formed by calculating the equivalent worth of the project's benefits divided by the equivalent worth of the project's costs. The benefit–cost ratio can be shown as follows:

$$B/C \text{ ratio} = \frac{\text{Equivalent worth of net benefits}}{\text{Equivalent worth of costs}}$$

$$= \frac{\text{PW benefits}}{\text{PW costs}} = \frac{\text{FW benefits}}{\text{FW costs}} = \frac{\text{AW benefits}}{\text{AW costs}}$$

Notice that *any* of the equivalent worth methods (present, future, and annual) can be used to calculate this ratio. Each formulation of the ratio will produce an identical result, as illustrated in Example 16-2.

EXAMPLE 16-2

Demonstrate that for this highway expansion project, the same B/C ratio is obtained using the present, future, and annual worth formulations.

Initial costs of expansion	$1,500,000
Annual costs for operating/maintenance	65,000
Annual savings and benefits to travelers	225,000
Scrap value after useful life	300,000
Useful life of investment	30 years
Interest rate	8%

SOLUTION

Using Present Worth

$$\text{PW benefits} = 225,000(P/A, 8\%, 30) + 300,000(P/F, 8\%, 30) = \$2,563,000$$
$$\text{PW costs} = 1,500,000 + 65,000(P/A, 8\%, 30) = \$2,232,000$$

Using Future Worth

$$\text{FW benefits} = 225{,}000(F/A, 8\%, 30) + 300{,}000 = \$25{,}790{,}000$$
$$\text{FW costs} = 1{,}500{,}000(F/P, 8\%, 30) + 65{,}000(F/A, 8\%, 30) = \$22{,}460{,}000$$

Using Annual Worth

$$\text{AW benefits} = 225{,}000 + 300{,}000(A/F, 8\%, 30) = \$227{,}600$$
$$\text{AW costs} = 1{,}500{,}000(A/P, 8\%, 30) + 65{,}000 = \$198{,}200$$

$$\text{B/C ratio} = \frac{2{,}563{,}000}{2{,}232{,}000} = \frac{25{,}790{,}000}{22{,}460{,}000} = \frac{227{,}600}{198{,}200} = 1.15$$

One can see that the ratio provided by each of these methods produces the same result: 1.15.

An economic analysis is performed to assist in the objective of making a decision. When one is using the B/C ratio, the decision rule is in two parts:

If the B/C ratio is > 1.0, then the decision should be to invest.

If the B/C ratio is < 1.0, then the decision should be not to invest.

Cases of a B/C ratio *near* 1.0 are analogous to the case of a calculated net present worth of $0 or an IRR analysis that yields $i = \text{MARR}\%$. In other words, the decision measure is *near* the breakeven criteria. In such cases a detailed analysis of the input variables and their estimates is necessary, and one should consider the merits of other available opportunities for the targeted funds. But, if the B/C ratio is clearly greater than or less than 1.0, the recommendation is clear.

The B/C ratio is a numerator/denominator relationship between the equivalent worths (EW) of *benefits* and *costs:*

$$\text{B/C ratio} = \frac{\text{EW of net benefits to whomsoever they may accrue}}{\text{EW of costs to the sponsors of the project}}$$

The numerator and denominator aspects of the ratio are sometimes interpreted and used in different fashions. For instance, the **conventional B/C ratio** (see Example 16-2) can be restated as follows:

$$\text{Conventional B/C ratio} = \frac{\text{EW of net benefits}}{\text{EW of initial costs} + \text{EW of operating and maintenance costs}}$$

However, there is also the **modified B/C ratio.** To use a *modified* version, the *annual operating and maintenance costs* are subtracted in the numerator, rather than being added as a cost in the denominator. This *modified B/C ratio* is mathematically similar to the present worth index defined in Chapter 9. The ratio becomes:

$$\text{Modified B/C ratio} = \frac{\text{EW of net benefits} - \text{EW of operating and maintenance costs}}{\text{EW of initial costs}}$$

For decision making, the two versions of the benefit–cost ratio will produce the same recommendation on whether to *invest* or *not invest* in the project being considered. The *numeric values for the B/C ratio* will not be the same, but the recommendation will be. This is illustrated in Example 16-3.

EXAMPLE 16-3

Consider the highway expansion project from Example 16-2. Let us use the present worth formulation of *conventional* and *modified* versions to calculate the B/C ratio.

SOLUTION

Using the *conventional B/C ratio*

$$\text{B/C ratio} = \frac{225,000(P/A, 8\%, 30) + 300,000(P/F, 8\%, 30)}{1,500,000 + 65,000(P/A, 8\%, 30)} = 1.15$$

Using the *modified B/C ratio*

$$\text{B/C ratio} = \frac{225,000(P/A, 8\%, 30) + 300,000(P/F, 8\%, 30) - 65,000(P/A, 8\%, 30)}{1,500,000} = 1.22$$

Whether the conventional or the modified ratio is used, the recommendation is to invest in the highway expansion project. The ratios are not identical in magnitude (1.15 vs. 1.22), but the decision is the same.

It is important when one is using the conventional and modified B/C ratios not to directly compare the magnitudes of the two versions. Evaluating a project with one version may produce a higher ratio than is produced with the other version, but this does not imply that the project is somehow better.

The *net benefits to the users* of government projects are the difference between the expected *benefits* from investment minus the expected *disbenefits*. Disbenefits are the negative effects of government projects felt by some individuals or groups. For example, consider the U.S. National Park System. Development projects by the skiing or lumber industries might provide enormous benefits to the recreation or construction sectors while creating simultaneous disbenefits for environmental groups.

INCREMENTAL BENEFIT–COST ANALYSIS

In Chapter 9 we discussed using the incremental benefit–cost ratio in economic decision analysis. As with the internal rate of return (IRR) measure, the incremental B/C ratio should be used in comparing *mutually exclusive alternatives*. Incremental B/C ratio analysis is consistent with maximizing the present worth of the alternatives. As with the incremental IRR method, it is *not* proper to simply calculate the B/C ratio for each alternative and choose the one with the highest value. Rather, an *incremental* approach is called for.

Elements of the Incremental Benefit–Cost Ratio Method

1. *Identify all relevant alternatives.* Decision rules or models can recommend a *best* course of action *only* from the set of identified alternatives. If a better alternative exists but is not considered, then it will never be selected, and when available, the solution will be suboptimal. For benefit–cost ratio problems, the "do-nothing" option, is always the "base case" from which the incremental methodology proceeds.

2. *(Optional) Calculate the B/C ratio of each alternative.* Once the individual B/C ratios have been calculated, the alternatives with a ratio *less than 1.0* are eliminated from further consideration. This step gets the "poor performers" out of the way before the incremental procedure is initiated. This step may be omitted, however, because the incremental analysis method will eliminate the subpar alternatives in due time.

Note: There is a case where this step *must* be skipped. If doing nothing is not an a alternative and if *all* alternatives have a B/C ratio less than 1.0, then the best of the unattractive alternatives will be selected through incremental analysis.

3. *Rank-order the projects.* The alternatives must be ordered from smallest to largest size of the *denominator of the B/C ratio.* (The rank order will be the same regardless of whether one uses the present worth, annual worth, or future worth of costs to calculate the *denominator.*) When available, the "do-nothing" alternative always becomes the first on the ordered list.

4. *Identify the increment under consideration.* The first increment considered is always that of going from the lowest cost alternative (when it is available, this is the do-nothing option) to the next higher cost alternative. As the analysis proceeds, any identified increment is always in reference to some previously justified alternative.

5. *Calculate the B/C ratio for the incremental cash flows.* First, calculate the *incremental benefits* and the *incremental costs.* This is done by finding the cash flows that represent the difference (Δ) between the two alternatives under consideration. For two alternatives X and Y, where X is the defender, or base case, and Y is the challenger, the increment can be written as $(X \rightarrow Y)$ to signify *going from X to Y* or as $(Y - X)$ to signify the *cash flows of Y minus* cash flows of X. Both modes identify the incremental costs (ΔC) and benefits (ΔB) of investing in Alternative Y, where X is a previously justified (or base) alternative. The *incremental B/C ratio* equals $\Delta B / \Delta C$.

6. *Use the incremental B/C ratio to decide which alternative is better.* If the incremental B/C ratio ($\Delta B / \Delta C$) calculated in Step 5 is greater than 1.0, then the increment is desirable or justified; if the ratio is less than 1.0, it is not desirable, or is not justified. If an increment is accepted, the alternative (the challenger) associated with that increment of investment becomes the base from which the next increment is formed. When the increment is not justified, the alternative associated with the additional increment is rejected and the previously justified alternative (the defender) is maintained as the base for formation of the next increment.

7. *Iterate to Step 4 until all increments (projects) have been considered.* The incremental method requires that the entire list of ranked feasible alternatives be evaluated. All pairwise comparisons are made such that the additional increment being considered is examined with respect to a previously justified alternative. The incremental method continues until all alternatives have been evaluated.

8. *Select the best alternative from the set of mutually exclusive competing projects.* After all alternatives (and associated increments) have been considered, the incremental B/C ratio method calls for selecting the alternative *associated with the last justified increment.* This assures that a maximum investment is made such that each ratio of equivalent worth of incremental benefits to equivalent worth of incremental costs is greater than 1.0. (A common error in applying the incremental B/C method is selecting the alternative with the *largest* incremental B/C ratio, which is inconsistent with the objective of maximizing investment size with incremental B/C ratios above 1.0.)

Both the conventional and modified versions of the B/C ratio can be used with the incremental B/C ratio methodology just described, but the two versions should not be mixed in the same problem. Such an approach could affect the rank order and cause confusion and errors. Instead, *one* of the two versions should be *consistently* used throughout the analysis.

In Examples 16-4 and 16-5 the incremental B/C ratio (conventional and modified version) is used to evaluate a set of mutually exclusive alternatives.

EXAMPLE 16-4

A midwestern industrial state may construct and operate two coal-burning power plants and a distribution network to provide electricity to several state-owned properties. The following costs and benefits have been identified.

Primary costs: Construction of the power plant facilities; cost of installing the power distribution network; life-cycle maintenance and operating costs.

Primary benefits: Elimination of payments to the current electricity provider; creation of jobs for construction, operation, and maintenance of the facilities and distribution network; revenue from selling excess power to utility companies; increased employment at coal mines in the state.

There have been four competing designs identified for the power plants. Each has a life of 45 years. Use the *conventional* B/C ratio with an interest rate of 8% to recommend a course of action.

	Values ($\times$ 10^4) for Competing Design Alternatives			
	I	II	III	IV
Project costs				
Plant construction cost	$12,500	$11,000	$12,500	$16,800
Annual operating and maintenance cost	120	480	450	145
Project benefits				
Annual savings from utility payments	580	700	950	1,300
Revenue from overcapacity	700	550	200	250
Annual effect of jobs created	400	750	150	500

SOLUTION

Alternatives I through IV and the do-nothing alternative are *mutually exclusive* choices because one and only one will be selected. Therefore, an incremental B/C ratio method is used to obtain the solution.

Step 1 *Identify alternatives.* The alternatives are do nothing, and designs I, II, III, and IV.

Step 2 *Calculate B/C ratio for each alternative.*

$$\text{B/C ratio (I)} = (580 + 700 + 400)\,(P/A, 8\%, 45)/[12{,}500 + 120\,(P/A, 8\%, 45)] = 1.46$$

$$\text{B/C ratio (II)} = (700 + 550 + 750)\,(P/A, 8\%, 45)/[11{,}000 + 480\,(P/A, 8\%, 45)] = 1.44$$

$$\text{B/C ratio (III)} = (200 + 950 + 150)\,(P/A, 8\%, 45)/[12{,}500 + 325\,(P/A, 8\%, 45)] = 0.96$$

$$\text{B/C ratio (IV)} = (1300 + 250 + 500)(P/A, 8\%, 45)/[16{,}800 + 145(P/A, 8\%, 45)] = 1.34$$

Alternatives I, II, and IV all have B/C ratios greater than 1.0 and thus merit further consideration. Alternative III does not meet the acceptability criterion and could be eliminated from further consideration. However, to illustrate that Step 2 is optional, all four design alternatives will be analyzed incrementally.

Step 3 *Rank-order projects.* Here we calculate the PW of costs for each alternative. The denominator of the *conventional* B/C ratio includes first cost and annual O&M costs, so the PW of costs for the alternatives are:

$$\text{PW costs (I)} = 12{,}500 + 120(P/A, 8\%, 45) = \$13{,}953$$

$$\text{PW costs (II)} = 11{,}000 + 480(P/A, 8\%, 45) = \$16{,}812$$

$$\text{PW costs (III)} = 12{,}500 + 325(P/A, 8\%, 45) = \$16{,}435$$

$$\text{PW costs (IV)} = 16{,}800 + 145(P/A, 8\%, 45) = \$18{,}556$$

The rank order from low to high value of the B/C ratio *denominator* is as follows: do nothing, I, III, II, IV.

Step 4 *Identify increment under consideration.*

Step 5 *Calculate B/C ratio.*

Step 6 *Which alternative is better?*

	1st Iteration	2nd Iteration	3rd Iteration	4th Iteration
Step 4 Increment under Consideration	(Do Nothing → I)	(I → III)	(I → II)	(II → IV)
ΔPlant construction cost	$12,500	$ 0	$−1500	$ 5800
ΔAnnual O&M cost	120	205	360	−335
PW of ΔCosts	13,953	2482	2859	1744
ΔAnnual utility payment savings	580	370	120	600
ΔAnnual overcapacity revenue	700	500	−150	−300
ΔAnnual benefits of new jobs	400	−250	350	−250
PW of ΔBenefits	20,342	−4601	3875	605
Step 5 ΔB/C ratio (PW ΔB)/(PW ΔC)	1.46	−1.15	1.36	0.35
Step 6 Is increment justified?	Yes	No	Yes	No

As an example of these calculations, consider the third increment (I → II).

ΔPlant construction cost	$= 11,000 - 12,500 = -\$1500$
ΔAnnual O&M cost	$= 480 - 120 = \$360$
PW of ΔCosts	$= -1500 + 360(P/A, 8\%, 45) = \$2859$
or	$= 16,812 - 13,953 = \$2859$
ΔAnnual utility payment savings	$= 700 - 580 = \$120$
ΔAnnual overcapacity revenue	$= 550 - 700 = -\$150$
ΔAnnual benefits of new jobs	$= 750 - 400 = \$350$
PW of ΔBenefits	$= (120 - 150 + 350)(P/A, 8\%, 45)$
	$= \$3875$
ΔB/C ratio (PW ΔB)/(PW ΔC)	$= 3875/2850 = 1.36$

The analysis in the table proceeded as follows: do nothing to Alternative I was justified ($\Delta B/C$ ratio = 1.46), Alternative I became the new base; Alternative I to Alternative III was not justified (ΔB/C ratio = −1.15), Alternative I remained base; Alternative I to Alternative II was justified (ΔB/C ratio = 1.36), Alternative II became the base; Alternative II to Alternative IV was not justified (ΔB/C ratio = 0.35).

Step 7 *Select best alternative.* Alternative II became the recommended power plant design alternative because it is the one associated with the last justified increment. Notice that Alternative III, did not affect the recommendation and was eliminated through the incremental method. Notice also that the first increment considered (do nothing → I) was not selected even though it had the *largest* ΔB/C ratio (1.45). The alternative associated with the *last justified increment* (in this case, Alt. II) should be selected.

EXAMPLE 16-5

Let us reconsider Example 16-4, this time using the modified B/C ratio to analyze the alternatives. Again we will use the present worth method.

SOLUTION

Here we use the modified B/C ratio.

Step 1 *Identify alternatives.* The alternatives are still do nothing and designs I, II, III, and IV.

Step 2 *Calculate modified B/C ratio for each alternative.*

$$\text{B/C ratio(I)} = (580 + 700 + 400 - 120)\,(P/A, 8\%, 45)/12,500 = 1.51$$

$$\text{B/C ratio(II)} = (700 + 550 + 750 - 480)\,(P/A, 8\%, 45)/11,000 = 1.67$$

$$\text{B/C ratio (III)} = (200 + 950 + 150 - 325)\,(P/A, 8\%, 45)/12,500 = 0.95$$

$$\text{B/C ratio (IV)} = (1300 + 250 + 500 - 145)\,(P/A, 8\%, 45)/16,800 = 1.37$$

Again Alternative III can be eliminated from further consideration because its B/C ratio is *less than 1.0.* In this case we will eliminate it. The remaining alternatives are do nothing and alternative designs I, II, and IV.

Step 3 *Rank-order projects.* The PW of costs for each alternative:

$$\text{PW Costs (I)} \ = \$12,500$$

$$\text{PW Costs (II)} = \$11,000$$

$$\text{PW Costs (IV)} = \$16,800$$

The correct rank order is now do nothing, II, I, IV. Notice that the *modified* B/C ratio produces a rank order different from that yielded by the *conventional* version in Example 16-4.

Step 4 *Identify increment being considered.*

Step 5 *Calculate B/C ratio.*

Step 6 *Which alternative is better?*

		1st Iteration	2nd Iteration	3rd Iteration
Step 4	**Incremental Effects**	**(Do Nothing → II)**	**(II → I)**	**(II → IV)**
	ΔPlant construction cost	$11,000	$ 1500	$ 5800
	PW of ΔCosts	11,000	1500	5800
	ΔAnnual utility payment savings	700	−120	600
	ΔAnnual overcapacity revenue	550	150	−300
	ΔAnnual benefits of new jobs	750	−350	−250
	ΔAnnual O&M disbenefit	480	−360	−335
	PW of ΔBenefits	18,405	484	4662
Step 5	ΔB/C ratio (PW ΔB)/(PW ΔC)	1.67	0.32	0.80
Step 6	Is increment justified?	Yes	No	No

As an example of the calculations in the foregoing table, consider the third increment (II → IV).

ΔPlant construction cost	$= 16,800 - 11,000 = \$5800$
PW of ΔCosts	$= \$5800$
ΔAnnual utility payment savings	$= 1300 - 700 = \$600$
ΔAnnual overcapacity revenue	$= 250 - 550 = -\$300$
ΔAnnual benefits of new jobs	$= 500 - 750 = -\$250$
ΔAnnual O&M disbenefit	$= 145 - 480 = -\$335$
PW of ΔBenefits	$= (600 - 300 - 250 + 335)(P/A, 8\%, 45) = \$4662$
ΔB/C ratio, (PW ΔB)/(PW ΔC)	$= 4662/5800 = 0.80$

When the modified version of the B/C ratio is used, Alt. II emerges as the recommended power plant design—just as it did when we used the conventional B/C ratio.

OTHER EFFECTS OF PUBLIC PROJECTS

Four areas remain that merit discussion in describing the differences between government and nongovernment economic analysis: (1) financing government versus nongovernment projects, (2) the typical length of government versus nongovernment project lives, (3) quantifying and valuing benefits and disbenefits, and (4) the general effects of politics on economic analysis.

Project Financing

Governmental organizations and market-driven firms differ in the way investments in equipment, facilities, and other projects are financed. In general, firms rely on monies from individual investors (through stock and bond issuance), private lenders, and retained earnings from operations. These sources serve as the pool from which investment dollars for projects come. Management's job is to match financial resources with projects in a way that keeps the firm growing, results in an efficient and productive environment, and continues to attract investors and future lenders of capital.

On the other hand, the government sector often uses taxation and municipal bond issuance as the source of investment capital. In government, taxation and revenue from operations is adequate to finance only modest projects. However, public projects tend to be large in scale (roadways, bridges, etc.), which means that for many public projects 100% of the investment costs must be borrowed. To prevent excessive public borrowing and to assure timely debt repayment, the U.S. government, through constitutional and legislative channels, has restricted government debt. These restrictions include:

1. Local government bodies are limited in their borrowing to a specified percentage of the assessed property value in their taxation district.
2. For new construction, borrowed funds attained through the sale of bonds require the approval of local voters (sometimes by a two-thirds majority). For example, a $2 million bond proposition for a new municipal jail might increase property taxes in the city's tax district by $1.50 for every $1000 of assessed property value. These added tax revenues would then be used to retire the debt on the bonds.
3. Repayment of public debt must be made following a specific plan over a preset period of time. For monies borrowed by issuing bonds interest payments and maturity dates are set at the time of issuance.

Limitations on the use and sources of borrowed monies make funding public sector projects much different from in the private sector. Private sector firms are seldom able to borrow 100% of required funds for projects, as can be done in the public sector, but at the same time, private entities do not face restrictions on debt retirement or the uncertainty of voter approval. Passing the bond proposition is the public *go-ahead*.

Project Duration

Government projects often have longer lives than those in the private sector. In the private sector, projects most often have a projected or intended life ranging between 5 and 15 years. Some markets and technologies change more rapidly and some more slowly, but a majority of projects fall in this interval. Complex advanced manufacturing technologies, like computer-aided manufacturing or flexible automated manufacturing cells, typically have project lives at the longer end of this range.

Government projects typically have lives in the range of 20 to 50 years (or longer). Typical projects include federal highways, city water/sewer infrastructure, county dumps, and state libraries and museums. These projects, by nature, have a longer useful life than a typical private sector project. There are exceptions to this rule because private firms invest in facilities and other long-range projects, and government entities also invest in projects with shorter-term lives. But, in general, investment duration in the government sector is longer.

Government projects, because they tend to be long range and large scale, usually require substantial funding in the early stages. Highway, water/sewer, and library projects can require millions of dollars in design, surveying, and construction costs. Therefore, it is in the best interest of decision makers who are advocates of such projects to spread that first cost over as many years as possible to reduce the annual cost of capital recovery. Using longer project lives to downplay the effects of a large first cost increases the desirability of the project, as measured by the B/C ratio. Another aspect closely associated with managing the size of the capital recovery cost in a B/C ratio analysis is the interest rate used for discounting. Lower interest rates reduce the capital recovery cost of having money tied up in a project. Example 16-6 illustrates the effects that project life and interest rate can have on the analysis and acceptability of a project.

EXAMPLE 16-6

Consider a project that has been approved by local voters to build a new junior high school, needed because of increased (and projected) population growth. Analyze the project with interest rates of 3, 10, and 15% and with horizons of 15, 30, and 60 years.

Building first costs (design, planning, and construction)	$10,000,000
Initial cost for roadway and parking facilities	5,500,000
First cost to equip and furnish facility	500,000
Annual operating and maintenance costs	350,000
Annual savings from rented space	400,000
Annual benefits to community	1,600,000

SOLUTION

With this project we examine the effect that varying project lives and interest rates have on the economic value of a public project. In each case, the formula is:

$$\text{Conventional B/C ratio} = \frac{1,600,000 + 400,000}{(10,000 + 5,500,000 + 500,000)(A/P, i, n) + 350,000}$$

The B/C ratio for each combination of project life and interest rate is tabulated as follows:

Conventional Benefit–Cost Ratio for Various Combinations of Project Life and Interest Rate

Project Life (years)	Interest		
	3%	10%	15%
15	1.24	0.86	0.69
30	1.79	1.03	0.76
60	2.24	1.08	0.77

From these numbers one can see the effect of project life and interest on the analysis and recommendation. At the lower interest rate, the project has B/C ratios above 1.0 in all cases of project life, while at the higher rate the ratios are all less than 1.0. At an interest rate of 10% the recommendation to invest changes from *no* at a life of 15 years to *yes* at 30 and 60 years. A higher interest rate discounts the benefits in later years more heavily, so that they may not matter. In this case, the benefits from Years 31 to 60 add only 0.01 to the B/C ratio at 15% and 0.05 at 10%. At 3% those benefits add 0.45 to the ratio. By manipulating these two parameters (project life and interest rate), it is possible to reach entirely different conclusions regarding the desirability of the project.

Example 16-6 demonstrates that we must ensure that a long life and a low interest rate for a public project are truly appropriate and not chosen solely to make a marginal project look better.

Quantifying and Valuing Benefits and Disbenefits

The junior high school in Example 16-6 included annual benefits to the community of $1.6 million. If you were evaluating the high school, how would you estimate this? Many public sector projects like the junior high school and the examples in Table 16-1 have consequences that are difficult to state in monetary terms.

First the number of people affected by the project have to be counted—now and through the project's horizon. Then a dollar value for each person is required. For the junior high school it may be easy to estimate the number of students. But how much better will the educational outcomes be with the new school, and how valuable is that improvement?

TABLE 16-1 Example Benefits and Disbenefits for Public Investments

Public Project	Primary Benefits	Primary Disbenefits
New airport outside city	More flights, new businesses	Increased travel time to airport, more traffic on outer belt
Interstate bypass around town	Quicker commute times, reduced congestion on surface roads	Lost sales to businesses on surface roads, lost agricultural lands
New metro subway system	Faster commute times, less pollution	Lost jobs due to bus line closing, less access to service (fewer stops)
Creation of a city waste disposal facility versus sending waste out of state	Less costly; faster and more responsive to customers	Objectionable sight and smells, lost market value to homeowners, lost forestland
Construction of a nuclear power plant	Cheaper energy costs, new industry in area	Environmental risks

On the other hand, consider the levees around New Orleans that must be rebuilt in the aftermath of Katrina. An economic evaluation of the different alternatives requires estimating the number of residents that would be protected by improving the levees in New Orleans, which is extremely difficult. This requires estimating not only how the rebuilding of the city will progress but also future storm surges from hurricanes, whose frequency and intensity may be changing. Once the number of people and homes has been estimated over the next 30 to 100 years, it will be necessary to put a value on property, on disrupted lives, and on human lives.

While many individuals find it difficult to put a dollar value on a human life, there are many public projects whose main intent is to reduce the number of deaths due to floods, cancer, auto accidents, and other causes. Those projects are often justified by the value of preventing deaths. Thus, valuing human lives is an inescapable part of public sector engineering economy.

Because the benefits and disbenefits of public projects are often difficult to quantify and value, the estimated values will have more uncertainty than is typical for private sector projects. Thus those who favor a project and those who oppose it will often push to have values used that support their position.

Project Politics

To some degree political influences are felt in nearly every decision made in any organization. Predictably, some individual or group will support particular interests over competing views. This actuality exists in both firms and government organizations. In government the effects of politics are continuously felt at all levels because of the large-scale and multi-purpose nature of projects, because government decision making involves the use of the citizenry's common pool of money, and because individuals and groups have different values and views. For example, how important is economic development relative to environmental protection?

The guideline for public decision making, as set forward in the Preamble to the Constitution, is to *promote the general welfare* of citizens. However, it is impossible to please everyone all the time. The term "general welfare" implies that the architects of this document understood that the political process would produce opposing viewpoints, but at the same time they empowered decision makers to act in a representative way.

Since government projects tend to be large in scale, the time required to plan, design, fund, and construct them is usually several years. However, the political process tends to produce government leaders who support short-term decision making (because many government terms of office, either elected or appointed, are relatively short). Therein lies another difference between firm and government decision making—short-term decision making, long-term projects.

Because government decision makers are in the public eye more than those in the private sector, governmental decisions are generally more affected by "politics." Thus, the decisions that public officials make may not always be the best from an *overall* perspective. If a particular situation exposes a public official to ridicule, he may choose an expedient action to eliminate negative exposure (whereas a more careful analysis might have been better). Or, such a decision maker may placate a small, but vocal, political group over the interest of the majority of citizens by committing funds to a favored project (at the expense of other better projects). Or, a public decision maker may avoid controversy by declining to make a decision on an important, but politically charged, issue (whereas it would be in the overall interest of the citizenry if action were taken). Indeed, the role of politics in government decision making is more complex and far ranging than in the private sector.

EXAMPLE 16-7

Consider again Example 16-4, where we evaluated power plants designs. Remember that government projects are often opposed and supported by different groups in the populace. Thus, decision makers become very aware of potential political aspects when they are considering such projects. For the electric power plant decision, several political considerations may affect any evaluation of funding this project.

- The governor has been a strong advocate of workers' rights and has received abundant campaign support from organized labor (which is especially important in an industrialized state). By championing this project, the governor should be seen as pro-labor, thereby benefiting his bid for reelection, even if the project is not funded.
- The regulated electric utility providers in the state are strongly against this project, claiming that it would directly compete with them and take away some of their biggest customers. The providers have a strong lobby and key contacts with the state's utilities commission. A senior state senator has already protested that this project is the first step toward "rampant socialism in this great state."
- Business leaders in the municipalities where the two facilities would be constructed favor the project because it would create more jobs and increase the tax base. These leaders promote the project as a win–win opportunity for government and industry, where the state can benefit by reducing costs, and the electric utilities can improve their service by focusing more effectively on residential customers and their needs.
- The lieutenant governor is promoting this project, proclaiming that it is an excellent example of "initiating proactive and creative solutions to the problems that this state faces."
- Federal and state regulatory agencies are closely watching this project with respect to the Clean Air Act. Speculation is that the state plans to use a high-sulfur grade of local coal exclusively. Thus "stack scrubbers" would be required, or the high-sulfur coal would have to be mixed with lower-sulfur coal imported from other states to bring the overall air emissions in line with federal standards. The governor is using this opportunity to make the point that "the people of this state don't need regulators to tell us if we can use our own coal!"
- The state's coal operators and mining unions strongly support this project. They see the increased demands for coal and the governor's pro-labor advocacy as very positive. They plan to lobby the legislature strongly in favor of the project.
- Land preservation and environmental groups are strongly opposing the proposed project. They have studied the potential negative impacts of this project on the land and on water and air quality, as well as on the ecosystem and wildlife, in the areas where the two facilities would be constructed. Environmentalists have started a public awareness campaign urging the governor to act as the "chief steward" of the state's natural beauty and resources.

Will the project be funded? We can only guess. Clearly, however, we can see the competing influences that can be, and often are, part of decision making in the public sector.

SUMMARY

Economic analysis and decision making in government is notably different from these processes in the private sector because the basic objectives of the public and private sectors are fundamentally different. Government investments in projects seek to maximize benefits to the *greatest number of citizens,* while minimizing the *disbenefits to citizens* and *costs to the government.* Private firms, on the other hand, are focused primarily on maximizing stockholder wealth.

Several factors, not affecting private firms, enter into the decision-making process in government. The source of capital for public projects is limited primarily to taxes and bonds. Government bonds issued for project construction are subject to legislative restrictions on debt not required for private firms. Also, raising tax and bond monies involves sometimes long and politically charged processes not present in the private sector. In addition, government projects tend to be larger in scale than the projects of competitive firms and the government projects affect many more people and groups in the population. The benefits and disbenefits to the many people affected are difficult to quantify and value, which is unlike the private sector, where products and services are sold and the revenue to the firm is clearly defined. All these factors slow down the process and make investment decision analysis more difficult for government decision makers than for those in the private sector. Another difference between the public and private sectors lies in how the interest rate (MARR) is set for economic studies. In the private sector, considerations for setting the rate include the cost of capital and opportunity costs. In government, establishing the interest rate is complicated by uncertainty in specifying the cost of capital and the issue of assigning opportunity costs to taxpayers or to the government.

The benefit–cost ratio is widely used to evaluate and justify government-funded projects. This measure of merit is the ratio of the equivalent worth of benefits to the equivalent worth of costs. This ratio can be calculated by PW, AW, or FW methods. A B/C ratio *greater than 1.0* indicates that a project should be invested in if funding sources are available. For considering *mutually exclusive alternatives,* an incremental analysis is required. This method results in the recommendation of the project with the highest investment cost that can be incrementally justified. Two versions of the B/C ratio, the *conventional* and *modified* B/C ratios, produce identical recommendations. The conventional B/C ratio treats annual operating and maintenance costs as a cost in the denominator, while the modified B/C ratio subtracts those costs from the benefits in the numerator.

PROBLEMS

Objective and Viewpoint

16-1 Public sector economic analysis and decision making is often called "a multi-actor or multi-stakeholder decision problem." Explain.

16-2 Compare the general underlying objective of public decision making versus private decision making.

16-3 The text recommends a viewpoint that is appropriate in public decision making. What is suggested? What example is given to highlight the dilemma of viewpoint in public decision making? Provide another example.

16-4 In government projects, what is meant by the phrase "most of the benefits are local"? What conflict does this create for the federal government in the funding projects from public monies.

16-5 List the potential costs, benefits, and disbenefits that should be considered in evaluating a potential nuclear power plant.

16-6 Improvements at a congested intersection require that the government acquire the properties on the four corners: two gas stations, a church, and a bank. Construction will take a year, and the costs will be shared 70% by the state and 30% by the city. Traffic through the intersection is mainly commuters, local residents, and deliveries to and by local businesses. There is some through traffic from other parts of the metropolitan area.

Identify the benefits, disbenefits, and costs that must be considered in evaluating this project. From the city's viewpoint, which must be included? From the state's viewpoint? What viewpoint should be used to evaluate this project?

16-7 The state may eliminate a railroad grade crossing by building an overpass. The new structure, together with the needed land, would cost $1.8 million. The analysis period is assumed to be 30 years because either the railroad or the highway above it will be relocated by then. Salvage value of the bridge (actually, the net value of the land on either side of the railroad tracks) 30 years hence is estimated to be $100,000. A 6% interest rate is to be used.

About 1000 vehicles per day are delayed by trains at the grade crossing. Trucks represent 40%, and 60% are other vehicles. Time for truck drivers is valued at $18 per hour and for other drivers at $5 per hour. Average time saving per vehicle will be 2 minutes if the overpass is built. No time saving occurs for the railroad.

The railroad spends $48,000 annually for crossing guards. During the preceding 10-year period, the railroad has paid out $600,000 in settling lawsuits and accident cases related to the grade crossing. The proposed project will entirely eliminate both these expenses. The state estimates that the new overpass will save it about $6000 per year in expenses directly due to the accidents. The overpass, if built, will belong to the state.

Should the overpass be built? If the overpass is built, how much should the railroad be asked to contribute to the state as its share of the $1.8 million construction cost?

Selecting Rate

16-8 Discuss the alternative concepts that can be employed when setting the discounting rate for economic analysis in the public sector. What is the authors' final recommendation for setting this rate?

16-9 The city's landfill department has a capital budget of $600,000. What is the government's opportunity cost of capital if it has the following projects to consider? What does this indicate about which projects should be done?

Project	First Cost	Rate of Return (%)	B/C Ratio at 7%
A	$100,000	23	1.30
B	200,000	22	1.40
C	300,000	17	1.50
D	200,000	19	1.35
E	100,000	18	1.56

16-10 The state's fish and game department has a capital budget of $9 million. What is the government's opportunity cost of capital if it has the following projects to consider? What does this indicate about which projects should be done?

Project	First Cost	Rate of Return (%)	B/C Ratio at 7%
A	$2,000,000	9	1.23
B	1,000,000	14	1.42
C	2,000,000	10	1.17
D	3,000,000	16	1.45
E	2,000,000	13	1.56
F	3,000,000	15	1.35
G	3,000,000	12	1.32
H	1,000,000	11	1.26

16-11 Is the 7% interest rate specified in OMB A94 a real or a nominal interest rate? Should it be used with costs expressed as constant-value dollars or with costs inflated using carefully selected inflation rates?

16-12 A municipal bond has a face value of $10,000. Interest of $400 is paid every 6 months. The bond has a life of 20 years. What is the effective rate of interest on this bond? Is this rate adjusted for inflation? Estimate the government's cost of capital for this bond?

B/C and Modified Ratio

16-13 Consider the following investment opportunity:

Initial cost	$100,000
Additional cost at end of Year 1	150,000
Benefit at end of Year 1	0
Annual benefit per year at end of Years 2–10	20,000

With interest at 7%, what is the benefit–cost ratio for this project?   (*Answer:* 0.51)

16-14 Calculate the conventional and modified benefit–cost ratios for the following project.

Required first costs	$1,200,000
Annual benefits to users	$500,000
Annual disbenefits to users	$25,000
Annual cost to government	$125,000
Project life	35 years
Interest rate	10%

16-15 For the data given in Problem 16-14, for handling benefits and costs, demonstrate that the calculated B/C ratio is the same using each of the following methods: present worth, annual worth, and future worth.

16-16 What is the essential difference between the *conventional* and *modified* versions of the benefit–cost ratio? Is it possible for these two measures to provide conflicting recommendations regarding invest/do-not-invest decisions?

16-17 A government agency has estimated that a flood control project has costs and benefits that are parabolic, according to the equation

$$(\text{Present worth of benefits})^2$$
$$- 22(\text{Present worth of cost}) + 44 = 0$$

where both benefits and costs are stated in millions of dollars. What is the present worth of cost for the optimal size project?

Incremental Analysis

16-18 The Highridge Water District needs an additional supply of water from Steep Creek. The engineer has selected two plans for comparison:

Gravity plan: Divert water at a point 10 miles up Steep Creek and carry it through a pipeline by gravity.

Pumping plan: Divert water at a point near the district and pump it through 2 miles of pipeline. The pumping plant can be built in two stages: half now and half 10 years later.

	Gravity	Pumping
Initial investment	$2,800,000	$1,400,000
Investment in 10th year	0	200,000
Operation, maintenance, replacements, per year	10,000	25,000
Average annual power cost		
First 10 years	0	50,000
Next 30 years	0	100,000

Use a 40-year analysis period and 8% interest. Salvage values can be ignored. Use the conventional benefit–cost ratio method to select the more economical plan. (*Answer:* Pumping plan)

16-19 Two different routes are being considered for a mountain highway. The **high road** will require building several bridges and navigates around the highest mountain points, thus requiring more roadway. The **low road** constructs several tunnels for a more direct route through the mountains. Projected travel volume for this new section of road is 2500 cars per day. Use the *modified* B/C ratio to determine which alternative should be recommended. Assume that project life is 45 years and $i = 6\%$.

	High Road	Low Road
Construction cost/ mile	$200,000	$450,000
Number of miles required	35	10
Annual benefit/ car-mile	$0.015	$0.045
Annual O&M costs/ mile	$2000	$10,000

16-20 The city engineer has prepared two plans for roads in the city park. Both plans meet anticipated requirements for the next 40 years. The city's minimum attractive rate of return is 7%.

Plan *A* is a three-stage development program: $300,000 is to be spent now followed by $250,000 at the end of 15 years and $300,000 at the end of 30 years. Annual maintenance will be $75,000 for the first 15 years, $125,000 for the next 15 years, and $250,000 for the final 10 years.

Plan *B* is a two-stage program: $450,000 is required now, followed by $50,000 at the end of 15 years. Annual maintenance will be $100,000 for the first 15 years and $125,000 for the subsequent years. At the end of 40 years, this plan has a salvage value of $150,000.

Use a conventional benefit–cost ratio analysis to determine which plan should be chosen.

16-21 A two-lane highway between two cities, 10 miles apart, is to be converted to a four-lane divided freeway. The average daily traffic (ADT) on the new freeway is forecast to average 20,000 vehicles per day over the next 20 years. Trucks represent 5% of the total traffic. Annual maintenance on the existing highway is $1500 per lane-mile. The existing accident rate is 4.58 per million vehicle-miles (MVM). Three alternate plans are under consideration.

Plan A: Improve along the existing development by adding two lanes adjacent to the existing lanes at a cost of $450,000 per mile. This will reduce auto travel time by 2 minutes and truck travel time by 1 minute. The estimated accident rate is 2.50 per MVM. Annual maintenance is $1250 per lane-mile.

Plan B: Improve the highway along the existing alignment with grade improvements at a cost of $650,000 per mile. Plan *B* adds two lanes and would reduce auto and truck travel time by 3 minutes each. The accident rate on this improved road is estimated to be 2.40 per MVM. Annual maintenance is $1000 per lane-mile.

Plan C: Construct a new freeway on a new alignment at a cost of $800,000 per mile. This plan would reduce auto travel time by 5 minutes and truck travel time by 4 minutes. Plan *C* is 0.3 mile longer than *A* or *B*. The estimated accident rate for *C* is 2.30 per MVM. Annual maintenance is $1000 per lane-mile. Plan *C* includes abandoning the existing highway with no salvage value.

Incremental operating cost
Autos	6¢ per mile
Trucks	18¢ per mile

Time saving
Autos	3¢ per minute
Trucks	15¢ per minute
Average accident cost	$1200

If a 5% interest rate is used, which of the three proposed plans should be adopted? (*Answer:* Plan *C*)

16-22 Evaluate these mutually exclusive alternatives with a horizon of 15 years and a MARR of 12%.

	A	*B*	*C*
Initial investment	$9500	$18,500	$22,000
Annual savings	3200	5,000	9,800
Annual costs	1000	2,750	6,400
Salvage value	6000	4,200	14,000

Use each of these approaches:

(*a*) *Conventional* B/C ratio

(*b*) *Modified* B/C ratio

(*c*) Present worth analysis

(*d*) Internal rate of return analysis

(*e*) Payback period

16-23 A 50-meter tunnel must be constructed for a new city aqueduct. One alternative is to build a full-capacity tunnel now for $500,000. The other alternative is to build a half-capacity tunnel now for $300,000 and then to build a second parallel half-capacity tunnel 20 years hence for $400,000. The cost to repair the tunnel lining every 10 years is $20,000 for the full-capacity tunnel and $16,000 for each half-capacity tunnel.

Determine whether the full-capacity tunnel or the half-capacity tunnel should be constructed now. Solve the problem by the conventional benefit–cost ratio analysis, using a 5% interest rate and a 50-year analysis period. There will be no tunnel lining repair at the end of the 50 years.

16-24 The Fishery and Wildlife Agency of Ireland is considering four mutually exclusive design alternatives (Table P16-24) for a major salmon hatchery. This agency of the Irish government uses the following B/C ratio for decision making:

$$\text{B/C ratio} = \frac{\text{EW(Net benefits)}}{\text{EW(Capital recovery cost)} + \text{EW(O\&M cost)}}$$

TABLE P16-24 Data (Values in 1000s)

	Irish Fishery Design Alternatives			
	A	B	C	D
First cost	$9,500	$12,500	$14,000	$15,750
Annual benefits	2,200	1,500	1,000	2,500
Annual O&M costs	550	175	325	145
Annual disbenefits	350	150	75	700
Salvage value	1,000	6,000	3,500	7,500

TABLE P16-26 Data (Costs in 1000s)

Year	Repair Now	Repair in 2 Years	Repair in 4 Years	Repair in 5 Years
0	−$150			
1	5			
2	10	−$150		
3	20	20		
4	30	30	−$150	
5	40	40	40	−$150
6	50	50	50	50
7	50	50	50	50
8	50	50	50	50
9	50	50	50	50
10	50	50	50	50
11	0	50	50	50
12	0	50	50	50
13	0	0	50	50
14	0	0	50	50
15	0	0	0	50

Using an interest rate of 8% and a project life of 30 years, recommend which design is best.

16-25 Six mutually exclusive investments have been identified for evaluation by means of the benefit–cost ratio method. Assume a MARR of 10% and? an equal project life of 25 years for all alternatives.

	Mutually Exclusive Alternatives					
Annualized	1	2	3	4	5	6
Net costs to sponsor ($M)	15.5	13.7	16.8	10.2	17.0	23.3
Net benefits to users ($M)	20.0	16.0	15.0	13.7	22.0	25.0

(a) Use annual worth and the B/C ratio to identify the better alternative.

(b) If this were a set of *independent* alternatives, how would you conduct a comparison?

16-26 A section of state highway needs repair. At present, the traffic volume is low and few motorists would benefit. However, traffic is expected to increase, with resulting increased motorist benefits. The repair work will produce benefits for 10 years after it is completed. (See Table P16.26.)

Should the road be repaired and, if so, when should the work be done? Use a 15% MARR.

16-27 The local highway department is analyzing reconstruction of a mountain road. The vehicle traffic increases each year, hence the benefits to the motoring public also increase. Based on a traffic count, the benefits are projected as follows:

Year	End-of-Year Benefit
2011	$10,000
2012	12,000
2013	14,000
2014	16,000
2015	18,000
2016	20,000
	and so on, increasing $2000 per year

The reconstructed pavement will cost $275,000 when it is installed and will have a 15-year useful life. The construction period is short, hence a beginning-of-year reconstruction will result in the listed end-of-year benefits. Assume a 6% interest rate. The reconstruction, if done at all, must be done not later than 2016. Should it be done, and if so, in what year?

Challenges for Public Sector

16-28 Big City Carl, a local politician, is pushing a new dock and pier system at the river to attract commerce. A committee appointed by the mayor (an opponent of Carl's) has developed the following estimates.

Cost to remove current facilities	$ 750,000
Material, labor, and overhead for new construction	2,750,000
Annual operating & maintenance costs	185,000
Annual benefits from new commerce	550,000

Annual disbenefits to sportsmen	35,000
Project life	20 years
Interest rate	8%

(a) Using the *conventional* B/C ratio, determine whether the project should be funded.

(b) After studying the numbers given by the committee, Big City Carl argued that the project life should be *at least* 25 years and more likely closer to 30 years. How did he arrive at this estimate, and why is he making this statement?

16-29 The federal government proposes to construct a multipurpose water project to provide water for irrigation and municipal use. In addition, there are flood control and recreation benefits. The benefits are given in Table P16-29. The annual benefits are one-tenth of the decade benefits. The operation and maintenance cost is $15,000 per year. Assume a 50-year analysis period with no net project salvage value.

(a) If an interest rate of 5% is used, and a benefit–cost ratio of unity, what capital expenditure can be justified to build the water project now?

(b) If the interest rate is changed to 8%, how does this change the justified capital expenditure?

16-30 If you favored Plan *B* in Problem 16-20, what value of MARR would you use in the computations? Explain.

16-31 Describe how a decision maker can use each of the following to "skew" the results of a B/C ratio analysis in favor of his or her own position on funding projects:

(a) Conventional versus modified ratios

(b) Interest rates

TABLE P16-29 Data (thousands)

Purpose	Decades				
	First	Second	Third	Fourth	Fifth
Municipal	$ 40	$ 50	$ 60	$ 70	$110
Irrigation	350	370	370	360	350
Flood control	150	150	150	150	150
Recreation	60	70	80	80	90
Totals	$600	$640	$660	$660	$700

(c) Project duration

(d) Benefits, costs, and disbenefits

16-32 Briefly describe your sources and methods for estimating the value of:

(a) a saved hour of commuting time

(b) converting 15 miles of unused railroad tracks near a community of 300,000 into a new bike path

(c) reducing annual flood risks from the Mississippi River for St. Louis by 5%.

(d) a human life

16-33 Discuss potential data sources and methods for estimating each of the costs, benefits, and disbenefits identified in Problem 16-5 for the nuclear power plant.

16-34 Discuss potential data sources and methods for estimating each of the costs, benefits, and disbenefits identified in Problem 16-6 for the congested intersection.

16-35 Think about a major government construction project under way in your state, city, or region. Are the decision makers who originally analyzed and initiated the project currently in office? How can politicians use "political posturing" with respect to government projects?

Minicase

The data supplied adds detail to the chapter-opening vignette.

The three problems which follow the data were contributed by *William R. Truran, Stevens Institute of Technology and Peter A. Cerenzio, Cerenzio & Panaro Consulting Engineers*

Daily and intermediate cover material is 20% of the landfill's usable space. Density of solid waste is 1500 pounds per cubic yard. LFG recovery rates:

3000 cubic feet per ton for municipal solid waste (MSW)

1500 cubic feet per ton for construction and demolition waste (C&D)

Assume waste composition is 80% MSW and 20% C&D.

Methane content in the LFG is 50% for MSW and 20% for C&D.

Heating value
For methane: 1,030,000 BTU/1000 cubic feet
For fuel oil: 138,800 BTU/gallon
Assumed furnace efficiency is 88% for methane and 82% for fuel oil.
Cost of fuel oil is $2.50/gallon.
Heating load for residential dwelling is 100 million BTU per year.
1.7×10^4 BTU per kilowatt-hours
Value = $0.05/kWh

16-36 An economic analysis is needed for a new municipal solid waste landfill. This includes determining the potential economic benefit from the gas that is generated when solid waste decomposes. The landfill is proposed at 14 acres with a design capacity of 1 million cubic yards of capacity. The final capping system will require a 3-foot layer over the 14 acres, and the waste flow rate is 120,000 tons per year. Calculate:

(a) The life of the landfill.

(b) The average annual methane production (assume all methane production has ceased by 15 years after landfill closure).

(c) The dollar value of annual methane converted to electricity (neglect collection and energy production costs).

16-37 A developer has proposed a 650-unit residential development adjacent to the landfill. The proposal includes using the gas generated from the landfill to heat the homes (and mitigate odors). To determine the economic feasibility of the proposal, the value of the gas for heating purposes must be determined. Determine whether the quantity is sufficient for heating the development. Is this economically more attractive than using fuel oil? Is it operationally feasible?

16-38 A 4.6-acre landfill must be evaluated for economic viability. As part of the cost–benefit analysis, the cost to extract and treat the landfill gas must be determined. Using the following, design a landfill gas extraction system and estimate the cost to implement such a system:

>Landfill dimensions are 1000 feet by 200 feet.
>Area of influence of a landfill gas extraction well is a 50-foot radius.
>Cost of well construction is $3000 each.
>Cost of wellhead (necessary for each well) is $2500 each.

Cost of collection header piping, diameter HDPE is $35/linear 8-inch foot.
Cost of condensate knockouts (located at low points) is $5000 each.
Cost of blower/flare station is $500,000.

CHAPTER 17

ACCOUNTING AND ENGINEERING ECONOMY

The Two Faces of ABB

In the late 1990s, ABB Ltd was flying high. The European conglomerate was an engineering giant with a global network of operations and forward-looking management. Unlike many of its competitors, ABB was also committed to modernizing its accounting system. ABB had followed the traditional practice of assigning overhead costs to its divisions on a roughly equal basis. But this practice tended to obscure the fact that some activities incurred far more costs than others.

So the company spent substantial time and resources switching over to an activity-based costing (ABC) system, which assigns costs to the activities that actually produce them. The results were quite positive, allowing ABB to zero in on areas where it could cut costs most effectively.

But in other respects, ABB's accounting practices were less effective. One of its most serious problems arose at Combustion Engineering, an American subsidiary that was exposed to numerous asbestos liability claims. For years, ABB had downplayed this potential liability, despite warnings from outside analysts. Finally, in late 2002, ABB admitted that its asbestos liability exceeded the subsidiary's total asset value.

But there was more bad news still. The company also issued a very poor third-quarter earnings report, despite having earlier assured investors that ABB was on target to improve its earnings and decrease its debt. When incredulous investors asked how this could have happened, ABB management blamed "poor internal reporting."

By this point, ABB's stock price had nose-dived, and credit rating agencies viewed the company's bonds as little better than junk. In 2003, Combustion Engineering went through bankruptcy and reached a settlement for the asbestos claims.

ABB has recovered, and by 2007 its stock price had climbed sevenfold over its low.

QUESTIONS TO CONSIDER

1. ABB's adoption of activity-based costing received widespread publicity and boosted the company's reputation for innovative management. How might this have affected investors' assumptions about the company's other accounting practices?
2. Outside analysts estimated that ABB lost over $690 million in 2001. Yet many investors were still stunned by its poor showing in late 2002. What does this say about the relationship between financial accounting and investor confidence?
3. ABB was not alone in its financial misery, of course. How did ABB's accounting problems compare with those of well-known American companies such as Enron and WorldCom?
4. In each of these well-publicized situations, were there ethical lapses? What were the lapses, and who was responsible?

After Completing This Chapter...
The student should be able to:

- Describe the links between engineering economy and accounting.
- Describe the objectives of general accounting, explain what financial transactions are, and show how they are important.
- Use a firm's balance sheet and associated financial ratios to evaluate the firm's health.
- Use a firm's income statement and associated financial ratios to evaluate the firm's performance.
- Use traditional absorption costing to calculate product costs.
- Understand the greater accuracy in product costs available with activity-based costing (ABC).

ngineering economy focuses on the financial aspects of projects, while accounting focuses on the financial aspects of firms. Thus the application of engineering economy is much easier if one has some understanding of accounting principles. In fact, one important accounting topic, depreciation, was the subject of an earlier chapter.

THE ROLE OF ACCOUNTING

Accounting data are used to value capital equipment, to decide whether to make or buy a part, to determine costs and set prices, to set indirect cost rates, and to make product mix decisions. Accounting is used in private-sector firms and public sector agencies, but for simplicity this chapter uses "the firm" to designate both. Accountants track the costs of projects and products, which are the basis for estimating future costs and revenues.

The engineering economy, accounting, and managerial functions are interdependent. As shown in Figure 17-1, data and communications flow between them. Whether carried out by a single person in a small firm or by distinct divisions in a large firm, all are needed.

- Engineering economy analyzes the economic impact of design alternatives and projects over their life cycles.
- Accounting determines the dollar impact of past decisions, reports on the economic viability of a unit or firm, and evaluates potential funding sources.
- Management allocates available investment funds to projects, evaluates unit and firm performance, allocates resources, and selects and directs personnel.

Accounting for Business Transactions

A business transaction involves two parties and the exchange of dollars (or the promise of dollars) for a product or service. Each day, millions of transactions occur between firms and their customers, suppliers, vendors, and employees. Transactions are the lifeblood of the business world and are most often stated in monetary terms. The accounting function records, analyzes, and reports these exchanges.

Transactions can be as simple as payment for a water bill, or as complex as the international transfer of millions of dollars of buildings, land, equipment, inventory, and other

Accounting	Management	Engineering Economy
About past	About past and future	About future
Analyzing	Capital budgeting	Feasibility of alternatives
Summarizing	Decision making	Collect & analyze data
Reporting	Setting goals	Estimate
Financial indicators	Assessing impacts	Evaluate projects
Economic trends	Analyzing risk	Recommend
Cost acquisitions	Planning	Audit
	Controlling	Identify needs
	Record keeping	Trade-offs & constraints

Data and Communication   Data and Communication

Budgeting   Data and Communication   Estimating

FIGURE 17-1 The accounting, managerial, and engineering economy functions.

assets. Also, with transactions, one business event may lead to another—all of which need to be accounted for. Consider, for example, the process of selling a robot or bull-dozer. This simple act involves several related transactions: (1) releasing equipment from inventory, (2) shipping equipment to the purchaser, (3) invoicing the purchaser, and finally (4) collecting from the purchaser.

Transaction accounting involves more than just reporting: it includes finding, synthesizing, summarizing, and analyzing data. For the engineering economist, historical data housed in the accounting function are the foundation for estimates of future costs and revenues.

Most accounting is done in nominal or *stable* dollars. Higher market values and costs due to inflation are less objective than cost data, and with a going concern, accountants have decided that objectivity should be maintained. Similarly, most assets are valued at their acquisition cost adjusted for depreciation and improvements. To be conservative, when market value is lower than this adjusted cost, the lower value is used. This restrains the interests of management in maximizing a firm's apparent value. If a firm must be liquidated, then current market value must be estimated.

The accounting function provides data for *general accounting* and *cost accounting*. This chapter's presentation begins with the balance sheet and income statement, which are the two key summaries of financial transactions for general accounting. This discussion includes some of the basic financial ratios used for short- and long-term evaluations. The chapter concludes with a key topic in cost accounting—allocating indirect expenses.

THE BALANCE SHEET

The primary accounting statements are the **balance sheet** and the **income statement.** The **balance sheet** describes the firm's financial condition at a specific time, while the income statement describes the firm's performance over a period of time—usually a year.

The balance sheet lists the firm's assets, liabilities, and equity on a specified date. This is a picture of the organization's financial health or a snapshot in time. Usually, balance sheets are taken at the end of the quarter and fiscal year. The balance sheet is based on the **fundamental accounting equation:**

$$\text{Assets} = \text{Liabilities} + \text{Equity} \qquad (17\text{-}1)$$

Figure 17-2 illustrates the basic format of the balance sheet. Notice in the balance sheet, as in Equation 17-1, that **assets** are listed on the left-hand side and **liabilities** and **equity** are on the right-hand side. The fact that the firm's resources are *balanced* by the sources of funds is the basis for the name of the balance sheet.

Assets

In Equation 17-1 and Figure 17-2, **assets** are owned by the firm and have monetary value. **Liabilities** are the dollar claims against the firm. **Equity** represents funding from the firm and its owners (the shareholders). In Equation 17-1, assets are always balanced by the sum of the liabilities and the equity. The value for retained earnings is set so that equity equals assets minus liabilities.

On a balance sheet, assets are listed in order of decreasing liquidity, that is, according to how quickly each one can be converted to cash. Thus, *current assets* are listed first, and within that category in order of decreasing liquidity are listed cash, receivables, securities,

Balance Sheet for Engineered Industries, December 31, 2007 (all amounts in $1000s)

Assets		Liabilities	
Current assets		Current liabilities	
Cash	$1940	Accounts payable	$1150
Accounts receivable	950	Notes payable	80
Securities	4100		
Inventories	1860	Accrued expense	950
(*minus*) Bad debt provision	−80	Total current liabilities	2180
Total current assets	8770		
		Long-term liabilities	1200
		Total liabilities	**3380**
Fixed assets			
Land	335		
Plant and equipment	6500		
(*minus*) Accumulated depr.	−2350		
Total fixed assets	4485	**Equity**	
		Preferred stock	110
Other assets		Common stock	650
Prepays/deferred charges	140	Capital surplus	930
Intangibles	420	Retained earnings	8745
Total other assets	560	Total equity	10,435
Total assets	**13,815**	**Total liabilities and equity**	**13,815**

FIGURE 17-2 Sample balance sheet.

and inventories. *Fixed assets,* or *property, plant, and equipment,* are used to produce and deliver goods and/or services, and they are not intended for sale. Items such as prepayments and intangibles such as patents are listed last.

The term "receivables" comes from the manner of handling billing and payment for most business sales. Rather than requesting immediate payment for every transaction by check or credit card, most businesses record each transaction and then once a month bill for all transactions. The total that has been billed less payments already received is called accounts receivable, or receivables.

Liabilities

On the balance sheet, liabilities are divided into two major classifications—short term and long term. The *short-term* or *current liabilities* are expenses, notes, and other payable accounts that are due within one year from the balance sheet date. *Long-term liabilities* include mortgages, bonds, and loans with later due dates. For Engineered Industries in Figure 17-2, total current and long-term liabilities are $2,180,000 and $1,200,000, respectively. Often in performing engineering economic analyses, the **working capital** for a project must be estimated. The total amount of working capital available may be calculated with Equation 17-2 as the difference between current assets and current liabilities.

$$\text{Working capital} = \text{Current assets} - \text{Current liabilities} \qquad (17\text{-}2)$$

For Engineered Industries, there would be $8,770,000 - $2,180,000 = $6,590,000 available in working capital.

Equity

Equity is also called *owner's equity* or *net worth*. It includes the par value of the owners' stockholdings and the capital surplus, which are the excess dollars brought in over par value when the stock was issued. The capital surplus can also be called *additional paid-in capital*, or APIC. Retained earnings are dollars a firm chooses to retain rather than paying out as dividends to stockholders.

Retained earnings within the equity component is the dollar quantity that always brings the balance sheet, and thus the fundamental accounting equation, into balance. For Engineered Industries, *total equity* value is listed at $10,435,000. From Equation 17-1 and the assets, liabilities, and equity values in Figure 17-2, we can write the balance as follows:

$$\text{Assets} = \text{Liabilities} + \text{Equity}$$

$$\text{Assets (current, fixed, other)} = \text{Liabilities (current and long-term)} + \text{Equity}$$

$$8,770,000 + 4,485,000 + 560,000 = 2,180,000 + 1,200,000 + 10,435,000$$

$$\$13,815,000 = \$13,815,000$$

An example of owner's equity is ownership of a home. Most homes are purchased by means of a mortgage loan that is paid off at a certain interest rate over 15 to 30 years. At any point in time, the difference between what is owed to the bank (the remaining balance on the mortgage) and what the house is worth (its appraised market value) is the *owner's equity*. In this case, the loan balance is the *liability*, and the home's value is the *asset*—with *equity* being the difference. Over time, as the house loan is paid off, the owner's equity increases.

The balance sheet is a very useful tool that shows one view of the firm's financial condition at a particular point in time.

Financial Ratios Derived from Balance Sheet Data

One common way to evaluate the firm's health is through ratios of quantities on the balance sheet. Firms in a particular industry will typically have similar values, and exceptions will often indicate firms with better or worse performance. Two common ratios used to analyze the firm's current position are the current ratio and the acid-test ratio.

A firm's **current ratio** is the ratio of current assets to current liabilities, as in Equation 17-3.

$$\text{Current Ratio} = \text{Current assets/Current liabilities} \tag{17-3}$$

This ratio provides insight into the firm's solvency over the short term by indicating its ability to cover current liabilities. Historically, firms aim to be at or above a ratio of 2.0; however, this depends heavily on the industry as well as the individual firm's management practices and philosophies. The current ratio for Engineered Industries in Figure 17-2 is above 2 ($8,770,000/2,180,000 = 4.02$).

Both working capital and the current ratio indicate the firm's ability to meet currently maturing obligations. However, neither describes the type of assets owned. The **acid-test ratio** or **quick ratio** becomes important when one wishes to consider the firm's ability to pay debt "instantly." The acid-test ratio is computed by dividing a firm's **quick assets** (cash, receivables, and market securities) by total current liabilities, as in Equation 17-4.

$$\text{Acid-test ratio} = \text{Quick assets/Total current liabilities} \quad\quad (17\text{-}4)$$

Current inventories are excluded from quick assets because of the time required to sell these inventories, collect the receivables, and subsequently have the cash on hand to reduce debt. For Engineered Industries in Figure 17-2, the calculated acid-test ratio is well above 1[(1,940,000+ 950,000 + 4,100,000)/2,180,000 = 3.21].

Working capital, current ratio, and acid-test ratio are all indications of the firm's financial health (status). A thorough financial evaluation would consider all three, including comparisons with values from previous periods and with broad-based industry standards. When trends extend over multiple periods, the trends may be more important than the current values.

THE INCOME STATEMENT

The **income statement** or **profit and loss statement** summarizes the firm's revenues and expenses over a month, quarter, or year. Rather than being a snapshot like the balance sheet, the income statement encompasses a *period* of business activity. The income statement is used to evaluate revenue and expenses that occur in the interval *between* consecutive balance sheet statements. The income statement reports the firm's *net income (profit)* or *loss* by subtracting expenses from revenues. If revenues minus expenses is positive in Equation 17-5, there has been a profit, if negative a loss has occurred.

$$\text{Revenues} - \text{Expenses} = \text{Net profit (Loss)} \quad\quad (17\text{-}5)$$

Revenue, as in Equation 17-5, serves to increase ownership in a firm, while expenses serve to decrease ownership. Figure 17-3 is an example income statement.

To aid in analyzing performance, the income statement in Figure 17-3 separates operating and nonoperating activities and shows revenues and expenses for each. Operating revenues are made up of sales revenues (minus returns and allowances), while nonoperating revenues come from rents and interest receipts.

Operating expenses produce the products and services that generate the firm's revenue stream of cash flows. Typical operating expenses include cost of goods sold, selling and promotion costs, depreciation, general and administrative costs, and lease payments. *Cost of goods sold (COGS)* includes the labor, materials, and indirect costs of production.

Engineers design production systems, and they are involved in labor loading, specifying materials, and make/buy decisions. All these items affect a firm's cost of goods sold. Good engineering design focuses not only on technical functionality but also on cost-effectiveness as the design *integrates* the entire production system. Also of interest to the engineering economist is *depreciation* (see Chapter 11)—which is the systematic "writing off" of a capital expense over a period of years. This noncash expense is important because it represents a decrease in value of the firm's capital assets.

Income Statement for Engineered Industries for End of Year 2008 (all amounts in $1000)

Operating revenues and expenses	
Operating revenues	
Sales	$ 18,900
(*minus*) Returns and allowances	−870
Total operating revenues	**18,030**
Operating expenses	
Cost of goods and services sold	
Labor	6140
Materials	4640
Indirect cost	2280
Selling and promotion	930
Depreciation	450
General and administrative	2160
Lease payments	510
Total operating expense	**17,110**
Total operating income	**920**
Nonoperating revenues and expenses	
Rents	20
Interest receipts	300
Interest payments	−120
Total nonoperating income	**200**
Net income before taxes	**1120**
Income taxes	−390
Net profit (loss) for 2008	**730**

FIGURE 17-3 Sample income statement.

The operating revenues and expenses are shown first, so that the firm's operating income from its products and services can be calculated. Also shown on the income statement are nonoperating expenses such as interest payments on debt in the form of loans or bonds.

From the data in Figure 17-3, Engineered Industries has total expenses (operating = $17,110,000 and nonoperating = $120,000) of $17,230,000. Total revenues are $18,350,000 (= $18,030,000 + $20,000 + $300,000). The net after-tax profit for year 2008 shown in Figure 17-3 as $730,000, but it can also be calculated using Equation 17-5 as:

$$\text{Net profits (Loss)} = \text{Revenues} - \text{Expenses [before taxes]}$$

$$\$1,120,000 = 18,350,000 - 17,230,000 \text{ [before taxes] and with}$$

$$\$390,000 \text{ taxes paid}$$

thus

$$\$730,000 = 1,120,000 - 390,000 \text{ [after taxes]}$$

Financial Ratios Derived from Income Statement Data

Interest coverage, as given in Equation 17-6, is calculated as the ratio of total income to interest payments—where *total income* is total revenues minus all expenses except interest payments.

$$\text{Interest coverage} = \text{Total income}/\text{Interest payments} \qquad (17\text{-}6)$$

The interest coverage ratio (which for industrial firms should be at least 3.0) indicates how much revenue must drop to affect the firm's ability to finance its debt. With an interest coverage ratio of 3.0, a firm's revenue would have to decrease by two-thirds (unlikely) before it became impossible to pay the interest on the debt. The larger the interest coverage ratio the better. Engineered Industries in Figure 17-3 has an interest coverage ratio of

$$(18,350,000 - 17,110,000)/120,000 = 10.3$$

Another important financial ratio based on the income statement is the **net profit ratio.** This ratio (Equation 17-7) equals net profits divided by net sales revenue. Net sales revenue equals sales minus returns and allowances.

$$\text{Net profit ratio} = \text{Net profit}/\text{Net sales revenue} \qquad (17\text{-}7)$$

This ratio provides insight into the cost efficiency of operations as well as a firm's ability to convert sales into profits. For Engineered Industries in Figure 17-3, the net profit ratio is $730,000/18,030,000 = 0.040 = 4.0\%$. Like other financial measures, the net profit ratio is best evaluated by comparisons with other time periods and industry benchmarks, and trends may be more significant than individual values.

Linking the Balance Sheet, Income Statement, and Capital Transactions

The balance sheet and the income statement are separate but linked documents. Understanding how the two are linked together helps clarify each. Accounting describes such links as the *articulation* between these reports.

The balance sheet shows a firm's assets, liabilities, and equity at a particular point in time, whereas the income statement summarizes revenues and expenses over a time interval. These tabulations can be visualized as a snapshot at the period's beginning (a balance sheet), a video summary over the period (the income statement), and a snapshot at the period's end (another balance sheet). The income statement and changes in the balance sheets summarize the business transactions that have occurred during that period.

There are many links between these statements and the cash flows that make up business transactions, but for engineering economic analysis the following are the most important.

1. Overall profit or loss (income statement) and the starting and ending equity (balance sheets).
2. Acquisition of capital assets.
3. Depreciation of capital assets.

The overall profit or loss during the year (shown on the income statement) is reflected in the change in retained earnings between the balance sheets at the beginning and end of the year. To find the change in retained earnings (RE), one must also subtract any dividends distributed to the owners and add the value of any new capital stock sold:

$$RE_{beg} + \text{Net income/Loss} + \text{New stock} - \text{Dividends} = RE_{end}$$

When capital equipment is purchased, the balance sheet changes, but the income statement does not. If cash is paid, then the cash asset account decrease equals the increase in the capital equipment account—there is no change in total assets. If a loan is used, then the capital equipment account increases, and so does the liability item for loans. In both cases the equity accounts and the income statement are unchanged.

The depreciation of capital equipment is shown as a line on the income statement. The depreciation for that year equals the change in accumulated depreciation between the beginning and the end of the year—after subtraction of the accumulated depreciation for any asset that is sold or disposed of during that year.

Example 17-1 applies these relationships to the data in Figures 17-1 and 17-2.

EXAMPLE 17-1

For simplicity, assume that Engineered Industries will not pay dividends in 2008 and did not sell any capital equipment. It did purchase $400,000 in capital equipment. What can be said about the values on the balance sheet at the end of 2008, using the linkages just described?

SOLUTION

First, the net profit of $730,000 will be added to the retained earnings from the end of 2007 to find the new retained earnings at the end of 2008:

$$RE_{12/31/2008} = \$730,000 + \$8,745,000 = \$9,475,000$$

Second, the fixed assets shown at the end of 2008 would increase from $6,500,000 to $6,900,000.

Third, the accumulated depreciation would increase by the $450,000 in depreciation shown in the 2008 income statement from the $2,350,000 posted in 2007. The new accumulated depreciation on the 2008 balance sheet would be $2,800,000. Combined with the change in the amount of capital equipment, the new fixed asset total for 2008 would equal:

$$\$335,000 + \$6,900,000 - \$2,800,000 = \$4,435,000$$

TRADITIONAL COST ACCOUNTING

A firm's *cost-accounting system* collects, analyzes, and reports operational performance data (costs, utilization rates, etc.). Cost-accounting data are used to develop product costs, to determine the mix of labor, materials, and other costs in a production setting, and to evaluate outsourcing and subcontracting possibilities.

Direct and Indirect Costs

Costs incurred to produce a product or service are traditionally classified as either *direct* or *indirect (overhead)*. Direct costs come from activities directly associated with the final product or service produced. Examples include material costs and labor costs for engineering design, component assembly, painting, and drilling.

Some organizational activities are difficult to link to specific projects, products, or services. For example, the receiving and shipping areas of a manufacturing plant are used by all incoming materials and all outgoing products. Materials and products differ in their weight, size, fragility, value, number of units, packaging, and so on, and the receiving and shipping costs depend on all these factors. Also, different materials arrive together and different products are shipped together, so these costs are intermingled and often cannot be tied directly to each product or material.

Other costs, such as the organization's management, sales, and administrative expenses, are difficult to link directly to individual products or services. These indirect or overhead expenses also include machine depreciation, engineering and technical support, and customer warranties.

Indirect Cost Allocation

To allocate indirect costs to different departments, products, and services, accountants use quantities such as direct-labor hours, direct-labor costs, material costs, and total direct cost. One of these is chosen to be the burden vehicle. The total of all indirect or overhead costs is divided by the total for the burden vehicle. For example, if direct-labor hours is the burden vehicle, then overhead will be allocated based on overhead dollar per direct-labor hour. Then each product, project, or department will *absorb* (or be allocated) overhead costs, based on the number of direct-labor hours each has.

This is the basis for calling traditional costing systems **absorption costing.** For decision making, the problem is that the absorbed costs represent average, not incremental, performance.

Four common ways of allocating overhead are direct-labor hours, direct-labor cost, direct-materials cost, and total direct cost. The first two differ significantly only if the cost per hour of labor differs for different products. Example 17-2 uses direct-labor and direct-materials cost to illustrate the different choices of burden vehicle.

EXAMPLE 17-2

Industrial Robots does not manufacture its own motors or computer chips. Its premium product differs from its standard product in having heavier-duty motors and more computer chips for greater flexibility.

As a result, Industrial Robots manufactures a higher fraction of the standard product's value itself, and it purchases a higher fraction of the premium product's value. Use the following data to allocate $850,000 in overhead on the basis of labor cost and materials cost.

	Standard	Premium
Number of units per year	750	400
Labor cost (each)	$400	$500
Materials cost (each)	$550	$900

SOLUTION

First, the labor and material costs for the standard product, the premium product, and in total are calculated.

	Standard	Premium	Total
Number of units per year	750	400	
Labor cost (each)	$ 400	$ 500	
Materials cost (each)	550	900	
Labor cost	300,000	200,000	$500,000
Materials cost	412,500	360,000	772,500

Then the allocated cost per labor dollar, $1.70, is found by dividing the $850,000 in overhead by the $500,000 in total labor cost. The allocated cost per material dollar, $1.100324, is found by dividing the $850,000 in overhead by the $772,500 in materials cost. Now, the $850,000 in allocated overhead is split between the two products using labor costs and material costs.

	Standard	Premium	Total
Labor cost	$300,000	$200,000	$500,000
Overhead/labor	1.70	1.70	
Allocation by labor	510,000	340,000	850,000
Material cost	412,500	360,000	0
Overhead/material	1.100324	1.100324	
Allocation by material	453,884	396,117	850,000

If labor cost is the burden vehicle, then 60% of the $850,000 in overhead is allocated to the standard product. If material cost is the burden vehicle, then 53.4% is allocated to the standard product. In both cases, the $850,000 has been split between the two products. Using total direct costs would produce another overhead allocation between these two values. However, for decision making about product mix and product prices, incremental overhead costs must be analyzed. All the allocation or burden vehicles are based on an average cost of overhead per unit of burden vehicle.

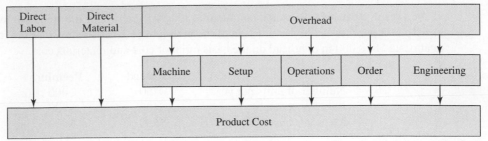

FIGURE 17-4 Activity-based costing versus traditional overhead allocation. (Based on an example by Kim LaScola Needy.)

Problems with Traditional Cost Accounting

Allocation of indirect costs can distort product costs and the decisions based on those costs. To be accurate, the analyst must determine which indirect or overhead expenses will be changed because of an engineering project. In other words, what are the incremental cash flows? For example, vacation and sick leave accrual may be part of overhead, but will they change if the labor content is changed? The changes in costs incurred must be estimated. Loadings, or allocations, of overhead expenses cannot be used.

This issue has become very important because in some firms, automation has reduced direct-labor content to less than 5% of the product's cost. Yet in some of these firms, the basis for allocating overhead is still direct-labor hours or cost.

Other firms are shifting to activity-based costing (ABC) where each activity is linked to specific cost drivers, and the number of dollars allocated as overhead is minimized (see Liggett, Trevino, and Lavelle, 1992). Figure 17-4 illustrates the difference between activity-based costing and traditional overhead allocations (see Tippet and Hoekstra, 1993).

Other Problems to Watch For

Project managers have often accused centralized accounting systems of being too slow or being "untimely." Because engineering economy is not concerned with the problem of daily project control, this is a less critical issue. However, if an organization establishes multiple files and systems so that project managers (and others) have the timely data they need, then the level of accuracy in one or all systems may be low. As a result, analysts making cost estimates will have to consider other internal data sources.

There are several cases in which data on equipment or inventory values may be questionable. When inventory is valued on a "last in, first out" basis, the remaining inventory may be valued too low. Similarly, land valued at its acquisition cost is likely to be significantly undervalued. Finally, capital equipment may be valued at either a low or a high value, depending on allowable depreciation techniques and company policy.

PROBLEMS

Accounting

17-1 Why is it important for engineers and managers to understand accounting principles? Name a few ways that they can do so.

17-2 Explain the accounting function within a firm. What does this function do, and why is it important? What types of data does it provide?

Balance Sheet

17-3 Develop short definitions for the following terms: balance sheet, income statement, and fundamental accounting equation.

17-4 Explain the difference between short-term and long-term liabilities.

17-5 Calculate the equity of the Gravel Construction Company if it has $1 million worth of assets. Gravel has $127,000 in current liabilities and $210,000 in long-term liabilities.

17-6 Matbach Industries has $870,000 in current assets and $430,000 in fixed assets less $180,000 in accumulated depreciation. The firm's current liabilities total $330,000, and the long-term liabilities $115,000.

(a) What is the firm's equity?

(b) If the firm's stock and capital surplus total $305,000, what is the value for retained earnings?

17-7 CalcTech has $930,000 in current assets and $320,000 in fixed assets less $108,000 in accumulated depreciation. The firm's current liabilities total $350,000, and the long-term liabilities $185,000.

(a) What is the firm's equity?

(b) If the firm's stock and capital surplus total $402,000, what is the value for retained earnings?

17-8 Mama L's Baby Monitor Company has current assets of $5 million and current liabilities of $2 million. Give the company's working capital and current ratio. (*Answers:* $3 million; 2.50)

17-9 For Gee-Whiz Devices calculate the following: working capital, current ratio, and acid-test ratio.

Gee-Whiz Devices Balance Sheet Data

Cash	$100,000
Market securities	45,000
Net accounts and notes receivable	150,000
Retailers' inventories	200,000
Prepaid expenses	8,000
Accounts and notes payable (short term)	315,000
Accrued expenses	90,000

(*Answers*: $90,000; 1.22; 0.73)

17-10 From the following data, taken from the balance sheet of Petey's Widget Factory, determine the working capital, current ratio, and acid-test ratio.

Cash	$ 90,000
Net accounts and notes receivable	175,000
Retailers' inventories	210,000
Prepaid expenses	6,000
Accounts and notes payable (short term)	322,000
Accrued expenses	87,000

17-11 (a) For Evergreen Environmental Engineering (EEE) determine the working capital, current ratio, and acid-test ratio. Evaluate the company's economic situation with respect to its ability to pay off debt.

EEE Balance Sheet Data ($1000s)

Cash	$110,000
Securities	40,000
Accounts receivable	160,000
Inventories	250,000
Prepaid expenses	3,000
Accounts payable	351,000
Accrued expenses	89,000

(b) The entries to complete EEE's balance sheet include:

More EEE Balance Sheet Data ($1000s)

Long-term liabilities	$ 220,000
Land	25,000
Plant and equipment	510,000
Accumulated depreciation	210,000
Stock	81,000
Capital surplus	15,000
Retained earnings	Value not given

Construct EEE's balance sheet.

(c) What are EEE's values for total assets, total liabilities, and retained earnings?

17-12 Turbo Start has current assets totaling $1.5 million (this includes $500,000 in current inventory) and current liabilities totaling $50,000. Find the current ratio and acid-test ratio. Are the ratios at desirable levels? Explain.

17-13 (a) For J&W Graphics Supply, compute the current ratio. Is this a financially healthy company? Explain.

J&W Graphics Supply Balance Sheet Data ($1000s)

Assets	
Cash	$1740
Accounts receivable	2500
Inventories	900
Bad debt provision	−75
Liabilities	
Accounts payable	1050
Notes payable	500
Accrued expenses	125

(b) The entries of complete J&W's balance sheet include:

More J&W Balance Sheet Data ($1000s)

Long-term liabilities	$950
Land	475
Plant and equipment	3100
Accumulated depreciation	1060
Stock	680
Capital surplus	45
Retained earnings	Value not given

Construct J&W's balance sheet.

(c) What J&W's values for total assets, total liabilities, and total earnings?

17-14 For Sutton Manufacturing, determine the current ratio and the acid-test ratio. Are these values acceptable? Why or why not?

Sutton Manufacturing Balance Sheet Data ($1000s)

Assets		Liabilities	
Current assets		Current liabilities	
Cash	$870	Notes payable	$500
Accounts receivable	450	Accounts payable	600
Inventory	1200	Accruals	200
Prepaid expenses	60	Taxes payable	30
		Current portion long-term debt	100
Total current assets	2670		
		Total current liabilities	1430
		Long-term debt	2000
		Officer debt (subordinated)	200
Net fixed assets			
Land	1200	Total liabilities	3630
Plant and equipment	3800	Equity	
(less accumulated depreciation)	−1000	Common stock	1670
		Capital surplus	400
		Retained earnings	1200
Other assets			
Notes receivable	200	Total equity	3270
Intangibles	120	Total liabilities and equity	6900
Total assets	6900		

17-15 If a firm has a current ratio less than 2.0 and an acid-test ratio less than 1.0, will the company eventually go bankrupt and out of business? Explain your answer.

17-16 What is the advantage of comparing financial statements across periods or against industry benchmarks over looking at statements associated with a single date or period?

Income Statement

17-17 List the two primary general accounting statements. What is each used for, and how do the two differ? Which is most important?

17-18 Scarmack's Paint Company has annual sales of $500,000 per year. If there is a profit of $1000 per day, 6 days per week operation, what is the total yearly business expense? All calculations are on a before-tax basis. (*Answer:* $188,000)

17-19 Find the net income of Turbo Start (Problem 17-12) given the following data from the balance sheet and income statement.

Turbo Start Data ($1000s)

Accounts payable	$ 1,000
Selling expense	5,000
Sales revenue	50,000
Owner's equity	4,500
Income taxes	2,000
Cost of goods sold	30,000
Accounts receivable	15,000

(*Answer:* $13,000,000)

17-20 Laila's Surveying Inc. had revenues of $100,000 in 2004. Expenses totaled $60,000. What was her net profit (or loss)?

17-21 The general ledger of the Fly-Buy-Nite (FBN) Engineering Company contained the following account balances. Construct an income statement. What is the net income before taxes and the net profit (or loss) after taxes? FBN has a tax rate of 27%.

	Amount ($1000s)
Administrative expenses	$ 2,750
Subcontracted services	18,000
Development expenses	900
Interest expense	200
Sales revenue	30,000
Selling expenses	4,500

17-22 For Magdalen Industries, compute the net income before taxes and net profit (or loss). Taxes for the year were $1 million.

(*a*) Calculate net profit for the year.

(*b*) Construct the income statement.

(*c*) Calculate the interest coverage and net profit ratio. Is the interest coverage acceptable? Explain why or why not.

Magdalen Industries Income Statement Data ($M)

Revenues	
Total operating revenue	$81
(including sales of $48 million)	
Total nonoperating revenue	5
Expenses	
Total operating expenses	70
Total nonoperating expenses	7
(interest payments)	

17-23 For Andrew's Electronic Instruments, calculate the interest coverage and net profit ratio. Is Andrew's business healthy?

Income Statement for Andrew's Electronics for End of Year 2008 ($1000s)

Revenues	
Operating revenues	
Sales	$395
(*minus*) Returns	−15
Total operating revenues	380
Nonoperating revenues	
Interest receipts	50
Stock revenues	25
Total nonoperating revenues	75
Total revenues, R	455
Expenses	
Operating expenses	
Cost of goods and services sold	
Labor	200
Materials	34
Indirect cost	68
Selling and promotion	20
Depreciation	30
General and administrative	10
Lease payments	10
Total operating expenses	372
Nonoperating expenses	
Interest payments	22
Total nonoperating expenses	22
Total expenses, E	394
Net income before taxes, R − E	61
Incomes taxes	30
Net profit (Loss) for the year 2008	31

Linking Balance Sheet and Income Statement

17-24 Sutton Manufacturing (balance sheet at the end of last year in Problem 17-14) had the following entries in this year's income statement.

Depreciation $420,000
Profit   480,000

In addition, we also know the firm purchased $800,000 of equipment with cash. The firm paid $200,000 in dividends this year.

What are the entries in the balance sheet at the end of this year for:

(a) plant and equipment?

(b) accumulated depreciation?

(c) retained earnings?

17-25 Magdalen Industries (Problem 17-22) had the following entries in its balance sheet at the end of last year.

Plant and equipment $15 million
(less accumulated depreciation)   8 million
Retained earnings   60 million

In addition to the income statement data for this year in Problem 17-22, we also know that the firm purchased $3 million of equipment with cash and that depreciation expenses were $2 million of the $70 million in operating expenses listed in Problem 17-22. The firm paid no dividends this year.

What are the entries in the balance sheet at the end of this year for:

(a) plant and equipment?

(b) accumulated depreciation?

(c) retained earnings?

Allocating Costs

17-26 Categorize each of the following costs as direct or indirect. Assume that a traditional costing system is in place.

Machine run costs	Cost to market the product
Machine depreciation	Cost of storage
Material handling costs	Insurance costs
Cost of materials	Cost of product sales force
Overtime expenses	Engineering drawings
Machine operator wages	Machine labor
Utility costs	Cost of tooling and fixtures
Support (administrative) staff salaries	

17-27 LeGaroutte Industries makes industrial pipe manufacturing equipment. Use direct-labor hours as the burden vehicle, and compute the total cost per unit for each model given in the table. Total manufacturing indirect costs are $15,892,000, and there are 100,000 units manufactured per year for Model S, 50,000 for Model M, and 82,250 for Model G.

Item	Model S	Model M	Model G
Direct-material costs	$3,800,000	$1,530,000	$2,105,000
Direct-labor costs	600,000	380,000	420,000
Direct-labor hours	64,000	20,000	32,000

(*Answers:* $132; $93; $84)

17-28 RLW-II Enterprises estimated that indirect manufacturing costs for the year would be $60 million and that 12,000 machine-hours would be used.

(a) Compute the predetermined indirect cost application rate using machine hours as the burden vehicle.

(b) Determine the total cost of production for a product with direct material costs of $1 million, direct-labor costs of $600,000, and 200 machine-hours.

(*Answers:* $5000; $2.6 million)

17-29 Par Golf Equipment Company produces two types of golf bag: the standard and deluxe models. The total indirect cost to be allocated to the two bags is $35,000. Determine the net revenue that Par Golf can expect from the sale of each bag.

(a) Use direct-labor cost to allocate indirect costs.

(b) Use direct-materials cost to allocate indirect costs.

Data Item	Standard	Deluxe
Direct-labor cost	$50,000	$65,000
Direct-material cost	35,000	47,500
Selling price	60	95
Units produced	1800	1400

Appendix A

Introduction to Spreadsheets

Computerized spreadsheets are available nearly everywhere, and they can be easily applied to economic analysis. In fact, spreadsheets were originally developed to analyze financial data, and they are often credited with initiating the explosive growth in demand for desktop computing.

A spreadsheet is a two-dimensional table, whose cells can contain numerical values, labels, or formulas. The software automatically updates the table when an entry is changed, and there are powerful tools for copying formulas, creating graphs, and formatting results.

THE ELEMENTS OF A SPREADSHEET

A spreadsheet is a two-dimensional table that labels the columns in alphabetical order A to Z, AA to AZ, BA to BZ, etc. The rows are numbered from 1 to 65,536 or higher. Thus a *cell* of the spreadsheet is specified by its column letter and row number. For example, A3 is the third row in column A and AA6 is the sixth row in the twenty-seventh column. Each cell can contain a label, a numerical value, or a formula.

A *label* is any cell where the contents should be treated as text. Arithmetic cannot be performed on labels. Labels are used for variable names, row and column headings, and explanatory notes. In Excel any cell that contains more than a simple number, such as 3.14159, is treated as a label, unless it begins with an =, which is the signal for a formula. Thus 2*3 and B1+B2 are labels. Meaningful labels can be wider than a normal column. One solution is to allow those cells to "wrap" text, which is one of the "alignment" options. The table heading row (row 8) in Example A-1 has turned this on by selecting row 8, right-clicking on the row, and selecting wrap text under the alignment tab.

A *numerical value* is any number. Acceptable formats for entry or display include percentages, currency, accounting, scientific, fractions, date, and time. In addition the number of decimal digits, the display of $ symbols, and commas for "thousands" separators can be adjusted. The format for cells can be changed by selecting a cell, a block of cells, a row, a column, or the entire spreadsheet. Then right-click on the selected area, and a menu that includes "format cells" will appear. Then number formats, alignment, borders, fonts, and patterns can be selected.

Formulas must begin with an =, such as = 3*4^2 or =B1+B2. They can include many functions—financial, statistical, trigonometric, etc. (and others can be defined by the user). The formula for the "current" cell is displayed in the formula bar at the top of the spreadsheet. The value resulting from the formula is displayed in the cell in the spreadsheet.

Often the printed-out spreadsheet will be part of a report or a homework assignment and the formulas must be explained. Here is an easy way to place a copy of the formula in an adjacent or nearby cell. (1) Convert the cell with the formula to a label by inserting a space before the = sign. (2) Copy that label to an adjacent cell by using cut and paste. Do not drag the cell to copy it, as any formula ending with a number (even an address like B4) will have the number automatically incremented. (3) Convert the original formula back into a formula by deleting the space.

DEFINING VARIABLES IN A DATA BLOCK

The cell A1, top left corner, is the HOME cell for a spreadsheet. Thus, the top left area is where the data block should be placed. This data block should have every variable in the spreadsheet with an adjacent label for each. This data block supports a basic principle of good spreadsheet modeling, which is to use variables in your models.

The data block in Example A-1 contains *entered data*—the loan amount (A1), the number of payments (A2), and the interest rate (A3), and *computed data*—the payment (A4). Then instead of using the loan amount of $5000 in a formula, the cell reference A1 is used. Even if a value is referenced only once, it is better to include it in the data block. By using one location to define each variable, you can change any value at one place in the spreadsheet and have the entire spreadsheet instantly recomputed.

Even for simple homework problems you should use a data block.

1. You may be able to use it for another problem.
2. Solutions to simple problems may grow into solutions for complex problems.
3. Good habits, like using data blocks, are easy to maintain once they are established.
4. It makes the assumptions clear if you've estimated a value or for grading.

In the real world, data blocks are even more important. Most problems are solved more than once, as more and more accurate values are estimated. Often the spreadsheet is revised to add other variables, time periods, locations, etc. Without data blocks, it is hard to change a spreadsheet and the likelihood of missing a required change skyrockets.

If you want your formulas to be easier to read, you can name your variables. *Note*: In Excel, the cell's location or name is displayed at the left of the formula bar. Variable names can be entered here. They will then automatically be applied if cell addresses are entered by the point and click method. If cell addresses are entered as A2, then A2 is what is displayed. To change a displayed A2 to the name of the cell (LoanAmount), the process is to click on Insert, click on Name, click on Apply, and then select the names to be applied.

COPY COMMAND

The copy command and relative/absolute addressing make spreadsheet models easy to build. If the range of cells to be copied contains only labels, numbers, and functions, then the copy command is easy to use and understand. For example, the formula =EXP(1.9) would be copied unchanged to a new location. However, cell addresses are usually part of the range being copied, and their absolute and relative addresses are treated differently.

An *absolute address* is denoted by adding $ signs before the column and/or row. For example in Figure A-1a, $A$4 is the absolute address for the interest rate. When an absolute address is copied, the column and/or row that is fixed is copied unchanged. Thus $A$4 is completely fixed, $A4 fixes the column, and A$4 fixes the row. One common use for absolute addresses is any data block entry, such as the interest rate. When entering or editing a formula, changing between A4, $A$4, A$4, $A4, and A4 is most easily done using the F4 key, which scrolls an address through the choices.

In contrast a *relative address* is best interpreted as directions from one cell to another. For example in Figure A-1a, the balance due in year t equals the balance due in year $t - 1$ minus the principal payment in Year t. Specifically for the balance due in Year 1, D10 contains =D9−C10. From cell D10, cell D9 is one row up and C10 is one column to the left, so the formula is really (contents of 1 up) minus (contents of 1 to the left). When a cell containing a relative address is copied to a new location, it is these directions that are copied to determine any new relative addresses. So if cell D10 is copied to cell F14, the formula is =F13−E14.

Thus to calculate a loan repayment schedule, as in Figure A-1, the row of formulas is created and then copied for the remaining years.

EXAMPLE A-1

Four repayment schedules for a loan of $5000 to be repaid over 5 years at an interest rate of 8% were shown in Table 3-1. Use a spreadsheet to calculate the amortization schedule for the constant principal payment option.

SOLUTION

The first step is to enter the loan amount, number of periods, and interest rate into a data block in the top left part of the spreadsheet. The next step is to calculate the constant principal payment amount, which was given as $1252.28 in Table 3-1. The factor approach to finding this value is given in Chapter 3 and the spreadsheet function is explained in Chapter 4.

The next step is to identify the columns for the amortization schedule. These are the year, interest owed, principal payment, and balance due. Because some of these labels are wider than a normal column, the cells are formatted so that the text wraps (row height increases automatically). The initial balance is shown in the Year-0 row.

Next, the formulas for the first year are written, as shown in Figure A-1a. The interest owed (cell B10) equals the interest rate ($A$4) times the balance due for Year 0 (D9). The principal payment (cell C10) equals the annual payment ($A$6) minus the interest owed and paid (B10). Finally, the balance due (cell D10) equals the balance due for the previous year (D9) minus the principal payment (C10). The results are shown in Figure A-1a.

Now cells A10 to D10 are selected for Year 1. By dragging down on the right corner of D10, the entire row can be copied for Years 2 through 5. Note that if you use cut and paste, then you must complete the year column separately (dragging increments the year, but cutting and pasting does not). The results are shown in Figure A-1b.

	A	B	C	D	E
1	Entered Data				
2	5000	Loan Amount			
3	5	Number of Payments			
4	8%	Interest Rate			
5	Computed Data				
6	$1,252.28	Loan Payment			
7					
8	Year	Interest Owed	Principal Payment	Balance Due	
9	0			5000.00	
10	1	400.00	852.28	4147.72	=D9−C10
11					
12		=$A$4*D9		=$A$6−B10	

(a)

	A	B	C	D	E
8	Year	Interest Owed	Principal Payment	Balance Due	
9	0			5000.00	
10	1	400.00	852.28	4147.72	
11	2	331.82	920.46	3227.25	
12	3	258.18	994.10	2233.15	
13	4	178.65	1073.63	1159.52	
14	5	92.76	1159.52	0.00	=D13−C14
15					
16		=$A$4*D13		=$A$6−B14	

(b)

FIGURE A-1 (a) Year 1 amortization schedule. (b) Completed amortization schedule.

This appendix has introduced the basics of spreadsheets. Chapter 2 uses spreadsheets and simple bar charts to draw cash flow diagrams. Chapters 4 to 15 each have spreadsheet sections. These are designed to develop spreadsheet modeling skills and to reinforce your understanding of engineering economy. As spreadsheet packages are built around using the computer mouse to click on cells and items in charts, there is usually an intuitive connection between what you would like to do and how to do it. The best way to learn how to use the spreadsheet package is to simply play around with it. In addition, as you look at the menu choices, you will find new commands that you hadn't thought of but will find useful.

Appendix B

Compound Interest Tables

Values of Interest Factors When n Equals Infinity

Single Payment:

$(F/P, i, \infty) = \infty$

$(P/F, i, \infty) = 0$

Arithmetic Gradient Series:

$(A/G, i, \infty) = 1/i$

$(P/G, i, \infty) = 1/i^2$

Uniform Payment Series:

$(A/F, i, \infty) = 0$

$(A/P, i, \infty) = i$

$(F/A, i, \infty) = \infty$

$(P/A, i, \infty) = 1/i$

1/4%   Compound Interest Factors   **1/4%**

	Single Payment		Uniform Payment Series				Arithmetic Gradient		
	Compound Amount Factor	Present Worth Factor	Sinking Fund Factor	Capital Recovery Factor	Compound Amount Factor	Present Worth Factor	Gradient Uniform Series	Gradient Present Worth	
	Find *F* Given *P*	Find *P* Given *F*	Find *A* Given *F*	Find *A* Given *P*	Find *F* Given *A*	Find *P* Given *A*	Find *A* Given *G*	Find *P* Given *G*	
n	F/P	P/F	A/F	A/P	F/A	P/A	A/G	P/G	*n*
1	1.003	.9975	1.0000	1.0025	1.000	0.998	0	0	1
2	1.005	.9950	.4994	.5019	2.003	1.993	0.504	1.005	2
3	1.008	.9925	.3325	.3350	3.008	2.985	1.005	2.999	3
4	1.010	.9901	.2491	.2516	4.015	3.975	1.501	5.966	4
5	1.013	.9876	.1990	.2015	5.025	4.963	1.998	9.916	5
6	1.015	.9851	.1656	.1681	6.038	5.948	2.498	14.861	6
7	1.018	.9827	.1418	.1443	7.053	6.931	2.995	20.755	7
8	1.020	.9802	.1239	.1264	8.070	7.911	3.490	27.611	8
9	1.023	.9778	.1100	.1125	9.091	8.889	3.987	35.440	9
10	1.025	.9753	.0989	.1014	10.113	9.864	4.483	44.216	10
11	1.028	.9729	.0898	.0923	11.139	10.837	4.978	53.950	11
12	1.030	.9705	.0822	.0847	12.167	11.807	5.474	64.634	12
13	1.033	.9681	.0758	.0783	13.197	12.775	5.968	76.244	13
14	1.036	.9656	.0703	.0728	14.230	13.741	6.464	88.826	14
15	1.038	.9632	.0655	.0680	15.266	14.704	6.957	102.301	15
16	1.041	.9608	.0613	.0638	16.304	15.665	7.451	116.716	16
17	1.043	.9584	.0577	.0602	17.344	16.624	7.944	132.063	17
18	1.046	.9561	.0544	.0569	18.388	17.580	8.437	148.319	18
19	1.049	.9537	.0515	.0540	19.434	18.533	8.929	165.492	19
20	1.051	.9513	.0488	.0513	20.482	19.485	9.421	183.559	20
21	1.054	.9489	.0464	.0489	21.534	20.434	9.912	202.531	21
22	1.056	.9465	.0443	.0468	22.587	21.380	10.404	222.435	22
23	1.059	.9442	.0423	.0448	23.644	22.324	10.894	243.212	23
24	1.062	.9418	.0405	.0430	24.703	23.266	11.384	264.854	24
25	1.064	.9395	.0388	.0413	25.765	24.206	11.874	287.407	25
26	1.067	.9371	.0373	.0398	26.829	25.143	12.363	310.848	26
27	1.070	.9348	.0358	.0383	27.896	26.078	12.852	335.150	27
28	1.072	.9325	.0345	.0370	28.966	27.010	13.341	360.343	28
29	1.075	.9301	.0333	.0358	30.038	27.940	13.828	386.366	29
30	1.078	.9278	.0321	.0346	31.114	28.868	14.317	413.302	30
36	1.094	.9140	.0266	.0291	37.621	34.387	17.234	592.632	36
40	1.105	.9049	.0238	.0263	42.014	38.020	19.171	728.882	40
48	1.127	.8871	.0196	.0221	50.932	45.179	23.025	1 040.22	48
50	1.133	.8826	.0188	.0213	53.189	46.947	23.984	1 125.96	50
52	1.139	.8782	.0180	.0205	55.458	48.705	24.941	1 214.76	52
60	1.162	.8609	.0155	.0180	64.647	55.653	28.755	1 600.31	60
70	1.191	.8396	.0131	.0156	76.395	64.144	33.485	2 147.87	70
72	1.197	.8355	.0127	.0152	78.780	65.817	34.426	2 265.81	72
80	1.221	.8189	.0113	.0138	88.440	72.427	38.173	2 764.74	80
84	1.233	.8108	.0107	.0132	93.343	75.682	40.037	3 030.06	84
90	1.252	.7987	.00992	.0124	100.789	80.504	42.820	3 447.19	90
96	1.271	.7869	.00923	.0117	108.349	85.255	45.588	3 886.62	96
100	1.284	.7790	.00881	.0113	113.451	88.383	47.425	4 191.60	100
104	1.297	.7713	.00843	.0109	118.605	91.480	49.256	4 505.93	104
120	1.349	.7411	.00716	.00966	139.743	103.563	56.512	5 852.52	120
240	1.821	.5492	.00305	.00555	328.306	180.312	107.590	19 399.75	240
360	2.457	.4070	.00172	.00422	582.745	237.191	152.894	36 264.96	360
480	3.315	.3016	.00108	.00358	926.074	279.343	192.673	53 821.93	480

Compound Interest Factors

	Single Payment		Uniform Payment Series				Arithmetic Gradient		
	Compound Amount Factor	Present Worth Factor	Sinking Fund Factor	Capital Recovery Factor	Compound Amount Factor	Present Worth Factor	Gradient Uniform Series	Gradient Present Worth	
n	Find F Given P F/P	Find P Given F P/F	Find A Given F A/F	Find A Given P A/P	Find F Given A F/A	Find P Given A P/A	Find A Given G A/G	Find P Given G P/G	n
1	1.005	.9950	1.0000	1.0050	1.000	0.995	0	0	1
2	1.010	.9901	.4988	.5038	2.005	1.985	0.499	0.991	2
3	1.015	.9851	.3317	.3367	3.015	2.970	0.996	2.959	3
4	1.020	.9802	.2481	.2531	4.030	3.951	1.494	5.903	4
5	1.025	.9754	.1980	.2030	5.050	4.926	1.990	9.803	5
6	1.030	.9705	.1646	.1696	6.076	5.896	2.486	14.660	6
7	1.036	.9657	.1407	.1457	7.106	6.862	2.980	20.448	7
8	1.041	.9609	.1228	.1278	8.141	7.823	3.474	27.178	8
9	1.046	.9561	.1089	.1139	9.182	8.779	3.967	34.825	9
10	1.051	.9513	.0978	.1028	10.228	9.730	4.459	43.389	10
11	1.056	.9466	.0887	.0937	11.279	10.677	4.950	52.855	11
12	1.062	.9419	.0811	.0861	12.336	11.619	5.441	63.218	12
13	1.067	.9372	.0746	.0796	13.397	12.556	5.931	74.465	13
14	1.072	.9326	.0691	.0741	14.464	13.489	6.419	86.590	14
15	1.078	.9279	.0644	.0694	15.537	14.417	6.907	99.574	15
16	1.083	.9233	.0602	.0652	16.614	15.340	7.394	113.427	16
17	1.088	.9187	.0565	.0615	17.697	16.259	7.880	128.125	17
18	1.094	.9141	.0532	.0582	18.786	17.173	8.366	143.668	18
19	1.099	.9096	.0503	.0553	19.880	18.082	8.850	160.037	19
20	1.105	.9051	.0477	.0527	20.979	18.987	9.334	177.237	20
21	1.110	.9006	.0453	.0503	22.084	19.888	9.817	195.245	21
22	1.116	.8961	.0431	.0481	23.194	20.784	10.300	214.070	22
23	1.122	.8916	.0411	.0461	24.310	21.676	10.781	233.680	23
24	1.127	.8872	.0393	.0443	25.432	22.563	11.261	254.088	24
25	1.133	.8828	.0377	.0427	26.559	23.446	11.741	275.273	25
26	1.138	.8784	.0361	.0411	27.692	24.324	12.220	297.233	26
27	1.144	.8740	.0347	.0397	28.830	25.198	12.698	319.955	27
28	1.150	.8697	.0334	.0384	29.975	26.068	13.175	343.439	28
29	1.156	.8653	.0321	.0371	31.124	26.933	13.651	367.672	29
30	1.161	.8610	.0310	.0360	32.280	27.794	14.127	392.640	30
36	1.197	.8356	.0254	.0304	39.336	32.871	16.962	557.564	36
40	1.221	.8191	.0226	.0276	44.159	36.172	18.836	681.341	40
48	1.270	.7871	.0185	.0235	54.098	42.580	22.544	959.928	48
50	1.283	.7793	.0177	.0227	56.645	44.143	23.463	1 035.70	50
52	1.296	.7716	.0169	.0219	59.218	45.690	24.378	1 113.82	52
60	1.349	.7414	.0143	.0193	69.770	51.726	28.007	1 448.65	60
70	1.418	.7053	.0120	.0170	83.566	58.939	32.468	1 913.65	70
72	1.432	.6983	.0116	.0166	86.409	60.340	33.351	2 012.35	72
80	1.490	.6710	.0102	.0152	98.068	65.802	36.848	2 424.65	80
84	1.520	.6577	.00961	.0146	104.074	68.453	38.576	2 640.67	84
90	1.567	.6383	.00883	.0138	113.311	72.331	41.145	2 976.08	90
96	1.614	.6195	.00814	.0131	122.829	76.095	43.685	3 324.19	96
100	1.647	.6073	.00773	.0127	129.334	78.543	45.361	3 562.80	100
104	1.680	.5953	.00735	.0124	135.970	80.942	47.025	3 806.29	104
120	1.819	.5496	.00610	.0111	163.880	90.074	53.551	4 823.52	120
240	3.310	.3021	.00216	.00716	462.041	139.581	96.113	13 415.56	240
360	6.023	.1660	.00100	.00600	1 004.5	166.792	128.324	21 403.32	360
480	10.957	.0913	.00050	.00550	1 991.5	181.748	151.795	27 588.37	480

3/4% Compound Interest Factors **3/4%**

	Single Payment		Uniform Payment Series				Arithmetic Gradient		
	Compound Amount Factor	Present Worth Factor	Sinking Fund Factor	Capital Recovery Factor	Compound Amount Factor	Present Worth Factor	Gradient Uniform Series	Gradient Present Worth	
	Find F Given P	Find P Given F	Find A Given F	Find A Given P	Find F Given A	Find P Given A	Find A Given G	Find P Given G	
n	F/P	P/F	A/F	A/P	F/A	P/A	A/G	P/G	n
1	1.008	.9926	1.0000	1.0075	1.000	0.993	0	0	1
2	1.015	.9852	.4981	.5056	2.008	1.978	0.499	0.987	2
3	1.023	.9778	.3308	.3383	3.023	2.956	0.996	2.943	3
4	1.030	.9706	.2472	.2547	4.045	3.926	1.492	5.857	4
5	1.038	.9633	.1970	.2045	5.076	4.889	1.986	9.712	5
6	1.046	.9562	.1636	.1711	6.114	5.846	2.479	14.494	6
7	1.054	.9490	.1397	.1472	7.160	6.795	2.971	20.187	7
8	1.062	.9420	.1218	.1293	8.213	7.737	3.462	26.785	8
9	1.070	.9350	.1078	.1153	9.275	8.672	3.951	34.265	9
10	1.078	.9280	.0967	.1042	10.344	9.600	4.440	42.619	10
11	1.086	.9211	.0876	.0951	11.422	10.521	4.927	51.831	11
12	1.094	.9142	.0800	.0875	12.508	11.435	5.412	61.889	12
13	1.102	.9074	.0735	.0810	13.602	12.342	5.897	72.779	13
14	1.110	.9007	.0680	.0755	14.704	13.243	6.380	84.491	14
15	1.119	.8940	.0632	.0707	15.814	14.137	6.862	97.005	15
16	1.127	.8873	.0591	.0666	16.932	15.024	7.343	110.318	16
17	1.135	.8807	.0554	.0629	18.059	15.905	7.822	124.410	17
18	1.144	.8742	.0521	.0596	19.195	16.779	8.300	139.273	18
19	1.153	.8676	.0492	.0567	20.339	17.647	8.777	154.891	19
20	1.161	.8612	.0465	.0540	21.491	18.508	9.253	171.254	20
21	1.170	.8548	.0441	.0516	22.653	19.363	9.727	188.352	21
22	1.179	.8484	.0420	.0495	23.823	20.211	10.201	206.170	22
23	1.188	.8421	.0400	.0475	25.001	21.053	10.673	224.695	23
24	1.196	.8358	.0382	.0457	26.189	21.889	11.143	243.924	24
25	1.205	.8296	.0365	.0440	27.385	22.719	11.613	263.834	25
26	1.214	.8234	.0350	.0425	28.591	23.542	12.081	284.421	26
27	1.224	.8173	.0336	.0411	29.805	24.360	12.548	305.672	27
28	1.233	.8112	.0322	.0397	31.029	25.171	13.014	327.576	28
29	1.242	.8052	.0310	.0385	32.261	25.976	13.479	350.122	29
30	1.251	.7992	.0298	.0373	33.503	26.775	13.942	373.302	30
36	1.309	.7641	.0243	.0318	41.153	31.447	16.696	525.038	36
40	1.348	.7416	.0215	.0290	46.447	34.447	18.507	637.519	40
48	1.431	.6986	.0174	.0249	57.521	40.185	22.070	886.899	48
50	1.453	.6882	.0166	.0241	60.395	41.567	22.949	953.911	50
52	1.475	.6780	.0158	.0233	63.312	42.928	23.822	1 022.64	52
60	1.566	.6387	.0133	.0208	75.425	48.174	27.268	1 313.59	60
70	1.687	.5927	.0109	.0184	91.621	54.305	31.465	1 708.68	70
72	1.713	.5839	.0105	.0180	95.008	55.477	32.289	1 791.33	72
80	1.818	.5500	.00917	.0167	109.074	59.995	35.540	2 132.23	80
84	1.873	.5338	.00859	.0161	116.428	62.154	37.137	2 308.22	84
90	1.959	.5104	.00782	.0153	127.881	65.275	39.496	2 578.09	90
96	2.049	.4881	.00715	.0147	139.858	68.259	41.812	2 854.04	96
100	2.111	.4737	.00675	.0143	148.147	70.175	43.332	3 040.85	100
104	2.175	.4597	.00638	.0139	156.687	72.035	44.834	3 229.60	104
120	2.451	.4079	.00517	.0127	193.517	78.942	50.653	3 998.68	120
240	6.009	.1664	.00150	.00900	667.901	111.145	85.422	9 494.26	240
360	14.731	.0679	.00055	.00805	1 830.8	124.282	107.115	13 312.50	360
480	36.111	.0277	.00021	.00771	4 681.5	129.641	119.662	15 513.16	480

1% Compound Interest Factors **1%**

	Single Payment		Uniform Payment Series				Arithmetic Gradient		
	Compound Amount Factor	Present Worth Factor	Sinking Fund Factor	Capital Recovery Factor	Compound Amount Factor	Present Worth Factor	Gradient Uniform Series	Gradient Present Worth	
	Find F Given P	Find P Given F	Find A Given F	Find A Given P	Find F Given A	Find P Given A	Find A Given G	Find P Given G	
n	F/P	P/F	A/F	A/P	F/A	P/A	A/G	P/G	n
1	1.010	.9901	1.0000	1.0100	1.000	0.990	0	0	1
2	1.020	.9803	.4975	.5075	2.010	1.970	0.498	0.980	2
3	1.030	.9706	.3300	.3400	3.030	2.941	0.993	2.921	3
4	1.041	.9610	.2463	.2563	4.060	3.902	1.488	5.804	4
5	1.051	.9515	.1960	.2060	5.101	4.853	1.980	9.610	5
6	1.062	.9420	.1625	.1725	6.152	5.795	2.471	14.320	6
7	1.072	.9327	.1386	.1486	7.214	6.728	2.960	19.917	7
8	1.083	.9235	.1207	.1307	8.286	7.652	3.448	26.381	8
9	1.094	.9143	.1067	.1167	9.369	8.566	3.934	33.695	9
10	1.105	.9053	.0956	.1056	10.462	9.471	4.418	41.843	10
11	1.116	.8963	.0865	.0965	11.567	10.368	4.900	50.806	11
12	1.127	.8874	.0788	.0888	12.682	11.255	5.381	60.568	12
13	1.138	.8787	.0724	.0824	13.809	12.134	5.861	71.112	13
14	1.149	.8700	.0669	.0769	14.947	13.004	6.338	82.422	14
15	1.161	.8613	.0621	.0721	16.097	13.865	6.814	94.481	15
16	1.173	.8528	.0579	.0679	17.258	14.718	7.289	107.273	16
17	1.184	.8444	.0543	.0643	18.430	15.562	7.761	120.783	17
18	1.196	.8360	.0510	.0610	19.615	16.398	8.232	134.995	18
19	1.208	.8277	.0481	.0581	20.811	17.226	8.702	149.895	19
20	1.220	.8195	.0454	.0554	22.019	18.046	9.169	165.465	20
21	1.232	.8114	.0430	.0530	23.239	18.857	9.635	181.694	21
22	1.245	.8034	.0409	.0509	24.472	19.660	10.100	198.565	22
23	1.257	.7954	.0389	.0489	25.716	20.456	10.563	216.065	23
24	1.270	.7876	.0371	.0471	26.973	21.243	11.024	234.179	24
25	1.282	.7798	.0354	.0454	28.243	22.023	11.483	252.892	25
26	1.295	.7720	.0339	.0439	29.526	22.795	11.941	272.195	26
27	1.308	.7644	.0324	.0424	30.821	23.560	12.397	292.069	27
28	1.321	.7568	.0311	.0411	32.129	24.316	12.852	312.504	28
29	1.335	.7493	.0299	.0399	33.450	25.066	13.304	333.486	29
30	1.348	.7419	.0287	.0387	34.785	25.808	13.756	355.001	30
36	1.431	.6989	.0232	.0332	43.077	30.107	16.428	494.620	36
40	1.489	.6717	.0205	.0305	48.886	32.835	18.178	596.854	40
48	1.612	.6203	.0163	.0263	61.223	37.974	21.598	820.144	48
50	1.645	.6080	.0155	.0255	64.463	39.196	22.436	879.417	50
52	1.678	.5961	.0148	.0248	67.769	40.394	23.269	939.916	52
60	1.817	.5504	.0122	.0222	81.670	44.955	26.533	1 192.80	60
70	2.007	.4983	.00993	.0199	100.676	50.168	30.470	1 528.64	70
72	2.047	.4885	.00955	.0196	104.710	51.150	31.239	1 597.86	72
80	2.217	.4511	.00822	.0182	121.671	54.888	34.249	1 879.87	80
84	2.307	.4335	.00765	.0177	130.672	56.648	35.717	2 023.31	84
90	2.449	.4084	.00690	.0169	144.863	59.161	37.872	2 240.56	90
96	2.599	.3847	.00625	.0163	159.927	61.528	39.973	2 459.42	96
100	2.705	.3697	.00587	.0159	170.481	63.029	41.343	2 605.77	100
104	2.815	.3553	.00551	.0155	181.464	64.471	42.688	2 752.17	104
120	3.300	.3030	.00435	.0143	230.039	69.701	47.835	3 334.11	120
240	10.893	.0918	.00101	.0110	989.254	90.819	75.739	6 878.59	240
360	35.950	.0278	.00029	.0103	3 495.0	97.218	89.699	8 720.43	360
480	118.648	.00843	.00008	.0101	11 764.8	99.157	95.920	9 511.15	480

1¼% Compound Interest Factors **1¼%**

	Single Payment		Uniform Payment Series				Arithmetic Gradient		
	Compound Amount Factor	Present Worth Factor	Sinking Fund Factor	Capital Recovery Factor	Compound Amount Factor	Present Worth Factor	Gradient Uniform Series	Gradient Present Worth	
	Find F Given P	Find P Given F	Find A Given F	Find A Given P	Find F Given A	Find P Given A	Find A Given G	Find P Given G	
n	F/P	P/F	A/F	A/P	F/A	P/A	A/G	P/G	n
1	1.013	.9877	1.0000	1.0125	1.000	0.988	0	0	1
2	1.025	.9755	.4969	.5094	2.013	1.963	0.497	0.976	2
3	1.038	.9634	.3292	.3417	3.038	2.927	0.992	2.904	3
4	1.051	.9515	.2454	.2579	4.076	3.878	1.485	5.759	4
5	1.064	.9398	.1951	.2076	5.127	4.818	1.976	9.518	5
6	1.077	.9282	.1615	.1740	6.191	5.746	2.464	14.160	6
7	1.091	.9167	.1376	.1501	7.268	6.663	2.951	19.660	7
8	1.104	.9054	.1196	.1321	8.359	7.568	3.435	25.998	8
9	1.118	.8942	.1057	.1182	9.463	8.462	3.918	33.152	9
10	1.132	.8832	.0945	.1070	10.582	9.346	4.398	41.101	10
11	1.146	.8723	.0854	.0979	11.714	10.218	4.876	49.825	11
12	1.161	.8615	.0778	.0903	12.860	11.079	5.352	59.302	12
13	1.175	.8509	.0713	.0838	14.021	11.930	5.827	69.513	13
14	1.190	.8404	.0658	.0783	15.196	12.771	6.299	80.438	14
15	1.205	.8300	.0610	.0735	16.386	13.601	6.769	92.058	15
16	1.220	.8197	.0568	.0693	17.591	14.420	7.237	104.355	16
17	1.235	.8096	.0532	.0657	18.811	15.230	7.702	117.309	17
18	1.251	.7996	.0499	.0624	20.046	16.030	8.166	130.903	18
19	1.266	.7898	.0470	.0595	21.297	16.849	8.628	145.119	19
20	1.282	.7800	.0443	.0568	22.563	17.599	9.088	159.940	20
21	1.298	.7704	.0419	.0544	23.845	18.370	9.545	175.348	21
22	1.314	.7609	.0398	.0523	25.143	19.131	10.001	191.327	22
23	1.331	.7515	.0378	.0503	26.458	19.882	10.455	207.859	23
24	1.347	.7422	.0360	.0485	27.788	20.624	10.906	224.930	24
25	1.364	.7330	.0343	.0468	29.136	21.357	11.355	242.523	25
26	1.381	.7240	.0328	.0453	30.500	22.081	11.803	260.623	26
27	1.399	.7150	.0314	.0439	31.881	22.796	12.248	279.215	27
28	1.416	.7062	.0300	.0425	32.280	23.503	12.691	298.284	28
29	1.434	.6975	.0288	.0413	34.696	24.200	13.133	317.814	29
30	1.452	.6889	.0277	.0402	36.129	24.889	13.572	337.792	30
36	1.564	.6394	.0222	.0347	45.116	28.847	16.164	466.297	36
40	1.644	.6084	.0194	.0319	51.490	31.327	17.852	559.247	40
48	1.845	.5509	.0153	.0278	65.229	35.932	21.130	759.248	48
50	1.861	.5373	.0145	.0270	68.882	37.013	21.930	811.692	50
52	1.908	.5242	.0138	.0263	72.628	38.068	22.722	864.960	52
60	2.107	.4746	.0113	.0238	88.575	42.035	25.809	1 084.86	60
70	2.386	.4191	.00902	.0215	110.873	46.470	29.492	1 370.47	70
72	2.446	.4088	.00864	.0211	115.675	47.293	30.205	1 428.48	72
80	2.701	.3702	.00735	.0198	136.120	50.387	32.983	1 661.89	80
84	2.839	.3522	.00680	.0193	147.130	51.822	34.326	1 778.86	84
90	3.059	.3269	.00607	.0186	164.706	53.846	36.286	1 953.85	90
96	3.296	.3034	.00545	.0179	183.643	55.725	38.180	2 127.55	96
100	3.463	.2887	.00507	.0176	197.074	56.901	39.406	2 242.26	100
104	3.640	.2747	.00474	.0172	211.190	58.021	40.604	2 355.90	104
120	4.440	.2252	.00363	.0161	275.220	61.983	45.119	2 796.59	120
240	19.716	.0507	.00067	.0132	1 497.3	75.942	67.177	5 101.55	240
360	87.543	.0114	.00014	.0126	6 923.4	79.086	75.840	5 997.91	360
480	388.713	.00257	.00003	.0125	31 017.1	79.794	78.762	6 284.74	480

1½% **Compound Interest Factors** 1½%

n	Single Payment		Uniform Payment Series				Arithmetic Gradient		n
	Compound Amount Factor	Present Worth Factor	Sinking Fund Factor	Capital Recovery Factor	Compound Amount Factor	Present Worth Factor	Gradient Uniform Series	Gradient Present Worth	
	Find *F* Given *P* *F/P*	Find *P* Given *F* *P/F*	Find *A* Given *F* *A/F*	Find *A* Given *P* *A/P*	Find *F* Given *A* *F/A*	Find *P* Given *A* *P/A*	Find *A* Given *G* *A/G*	Find *P* Given *G* *P/G*	
1	1.015	.9852	1.0000	1.0150	1.000	0.985	0	0	**1**
2	1.030	.9707	.4963	.5113	2.015	1.956	0.496	0.970	**2**
3	1.046	.9563	.3284	.3434	3.045	2.912	0.990	2.883	**3**
4	1.061	.9422	.2444	.2594	4.091	3.854	1.481	5.709	**4**
5	1.077	.9283	.1941	.2091	5.152	4.783	1.970	9.422	**5**
6	1.093	.9145	.1605	.1755	6.230	5.697	2.456	13.994	**6**
7	1.110	.9010	.1366	.1516	7.323	6.598	2.940	19.400	**7**
8	1.126	.8877	.1186	.1336	8.433	7.486	3.422	25.614	**8**
9	1.143	.8746	.1046	.1196	9.559	8.360	3.901	32.610	**9**
10	1.161	.8617	.0934	.1084	10.703	9.222	4.377	40.365	**10**
11	1.178	.8489	.0843	.0993	11.863	10.071	4.851	48.855	**11**
12	1.196	.8364	.0767	.0917	13.041	10.907	5.322	58.054	**12**
13	1.214	.8240	.0702	.0852	14.237	11.731	5.791	67.943	**13**
14	1.232	.8118	.0647	.0797	15.450	12.543	6.258	78.496	**14**
15	1.250	.7999	.0599	.0749	16.682	13.343	6.722	89.694	**15**
16	1.269	.7880	.0558	.0708	17.932	14.131	7.184	101.514	**16**
17	1.288	.7764	.0521	.0671	19.201	14.908	7.643	113.937	**17**
18	1.307	.7649	.0488	.0638	20.489	15.673	8.100	126.940	**18**
19	1.327	.7536	.0459	.0609	21.797	16.426	8.554	140.505	**19**
20	1.347	.7425	.0432	.0582	23.124	17.169	9.005	154.611	**20**
21	1.367	.7315	.0409	.0559	24.470	17.900	9.455	169.241	**21**
22	1.388	.7207	.0387	.0537	25.837	18.621	9.902	184.375	**22**
23	1.408	.7100	.0367	.0517	27.225	19.331	10.346	199.996	**23**
24	1.430	.6995	.0349	.0499	28.633	20.030	10.788	216.085	**24**
25	1.451	.6892	.0333	.0483	30.063	20.720	11.227	232.626	**25**
26	1.473	.6790	.0317	.0467	31.514	21.399	11.664	249.601	**26**
27	1.495	.6690	.0303	.0453	32.987	22.068	12.099	266.995	**27**
28	1.517	.6591	.0290	.0440	34.481	22.727	12.531	284.790	**28**
29	1.540	.6494	.0278	.0428	35.999	23.376	12.961	302.972	**29**
30	1.563	.6398	.0266	.0416	37.539	24.016	13.388	321.525	**30**
36	1.709	.5851	.0212	.0362	47.276	27.661	15.901	439.823	**36**
40	1.814	.5513	.0184	.0334	54.268	29.916	17.528	524.349	**40**
48	2.043	.4894	.0144	.0294	69.565	34.042	20.666	703.537	**48**
50	2.105	.4750	.0136	.0286	73.682	35.000	21.428	749.955	**50**
52	2.169	.4611	.0128	.0278	77.925	35.929	22.179	796.868	**52**
60	2.443	.4093	.0104	.0254	96.214	39.380	25.093	988.157	**60**
70	2.835	.3527	.00817	.0232	122.363	43.155	28.529	1 231.15	**70**
72	2.921	.3423	.00781	.0228	128.076	43.845	29.189	1 279.78	**72**
80	3.291	.3039	.00655	.0215	152.710	46.407	31.742	1 473.06	**80**
84	3.493	.2863	.00602	.0210	166.172	47.579	32.967	1 568.50	**84**
90	3.819	.2619	.00532	.0203	187.929	49.210	34.740	1 709.53	**90**
96	4.176	.2395	.00472	.0197	211.719	50.702	36.438	1 847.46	**96**
100	4.432	.2256	.00437	.0194	228.802	51.625	37.529	1 937.43	**100**
104	4.704	.2126	.00405	.0190	246.932	52.494	38.589	2 025.69	**104**
120	5.969	.1675	.00302	.0180	331.286	55.498	42.518	2 359.69	**120**
240	35.632	.0281	.00043	.0154	2 308.8	64.796	59.737	3 870.68	**240**
360	212.700	.00470	.00007	.0151	14 113.3	66.353	64.966	4 310.71	**360**
480	1 269.7	.00079	.00001	.0150	84 577.8	66.614	66.288	4 415.74	**480**

1³/4% Compound Interest Factors **1³/4%**

	Single Payment		Uniform Payment Series				Arithmetic Gradient		
	Compound Amount Factor	Present Worth Factor	Sinking Fund Factor	Capital Recovery Factor	Compound Amount Factor	Present Worth Factor	Gradient Uniform Series	Gradient Present Worth	
	Find F Given P	Find P Given F	Find A Given F	Find A Given P	Find F Given A	Find P Given A	Find A Given G	Find P Given G	
n	F/P	P/F	A/F	A/P	F/A	P/A	A/G	P/G	n
1	1.018	.9828	1.0000	1.0175	1.000	0.983	0	0	1
2	1.035	.9659	.4957	.5132	2.018	1.949	0.496	0.966	2
3	1.053	.9493	.3276	.3451	3.053	2.898	0.989	2.865	3
4	1.072	.9330	.2435	.2610	4.106	3.831	1.478	5.664	4
5	1.091	.9169	.1931	.2106	5.178	4.748	1.965	9.332	5
6	1.110	.9011	.1595	.1770	6.269	5.649	2.450	13.837	6
7	1.129	.8856	.1355	.1530	7.378	6.535	2.931	19.152	7
8	1.149	.8704	.1175	.1350	8.508	7.405	3.409	25.245	8
9	1.169	.8554	.1036	.1211	9.656	8.261	3.885	32.088	9
10	1.189	.8407	.0924	.1099	10.825	9.101	4.357	39.655	10
11	1.210	.8263	.0832	.1007	12.015	9.928	4.827	47.918	11
12	1.231	.8121	.0756	.0931	13.225	10.740	5.294	56.851	12
13	1.253	.7981	.0692	.0867	14.457	11.538	5.758	66.428	13
14	1.275	.7844	.0637	.0812	15.710	12.322	6.219	76.625	14
15	1.297	.7709	.0589	.0764	16.985	13.093	6.677	87.417	15
16	1.320	.7576	.0547	.0722	18.282	13.851	7.132	98.782	16
17	1.343	.7446	.0510	.0685	19.602	14.595	7.584	110.695	17
18	1.367	.7318	.0477	.0652	20.945	15.327	8.034	123.136	18
19	1.390	.7192	.0448	.0623	22.311	16.046	8.481	136.081	19
20	1.415	.7068	.0422	.0597	23.702	16.753	8.924	149.511	20
21	1.440	.6947	.0398	.0573	25.116	17.448	9.365	163.405	21
22	1.465	.6827	.0377	.0552	26.556	18.130	9.804	177.742	22
23	1.490	.6710	.0357	.0532	28.021	18.801	10.239	192.503	23
24	1.516	.6594	.0339	.0514	29.511	19.461	10.671	207.671	24
25	1.543	.6481	.0322	.0497	31.028	20.109	11.101	223.225	25
26	1.570	.6369	.0307	.0482	32.571	20.746	11.528	239.149	26
27	1.597	.6260	.0293	.0468	34.141	21.372	11.952	255.425	27
28	1.625	.6152	.0280	.0455	35.738	21.987	12.373	272.036	28
29	1.654	.6046	.0268	.0443	37.363	22.592	12.791	288.967	29
30	1.683	.5942	.0256	.0431	39.017	23.186	13.206	306.200	30
36	1.867	.5355	.0202	.0377	49.566	26.543	15.640	415.130	36
40	2.002	.4996	.0175	.0350	57.234	28.594	17.207	492.017	40
48	2.300	.4349	.0135	.0310	74.263	32.294	20.209	652.612	48
50	2.381	.4200	.0127	.0302	78.903	33.141	20.932	693.708	50
52	2.465	.4057	.0119	.0294	83.706	33.960	21.644	735.039	52
60	2.832	.3531	.00955	.0271	104.676	36.964	24.389	901.503	60
70	3.368	.2969	.00739	.0249	135.331	40.178	27.586	1 108.34	70
72	3.487	.2868	.00704	.0245	142.127	40.757	28.195	1 149.12	72
80	4.006	.2496	.00582	.0233	171.795	42.880	30.533	1 309.25	80
84	4.294	.2329	.00531	.0228	188.246	43.836	31.644	1 387.16	84
90	4.765	.2098	.00465	.0221	215.166	45.152	33.241	1 500.88	90
96	5.288	.1891	.00408	.0216	245.039	46.337	34.756	1 610.48	96
100	5.668	.1764	.00375	.0212	266.753	47.062	35.721	1 681.09	100
104	6.075	.1646	.00345	.0209	290.028	47.737	36.652	1 749.68	104
120	8.019	.1247	.00249	.0200	401.099	50.017	40.047	2 003.03	120
240	64.308	.0156	.00028	.0178	3 617.6	56.254	53.352	3 001.27	240
360	515.702	.00194	.00003	.0175	29 411.5	57.032	56.443	3 219.08	360
480	4 135.5	.00024		.0175	236 259.0	57.129	57.027	3 257.88	480

2% Compound Interest Factors **2%**

	Single Payment		Uniform Payment Series				Arithmetic Gradient		
	Compound Amount Factor	Present Worth Factor	Sinking Fund Factor	Capital Recovery Factor	Compound Amount Factor	Present Worth Factor	Gradient Uniform Series	Gradient Present Worth	
	Find *F* Given *P*	Find *P* Given *F*	Find *A* Given *F*	Find *A* Given *P*	Find *F* Given *A*	Find *P* Given *A*	Find *A* Given *G*	Find *P* Given *G*	
n	*F/P*	*P/F*	*A/F*	*A/P*	*F/A*	*P/A*	*A/G*	*P/G*	*n*
1	1.020	.9804	1.0000	1.0200	1.000	0.980	0	0	1
2	1.040	.9612	.4951	.5151	2.020	1.942	0.495	0.961	2
3	1.061	.9423	.3268	.3468	3.060	2.884	0.987	2.846	3
4	1.082	.9238	.2426	.2626	4.122	3.808	1.475	5.617	4
5	1.104	.9057	.1922	.2122	5.204	4.713	1.960	9.240	5
6	1.126	.8880	.1585	.1785	6.308	5.601	2.442	13.679	6
7	1.149	.8706	.1345	.1545	7.434	6.472	2.921	18.903	7
8	1.172	.8535	.1165	.1365	8.583	7.325	3.396	24.877	8
9	1.195	.8368	.1025	.1225	9.755	8.162	3.868	31.571	9
10	1.219	.8203	.0913	.1113	10.950	8.983	4.337	38.954	10
11	1.243	.8043	.0822	.1022	12.169	9.787	4.802	46.996	11
12	1.268	.7885	.0746	.0946	13.412	10.575	5.264	55.669	12
13	1.294	.7730	.0681	.0881	14.680	11.348	5.723	64.946	13
14	1.319	.7579	.0626	.0826	15.974	12.106	6.178	74.798	14
15	1.346	.7430	.0578	.0778	17.293	12.849	6.631	85.200	15
16	1.373	.7284	.0537	.0737	18.639	13.578	7.080	96.127	16
17	1.400	.7142	.0500	.0700	20.012	14.292	7.526	107.553	17
18	1.428	.7002	.0467	.0667	21.412	14.992	7.968	119.456	18
19	1.457	.6864	.0438	.0638	22.840	15.678	8.407	131.812	19
20	1.486	.6730	.0412	.0612	24.297	16.351	8.843	144.598	20
21	1.516	.6598	.0388	.0588	25.783	17.011	9.276	157.793	21
22	1.546	.6468	.0366	.0566	27.299	17.658	9.705	171.377	22
23	1.577	.6342	.0347	.0547	28.845	18.292	10.132	185.328	23
24	1.608	.6217	.0329	.0529	30.422	18.914	10.555	199.628	24
25	1.641	.6095	.0312	.0512	32.030	19.523	10.974	214.256	25
26	1.673	.5976	.0297	.0497	33.671	20.121	11.391	229.196	26
27	1.707	.5859	.0283	.0483	35.344	20.707	11.804	244.428	27
28	1.741	.5744	.0270	.0470	37.051	21.281	12.214	259.936	28
29	1.776	.5631	.0258	.0458	38.792	21.844	12.621	275.703	29
30	1.811	.5521	.0247	.0447	40.568	22.396	13.025	291.713	30
36	2.040	.4902	.0192	.0392	51.994	25.489	15.381	392.036	36
40	2.208	.4529	.0166	.0366	60.402	27.355	16.888	461.989	40
48	2.587	.3865	.0126	.0326	79.353	30.673	19.755	605.961	48
50	2.692	.3715	.0118	.0318	84.579	31.424	20.442	642.355	50
52	2.800	.3571	.0111	.0311	90.016	32.145	21.116	678.779	52
60	3.281	.3048	.00877	.0288	114.051	34.761	23.696	823.692	60
70	4.000	.2500	.00667	.0267	149.977	37.499	26.663	999.829	70
72	4.161	.2403	.00633	.0263	158.056	37.984	27.223	1 034.050	72
80	4.875	.2051	.00516	.0252	193.771	39.744	29.357	1 166.781	80
84	5.277	.1895	.00468	.0247	213.865	40.525	30.361	1 230.413	84
90	5.943	.1683	.00405	.0240	247.155	41.587	31.793	1 322.164	90
96	6.693	.1494	.00351	.0235	284.645	42.529	33.137	1 409.291	96
100	7.245	.1380	.00320	.0232	312.230	43.098	33.986	1 464.747	100
104	7.842	.1275	.00292	.0229	342.090	43.624	34.799	1 518.082	104
120	10.765	.0929	.00205	.0220	488.255	45.355	37.711	1 710.411	120
240	115.887	.00863	.00017	.0202	5 744.4	49.569	47.911	2 374.878	240
360	1 247.5	.00080	.00002	.0200	62 326.8	49.960	49.711	2 483.567	360
480	13 429.8	.00007		.0200	671 442.0	49.996	49.964	2 498.027	480

2¹/₂%   Compound Interest Factors   **2¹/₂%**

	Single Payment		Uniform Payment Series				Arithmetic Gradient		
	Compound Amount Factor	Present Worth Factor	Sinking Fund Factor	Capital Recovery Factor	Compound Amount Factor	Present Worth Factor	Gradient Uniform Series	Gradient Present Worth	
	Find F Given P	Find P Given F	Find A Given F	Find A Given P	Find F Given A	Find P Given A	Find A Given G	Find P Given G	
n	F/P	P/F	A/F	A/P	F/A	P/A	A/G	P/G	n
1	1.025	.9756	1.0000	1.0250	1.000	0.976	0	0	1
2	1.051	.9518	.4938	.5188	2.025	1.927	0.494	0.952	2
3	1.077	.9286	.3251	.3501	3.076	2.856	0.984	2.809	3
4	1.104	.9060	.2408	.2658	4.153	3.762	1.469	5.527	4
5	1.131	.8839	.1902	.2152	5.256	4.646	1.951	9.062	5
6	1.160	.8623	.1566	.1816	6.388	5.508	2.428	13.374	6
7	1.189	.8413	.1325	.1575	7.547	6.349	2.901	18.421	7
8	1.218	.8207	.1145	.1395	8.736	7.170	3.370	24.166	8
9	1.249	.8007	.1005	.1255	9.955	7.971	3.835	30.572	9
10	1.280	.7812	.0893	.1143	11.203	8.752	4.296	37.603	10
11	1.312	.7621	.0801	.1051	12.483	9.514	4.753	45.224	11
12	1.345	.7436	.0725	.0975	13.796	10.258	5.206	53.403	12
13	1.379	.7254	.0660	.0910	15.140	10.983	5.655	62.108	13
14	1.413	.7077	.0605	.0855	16.519	11.691	6.100	71.309	14
15	1.448	.6905	.0558	.0808	17.932	12.381	6.540	80.975	15
16	1.485	.6736	.0516	.0766	19.380	13.055	6.977	91.080	16
17	1.522	.6572	.0479	.0729	20.865	13.712	7.409	101.595	17
18	1.560	.6412	.0447	.0697	22.386	14.353	7.838	112.495	18
19	1.599	.6255	.0418	.0668	23.946	14.979	8.262	123.754	19
20	1.639	.6103	.0391	.0641	25.545	15.589	8.682	135.349	20
21	1.680	.5954	.0368	.0618	27.183	16.185	9.099	147.257	21
22	1.722	.5809	.0346	.0596	28.863	16.765	9.511	159.455	22
23	1.765	.5667	.0327	.0577	30.584	17.332	9.919	171.922	23
24	1.809	.5529	.0309	.0559	32.349	17.885	10.324	184.638	24
25	1.854	.5394	.0293	.0543	34.158	18.424	10.724	197.584	25
26	1.900	.5262	.0278	.0528	36.012	18.951	11.120	210.740	26
27	1.948	.5134	.0264	.0514	37.912	19.464	11.513	224.088	27
28	1.996	.5009	.0251	.0501	39.860	19.965	11.901	237.612	28
29	2.046	.4887	.0239	.0489	41.856	20.454	12.286	251.294	29
30	2.098	.4767	.0228	.0478	43.903	20.930	12.667	265.120	30
31	2.150	.4651	.0217	.0467	46.000	21.395	13.044	279.073	31
32	2.204	.4538	.0208	.0458	48.150	24.849	13.417	293.140	32
33	2.259	.4427	.0199	.0449	50.354	22.292	13.786	307.306	33
34	2.315	.4319	.0190	.0440	52.613	22.724	14.151	321.559	34
35	2.373	.4214	.0182	.0432	54.928	23.145	14.512	335.886	35
40	2.685	.3724	.0148	.0398	67.402	25.103	16.262	408.221	40
45	3.038	.3292	.0123	.0373	81.516	26.833	17.918	480.806	45
50	3.437	.2909	.0103	.0353	97.484	28.362	19.484	552.607	50
55	3.889	.2572	.00865	.0337	115.551	29.714	20.961	622.827	55
60	4.400	.2273	.00735	.0324	135.991	30.909	22.352	690.865	60
65	4.978	.2009	.00628	.0313	159.118	31.965	23.660	756.280	65
70	5.632	.1776	.00540	.0304	185.284	32.898	24.888	818.763	70
75	6.372	.1569	.00465	.0297	214.888	33.723	26.039	878.114	75
80	7.210	.1387	.00403	.0290	248.382	34.452	27.117	934.217	80
85	8.157	.1226	.00349	.0285	286.278	35.096	28.123	987.026	85
90	9.229	.1084	.00304	.0280	329.154	35.666	29.063	1 036.54	90
95	10.442	.0958	.00265	.0276	377.663	36.169	29.938	1 082.83	95
100	11.814	.0846	.00231	.0273	432.548	36.614	30.752	1 125.97	100

3%   **Compound Interest Factors**   3%

	Single Payment		Uniform Payment Series				Arithmetic Gradient		
	Compound Amount Factor	Present Worth Factor	Sinking Fund Factor	Capital Recovery Factor	Compound Amount Factor	Present Worth Factor	Gradient Uniform Series	Gradient Present Worth	
	Find F Given P	Find P Given F	Find A Given F	Find A Given P	Find F Given A	Find P Given A	Find A Given G	Find P Given G	
n	F/P	P/F	A/F	A/P	F/A	P/A	A/G	P/G	n
1	1.030	.9709	1.0000	1.0300	1.000	0.971	0	0	1
2	1.061	.9426	.4926	.5226	2.030	1.913	0.493	0.943	2
3	1.093	.9151	.3235	.3535	3.091	2.829	0.980	2.773	3
4	1.126	.8885	.2390	.2690	4.184	3.717	1.463	5.438	4
5	1.159	.8626	.1884	.2184	5.309	4.580	1.941	8.889	5
6	1.194	.8375	.1546	.1846	6.468	5.417	2.414	13.076	6
7	1.230	.8131	.1305	.1605	7.662	6.230	2.882	17.955	7
8	1.267	.7894	.1125	.1425	8.892	7.020	3.345	23.481	8
9	1.305	.7664	.0984	.1284	10.159	7.786	3.803	29.612	9
10	1.344	.7441	.0872	.1172	11.464	8.530	4.256	36.309	10
11	1.384	.7224	.0781	.1081	12.808	9.253	4.705	43.533	11
12	1.426	.7014	.0705	.1005	14.192	9.954	5.148	51.248	12
13	1.469	.6810	.0640	.0940	15.618	10.635	5.587	59.419	13
14	1.513	.6611	.0585	.0885	17.086	11.296	6.021	68.014	14
15	1.558	.6419	.0538	.0838	18.599	11.938	6.450	77.000	15
16	1.605	.6232	.0496	.0796	20.157	12.561	6.874	86.348	16
17	1.653	.6050	.0460	.0760	21.762	13.166	7.294	96.028	17
18	1.702	.5874	.0427	.0727	23.414	13.754	7.708	106.014	18
19	1.754	.5703	.0398	.0698	25.117	14.324	8.118	116.279	19
20	1.806	.5537	.0372	.0672	26.870	14.877	8.523	126.799	20
21	1.860	.5375	.0349	.0649	28.676	15.415	8.923	137.549	21
22	1.916	.5219	.0327	.0627	30.537	15.937	9.319	148.509	22
23	1.974	.5067	.0308	.0608	32.453	16.444	9.709	159.656	23
24	2.033	.4919	.0290	.0590	34.426	16.936	10.095	170.971	24
25	2.094	.4776	.0274	.0574	36.459	17.413	10.477	182.433	25
26	2.157	.4637	.0259	.0559	38.553	17.877	10.853	194.026	26
27	2.221	.4502	.0246	.0546	40.710	18.327	11.226	205.731	27
28	2.288	.4371	.0233	.0533	42.931	18.764	11.593	217.532	28
29	2.357	.4243	.0221	.0521	45.219	19.188	11.956	229.413	29
30	2.427	.4120	.0210	.0510	47.575	19.600	12.314	241.361	30
31	2.500	.4000	.0200	.0500	50.003	20.000	12.668	253.361	31
32	2.575	.3883	.0190	.0490	52.503	20.389	13.017	265.399	32
33	2.652	.3770	.0182	.0482	55.078	20.766	13.362	277.464	33
34	2.732	.3660	.0173	.0473	57.730	21.132	13.702	289.544	34
35	2.814	.3554	.0165	.0465	60.462	21.487	14.037	301.627	35
40	3.262	.3066	.0133	.0433	75.401	23.115	15.650	361.750	40
45	3.782	.2644	.0108	.0408	92.720	24.519	17.156	420.632	45
50	4.384	.2281	.00887	.0389	112.797	25.730	18.558	477.480	50
55	5.082	.1968	.00735	.0373	136.072	26.774	19.860	531.741	55
60	5.892	.1697	.00613	.0361	163.053	27.676	21.067	583.052	60
65	6.830	.1464	.00515	.0351	194.333	28.453	22.184	631.201	65
70	7.918	.1263	.00434	.0343	230.594	29.123	23.215	676.087	70
75	9.179	.1089	.00367	.0337	272.631	29.702	24.163	717.698	75
80	10.641	.0940	.00311	.0331	321.363	30.201	25.035	756.086	80
85	12.336	.0811	.00265	.0326	377.857	30.631	25.835	791.353	85
90	14.300	.0699	.00226	.0323	443.349	31.002	26.567	823.630	90
95	16.578	.0603	.00193	.0319	519.272	31.323	27.235	853.074	95
100	19.219	.0520	.00165	.0316	607.287	31.599	27.844	879.854	100

3¹/₂%   Compound Interest Factors   **3¹/₂%**

	Single Payment		Uniform Payment Series				Arithmetic Gradient		
	Compound Amount Factor	Present Worth Factor	Sinking Fund Factor	Capital Recovery Factor	Compound Amount Factor	Present Worth Factor	Gradient Uniform Series	Gradient Present Worth	
	Find F Given P	Find P Given F	Find A Given F	Find A Given P	Find F Given A	Find P Given A	Find A Given G	Find P Given G	
n	F/P	P/F	A/F	A/P	F/A	P/A	A/G	P/G	n
1	1.035	.9662	1.0000	1.0350	1.000	0.966	0	0	1
2	1.071	.9335	.4914	.5264	2.035	1.900	0.491	0.933	2
3	1.109	.9019	.3219	.3569	3.106	2.802	0.977	2.737	3
4	1.148	.8714	.2373	.2723	4.215	3.673	1.457	5.352	4
5	1.188	.8420	.1865	.2215	5.362	4.515	1.931	8.719	5
6	1.229	.8135	.1527	.1877	6.550	5.329	2.400	12.787	6
7	1.272	.7860	.1285	.1635	7.779	6.115	2.862	17.503	7
8	1.317	.7594	.1105	.1455	9.052	6.874	3.320	22.819	8
9	1.363	.7337	.0964	.1314	10.368	7.608	3.771	28.688	9
10	1.411	.7089	.0852	.1202	11.731	8.317	4.217	35.069	10
11	1.460	.6849	.0761	.1111	13.142	9.002	4.657	41.918	11
12	1.511	.6618	.0685	.1035	14.602	9.663	5.091	49.198	12
13	1.564	.6394	.0621	.0971	16.113	10.303	5.520	56.871	13
14	1.619	.6178	.0566	.0916	17.677	10.921	5.943	64.902	14
15	1.675	.5969	.0518	.0868	19.296	11.517	6.361	73.258	15
16	1.734	.5767	.0477	.0827	20.971	12.094	6.773	81.909	16
17	1.795	.5572	.0440	.0790	22.705	12.651	7.179	90.824	17
18	1.857	.5384	.0408	.0758	24.500	13.190	7.580	99.976	18
19	1.922	.5202	.0379	.0729	26.357	13.710	7.975	109.339	19
20	1.990	.5026	.0354	.0704	28.280	14.212	8.365	118.888	20
21	2.059	.4856	.0330	.0680	30.269	14.698	8.749	128.599	21
22	2.132	.4692	.0309	.0659	32.329	15.167	9.128	138.451	22
23	2.206	.4533	.0290	.0640	34.460	15.620	9.502	148.423	23
24	2.283	.4380	.0273	.0623	36.666	16.058	9.870	158.496	24
25	2.363	.4231	.0257	.0607	38.950	16.482	10.233	168.652	25
26	2.446	.4088	.0242	.0592	41.313	16.890	10.590	178.873	26
27	2.532	.3950	.0229	.0579	43.759	17.285	10.942	189.143	27
28	2.620	.3817	.0216	.0566	46.291	17.667	11.289	199.448	28
29	2.712	.3687	.0204	.0554	48.911	18.036	11.631	209.773	29
30	2.807	.3563	.0194	.0544	51.623	18.392	11.967	220.105	30
31	2.905	.3442	.0184	.0534	54.429	18.736	12.299	230.432	31
32	3.007	.3326	.0174	.0524	57.334	19.069	12.625	240.742	32
33	3.112	.3213	.0166	.0516	60.341	19.390	12.946	251.025	33
34	3.221	.3105	.0158	.0508	63.453	19.701	13.262	261.271	34
35	3.334	.3000	.0150	.0500	66.674	20.001	13.573	271.470	35
40	3.959	.2526	.0118	.0468	84.550	21.355	15.055	321.490	40
45	4.702	.2127	.00945	.0445	105.781	22.495	16.417	369.307	45
50	5.585	.1791	.00763	.0426	130.998	23.456	17.666	414.369	50
55	6.633	.1508	.00621	.0412	160.946	24.264	18.808	456.352	55
60	7.878	.1269	.00509	.0401	196.516	24.945	19.848	495.104	60
65	9.357	.1069	.00419	.0392	238.762	25.518	20.793	530.598	65
70	11.113	.0900	.00346	.0385	288.937	26.000	21.650	562.895	70
75	13.199	.0758	.00287	.0379	348.529	26.407	22.423	592.121	75
80	15.676	.0638	.00238	.0374	419.305	26.749	23.120	618.438	80
85	18.618	.0537	.00199	.0370	503.365	27.037	23.747	642.036	85
90	22.112	.0452	.00166	.0367	603.202	27.279	24.308	663.118	90
95	26.262	.0381	.00139	.0364	721.778	27.483	24.811	681.890	95
100	31.191	.0321	.00116	.0362	862.608	27.655	25.259	698.554	100

4% — Compound Interest Factors — **4%**

	Single Payment		Uniform Payment Series				Arithmetic Gradient		
	Compound Amount Factor	Present Worth Factor	Sinking Fund Factor	Capital Recovery Factor	Compound Amount Factor	Present Worth Factor	Gradient Uniform Series	Gradient Present Worth	
	Find F Given P	Find P Given F	Find A Given F	Find A Given P	Find F Given A	Find P Given A	Find A Given G	Find P Given G	
n	F/P	P/F	A/F	A/P	F/A	P/A	A/G	P/G	n
1	1.040	.9615	1.0000	1.0400	1.000	0.962	0	0	1
2	1.082	.9246	.4902	.5302	2.040	1.886	0.490	0.925	2
3	1.125	.8890	.3203	.3603	3.122	2.775	0.974	2.702	3
4	1.170	.8548	.2355	.2755	4.246	3.630	1.451	5.267	4
5	1.217	.8219	.1846	.2246	5.416	4.452	1.922	8.555	5
6	1.265	.7903	.1508	.1908	6.633	5.242	2.386	12.506	6
7	1.316	.7599	.1266	.1666	7.898	6.002	2.843	17.066	7
8	1.369	.7307	.1085	.1485	9.214	6.733	3.294	22.180	8
9	1.423	.7026	.0945	.1345	10.583	7.435	3.739	27.801	9
10	1.480	.6756	.0833	.1233	12.006	8.111	4.177	33.881	10
11	1.539	.6496	.0741	.1141	13.486	8.760	4.609	40.377	11
12	1.601	.6246	.0666	.1066	15.026	9.385	5.034	47.248	12
13	1.665	.6006	.0601	.1001	16.627	9.986	5.453	54.454	13
14	1.732	.5775	.0547	.0947	18.292	10.563	5.866	61.962	14
15	1.801	.5553	.0499	.0899	20.024	11.118	6.272	69.735	15
16	1.873	.5339	.0458	.0858	21.825	11.652	6.672	77.744	16
17	1.948	.5134	.0422	.0822	23.697	12.166	7.066	85.958	17
18	2.026	.4936	.0390	.0790	25.645	12.659	7.453	94.350	18
19	2.107	.4746	.0361	.0761	27.671	13.134	7.834	102.893	19
20	2.191	.4564	.0336	.0736	29.778	13.590	8.209	111.564	20
21	2.279	.4388	.0313	.0713	31.969	14.029	8.578	120.341	21
22	2.370	.4220	.0292	.0692	34.248	14.451	8.941	129.202	22
23	2.465	.4057	.0273	.0673	36.618	14.857	9.297	138.128	23
24	2.563	.3901	.0256	.0656	39.083	15.247	9.648	147.101	24
25	2.666	.3751	.0240	.0640	41.646	15.622	9.993	156.104	25
26	2.772	.3607	.0226	.0626	44.312	15.983	10.331	165.121	26
27	2.883	.3468	.0212	.0612	47.084	16.330	10.664	174.138	27
28	2.999	.3335	.0200	.0600	49.968	16.663	10.991	183.142	28
29	3.119	.3207	.0189	.0589	52.966	16.984	11.312	192.120	29
30	3.243	.3083	.0178	.0578	56.085	17.292	11.627	201.062	30
31	3.373	.2965	.0169	.0569	59.328	17.588	11.937	209.955	31
32	3.508	.2851	.0159	.0559	62.701	17.874	12.241	218.792	32
33	3.648	.2741	.0151	.0551	66.209	18.148	12.540	227.563	33
34	3.794	.2636	.0143	.0543	69.858	18.411	12.832	236.260	34
35	3.946	.2534	.0136	.0536	73.652	18.665	13.120	244.876	35
40	4.801	.2083	.0105	.0505	95.025	19.793	14.476	286.530	40
45	5.841	.1712	.00826	.0483	121.029	20.720	15.705	325.402	45
50	7.107	.1407	.00655	.0466	152.667	21.482	16.812	361.163	50
55	8.646	.1157	.00523	.0452	191.159	22.109	17.807	393.689	55
60	10.520	.0951	.00420	.0442	237.990	22.623	18.697	422.996	60
65	12.799	.0781	.00339	.0434	294.968	23.047	19.491	449.201	65
70	15.572	.0642	.00275	.0427	364.290	23.395	20.196	472.479	70
75	18.945	.0528	.00223	.0422	448.630	23.680	20.821	493.041	75
80	23.050	.0434	.00181	.0418	551.244	23.915	21.372	511.116	80
85	28.044	.0357	.00148	.0415	676.089	24.109	21.857	526.938	85
90	34.119	.0293	.00121	.0412	827.981	24.267	22.283	540.737	90
95	41.511	.0241	.00099	.0410	1 012.8	24.398	22.655	552.730	95
100	50.505	.0198	.00081	.0408	1 237.6	24.505	22.980	563.125	100

4½%   Compound Interest Factors   **4½%**

	Single Payment		Uniform Payment Series				Arithmetic Gradient		
	Compound Amount Factor	Present Worth Factor	Sinking Fund Factor	Capital Recovery Factor	Compound Amount Factor	Present Worth Factor	Gradient Uniform Series	Gradient Present Worth	
	Find F Given P	Find P Given F	Find A Given F	Find A Given P	Find F Given A	Find P Given A	Find A Given G	Find P Given G	
n	F/P	P/F	A/F	A/P	F/A	P/A	A/G	P/G	n
1	1.045	.9569	1.0000	1.0450	1.000	0.957	0	0	1
2	1.092	.9157	.4890	.5340	2.045	1.873	0.489	0.916	2
3	1.141	.8763	.3188	.3638	3.137	2.749	0.971	2.668	3
4	1.193	.8386	.2337	.2787	4.278	3.588	1.445	5.184	4
5	1.246	.8025	.1828	.2278	5.471	4.390	1.912	8.394	5
6	1.302	.7679	.1489	.1939	6.717	5.158	2.372	12.233	6
7	1.361	.7348	.1247	.1697	8.019	5.893	2.824	16.642	7
8	1.422	.7032	.1066	.1516	9.380	6.596	3.269	21.564	8
9	1.486	.6729	.0926	.1376	10.802	7.269	3.707	26.948	9
10	1.553	.6439	.0814	.1264	12.288	7.913	4.138	32.743	10
11	1.623	.6162	.0722	.1172	13.841	8.529	4.562	38.905	11
12	1.696	.5897	.0647	.1097	15.464	9.119	4.978	45.391	12
13	1.772	.5643	.0583	.1033	17.160	9.683	5.387	52.163	13
14	1.852	.5400	.0528	.0978	18.932	10.223	5.789	59.182	14
15	1.935	.5167	.0481	.0931	20.784	10.740	6.184	66.416	15
16	2.022	.4945	.0440	.0890	22.719	11.234	6.572	73.833	16
17	2.113	.4732	.0404	.0854	24.742	11.707	6.953	81.404	17
18	2.208	.4528	.0372	.0822	26.855	12.160	7.327	89.102	18
19	2.308	.4333	.0344	.0794	29.064	12.593	7.695	96.901	19
20	2.412	.4146	.0319	.0769	31.371	13.008	8.055	104.779	20
21	2.520	.3968	.0296	.0746	33.783	13.405	8.409	112.715	21
22	2.634	.3797	.0275	.0725	36.303	13.784	8.755	120.689	22
23	2.752	.3634	.0257	.0707	38.937	14.148	9.096	128.682	23
24	2.876	.3477	.0240	.0690	41.689	14.495	9.429	136.680	24
25	3.005	.3327	.0224	.0674	44.565	14.828	9.756	144.665	25
26	3.141	.3184	.0210	.0660	47.571	15.147	10.077	152.625	26
27	3.282	.3047	.0197	.0647	50.711	15.451	10.391	160.547	27
28	3.430	.2916	.0185	.0635	53.993	15.743	10.698	168.420	28
29	3.584	.2790	.0174	.0624	57.423	16.022	10.999	176.232	29
30	3.745	.2670	.0164	.0614	61.007	16.289	11.295	183.975	30
31	3.914	.2555	.0154	.0604	64.752	16.544	11.583	191.640	31
32	4.090	.2445	.0146	.0596	68.666	16.789	11.866	199.220	32
33	4.274	.2340	.0137	.0587	72.756	17.023	12.143	206.707	33
34	4.466	.2239	.0130	.0580	77.030	17.247	12.414	214.095	34
35	4.667	.2143	.0123	.0573	81.497	17.461	12.679	221.380	35
40	5.816	.1719	.00934	.0543	107.030	18.402	13.917	256.098	40
45	7.248	.1380	.00720	.0522	138.850	19.156	15.020	287.732	45
50	9.033	.1107	.00560	.0506	178.503	19.762	15.998	316.145	50
55	11.256	.0888	.00439	.0494	227.918	20.248	16.860	341.375	55
60	14.027	.0713	.00345	.0485	289.497	20.638	17.617	363.571	60
65	17.481	.0572	.00273	.0477	366.237	20.951	18.278	382.946	65
70	21.784	.0459	.00217	.0472	461.869	21.202	18.854	399.750	70
75	27.147	.0368	.00172	.0467	581.043	21.404	19.354	414.242	75
80	33.830	.0296	.00137	.0464	729.556	21.565	19.785	426.680	80
85	42.158	.0237	.00109	.0461	914.630	21.695	20.157	437.309	85
90	52.537	.0190	.00087	.0459	1 145.3	21.799	20.476	446.359	90
95	65.471	.0153	.00070	.0457	1 432.7	21.883	20.749	454.039	95
100	81.588	.0123	.00056	.0456	1 790.9	21.950	20.981	460.537	100

5%

Compound Interest Factors

5%

	Single Payment		Uniform Payment Series				Arithmetic Gradient		
	Compound Amount Factor	Present Worth Factor	Sinking Fund Factor	Capital Recovery Factor	Compound Amount Factor	Present Worth Factor	Gradient Uniform Series	Gradient Present Worth	
	Find F Given P	Find P Given F	Find A Given F	Find A Given P	Find F Given A	Find P Given A	Find A Given G	Find P Given G	
n	F/P	P/F	A/F	A/P	F/A	P/A	A/G	P/G	n
1	1.050	.9524	1.0000	1.0500	1.000	0.952	0	0	1
2	1.102	.9070	.4878	.5378	2.050	1.859	0.488	0.907	2
3	1.158	.8638	.3172	.3672	3.152	2.723	0.967	2.635	3
4	1.216	.8227	.2320	.2820	4.310	3.546	1.439	5.103	4
5	1.276	.7835	.1810	.2310	5.526	4.329	1.902	8.237	5
6	1.340	.7462	.1470	.1970	6.802	5.076	2.358	11.968	6
7	1.407	.7107	.1228	.1728	8.142	5.786	2.805	16.232	7
8	1.477	.6768	.1047	.1547	9.549	6.463	3.244	20.970	8
9	1.551	.6446	.0907	.1407	11.027	7.108	3.676	26.127	9
10	1.629	.6139	.0795	.1295	12.578	7.722	4.099	31.652	10
11	1.710	.5847	.0704	.1204	14.207	8.306	4.514	37.499	11
12	1.796	.5568	.0628	.1128	15.917	8.863	4.922	43.624	12
13	1.886	.5303	.0565	.1065	17.713	9.394	5.321	49.988	13
14	1.980	.5051	.0510	.1010	19.599	9.899	5.713	56.553	14
15	2.079	.4810	.0463	.0963	21.579	10.380	6.097	63.288	15
16	2.183	.4581	.0423	.0923	23.657	10.838	6.474	70.159	16
17	2.292	.4363	.0387	.0887	25.840	11.274	6.842	77.140	17
18	2.407	.4155	.0355	.0855	28.132	11.690	7.203	84.204	18
19	2.527	.3957	.0327	.0827	30.539	12.085	7.557	91.327	19
20	2.653	.3769	.0302	.0802	33.066	12.462	7.903	98.488	20
21	2.786	.3589	.0280	.0780	35.719	12.821	8.242	105.667	21
22	2.925	.3419	.0260	.0760	38.505	13.163	8.573	112.846	22
23	3.072	.3256	.0241	.0741	41.430	13.489	8.897	120.008	23
24	3.225	.3101	.0225	.0725	44.502	13.799	9.214	127.140	24
25	3.386	.2953	.0210	.0710	47.727	14.094	9.524	134.227	25
26	3.556	.2812	.0196	.0696	51.113	14.375	9.827	141.258	26
27	3.733	.2678	.0183	.0683	54.669	14.643	10.122	148.222	27
28	3.920	.2551	.0171	.0671	58.402	14.898	10.411	155.110	28
29	4.116	.2429	.0160	.0660	62.323	15.141	10.694	161.912	29
30	4.322	.2314	.0151	.0651	66.439	15.372	10.969	168.622	30
31	4.538	.2204	.0141	.0641	70.761	15.593	11.238	175.233	31
32	4.765	.2099	.0133	.0633	75.299	15.803	11.501	181.739	32
33	5.003	.1999	.0125	.0625	80.063	16.003	11.757	188.135	33
34	5.253	.1904	.0118	.0618	85.067	16.193	12.006	194.416	34
35	5.516	.1813	.0111	.0611	90.320	16.374	12.250	200.580	35
40	7.040	.1420	.00828	.0583	120.799	17.159	13.377	229.545	40
45	8.985	.1113	.00626	.0563	159.699	17.774	14.364	255.314	45
50	11.467	.0872	.00478	.0548	209.347	18.256	15.223	277.914	50
55	14.636	.0683	.00367	.0537	272.711	18.633	15.966	297.510	55
60	18.679	.0535	.00283	.0528	353.582	18.929	16.606	314.343	60
65	23.840	.0419	.00219	.0522	456.795	19.161	17.154	328.691	65
70	30.426	.0329	.00170	.0517	588.525	19.343	17.621	340.841	70
75	38.832	.0258	.00132	.0513	756.649	19.485	18.018	351.072	75
80	49.561	.0202	.00103	.0510	971.222	19.596	18.353	359.646	80
85	63.254	.0158	.00080	.0508	1 245.1	19.684	18.635	366.800	85
90	80.730	.0124	.00063	.0506	1 594.6	19.752	18.871	372.749	90
95	103.034	.00971	.00049	.0505	2 040.7	19.806	19.069	377.677	95
100	131.500	.00760	.00038	.0504	2 610.0	19.848	19.234	381.749	100

6%   Compound Interest Factors   6%

	Single Payment		Uniform Payment Series				Arithmetic Gradient		
	Compound Amount Factor	Present Worth Factor	Sinking Fund Factor	Capital Recovery Factor	Compound Amount Factor	Present Worth Factor	Gradient Uniform Series	Gradient Present Worth	
	Find F Given P	Find P Given F	Find A Given F	Find A Given P	Find F Given A	Find P Given A	Find A Given G	Find P Given G	
n	F/P	P/F	A/F	A/P	F/A	P/A	A/G	P/G	n
1	1.060	.9434	1.0000	1.0600	1.000	0.943	0	0	1
2	1.124	.8900	.4854	.5454	2.060	1.833	0.485	0.890	2
3	1.191	.8396	.3141	.3741	3.184	2.673	0.961	2.569	3
4	1.262	.7921	.2286	.2886	4.375	3.465	1.427	4.945	4
5	1.338	.7473	.1774	.2374	5.637	4.212	1.884	7.934	5
6	1.419	.7050	.1434	.2034	6.975	4.917	2.330	11.459	6
7	1.504	.6651	.1191	.1791	8.394	5.582	2.768	15.450	7
8	1.594	.6274	.1010	.1610	9.897	6.210	3.195	19.841	8
9	1.689	.5919	.0870	.1470	11.491	6.802	3.613	24.577	9
10	1.791	.5584	.0759	.1359	13.181	7.360	4.022	29.602	10
11	1.898	.5268	.0668	.1268	14.972	7.887	4.421	34.870	11
12	2.012	.4970	.0593	.1193	16.870	8.384	4.811	40.337	12
13	2.133	.4688	.0530	.1130	18.882	8.853	5.192	45.963	13
14	2.261	.4423	.0476	.1076	21.015	9.295	5.564	51.713	14
15	2.397	.4173	.0430	.1030	23.276	9.712	5.926	57.554	15
16	2.540	.3936	.0390	.0990	25.672	10.106	6.279	63.459	16
17	2.693	.3714	.0354	.0954	28.213	10.477	6.624	69.401	17
18	2.854	.3503	.0324	.0924	30.906	10.828	6.960	75.357	18
19	3.026	.3305	.0296	.0896	33.760	11.158	7.287	81.306	19
20	3.207	.3118	.0272	.0872	36.786	11.470	7.605	87.230	20
21	3.400	.2942	.0250	.0850	39.993	11.764	7.915	93.113	21
22	3.604	.2775	.0230	.0830	43.392	12.042	8.217	98.941	22
23	3.820	.2618	.0213	.0813	46.996	12.303	8.510	104.700	23
24	4.049	.2470	.0197	.0797	50.815	12.550	8.795	110.381	24
25	4.292	.2330	.0182	.0782	54.864	12.783	9.072	115.973	25
26	4.549	.2198	.0169	.0769	59.156	13.003	9.341	121.468	26
27	4.822	.2074	.0157	.0757	63.706	13.211	9.603	126.860	27
28	5.112	.1956	.0146	.0746	68.528	13.406	9.857	132.142	28
29	5.418	.1846	.0136	.0736	73.640	13.591	10.103	137.309	29
30	5.743	.1741	.0126	.0726	79.058	13.765	10.342	142.359	30
31	6.088	.1643	.0118	.0718	84.801	13.929	10.574	147.286	31
32	6.453	.1550	.0110	.0710	90.890	14.084	10.799	152.090	32
33	6.841	.1462	.0103	.0703	97.343	14.230	11.017	156.768	33
34	7.251	.1379	.00960	.0696	104.184	14.368	11.228	161.319	34
35	7.686	.1301	.00897	.0690	111.435	14.498	11.432	165.743	35
40	10.286	.0972	.00646	.0665	154.762	15.046	12.359	185.957	40
45	13.765	.0727	.00470	.0647	212.743	15.456	13.141	203.109	45
50	18.420	.0543	.00344	.0634	290.335	15.762	13.796	217.457	50
55	24.650	.0406	.00254	.0625	394.171	15.991	14.341	229.322	55
60	32.988	.0303	.00188	.0619	533.126	16.161	14.791	239.043	60
65	44.145	.0227	.00139	.0614	719.080	16.289	15.160	246.945	65
70	59.076	.0169	.00103	.0610	967.928	16.385	15.461	253.327	70
75	79.057	.0126	.00077	.0608	1 300.9	16.456	15.706	258.453	75
80	105.796	.00945	.00057	.0606	1 746.6	16.509	15.903	262.549	80
85	141.578	.00706	.00043	.0604	2 343.0	16.549	16.062	265.810	85
90	189.464	.00528	.00032	.0603	3 141.1	16.579	16.189	268.395	90
95	253.545	.00394	.00024	.0602	4 209.1	16.601	16.290	270.437	95
100	339.300	.00295	.00018	.0602	5 638.3	16.618	16.371	272.047	100

7%   **Compound Interest Factors**   **7%**

	Single Payment		Uniform Payment Series				Arithmetic Gradient		
	Compound Amount Factor	Present Worth Factor	Sinking Fund Factor	Capital Recovery Factor	Compound Amount Factor	Present Worth Factor	Gradient Uniform Series	Gradient Present Worth	
	Find F Given P	Find P Given F	Find A Given F	Find A Given P	Find F Given A	Find P Given A	Find A Given G	Find P Given G	
n	F/P	P/F	A/F	A/P	F/A	P/A	A/G	P/G	n
1	1.070	.9346	1.0000	1.0700	1.000	0.935	0	0	1
2	1.145	.8734	.4831	.5531	2.070	1.808	0.483	0.873	2
3	1.225	.8163	.3111	.3811	3.215	2.624	0.955	2.506	3
4	1.311	.7629	.2252	.2952	4.440	3.387	1.416	4.795	4
5	1.403	.7130	.1739	.2439	5.751	4.100	1.865	7.647	5
6	1.501	.6663	.1398	.2098	7.153	4.767	2.303	10.978	6
7	1.606	.6227	.1156	.1856	8.654	5.389	2.730	14.715	7
8	1.718	.5820	.0975	.1675	10.260	5.971	3.147	18.789	8
9	1.838	.5439	.0835	.1535	11.978	6.515	3.552	23.140	9
10	1.967	.5083	.0724	.1424	13.816	7.024	3.946	27.716	10
11	2.105	.4751	.0634	.1334	15.784	7.499	4.330	32.467	11
12	2.252	.4440	.0559	.1259	17.888	7.943	4.703	37.351	12
13	2.410	.4150	.0497	.1197	20.141	8.358	5.065	42.330	13
14	2.579	.3878	.0443	.1143	22.551	8.745	5.417	47.372	14
15	2.759	.3624	.0398	.1098	25.129	9.108	5.758	52.446	15
16	2.952	.3387	.0359	.1059	27.888	9.447	6.090	57.527	16
17	3.159	.3166	.0324	.1024	30.840	9.763	6.411	62.592	17
18	3.380	.2959	.0294	.0994	33.999	10.059	6.722	67.622	18
19	3.617	.2765	.0268	.0968	37.379	10.336	7.024	72.599	19
20	3.870	.2584	.0244	.0944	40.996	10.594	7.316	77.509	20
21	4.141	.2415	.0223	.0923	44.865	10.836	7.599	82.339	21
22	4.430	.2257	.0204	.0904	49.006	11.061	7.872	87.079	22
23	4.741	.2109	.0187	.0887	53.436	11.272	8.137	91.720	23
24	5.072	.1971	.0172	.0872	58.177	11.469	8.392	96.255	24
25	5.427	.1842	.0158	.0858	63.249	11.654	8.639	100.677	25
26	5.807	.1722	.0146	.0846	68.677	11.826	8.877	104.981	26
27	6.214	.1609	.0134	.0834	74.484	11.987	9.107	109.166	27
28	6.649	.1504	.0124	.0824	80.698	12.137	9.329	113.227	28
29	7.114	.1406	.0114	.0814	87.347	12.278	9.543	117.162	29
30	7.612	.1314	.0106	.0806	94.461	12.409	9.749	120.972	30
31	8.145	.1228	.00980	.0798	102.073	12.532	9.947	124.655	31
32	8.715	.1147	.00907	.0791	110.218	12.647	10.138	128.212	32
33	9.325	.1072	.00841	.0784	118.934	12.754	10.322	131.644	33
34	9.978	.1002	.00780	.0778	128.259	12.854	10.499	134.951	34
35	10.677	.0937	.00723	.0772	138.237	12.948	10.669	138.135	35
40	14.974	.0668	.00501	.0750	199.636	13.332	11.423	152.293	40
45	21.002	.0476	.00350	.0735	285.750	13.606	12.036	163.756	45
50	29.457	.0339	.00246	.0725	406.530	13.801	12.529	172.905	50
55	41.315	.0242	.00174	.0717	575.930	13.940	12.921	180.124	55
60	57.947	.0173	.00123	.0712	813.523	14.039	13.232	185.768	60
65	81.273	.0123	.00087	.0709	1 146.8	14.110	13.476	190.145	65
70	113.990	.00877	.00062	.0706	1 614.1	14.160	13.666	193.519	70
75	159.877	.00625	.00044	.0704	2 269.7	14.196	13.814	196.104	75
80	224.235	.00446	.00031	.0703	3 189.1	14.222	13.927	198.075	80
85	314.502	.00318	.00022	.0702	4 478.6	14.240	14.015	199.572	85
90	441.105	.00227	.00016	.0702	6 287.2	14.253	14.081	200.704	90
95	618.673	.00162	.00011	.0701	8 823.9	14.263	14.132	201.558	95
100	867.720	.00115	.00008	.0701	12 381.7	14.269	14.170	202.200	100

8% Compound Interest Factors **8%**

	Single Payment		Uniform Payment Series				Arithmetic Gradient		
	Compound Amount Factor	Present Worth Factor	Sinking Fund Factor	Capital Recovery Factor	Compound Amount Factor	Present Worth Factor	Gradient Uniform Series	Gradient Present Worth	
	Find F Given P	Find P Given F	Find A Given F	Find A Given P	Find F Given A	Find P Given A	Find A Given G	Find P Given G	
n	F/P	P/F	A/F	A/P	F/A	P/A	A/G	P/G	n
1	1.080	.9259	1.0000	1.0800	1.000	0.926	0	0	1
2	1.166	.8573	.4808	.5608	2.080	1.783	0.481	0.857	2
3	1.260	.7938	.3080	.3880	3.246	2.577	0.949	2.445	3
4	1.360	.7350	.2219	.3019	4.506	3.312	1.404	4.650	4
5	1.469	.6806	.1705	.2505	5.867	3.993	1.846	7.372	5
6	1.587	.6302	.1363	.2163	7.336	4.623	2.276	10.523	6
7	1.714	.5835	.1121	.1921	8.923	5.206	2.694	14.024	7
8	1.851	.5403	.0940	.1740	10.637	5.747	3.099	17.806	8
9	1.999	.5002	.0801	.1601	12.488	6.247	3.491	21.808	9
10	2.159	.4632	.0690	.1490	14.487	6.710	3.871	25.977	10
11	2.332	.4289	.0601	.1401	16.645	7.139	4.240	30.266	11
12	2.518	.3971	.0527	.1327	18.977	7.536	4.596	34.634	12
13	2.720	.3677	.0465	.1265	21.495	7.904	4.940	39.046	13
14	2.937	.3405	.0413	.1213	24.215	8.244	5.273	43.472	14
15	3.172	.3152	.0368	.1168	27.152	8.559	5.594	47.886	15
16	3.426	.2919	.0330	.1130	30.324	8.851	5.905	52.264	16
17	3.700	.2703	.0296	.1096	33.750	9.122	6.204	56.588	17
18	3.996	.2502	.0267	.1067	37.450	9.372	6.492	60.843	18
19	4.316	.2317	.0241	.1041	41.446	9.604	6.770	65.013	19
20	4.661	.2145	.0219	.1019	45.762	9.818	7.037	69.090	20
21	5.034	.1987	.0198	.0998	50.423	10.017	7.294	73.063	21
22	5.437	.1839	.0180	.0980	55.457	10.201	7.541	76.926	22
23	5.871	.1703	.0164	.0964	60.893	10.371	7.779	80.673	23
24	6.341	.1577	.0150	.0950	66.765	10.529	8.007	84.300	24
25	6.848	.1460	.0137	.0937	73.106	10.675	8.225	87.804	25
26	7.396	.1352	.0125	.0925	79.954	10.810	8.435	91.184	26
27	7.988	.1252	.0114	.0914	87.351	10.935	8.636	94.439	27
28	8.627	.1159	.0105	.0905	95.339	11.051	8.829	97.569	28
29	9.317	.1073	.00962	.0896	103.966	11.158	9.013	100.574	29
30	10.063	.0994	.00883	.0888	113.283	11.258	9.190	103.456	30
31	10.868	.0920	.00811	.0881	123.346	11.350	9.358	106.216	31
32	11.737	.0852	.00745	.0875	134.214	11.435	9.520	108.858	32
33	12.676	.0789	.00685	.0869	145.951	11.514	9.674	111.382	33
34	13.690	.0730	.00630	.0863	158.627	11.587	9.821	113.792	34
35	14.785	.0676	.00580	.0858	172.317	11.655	9.961	116.092	35
40	21.725	.0460	.00386	.0839	259.057	11.925	10.570	126.042	40
45	31.920	.0313	.00259	.0826	386.506	12.108	11.045	133.733	45
50	46.902	.0213	.00174	.0817	573.771	12.233	11.411	139.593	50
55	68.914	.0145	.00118	.0812	848.925	12.319	11.690	144.006	55
60	101.257	.00988	.00080	.0808	1 253.2	12.377	11.902	147.300	60
65	148.780	.00672	.00054	.0805	1 847.3	12.416	12.060	149.739	65
70	218.607	.00457	.00037	.0804	2 720.1	12.443	12.178	151.533	70
75	321.205	.00311	.00025	.0802	4 002.6	12.461	12.266	152.845	75
80	471.956	.00212	.00017	.0802	5 887.0	12.474	12.330	153.800	80
85	693.458	.00144	.00012	.0801	8 655.7	12.482	12.377	154.492	85
90	1 018.9	.00098	.00008	.0801	12 724.0	12.488	12.412	154.993	90
95	1 497.1	.00067	.00005	.0801	18 701.6	12.492	12.437	155.352	95
100	2 199.8	.00045	.00004	.0800	27 484.6	12.494	12.455	155.611	100

9% Compound Interest Factors **9%**

	Single Payment		Uniform Payment Series				Arithmetic Gradient		
	Compound Amount Factor	Present Worth Factor	Sinking Fund Factor	Capital Recovery Factor	Compound Amount Factor	Present Worth Factor	Gradient Uniform Series	Gradient Present Worth	
	Find F Given P	Find P Given F	Find A Given F	Find A Given P	Find F Given A	Find P Given A	Find A Given G	Find P Given G	
n	F/P	P/F	A/F	A/P	F/A	P/A	A/G	P/G	n
1	1.090	.9174	1.0000	1.0900	1.000	0.917	0	0	1
2	1.188	.8417	.4785	.5685	2.090	1.759	0.478	0.842	2
3	1.295	.7722	.3051	.3951	3.278	2.531	0.943	2.386	3
4	1.412	.7084	.2187	.3087	4.573	3.240	1.393	4.511	4
5	1.539	.6499	.1671	.2571	5.985	3.890	1.828	7.111	5
6	1.677	.5963	.1329	.2229	7.523	4.486	2.250	10.092	6
7	1.828	.5470	.1087	.1987	9.200	5.033	2.657	13.375	7
8	1.993	.5019	.0907	.1807	11.028	5.535	3.051	16.888	8
9	2.172	.4604	.0768	.1668	13.021	5.995	3.431	20.571	9
10	2.367	.4224	.0658	.1558	15.193	6.418	3.798	24.373	10
11	2.580	.3875	.0569	.1469	17.560	6.805	4.151	28.248	11
12	2.813	.3555	.0497	.1397	20.141	7.161	4.491	32.159	12
13	3.066	.3262	.0436	.1336	22.953	7.487	4.818	36.073	13
14	3.342	.2992	.0384	.1284	26.019	7.786	5.133	39.963	14
15	3.642	.2745	.0341	.1241	29.361	8.061	5.435	43.807	15
16	3.970	.2519	.0303	.1203	33.003	8.313	5.724	47.585	16
17	4.328	.2311	.0270	.1170	36.974	8.544	6.002	51.282	17
18	4.717	.2120	.0242	.1142	41.301	8.756	6.269	54.886	18
19	5.142	.1945	.0217	.1117	46.019	8.950	6.524	58.387	19
20	5.604	.1784	.0195	.1095	51.160	9.129	6.767	61.777	20
21	6.109	.1637	.0176	.1076	56.765	9.292	7.001	65.051	21
22	6.659	.1502	.0159	.1059	62.873	9.442	7.223	68.205	22
23	7.258	.1378	.0144	.1044	69.532	9.580	7.436	71.236	23
24	7.911	.1264	.0130	.1030	76.790	9.707	7.638	74.143	24
25	8.623	.1160	.0118	.1018	84.701	9.823	7.832	76.927	25
26	9.399	.1064	.0107	.1007	93.324	9.929	8.016	79.586	26
27	10.245	.0976	.00973	.0997	102.723	10.027	8.191	82.124	27
28	11.167	.0895	.00885	.0989	112.968	10.116	8.357	84.542	28
29	12.172	.0822	.00806	.0981	124.136	10.198	8.515	86.842	29
30	13.268	.0754	.00734	.0973	136.308	10.274	8.666	89.028	30
31	14.462	.0691	.00669	.0967	149.575	10.343	8.808	91.102	31
32	15.763	.0634	.00610	.0961	164.037	10.406	8.944	93.069	32
33	17.182	.0582	.00556	.0956	179.801	10.464	9.072	94.931	33
34	18.728	.0534	.00508	.0951	196.983	10.518	9.193	96.693	34
35	20.414	.0490	.00464	.0946	215.711	10.567	9.308	98.359	35
40	31.409	.0318	.00296	.0930	337.883	10.757	9.796	105.376	40
45	48.327	.0207	.00190	.0919	525.860	10.881	10.160	110.556	45
50	74.358	.0134	.00123	.0912	815.085	10.962	10.430	114.325	50
55	114.409	.00874	.00079	.0908	1 260.1	11.014	10.626	117.036	55
60	176.032	.00568	.00051	.0905	1 944.8	11.048	10.768	118.968	60
65	270.847	.00369	.00033	.0903	2 998.3	11.070	10.870	120.334	65
70	416.731	.00240	.00022	.0902	4 619.2	11.084	10.943	121.294	70
75	641.193	.00156	.00014	.0901	7 113.3	11.094	10.994	121.965	75
80	986.555	.00101	.00009	.0901	10 950.6	11.100	11.030	122.431	80
85	1 517.9	.00066	.00006	.0901	16 854.7	11.104	11.055	122.753	85
90	2 335.5	.00043	.00004	.0900	25 939.3	11.106	11.073	122.976	90
95	3 593.5	.00028	.00003	.0900	39 916.8	11.108	11.085	123.129	95
100	5 529.1	.00018	.00002	.0900	61 422.9	11.109	11.093	123.233	100

10%   Compound Interest Factors   **10%**

	Single Payment		Uniform Payment Series				Arithmetic Gradient		
	Compound Amount Factor	Present Worth Factor	Sinking Fund Factor	Capital Recovery Factor	Compound Amount Factor	Present Worth Factor	Gradient Uniform Series	Gradient Present Worth	
	Find F Given P	Find P Given F	Find A Given F	Find A Given P	Find F Given A	Find P Given A	Find A Given G	Find P Given G	
n	F/P	P/F	A/F	A/P	F/A	P/A	A/G	P/G	*n*
1	1.100	.9091	1.0000	1.1000	1.000	0.909	0	0	1
2	1.210	.8264	.4762	.5762	2.100	1.736	0.476	0.826	2
3	1.331	.7513	.3021	.4021	3.310	2.487	0.937	2.329	3
4	1.464	.6830	.2155	.3155	4.641	3.170	1.381	4.378	4
5	1.611	.6209	.1638	.2638	6.105	3.791	1.810	6.862	5
6	1.772	.5645	.1296	.2296	7.716	4.355	2.224	9.684	6
7	1.949	.5132	.1054	.2054	9.487	4.868	2.622	12.763	7
8	2.144	.4665	.0874	.1874	11.436	5.335	3.004	16.029	8
9	2.358	.4241	.0736	.1736	13.579	5.759	3.372	19.421	9
10	2.594	.3855	.0627	.1627	15.937	6.145	3.725	22.891	10
11	2.853	.3505	.0540	.1540	18.531	6.495	4.064	26.396	11
12	3.138	.3186	.0468	.1468	21.384	6.814	4.388	29.901	12
13	3.452	.2897	.0408	.1408	24.523	7.103	4.699	33.377	13
14	3.797	.2633	.0357	.1357	27.975	7.367	4.996	36.801	14
15	4.177	.2394	.0315	.1315	31.772	7.606	5.279	40.152	15
16	4.595	.2176	.0278	.1278	35.950	7.824	5.549	43.416	16
17	5.054	.1978	.0247	.1247	40.545	8.022	5.807	46.582	17
18	5.560	.1799	.0219	.1219	45.599	8.201	6.053	49.640	18
19	6.116	.1635	.0195	.1195	51.159	8.365	6.286	52.583	19
20	6.728	.1486	.0175	.1175	57.275	8.514	6.508	55.407	20
21	7.400	.1351	.0156	.1156	64.003	8.649	6.719	58.110	21
22	8.140	.1228	.0140	.1140	71.403	8.772	6.919	60.689	22
23	8.954	.1117	.0126	.1126	79.543	8.883	7.108	63.146	23
24	9.850	.1015	.0113	.1113	88.497	8.985	7.288	65.481	24
25	10.835	.0923	.0102	.1102	98.347	9.077	7.458	67.696	25
26	11.918	.0839	.00916	.1092	109.182	9.161	7.619	69.794	26
27	13.110	.0763	.00826	.1083	121.100	9.237	7.770	71.777	27
28	14.421	.0693	.00745	.1075	134.210	9.307	7.914	73.650	28
29	15.863	.0630	.00673	.1067	148.631	9.370	8.049	75.415	29
30	17.449	.0573	.00608	.1061	164.494	9.427	8.176	77.077	30
31	19.194	.0521	.00550	.1055	181.944	9.479	8.296	78.640	31
32	21.114	.0474	.00497	.1050	201.138	9.526	8.409	80.108	32
33	23.225	.0431	.00450	.1045	222.252	9.569	8.515	81.486	33
34	25.548	.0391	.00407	.1041	245.477	9.609	8.615	82.777	34
35	28.102	.0356	.00369	.1037	271.025	9.644	8.709	83.987	35
40	45.259	.0221	.00226	.1023	442.593	9.779	9.096	88.953	40
45	72.891	.0137	.00139	.1014	718.905	9.863	9.374	92.454	45
50	117.391	.00852	.00086	.1009	1 163.9	9.915	9.570	94.889	50
55	189.059	.00529	.00053	.1005	1 880.6	9.947	9.708	96.562	55
60	304.482	.00328	.00033	.1003	3 034.8	9.967	9.802	97.701	60
65	490.371	.00204	.00020	.1002	4 893.7	9.980	9.867	98.471	65
70	789.748	.00127	.00013	.1001	7 887.5	9.987	9.911	98.987	70
75	1 271.9	.00079	.00008	.1001	12 709.0	9.992	9.941	99.332	75
80	2 048.4	.00049	.00005	.1000	20 474.0	9.995	9.961	99.561	80
85	3 299.0	.00030	.00003	.1000	32 979.7	9.997	9.974	99.712	85
90	5 313.0	.00019	.00002	.1000	53 120.3	9.998	9.983	99.812	90
95	8 556.7	.00012	.00001	.1000	85 556.9	9.999	9.989	99.877	95
100	13 780.6	.00007	.00001	.1000	137 796.3	9.999	9.993	99.920	100

12%   Compound Interest Factors   12%

	Single Payment		Uniform Payment Series				Arithmetic Gradient		
	Compound Amount Factor	Present Worth Factor	Sinking Fund Factor	Capital Recovery Factor	Compound Amount Factor	Present Worth Factor	Gradient Uniform Series	Gradient Present Worth	
	Find F Given P	Find P Given F	Find A Given F	Find A Given P	Find F Given A	Find P Given A	Find A Given G	Find P Given G	
n	F/P	P/F	A/F	A/P	F/A	P/A	A/G	P/G	n
1	1.120	.8929	1.0000	1.1200	1.000	0.893	0	0	1
2	1.254	.7972	.4717	.5917	2.120	1.690	0.472	0.797	2
3	1.405	.7118	.2963	.4163	3.374	2.402	0.925	2.221	3
4	1.574	.6355	.2092	.3292	4.779	3.037	1.359	4.127	4
5	1.762	.5674	.1574	.2774	6.353	3.605	1.775	6.397	5
6	1.974	.5066	.1232	.2432	8.115	4.111	2.172	8.930	6
7	2.211	.4523	.0991	.2191	10.089	4.564	2.551	11.644	7
8	2.476	.4039	.0813	.2013	12.300	4.968	2.913	14.471	8
9	2.773	.3606	.0677	.1877	14.776	5.328	3.257	17.356	9
10	3.106	.3220	.0570	.1770	17.549	5.650	3.585	20.254	10
11	3.479	.2875	.0484	.1684	20.655	5.938	3.895	23.129	11
12	3.896	.2567	.0414	.1614	24.133	6.194	4.190	25.952	12
13	4.363	.2292	.0357	.1557	28.029	6.424	4.468	28.702	13
14	4.887	.2046	.0309	.1509	32.393	6.628	4.732	31.362	14
15	5.474	.1827	.0268	.1468	37.280	6.811	4.980	33.920	15
16	6.130	.1631	.0234	.1434	42.753	6.974	5.215	36.367	16
17	6.866	.1456	.0205	.1405	48.884	7.120	5.435	38.697	17
18	7.690	.1300	.0179	.1379	55.750	7.250	5.643	40.908	18
19	8.613	.1161	.0158	.1358	63.440	7.366	5.838	42.998	19
20	9.646	.1037	.0139	.1339	72.052	7.469	6.020	44.968	20
21	10.804	.0926	.0122	.1322	81.699	7.562	6.191	46.819	21
22	12.100	.0826	.0108	.1308	92.503	7.645	6.351	48.554	22
23	13.552	.0738	.00956	.1296	104.603	7.718	6.501	50.178	23
24	15.179	.0659	.00846	.1285	118.155	7.784	6.641	51.693	24
25	17.000	.0588	.00750	.1275	133.334	7.843	6.771	53.105	25
26	19.040	.0525	.00665	.1267	150.334	7.896	6.892	54.418	26
27	21.325	.0469	.00590	.1259	169.374	7.943	7.005	55.637	27
28	23.884	.0419	.00524	.1252	190.699	7.984	7.110	56.767	28
29	26.750	.0374	.00466	.1247	214.583	8.022	7.207	57.814	29
30	29.960	.0334	.00414	.1241	241.333	8.055	7.297	58.782	30
31	33.555	.0298	.00369	.1237	271.293	8.085	7.381	59.676	31
32	37.582	.0266	.00328	.1233	304.848	8.112	7.459	60.501	32
33	42.092	.0238	.00292	.1229	342.429	8.135	7.530	61.261	33
34	47.143	.0212	.00260	.1226	384.521	8.157	7.596	61.961	34
35	52.800	.0189	.00232	.1223	431.663	8.176	7.658	62.605	35
40	93.051	.0107	.00130	.1213	767.091	8.244	7.899	65.116	40
45	163.988	.00610	.00074	.1207	1358.2	8.283	8.057	66.734	45
50	289.002	.00346	.00042	.1204	2400.0	8.304	8.160	67.762	50
55	509.321	.00196	.00024	.1202	4236.0	8.317	8.225	68.408	55
60	897.597	.00111	.00013	.1201	7471.6	8.324	8.266	68.810	60
65	1581.9	.00063	.00008	.1201	13173.9	8.328	8.292	69.058	65
70	2787.8	.00036	.00004	.1200	23223.3	8.330	8.308	69.210	70
75	4913.1	.00020	.00002	.1200	40933.8	8.332	8.318	69.303	75
80	8658.5	.00012	.00001	.1200	72145.7	8.332	8.324	69.359	80
85	15259.2	.00007	.00001	.1200	127151.7	8.333	8.328	69.393	85
90	26891.9	.00004		.1200	224091.1	8.333	8.330	69.414	90
95	47392.8	.00002		.1200	394931.4	8.333	8.331	69.426	95
100	83522.3	.00001		.1200	696010.5	8.333	8.332	69.434	100

15%   Compound Interest Factors   **15%**

	Single Payment		Uniform Payment Series				Arithmetic Gradient		
	Compound Amount Factor	Present Worth Factor	Sinking Fund Factor	Capital Recovery Factor	Compound Amount Factor	Present Worth Factor	Gradient Uniform Series	Gradient Present Worth	
	Find F Given P	Find P Given F	Find A Given F	Find A Given P	Find F Given A	Find P Given A	Find A Given G	Find P Given G	
n	F/P	P/F	A/F	A/P	F/A	P/A	A/G	P/G	n
1	1.150	.8696	1.0000	1.1500	1.000	0.870	0	0	1
2	1.322	.7561	.4651	.6151	2.150	1.626	0.465	0.756	2
3	1.521	.6575	.2880	.4380	3.472	2.283	0.907	2.071	3
4	1.749	.5718	.2003	.3503	4.993	2.855	1.326	3.786	4
5	2.011	.4972	.1483	.2983	6.742	3.352	1.723	5.775	5
6	2.313	.4323	.1142	.2642	8.754	3.784	2.097	7.937	6
7	2.660	.3759	.0904	.2404	11.067	4.160	2.450	10.192	7
8	3.059	.3269	.0729	.2229	13.727	4.487	2.781	12.481	8
9	3.518	.2843	.0596	.2096	16.786	4.772	3.092	14.755	9
10	4.046	.2472	.0493	.1993	20.304	5.019	3.383	16.979	10
11	4.652	.2149	.0411	.1911	24.349	5.234	3.655	19.129	11
12	5.350	.1869	.0345	.1845	29.002	5.421	3.908	21.185	12
13	6.153	.1625	.0291	.1791	34.352	5.583	4.144	23.135	13
14	7.076	.1413	.0247	.1747	40.505	5.724	4.362	24.972	14
15	8.137	.1229	.0210	.1710	47.580	5.847	4.565	26.693	15
16	9.358	.1069	.0179	.1679	55.717	5.954	4.752	28.296	16
17	10.761	.0929	.0154	.1654	65.075	6.047	4.925	29.783	17
18	12.375	.0808	.0132	.1632	75.836	6.128	5.084	31.156	18
19	14.232	.0703	.0113	.1613	88.212	6.198	5.231	32.421	19
20	16.367	.0611	.00976	.1598	102.444	6.259	5.365	33.582	20
21	18.822	.0531	.00842	.1584	118.810	6.312	5.488	34.645	21
22	21.645	.0462	.00727	.1573	137.632	6.359	5.601	35.615	22
23	24.891	.0402	.00628	.1563	159.276	6.399	5.704	36.499	23
24	28.625	.0349	.00543	.1554	184.168	6.434	5.798	37.302	24
25	32.919	.0304	.00470	.1547	212.793	6.464	5.883	38.031	25
26	37.857	.0264	.00407	.1541	245.712	6.491	5.961	38.692	26
27	43.535	.0230	.00353	.1535	283.569	6.514	6.032	39.289	27
28	50.066	.0200	.00306	.1531	327.104	6.534	6.096	39.828	28
29	57.575	.0174	.00265	.1527	377.170	6.551	6.154	40.315	29
30	66.212	.0151	.00230	.1523	434.745	6.566	6.207	40.753	30
31	76.144	.0131	.00200	.1520	500.957	6.579	6.254	41.147	31
32	87.565	.0114	.00173	.1517	577.100	6.591	6.297	41.501	32
33	100.700	.00993	.00150	.1515	664.666	6.600	6.336	41.818	33
34	115.805	.00864	.00131	.1513	765.365	6.609	6.371	42.103	34
35	133.176	.00751	.00113	.1511	881.170	6.617	6.402	42.359	35
40	267.864	.00373	.00056	.1506	1 779.1	6.642	6.517	43.283	40
45	538.769	.00186	.00028	.1503	3 585.1	6.654	6.583	43.805	45
50	1 083.7	.00092	.00014	.1501	7 217.7	6.661	6.620	44.096	50
55	2 179.6	.00046	.00007	.1501	14 524.1	6.664	6.641	44.256	55
60	4 384.0	.00023	.00003	.1500	29 220.0	6.665	6.653	44.343	60
65	8 817.8	.00011	.00002	.1500	58 778.6	6.666	6.659	44.390	65
70	17 735.7	.00006	.00001	.1500	118 231.5	6.666	6.663	44.416	70
75	35 672.9	.00003		.1500	237 812.5	6.666	6.665	44.429	75
80	71 750.9	.00001		.1500	478 332.6	6.667	6.666	44.436	80
85	144 316.7	.00001		.1500	962 104.4	6.667	6.666	44.440	85

18%   **Compound Interest Factors**   18%

	Single Payment		Uniform Payment Series				Arithmetic Gradient		
	Compound Amount Factor	Present Worth Factor	Sinking Fund Factor	Capital Recovery Factor	Compound Amount Factor	Present Worth Factor	Gradient Uniform Series	Gradient Present Worth	
	Find F Given P	Find P Given F	Find A Given F	Find A Given P	Find F Given A	Find P Given A	Find A Given G	Find P Given G	
n	F/P	P/F	A/F	A/P	F/A	P/A	A/G	P/G	n
1	1.180	.8475	1.0000	1.1800	1.000	0.847	0	0	1
2	1.392	.7182	.4587	.6387	2.180	1.566	0.459	0.718	2
3	1.643	.6086	.2799	.4599	3.572	2.174	0.890	1.935	3
4	1.939	.5158	.1917	.3717	5.215	2.690	1.295	3.483	4
5	2.288	.4371	.1398	.3198	7.154	3.127	1.673	5.231	5
6	2.700	.3704	.1059	.2859	9.442	3.498	2.025	7.083	6
7	3.185	.3139	.0824	.2624	12.142	3.812	2.353	8.967	7
8	3.759	.2660	.0652	.2452	15.327	4.078	2.656	10.829	8
9	4.435	.2255	.0524	.2324	19.086	4.303	2.936	12.633	9
10	5.234	.1911	.0425	.2225	23.521	4.494	3.194	14.352	10
11	6.176	.1619	.0348	.2148	28.755	4.656	3.430	15.972	11
12	7.288	.1372	.0286	.2086	34.931	4.793	3.647	17.481	12
13	8.599	.1163	.0237	.2037	42.219	4.910	3.845	18.877	13
14	10.147	.0985	.0197	.1997	50.818	5.008	4.025	20.158	14
15	11.974	.0835	.0164	.1964	60.965	5.092	4.189	21.327	15
16	14.129	.0708	.0137	.1937	72.939	5.162	4.337	22.389	16
17	16.672	.0600	.0115	.1915	87.068	5.222	4.471	23.348	17
18	19.673	.0508	.00964	.1896	103.740	5.273	4.592	24.212	18
19	23.214	.0431	.00810	.1881	123.413	5.316	4.700	24.988	19
20	27.393	.0365	.00682	.1868	146.628	5.353	4.798	25.681	20
21	32.324	.0309	.00575	.1857	174.021	5.384	4.885	26.300	21
22	38.142	.0262	.00485	.1848	206.345	5.410	4.963	26.851	22
23	45.008	.0222	.00409	.1841	244.487	5.432	5.033	27.339	23
24	53.109	.0188	.00345	.1835	289.494	5.451	5.095	27.772	24
25	62.669	.0160	.00292	.1829	342.603	5.467	5.150	28.155	25
26	73.949	.0135	.00247	.1825	405.272	5.480	5.199	28.494	26
27	87.260	.0115	.00209	.1821	479.221	5.492	5.243	28.791	27
28	102.966	.00971	.00177	.1818	566.480	5.502	5.281	29.054	28
29	121.500	.00823	.00149	.1815	669.447	5.510	5.315	29.284	29
30	143.370	.00697	.00126	.1813	790.947	5.517	5.345	29.486	30
31	169.177	.00591	.00107	.1811	934.317	5.523	5.371	29.664	31
32	199.629	.00501	.00091	.1809	1 103.5	5.528	5.394	29.819	32
33	235.562	.00425	.00077	.1808	1 303.1	5.532	5.415	29.955	33
34	277.963	.00360	.00065	.1806	1 538.7	5.536	5.433	30.074	34
35	327.997	.00305	.00055	.1806	1 816.6	5.539	5.449	30.177	35
40	750.377	.00133	.00024	.1802	4 163.2	5.548	5.502	30.527	40
45	1 716.7	.00058	.00010	.1801	9 531.6	5.552	5.529	30.701	45
50	3 927.3	.00025	.00005	.1800	21 813.0	5.554	5.543	30.786	50
55	8 984.8	.00011	.00002	.1800	49 910.1	5.555	5.549	30.827	55
60	20 555.1	.00005	.00001	.1800	114 189.4	5.555	5.553	30.846	60
65	47 025.1	.00002		.1800	261 244.7	5.555	5.554	30.856	65
70	107 581.9	.00001		.1800	597 671.7	5.556	5.555	30.860	70

20%   **Compound Interest Factors**   **20%**

	Single Payment		Uniform Payment Series				Arithmetic Gradient		
	Compound Amount Factor	Present Worth Factor	Sinking Fund Factor	Capital Recovery Factor	Compound Amount Factor	Present Worth Factor	Gradient Uniform Series	Gradient Present Worth	
	Find F Given P	Find P Given F	Find A Given F	Find A Given P	Find F Given A	Find P Given A	Find A Given G	Find P Given G	
n	F/P	P/F	A/F	A/P	F/A	P/A	A/G	P/G	n
1	1.200	.8333	1.0000	1.2000	1.000	0.833	0	0	1
2	1.440	.6944	.4545	.6545	2.200	1.528	0.455	0.694	2
3	1.728	.5787	.2747	.4747	3.640	2.106	0.879	1.852	3
4	2.074	.4823	.1863	.3863	5.368	2.589	1.274	3.299	4
5	2.488	.4019	.1344	.3344	7.442	2.991	1.641	4.906	5
6	2.986	.3349	.1007	.3007	9.930	3.326	1.979	6.581	6
7	3.583	.2791	.0774	.2774	12.916	3.605	2.290	8.255	7
8	4.300	.2326	.0606	.2606	16.499	3.837	2.576	9.883	8
9	5.160	.1938	.0481	.2481	20.799	4.031	2.836	11.434	9
10	6.192	.1615	.0385	.2385	25.959	4.192	3.074	12.887	10
11	7.430	.1346	.0311	.2311	32.150	4.327	3.289	14.233	11
12	8.916	.1122	.0253	.2253	39.581	4.439	3.484	15.467	12
13	10.699	.0935	.0206	.2206	48.497	4.533	3.660	16.588	13
14	12.839	.0779	.0169	.2169	59.196	4.611	3.817	17.601	14
15	15.407	.0649	.0139	.2139	72.035	4.675	3.959	18.509	15
16	18.488	.0541	.0114	.2114	87.442	4.730	4.085	19.321	16
17	22.186	.0451	.00944	.2094	105.931	4.775	4.198	20.042	17
18	26.623	.0376	.00781	.2078	128.117	4.812	4.298	20.680	18
19	31.948	.0313	.00646	.2065	154.740	4.843	4.386	21.244	19
20	38.338	.0261	.00536	.2054	186.688	4.870	4.464	21.739	20
21	46.005	.0217	.00444	.2044	225.026	4.891	4.533	22.174	21
22	55.206	.0181	.00369	.2037	271.031	4.909	4.594	22.555	22
23	66.247	.0151	.00307	.2031	326.237	4.925	4.647	22.887	23
24	79.497	.0126	.00255	.2025	392.484	4.937	4.694	23.176	24
25	95.396	.0105	.00212	.2021	471.981	4.948	4.735	23.428	25
26	114.475	.00874	.00176	.2018	567.377	4.956	4.771	23.646	26
27	137.371	.00728	.00147	.2015	681.853	4.964	4.802	23.835	27
28	164.845	.00607	.00122	.2012	819.223	4.970	4.829	23.999	28
29	197.814	.00506	.00102	.2010	984.068	4.975	4.853	24.141	29
30	237.376	.00421	.00085	.2008	1 181.9	4.979	4.873	24.263	30
31	284.852	.00351	.00070	.2007	1 419.3	4.982	4.891	24.368	31
32	341.822	.00293	.00059	.2006	1 704.1	4.985	4.906	24.459	32
33	410.186	.00244	.00049	.2005	2 045.9	4.988	4.919	24.537	33
34	492.224	.00203	.00041	.2004	2 456.1	4.990	4.931	24.604	34
35	590.668	.00169	.00034	.2003	2 948.3	4.992	4.941	24.661	35
40	1 469.8	.00068	.00014	.2001	7 343.9	4.997	4.973	24.847	40
45	3 657.3	.00027	.00005	.2001	18 281.3	4.999	4.988	24.932	45
50	9 100.4	.00011	.00002	.2000	45 497.2	4.999	4.995	24.970	50
55	22 644.8	.00004	.00001	.2000	113 219.0	5.000	4.998	24.987	55
60	56 347.5	.00002		.2000	281 732.6	5.000	4.999	24.994	60

25% Compound Interest Factors 25%

	Single Payment		Uniform Payment Series				Arithmetic Gradient		
	Compound Amount Factor Find F Given P F/P	Present Worth Factor Find P Given F P/F	Sinking Fund Factor Find A Given F A/F	Capital Recovery Factor Find A Given P A/P	Compound Amount Factor Find F Given A F/A	Present Worth Factor Find P Given A P/A	Gradient Uniform Series Find A Given G A/G	Gradient Present Worth Find P Given G P/G	
n									n
1	1.250	.8000	1.0000	1.2500	1.000	0.800	0	0	1
2	1.563	.6400	.4444	.6944	2.250	1.440	0.444	0.640	2
3	1.953	.5120	.2623	.5123	3.813	1.952	0.852	1.664	3
4	2.441	.4096	.1734	.4234	5.766	2.362	1.225	2.893	4
5	3.052	.3277	.1218	.3718	8.207	2.689	1.563	4.204	5
6	3.815	.2621	.0888	.3388	11.259	2.951	1.868	5.514	6
7	4.768	.2097	.0663	.3163	15.073	3.161	2.142	6.773	7
8	5.960	.1678	.0504	.3004	19.842	3.329	2.387	7.947	8
9	7.451	.1342	.0388	.2888	25.802	3.463	2.605	9.021	9
10	9.313	.1074	.0301	.2801	33.253	3.571	2.797	9.987	10
11	11.642	.0859	.0235	.2735	42.566	3.656	2.966	10.846	11
12	14.552	.0687	.0184	.2684	54.208	3.725	3.115	11.602	12
13	18.190	.0550	.0145	.2645	68.760	3.780	3.244	12.262	13
14	22.737	.0440	.0115	.2615	86.949	3.824	3.356	12.833	14
15	28.422	.0352	.00912	.2591	109.687	3.859	3.453	13.326	15
16	35.527	.0281	.00724	.2572	138.109	3.887	3.537	13.748	16
17	44.409	.0225	.00576	.2558	173.636	3.910	3.608	14.108	17
18	55.511	.0180	.00459	.2546	218.045	3.928	3.670	14.415	18
19	69.389	.0144	.00366	.2537	273.556	3.942	3.722	14.674	19
20	86.736	.0115	.00292	.2529	342.945	3.954	3.767	14.893	20
21	108.420	.00922	.00233	.2523	429.681	3.963	3.805	15.078	21
22	135.525	.00738	.00186	.2519	538.101	3.970	3.836	15.233	22
23	169.407	.00590	.00148	.2515	673.626	3.976	3.863	15.362	23
24	211.758	.00472	.00119	.2512	843.033	3.981	3.886	15.471	24
25	264.698	.00378	.00095	.2509	1 054.8	3.985	3.905	15.562	25
26	330.872	.00302	.00076	.2508	1 319.5	3.988	3.921	15.637	26
27	413.590	.00242	.00061	.2506	1 650.4	3.990	3.935	15.700	27
28	516.988	.00193	.00048	.2505	2 064.0	3.992	3.946	15.752	28
29	646.235	.00155	.00039	.2504	2 580.9	3.994	3.955	15.796	29
30	807.794	.00124	.00031	.2503	3 227.2	3.995	3.963	15.832	30
31	1 009.7	.00099	.00025	.2502	4 035.0	3.996	3.969	15.861	31
32	1 262.2	.00079	.00020	.2502	5 044.7	3.997	3.975	15.886	32
33	1 577.7	.00063	.00016	.2502	6 306.9	3.997	3.979	15.906	33
34	1 972.2	.00051	.00013	.2501	7 884.6	3.998	3.983	15.923	34
35	2 465.2	.00041	.00010	.2501	9 856.8	3.998	3.986	15.937	35
40	7 523.2	.00013	.00003	.2500	30 088.7	3.999	3.995	15.977	40
45	22 958.9	.00004	.00001	.2500	91 831.5	4.000	3.998	15.991	45
50	70 064.9	.00001		.2500	280 255.7	4.000	3.999	15.997	50
55	213 821.2			.2500	855 280.7	4.000	4.000	15.999	55

30% Compound Interest Factors **30%**

	Single Payment		Uniform Payment Series				Arithmetic Gradient		
	Compound Amount Factor	Present Worth Factor	Sinking Fund Factor	Capital Recovery Factor	Compound Amount Factor	Present Worth Factor	Gradient Uniform Series	Gradient Present Worth	
	Find F Given P	Find P Given F	Find A Given F	Find A Given P	Find F Given A	Find P Given A	Find A Given G	Find P Given G	
n	F/P	P/F	A/F	A/P	F/A	P/A	A/G	P/G	n
1	1.300	.7692	1.0000	1.3000	1.000	0.769	0	0	1
2	1.690	.5917	.4348	.7348	2.300	1.361	0.435	0.592	2
3	2.197	.4552	.2506	.5506	3.990	1.816	0.827	1.502	3
4	2.856	.3501	.1616	.4616	6.187	2.166	1.178	2.552	4
5	3.713	.2693	.1106	.4106	9.043	2.436	1.490	3.630	5
6	4.827	.2072	.0784	.3784	12.756	2.643	1.765	4.666	6
7	6.275	.1594	.0569	.3569	17.583	2.802	2.006	5.622	7
8	8.157	.1226	.0419	.3419	23.858	2.925	2.216	6.480	8
9	10.604	.0943	.0312	.3312	32.015	3.019	2.396	7.234	9
10	13.786	.0725	.0235	.3235	42.619	3.092	2.551	7.887	10
11	17.922	.0558	.0177	.3177	56.405	3.147	2.683	8.445	11
12	23.298	.0429	.0135	.3135	74.327	3.190	2.795	8.917	12
13	30.287	.0330	.0102	.3102	97.625	3.223	2.889	9.314	13
14	39.374	.0254	.00782	.3078	127.912	3.249	2.969	9.644	14
15	51.186	.0195	.00598	.3060	167.286	3.268	3.034	9.917	15
16	66.542	.0150	.00458	.3046	218.472	3.283	3.089	10.143	16
17	86.504	.0116	.00351	.3035	285.014	3.295	3.135	10.328	17
18	112.455	.00889	.00269	.3027	371.518	3.304	3.172	10.479	18
19	146.192	.00684	.00207	.3021	483.973	3.311	3.202	10.602	19
20	190.049	.00526	.00159	.3016	630.165	3.316	3.228	10.702	20
21	247.064	.00405	.00122	.3012	820.214	3.320	3.248	10.783	21
22	321.184	.00311	.00094	.3009	1 067.3	3.323	3.265	10.848	22
23	417.539	.00239	.00072	.3007	1 388.5	3.325	3.278	10.901	23
24	542.800	.00184	.00055	.3006	1 806.0	3.327	3.289	10.943	24
25	705.640	.00142	.00043	.3004	2 348.8	3.329	3.298	10.977	25
26	917.332	.00109	.00033	.3003	3 054.4	3.330	3.305	11.005	26
27	1 192.5	.00084	.00025	.3003	3 971.8	3.331	3.311	11.026	27
28	1 550.3	.00065	.00019	.3002	5 164.3	3.331	3.315	11.044	28
29	2 015.4	.00050	.00015	.3001	6 714.6	3.332	3.319	11.058	29
30	2 620.0	.00038	.00011	.3001	8 730.0	3.332	3.322	11.069	30
31	3 406.0	.00029	.00009	.3001	11 350.0	3.332	3.324	11.078	31
32	4 427.8	.00023	.00007	.3001	14 756.0	3.333	3.326	11.085	32
33	5 756.1	.00017	.00005	.3001	19 183.7	3.333	3.328	11.090	33
34	7 483.0	.00013	.00004	.3000	24 939.9	3.333	3.329	11.094	34
35	9 727.8	.00010	.00003	.3000	32 422.8	3.333	3.330	11.098	35
40	36 118.8	.00003	.00001	.3000	120 392.6	3.333	3.332	11.107	40
45	134 106.5	.00001		.3000	447 018.3	3.333	3.333	11.110	45

35%   **Compound Interest Factors**   35%

	Single Payment		Uniform Payment Series				Arithmetic Gradient		
	Compound Amount Factor	Present Worth Factor	Sinking Fund Factor	Capital Recovery Factor	Compound Amount Factor	Present Worth Factor	Gradient Uniform Series	Gradient Present Worth	
	Find F Given P	Find P Given F	Find A Given F	Find A Given P	Find F Given A	Find P Given A	Find A Given G	Find P Given G	
n	F/P	P/F	A/F	A/P	F/A	P/A	A/G	P/G	n
1	1.350	.7407	1.0000	1.3500	1.000	0.741	0	0	1
2	1.822	.5487	.4255	.7755	2.350	1.289	0.426	0.549	2
3	2.460	.4064	.2397	.5897	4.173	1.696	0.803	1.362	3
4	3.322	.3011	.1508	.5008	6.633	1.997	1.134	2.265	4
5	4.484	.2230	.1005	.4505	9.954	2.220	1.422	3.157	5
6	6.053	.1652	.0693	.4193	14.438	2.385	1.670	3.983	6
7	8.172	.1224	.0488	.3988	20.492	2.508	1.881	4.717	7
8	11.032	.0906	.0349	.3849	28.664	2.598	2.060	5.352	8
9	14.894	.0671	.0252	.3752	39.696	2.665	2.209	5.889	9
10	20.107	.0497	.0183	.3683	54.590	2.715	2.334	6.336	10
11	27.144	.0368	.0134	.3634	74.697	2.752	2.436	6.705	11
12	36.644	.0273	.00982	.3598	101.841	2.779	2.520	7.005	12
13	49.470	.0202	.00722	.3572	138.485	2.799	2.589	7.247	13
14	66.784	.0150	.00532	.3553	187.954	2.814	2.644	7.442	14
15	90.158	.0111	.00393	.3539	254.739	2.825	2.689	7.597	15
16	121.714	.00822	.00290	.3529	344.897	2.834	2.725	7.721	16
17	164.314	.00609	.00214	.3521	466.611	2.840	2.753	7.818	17
18	221.824	.00451	.00158	.3516	630.925	2.844	2.776	7.895	18
19	299.462	.00334	.00117	.3512	852.748	2.848	2.793	7.955	19
20	404.274	.00247	.00087	.3509	1 152.2	2.850	2.808	8.002	20
21	545.769	.00183	.00064	.3506	1 556.5	2.852	2.819	8.038	21
22	736.789	.00136	.00048	.3505	2 102.3	2.853	2.827	8.067	22
23	994.665	.00101	.00035	.3504	2 839.0	2.854	2.834	8.089	23
24	1 342.8	.00074	.00026	.3503	3 833.7	2.855	2.839	8.106	24
25	1 812.8	.00055	.00019	.3502	5 176.5	2.856	2.843	8.119	25
26	2 447.2	.00041	.00014	.3501	6 989.3	2.856	2.847	8.130	26
27	3 303.8	.00030	.00011	.3501	9 436.5	2.856	2.849	8.137	27
28	4 460.1	.00022	.00008	.3501	12 740.3	2.857	2.851	8.143	28
29	6 021.1	.00017	.00006	.3501	17 200.4	2.857	2.852	8.148	29
30	8 128.5	.00012	.00004	.3500	23 221.6	2.857	2.853	8.152	30
31	10 973.5	.00009	.00003	.3500	31 350.1	2.857	2.854	8.154	31
32	14 814.3	.00007	.00002	.3500	42 323.7	2.857	2.855	8.157	32
33	19 999.3	.00005	.00002	.3500	57 137.9	2.857	2.855	8.158	33
34	26 999.0	.00004	.00001	.3500	77 137.2	2.857	2.856	8.159	34
35	36 448.7	.00003	.00001	.3500	104 136.3	2.857	2.856	8.160	35

40%   Compound Interest Factors   **40%**

	Single Payment		Uniform Payment Series				Arithmetic Gradient		
	Compound Amount Factor	Present Worth Factor	Sinking Fund Factor	Capital Recovery Factor	Compound Amount Factor	Present Worth Factor	Gradient Uniform Series	Gradient Present Worth	
	Find F Given P	Find P Given F	Find A Given F	Find A Given P	Find F Given A	Find P Given A	Find A Given G	Find P Given G	
n	F/P	P/F	A/F	A/P	F/A	P/A	A/G	P/G	n
1	1.400	.7143	1.0000	1.4000	1.000	0.714	0	0	1
2	1.960	.5102	.4167	.8167	2.400	1.224	0.417	0.510	2
3	2.744	.3644	.2294	.6294	4.360	1.589	0.780	1.239	3
4	3.842	.2603	.1408	.5408	7.104	1.849	1.092	2.020	4
5	5.378	.1859	.0914	.4914	10.946	2.035	1.358	2.764	5
6	7.530	.1328	.0613	.4613	16.324	2.168	1.581	3.428	6
7	10.541	.0949	.0419	.4419	23.853	2.263	1.766	3.997	7
8	14.758	.0678	.0291	.4291	34.395	2.331	1.919	4.471	8
9	20.661	.0484	.0203	.4203	49.153	2.379	2.042	4.858	9
10	28.925	.0346	.0143	.4143	69.814	2.414	2.142	5.170	10
11	40.496	.0247	.0101	.4101	98.739	2.438	2.221	5.417	11
12	56.694	.0176	.00718	.4072	139.235	2.456	2.285	5.611	12
13	79.371	.0126	.00510	.4051	195.929	2.469	2.334	5.762	13
14	111.120	.00900	.00363	.4036	275.300	2.478	2.373	5.879	14
15	155.568	.00643	.00259	.4026	386.420	2.484	2.403	5.969	15
16	217.795	.00459	.00185	.4018	541.988	2.489	2.426	6.038	16
17	304.913	.00328	.00132	.4013	759.783	2.492	2.444	6.090	17
18	426.879	.00234	.00094	.4009	1 064.7	2.494	2.458	6.130	18
19	597.630	.00167	.00067	.4007	1 419.6	2.496	2.468	6.160	19
20	836.682	.00120	.00048	.4005	2 089.2	2.497	2.476	6.183	20
21	1 171.4	.00085	.00034	.4003	2 925.9	2.498	2.482	6.200	21
22	1 639.9	.00061	.00024	.4002	4 097.2	2.498	2.487	6.213	22
23	2 295.9	.00044	.00017	.4002	5 737.1	2.499	2.490	6.222	23
24	3 214.2	.00031	.00012	.4001	8 033.0	2.499	2.493	6.229	24
25	4 499.9	.00022	.00009	.4001	11 247.2	2.499	2.494	6.235	25
26	6 299.8	.00016	.00006	.4001	15 747.1	2.500	2.496	6.239	26
27	8 819.8	.00011	.00005	.4000	22 046.9	2.500	2.497	6.242	27
28	12 347.7	.00008	.00003	.4000	30 866.7	2.500	2.498	6.244	28
29	17 286.7	.00006	.00002	.4000	43 214.3	2.500	2.498	6.245	29
30	24 201.4	.00004	.00002	.4000	60 501.0	2.500	2.499	6.247	30
31	33 882.0	.00003	.00001	.4000	84 702.5	2.500	2.499	6.248	31
32	47 434.8	.00002	.00001	.4000	118 584.4	2.500	2.499	6.248	32
33	66 408.7	.00002	.00001	.4000	166 019.2	2.500	2.500	6.249	33
34	92 972.1	.00001		.4000	232 427.9	2.500	2.500	6.249	34
35	130 161.0	.00001		.4000	325 400.0	2.500	2.500	6.249	35

45% · Compound Interest Factors · **45%**

	Single Payment		Uniform Payment Series				Arithmetic Gradient		
	Compound Amount Factor	Present Worth Factor	Sinking Fund Factor	Capital Recovery Factor	Compound Amount Factor	Present Worth Factor	Gradient Uniform Series	Gradient Present Worth	
	Find F Given P	Find P Given F	Find A Given F	Find A Given P	Find F Given A	Find P Given A	Find A Given G	Find P Given G	
n	F/P	P/F	A/F	A/P	F/A	P/A	A/G	P/G	n
1	1.450	.6897	1.0000	1.4500	1.000	0.690	0	0	1
2	2.103	.4756	.4082	.8582	2.450	1.165	0.408	0.476	2
3	3.049	.3280	.2197	.6697	4.553	1.493	0.758	1.132	3
4	4.421	.2262	.1316	.5816	7.601	1.720	1.053	1.810	4
5	6.410	.1560	.0832	.5332	12.022	1.876	1.298	2.434	5
6	9.294	.1076	.0543	.5043	18.431	1.983	1.499	2.972	6
7	13.476	.0742	.0361	.4861	27.725	2.057	1.661	3.418	7
8	19.541	.0512	.0243	.4743	41.202	2.109	1.791	3.776	8
9	28.334	.0353	.0165	.4665	60.743	2.144	1.893	4.058	9
10	41.085	.0243	.0112	.4612	89.077	2.168	1.973	4.277	10
11	59.573	.0168	.00768	.4577	130.162	2.185	2.034	4.445	11
12	86.381	.0116	.00527	.4553	189.735	2.196	2.082	4.572	12
13	125.252	.00798	.00362	.4536	276.115	2.204	2.118	4.668	13
14	181.615	.00551	.00249	.4525	401.367	2.210	2.145	4.740	14
15	263.342	.00380	.00172	.4517	582.982	2.214	2.165	4.793	15
16	381.846	.00262	.00118	.4512	846.325	2.216	2.180	4.832	16
17	553.677	.00181	.00081	.4508	1 228.2	2.218	2.191	4.861	17
18	802.831	.00125	.00056	.4506	1 781.8	2.219	2.200	4.882	18
19	1 164.1	.00086	.00039	.4504	2 584.7	2.220	2.206	4.898	19
20	1 688.0	.00059	.00027	.4503	3 748.8	2.221	2.210	4.909	20
21	2 447.5	.00041	.00018	.4502	5 436.7	2.221	2.214	4.917	21
22	3 548.9	.00028	.00013	.4501	7 884.3	2.222	2.216	4.923	22
23	5 145.9	.00019	.00009	.4501	11 433.2	2.222	2.218	4.927	23
24	7 461.6	.00013	.00006	.4501	16 579.1	2.222	2.219	4.930	24
25	10 819.3	.00009	.00004	.4500	24 040.7	2.222	2.220	4.933	25
26	15 688.0	.00006	.00003	.4500	34 860.1	2.222	2.221	4.934	26
27	22 747.7	.00004	.00002	.4500	50 548.1	2.222	2.221	4.935	27
28	32 984.1	.00003	.00001	.4500	73 295.8	2.222	2.221	4.936	28
29	47 826.9	.00002	.00001	.4500	106 279.9	2.222	2.222	4.937	29
30	69 349.1	.00001	.00001	.4500	154 106.8	2.222	2.222	4.937	30
31	100 556.1	.00001		.4500	223 455.9	2.222	2.222	4.938	31
32	145 806.4	.00001		.4500	324 012.0	2.222	2.222	4.938	32
33	211 419.3			.4500	469 818.5	2.222	2.222	4.938	33
34	306 558.0			.4500	681 237.8	2.222	2.222	4.938	34
35	444 509.2			.4500	987 795.9	2.222	2.222	4.938	35

50% Compound Interest Factors **50%**

	Single Payment		Uniform Payment Series				Arithmetic Gradient		
	Compound Amount Factor	Present Worth Factor	Sinking Fund Factor	Capital Recovery Factor	Compound Amount Factor	Present Worth Factor	Gradient Uniform Series	Gradient Present Worth	
	Find F Given P F/P	Find P Given F P/F	Find A Given F A/F	Find A Given P A/P	Find F Given A F/A	Find P Given A P/A	Find A Given G A/G	Find P Given G P/G	
n									n
1	1.500	.6667	1.0000	1.5000	1.000	0.667	0	0	1
2	2.250	.4444	.4000	.9000	2.500	1.111	0.400	0.444	2
3	3.375	.2963	.2105	.7105	4.750	1.407	0.737	1.037	3
4	5.063	.1975	.1231	.6231	8.125	1.605	1.015	1.630	4
5	7.594	.1317	.0758	.5758	13.188	1.737	1.242	2.156	5
6	11.391	.0878	.0481	.5481	20.781	1.824	1.423	2.595	6
7	17.086	.0585	.0311	.5311	32.172	1.883	1.565	2.947	7
8	25.629	.0390	.0203	.5203	49.258	1.922	1.675	3.220	8
9	38.443	.0260	.0134	.5134	74.887	1.948	1.760	3.428	9
10	57.665	.0173	.00882	.5088	113.330	1.965	1.824	3.584	10
11	86.498	.0116	.00585	.5058	170.995	1.977	1.871	3.699	11
12	129.746	.00771	.00388	.5039	257.493	1.985	1.907	3.784	12
13	194.620	.00514	.00258	.5026	387.239	1.990	1.933	3.846	13
14	291.929	.00343	.00172	.5017	581.859	1.993	1.952	3.890	14
15	437.894	.00228	.00114	.5011	873.788	1.995	1.966	3.922	15
16	656.814	.00152	.00076	.5008	1 311.7	1.997	1.976	3.945	16
17	985.261	.00101	.00051	.5005	1 968.5	1.998	1.983	3.961	17
18	1 477.9	.00068	.00034	.5003	2 953.8	1.999	1.988	3.973	18
19	2 216.8	.00045	.00023	.5002	4 431.7	1.999	1.991	3.981	19
20	3 325.3	.00030	.00015	.5002	6 648.5	1.999	1.994	3.987	20
21	4 987.9	.00020	.00010	.5001	9 973.8	2.000	1.996	3.991	21
22	7 481.8	.00013	.00007	.5001	14 961.7	2.000	1.997	3.994	22
23	11 222.7	.00009	.00004	.5000	22 443.5	2.000	1.998	3.996	23
24	16 834.1	.00006	.00003	.5000	33 666.2	2.000	1.999	3.997	24
25	25 251.2	.00004	.00002	.5000	50 500.3	2.000	1.999	3.998	25
26	37 876.8	.00003	.00001	.5000	75 751.5	2.000	1.999	3.999	26
27	56 815.1	.00002	.00001	.5000	113 628.3	2.000	2.000	3.999	27
28	85 222.7	.00001	.00001	.5000	170 443.4	2.000	2.000	3.999	28
29	127 834.0	.00001		.5000	255 666.1	2.000	2.000	4.000	29
30	191 751.1	.00001		.5000	383 500.1	2.000	2.000	4.000	30
31	287 626.6			.5000	575 251.2	2.000	2.000	4.000	31
32	431 439.9			.5000	862 877.8	2.000	2.000	4.000	32

60%   **Compound Interest Factors**   60%

	Single Payment		Uniform Payment Series				Arithmetic Gradient		
	Compound Amount Factor	Present Worth Factor	Sinking Fund Factor	Capital Recovery Factor	Compound Amount Factor	Present Worth Factor	Gradient Uniform Series	Gradient Present Worth	
	Find F Given P	Find P Given F	Find A Given F	Find A Given P	Find F Given A	Find P Given A	Find A Given G	Find P Given G	
n	F/P	P/F	A/F	A/P	F/A	P/A	A/G	P/G	n
1	1.600	.6250	1.0000	1.6000	1.000	0.625	0	0	1
2	2.560	.3906	.3846	.9846	2.600	1.016	0.385	0.391	2
3	4.096	.2441	.1938	.7938	5.160	1.260	0.698	0.879	3
4	6.554	.1526	.1080	.7080	9.256	1.412	0.946	1.337	4
5	10.486	.0954	.0633	.6633	15.810	1.508	1.140	1.718	5
6	16.777	.0596	.0380	.6380	26.295	1.567	1.286	2.016	6
7	26.844	.0373	.0232	.6232	43.073	1.605	1.396	2.240	7
8	42.950	.0233	.0143	.6143	69.916	1.628	1.476	2.403	8
9	68.719	.0146	.00886	.6089	112.866	1.642	1.534	2.519	9
10	109.951	.00909	.00551	.6055	181.585	1.652	1.575	2.601	10
11	175.922	.00568	.00343	.6034	291.536	1.657	1.604	2.658	11
12	281.475	.00355	.00214	.6021	467.458	1.661	1.624	2.697	12
13	450.360	.00222	.00134	.6013	748.933	1.663	1.638	2.724	13
14	720.576	.00139	.00083	.6008	1 199.3	1.664	1.647	2.742	14
15	1 152.9	.00087	.00052	.6005	1 919.9	1.665	1.654	2.754	15
16	1 844.7	.00054	.00033	.6003	3 072.8	1.666	1.658	2.762	16
17	2 951.5	.00034	.00020	.6002	4 917.5	1.666	1.661	2.767	17
18	4 722.4	.00021	.00013	.6001	7 868.9	1.666	1.663	2.771	18
19	7 555.8	.00013	.00008	.6011	12 591.3	1.666	1.664	2.773	19
20	12 089.3	.00008	.00005	.6000	20 147.1	1.667	1.665	2.775	20
21	19 342.8	.00005	.00003	.6000	32 236.3	1.667	1.666	2.776	21
22	30 948.5	.00003	.00002	.6000	51 579.2	1.667	1.666	2.777	22
23	49 517.6	.00002	.00001	.6000	82 527.6	1.667	1.666	2.777	23
24	79 228.1	.00001	.00001	.6000	132 045.2	1.667	1.666	2.777	24
25	126 765.0	.00001		.6000	211 273.4	1.667	1.666	2.777	25
26	202 824.0			.6000	338 038.4	1.667	1.667	2.778	26
27	324 518.4			.6000	540 862.4	1.667	1.667	2.778	27
28	519 229.5			.6000	865 380.9	1.667	1.667	2.778	28

Continuous Compounding—Single Payment Factors

rn	Compound Amount Factor e^{rn} Find F Given P F/P	Present Worth Factor e^{-rn} Find P Given F P/F	rn	Compound Amount Factor e^{rn} Find F Given P F/P	Present Worth Factor e^{-rn} Find P Given F P/F
.01	1.0101	.9900	.51	1.6653	.6005
.02	1.0202	.9802	.52	1.6820	.5945
.03	1.0305	.9704	.53	1.6989	.5886
.04	1.0408	.9608	.54	1.7160	.5827
.05	1.0513	.9512	.55	1.7333	.5769
.06	1.0618	.9418	.56	1.7507	.5712
.07	1.0725	.9324	.57	1.7683	.5655
.08	1.0833	.9231	.58	1.7860	.5599
.09	1.0942	.9139	.59	1.8040	.5543
.10	1.1052	.9048	.60	1.8221	.5488
.11	1.1163	.8958	.61	1.8404	.5434
.12	1.1275	.8869	.62	1.8589	.5379
.13	1.1388	.8781	.63	1.8776	.5326
.14	1.1503	.8694	.64	1.8965	.5273
.15	1.1618	.8607	.65	1.9155	.5220
.16	1.1735	.8521	.66	1.9348	.5169
.17	1.1853	.8437	.67	1.9542	.5117
.18	1.1972	.8353	.68	1.9739	.5066
.19	1.2092	.8270	.69	1.9937	.5016
.20	1.2214	.8187	.70	2.0138	.4966
.21	1.2337	.8106	.71	2.0340	.4916
.22	1.2461	.8025	.72	2.0544	.4868
.23	1.2586	.7945	.73	2.0751	.4819
.24	1.2712	.7866	.74	2.0959	.4771
.25	1.2840	.7788	.75	2.1170	.4724
.26	1.2969	.7711	.76	2.1383	.4677
.27	1.3100	.7634	.77	2.1598	.4630
.28	1.3231	.7558	.78	2.1815	.4584
.29	1.3364	.7483	.79	2.2034	.4538
.30	1.3499	.7408	.80	2.2255	.4493
.31	1.3634	.7334	.81	2.2479	.4449
.32	1.3771	.7261	.82	2.2705	.4404
.33	1.3910	.7189	.83	2.2933	.4360
.34	1.4049	.7118	.84	2.3164	.4317
.35	1.4191	.7047	.85	2.3396	.4274
.36	1.4333	.6977	.86	2.3632	.4232
.37	1.4477	.6907	.87	2.3869	.4190
.38	1.4623	.6839	.88	2.4109	.4148
.39	1.4770	.6771	.89	2.4351	.4107
.40	1.4918	.6703	.90	2.4596	.4066
.41	1.5068	.6637	.91	2.4843	.4025
.42	1.5220	.6570	.92	2.5093	.3985
.43	1.5373	.6505	.93	2.5345	.3946
.44	1.5527	.6440	.94	2.5600	.3906
.45	1.5683	.6376	.95	2.5857	.3867
.46	1.5841	.6313	.96	2.6117	.3829
.47	1.6000	.6250	.97	2.6379	.3791
.48	1.6161	.6188	.98	2.6645	.3753
.49	1.6323	.6126	.99	2.6912	.3716
.50	1.6487	.6065	1.00	2.7183	.3679

Appendix C

Fundamentals of Engineering (FE) Exam Practice Problems

This set of homework problems has been developed in the multiple-choice style of the Fundamentals of Engineering (FE) examination. Passing the FE is an important step toward achieving professional licensure through the National Council of Examiners for Engineering and Surveying (NCEES). Problems have been divided by type and chapter to assist the student in learning and mastering the material in preparation for the examination. Bold italic signifies the topic covered in this book.

I. Engineering economy topic problems are found in the general morning session of the FE exam, as well as in several of the discipline-specific afternoon session exams. Engineering economy topics constitute 8% of the morning exam, and from 1 to 15% of some afternoon exams. Afternoon exams that include engineering economy topics include Chemical, Civil, Industrial, and Other/General. The following listing of *engineering economy topics* found in the FE exam is from the 2005 NCEES publication of exam topics.

Morning Session: (all examinees complete this section; there are 120 questions in 12 topical areas)

Topic VI. Engineering Economics (8% of questions)

A. *Discounted cash flow (e.g., equivalence, PW, equivalent annual FW, rate of return)*

B. *Cost (e.g., incremental, average, sunk, estimating)*

C. *Analyses (e.g., breakeven, benefit–cost)*

D. *Uncertainty (e.g., expected value and risk)*

Afternoon Session: (examinees chose one of seven possible modules: Chemical, Civil, Electrical, Environmental, Industrial, Mechanical, and Other/General)

Chemical Engineering Session (60 questions in 11 topical areas)

Topic VIII. Process Design and Economic Optimization (10% of questions)

A. Process flow diagrams (PFD)

B. Piping and instrumentation diagrams (P&ID)

  C. Scale-up

  D. *Comparison of economic alternatives (e.g., net present value, discounted cash flow, rate of return)*

  E. *Cost estimation*

Civil Engineering Session (60 questions in 9 topical areas)

  Topic VIII. Construction Management (10% of questions)

  A. Procurement methods (e.g., design–build, design–bid–build, qualifications based)

  B. Allocation of resources (e.g., labor, equipment, materials, money, time)

  C. Contracts/contract law

  D. Project scheduling (e.g., CPM, PERT)

  E. *Engineering economics*

  F. Project management (e.g., owner/contractor/client relations, safety)

  G. Construction estimating

Industrial Engineering Session (60 questions in 8 topical areas)

  Topic I. Engineering Economics (15% of questions)

  A. *Discounted cash flows (equivalence, PW, EAC, FW, IRR, loan amortization)*

  B. *Types and breakdown of costs (e.g., fixed, variable, direct and indirect labor, material, capitalized)*

  C. *Analyses (e.g., benefit–cost, breakeven, minimum cost, overhead, risk, incremental, life cycle)*

  D. *Accounting (financial statements and overhead cost allocation)*

  E. *Cost estimating*

  F. *Depreciation and taxes*

  G. *Capital budgeting*

Other/General Engineering Session (60 questions in 9 topical areas)

  Topic IV. Engineering Economics (10% of questions)

  A. *Cost estimating*

  B. *Project selection*

  C. *Lease/buy/make*

  D. *Replacement analysis (e.g., optimal economic life)*

II. Engineering ethics topic problems are found only in the general morning session of the FE exam. These questions constitute about 7% of the morning exam. The following list of engineering ethics topics found in the FE exam is from the 2005 NCEES publication of exam topics.

Morning Session: (all examinees complete this section; there are 120 questions in 12 topical areas)

  Topic V. Ethics and Business Practices (7% of questions)

  A. *Code of ethics (professional and technical societies)*

  B. Agreements and contracts

C. Ethical versus legal

D. Professional liability

E. Public protection issues (e.g., licensing boards)

PROBLEMS

Chapter 1
Decision Making

1   Engineering economic analysis provides useful input in all of the following situations except which one?

   (*a*) Determining how much we should pay for a machine that will provide a savings.

   (*b*) Determining the priority of investing our company's retained earnings.

   (*c*) Illustrating the economic advantages of one alternative over other feasible choices.

   (*d*) Convincing management that one person should be hired over another.

Engineering Ethics

2   Engineers are acting in an ethical way most in which of these situations?

   (*a*) They are making sure the company's interests are protected.

   (*b*) They feel good about the decisions that they've made.

   (*c*) They act to protect the interests of society in general.

   (*d*) They ensure that their own best interest is protected.

3   To act as an ethical engineer, you should accept fees for engineering work in which situation?

   (*a*) If you need the money to keep your business open and thriving.

   (*b*) If you are competent to complete all aspects of the job.

   (*c*) If the contract is a cost plus contract.

   (*d*) If there were no other engineers who bid on the job.

4   A registered professional engineer (PE) has as a primary obligation protection of which of the following entities?

   (*a*) The government

   (*b*) The PE's company

   (*c*) The PE's country

   (*d*) The general public

5   Engineers should act in ethical ways for what reason?

   (*a*) It creates a feel-good situation for them.

   (*b*) The engineering code of conduct requires it.

   (*c*) They may be considered for a raise because of it.

   (*d*) They may be violating the law if they don't.

6   A design engineer is responsible for an important subelement of a large project at a firm. The project has fallen behind schedule, and the important client is very angry and threatening to sue. The boss is expecting the engineer's design review to go well so that the project can be shipped by the end of the week. The engineer notices during final design review that an element of the design is wrong and will create a major safety issue for the entire system. To rework that portion of the design will take several months. The engineer should do which of the following?

   (*a*) Sign off on the drawings because of the threatened lawsuit and because the project is so far behind.

   (*b*) Do not sign off on the drawings, and let the boss know what is found at the final design review.

   (*c*) Tell the boss that to sign off on the design now, the engineer must have an immediate raise.

   (*d*) Sign off and keep an eye on how construction goes. Maybe the engineer can correct the safety issues before the project is fully operational.

7   As team leader for your unit, you function as both engineer and manager. One of your roles is to approve major purchases, and you have been contacted by a new supplier in your area. The new company has invited you to expensive dinners, has offered trips to vacation spots to attend "product shows," and has recently been sending to your private address personal items such as collectible art, coins, sports tickets, and

golf club memberships. You are unsure how to handle this situation. You should do which of the following?

(a) Accept all of the gifts because you know that everyone else is doing it and that this is your chance to get a share of the action.

(b) Decline the gifts and other offers that would be considered outside the scope of ordinary business or professional contact.

(c) Knowing that the gifts will not influence your purchasing decisions, accept the items with no guilt.

(d) Accept the gifts but make sure that your boss and others on your team share in the bounty.

Cost Problems

8   A company produces several product lines. One of those lines generates the following annual cost and production data:

Manufacturing/Materials costs	$200,000
General/Administrative costs	50,000
Direct-labor costs	170,000
Other overhead costs	60,000
Annual production demand	10,000 units

The company adds 40% to its production cost in selling to the retailer. The retailer in turn adds a 50% profit margin when selling to its customers. How much would it cost a retailer to buy 100 units of the product?

(a) $4800

(b) $6720

(c) $7200

(d) $1008

9   An agribusiness is deciding what crop to plant in this area for the next growing season. The local agricultural extension office has provided the following data (in $100):

Crop	Cost per acre	Income/acre at 100% Yield	Estimated Yield(%)
A	$30	$45	80
B	45	75	65
C	15	25	90

Using a 200-acre plot as an example (subtract 10% for unusable areas of the field), which crop should be planted this year, and what is the total profit?

(a) A; $108,000

(b) C; $135,000

(c) C; $150,000

(d) B; $540,000

Chapter 2
Cost Concepts

10   A firm bought a used machine 2 years ago for $1500. When new, the machine cost $8000. Today it could be sold for $500. Which of the following statements is true?

(a) The fixed cost for operating the machine can be ignored in any analysis.

(b) The $8000 purchase price is not included in the analysis.

(c) The $1500 paid 2 years ago is included in the analysis.

(d) The variable cost of ownership is the difference between what was paid and what the machine is now worth ($1500 − $500 = $1000).

11   When considering two alternatives that are described only in terms of the cost of ownership, the breakeven point cannot be described by which of the following statements?

(a) The difference in initial cost between the two alternatives.

(b) The level of production (or activity) of each alternative under consideration is equivalent.

(c) Fixed plus variable costs of each alternative is equivalent.

(d) A rational decision maker should be indifferent between the two alternatives.

12   If JMJ Industries realizes a profit of $4.00 per unit sold, what is the fixed-cost portion of their production costs? Their variable costs are $1.50 per unit, and they sell 1000 units per year at a price of $6.00 per unit.

(a) There are no fixed costs in this type of problem.

(b) $250

(c) $500

(d) $2000

13   Consider the following production data for Alternatives A and B in a firm that uses a 10% interest rate.

	Alt. A	Alt. B
Annual fixed cost per unit	$2 million	$3.5 million
Annual variable cost per unit	850	250

If the company is going to produce 4000 units annually, which alternative should be chosen?

(a) Neither alternative should be chosen because the negative cash flows are greater than the positive cash flows for both alternatives.

(b) This problem cannot be solved because there is not enough data given.

(c) Alt. A

(d) Alt. B

14 A company has annual fixed costs of $2,500,000 and variable costs of 0.15¢ per unit produced. For the firm to break even if they charge $1.85 for their product, the level of annual production is nearest to what value?

(a) 375,000 units

(b) 1,315,789 units

(c) 1,351,351 units

(d) 1,562,500 units

Estimating in Engineering Economy

15 Which statement is not true with respect to estimating the economic impacts of proposed engineering projects?

(a) Order-of-magnitude estimates are used for high-level planning.

(b) Order-of-magnitude estimates are the most accurate type at about –3 to 5%.

(c) Increasing the accuracy of estimates requires added time and resources.

(d) Estimators tend to underestimate the magnitude of costs and overestimate benefits.

16 The Department of Transportation is accepting bids for materials to provide "signage and safety" for a new 25-mile section of a 4-lane highway. DOT estimates are:

Lane paint	$500 per mile per lane
Reflective lane markers	$6 per 25 feet
Mile markers	$18 per unit
Flexible roadside delineators	$32 per unit; spaced at 1000 feet
Emergency boxes	$500 per unit spaced at 2.5 miles
Signage (various messages)	$1000 per mile; based on historical costs

The bids that DOT receives should be in what range of costs?

(a) Less than $100,000

(b) Between $100,000 and $200,000

(c) Between $200,000 and $250,000

(d) Greater than $250,000

17 A 250-gallon reactor cost $780,000 when it was constructed 20 years ago. What would a 750-gallon model cost today if the power-sizing exponent is 0.56 and the construction cost index for such facil-

ities has increased from 141 to 556 over the last 20 years? Choose the closest value.

(a) $0.37 million

(b) $1.66 million

(c) $1.44 million

(d) $5.69 million

18 A half-million-square-foot warehouse facility is being considered by your company for a location in Kansas City, KS. You have a bid for a similar type 25,000 ft^2 facility in Washington, DC, at $4,375,000. If the warehouse construction cost index value for KS is 0.75 and for DC it is 1.34, what range should the KS bid fall into if you assume that construction costs are linear across size?

(a) Less than $30 million

(b) Between $30 million and $60 million

(c) Between $60 million and $100 million

(d) Greater than $100 million

Chapter 3
Simple Interest

19 Which of the following statements is not true?

(a) Simple interest is to be used only in simple decision situations.

(b) Compounded interest involves computing interest on top of interest.

(c) Simple interest is rare in practical situations of borrowing and loaning.

(d) If the interest is not stated as being simple or compounded, we assume the latter.

20 If you borrow $1000 from the bank at 5% simple interest per month due back in 2 years, what is the size of your monthly payments?

(a) $25

(b) $50

(c) $500

(d) $1200

21 Your quarterly payments on a loan are $500 and the interest that you are paying is 1% per quarter simple interest. The size of the principal that you have borrowed is closest to which value?

(a) $5000

(b) $12,500

(c) $20,000

(d) $50,000

22 The principal that you borrowed for a recent purchase was $15,000. You will pay the purchase off through a simple interest loan at 8% per year

due in 3 years. The amount that is due at the end of 3 years is closest to what value?

(a) $1200

(b) $3600

(c) $16,200

(d) $18,900

Compound Interest, Single Cash Flows

23 A savings account's value today is $150 and it earns interest at 1% per month. How much will be in the account one year from today? Which of the following is correct to solve for the unknown value?

(a) $P = 150(1+0.01)^{12}$

(b) $150 = F(F/P, 12\%, 1)$

(c) $F = 150(1.12)^{-12}$

(d) $F = 150(F/P, 1\%, 12)$

24 To calculate how many years (n) an investment (P) must be kept in an account that earns interest at $i\%$, in order to triple in amount, which of the following expressions should be used?

(a) $n = -P + F(P/F, i\%, n)$

(b) $n = [\log (F/P)] / [\log (1+i\%)]$

(c) $n = [\ln (-P + F)] / [\ln i\%]$

(d) $n = -F(1+i\%)^{P}$

25 If you invest $40,000 in a stock whose value grows at 2% per year, your investment is nearest what value after 5 years?

(a) $40,800

(b) $43,296

(c) $44,164

(d) $64,420

26 An account pays interest at 1.5% per month. If you deposit $5000 at the beginning of this year, how much could you withdraw at the end of next year?

(a) $5151

(b) $7148

(c) $49,249

(d) $265,545

27 A machine will need to be replaced 15 years from today for $10,000. How much must be deposited now into an account that earns 5% per year to cover the replacement cost?

(a) $1486

(b) $4810

(c) $6139

(d) $10,000

28 You deposit $5000 into an account that will grow to $14,930 in 6 years. Your rate of return on this investment is closest to what value?

(a) 18%

(b) 20%

(c) 22%

(d) 25%

29 An investment company owns land now worth $500,000 that is increasing in value each year. If the land value doubles in 7 years, what is the yearly rate of return nearest to?

(a) 0.0%

(b) 2.0%

(c) 7.0%

(d) 10.5%

30 Your friend withdrew $630,315 from an account into which she had invested $350,000. If the account paid interest at 4% per year, she kept her money in the account for how many years?

(a) 1.8 years

(b) 6.5 years

(c) 12.5 years

(d) 15 years

31 Annual revenues in our company are $1.5 million this year. If they are expected to grow at a compounded rate of 20% per year, what will they be 10 years from now?

(a) $3.89 million

(b) $9.29 million

(c) $10.9 million

(d) $57.51 million

32 A student inherits $50,000 and invests it in government bonds that will average 3% annual interest. What is the value of the investment after 50 years?

(a) $67,195

(b) $219,195

(c) $355,335

(d) $5,869,545

33 A zero-coupon bond has a face (par) value of $10,000. The bond is sold at a discount for $6500 and held for 3 years, at which time it is sold. If an annual rate of 10% is earned over that 3-year period, how much was the bond sold for?

(a) $8652

(b) $13,310

(c) $16,859

(d) $25,937

Chapter 4

Uniform Cash Flow Series

34   You place $100 per month into an account that earns 1% per month. Which of the following expressions can be used to calculate the account's value after 3 years?

(a) $P = 100(P/A, 1\%, 3)$

(b) $F = 100(P/A, 1\%, 36)(F/P, 1\%, 36)$

(c) $F = 100[(1 + 0.01)^n - 1]/0.01$

(d) $F = 100(F/A, 12.68\%, 3)$

35   A machine must be replaced in 7 years at a cost of $7500. How much must be deposited at the end of each year into an account that earns 5% in order to have accumulated enough to pay for the replacement?

(a) $471

(b) $596

(c) $791

(d) $921

36   Winners of the PowerState Lottery can take $30 million now or payments of $2.5 million per year for the next 15 years. These are equivalent at what annual interest rate? The answer is closest to what value?

(a) 1%

(b) 2%

(c) 3%

(d) 5%

37   You deposit $1000 in a retirement investment account today that earns 1% per month. In addition, you deposit $50 at the end of every month starting this month and continue to do so for 30 years. The amount that has accumulated in this account at the end of 30 years is nearest to

(a) $35,949

(b) $42,027

(c) $174,748

(d) $210,698

38   Suppose $10,000 is deposited into an account that earns 10% per year for 5 years. At that point in time, uniform end-of-year withdrawals are made such that the account is emptied after the 15th withdrawal. The size of these annual withdrawals is closest to what value?

(a) $2118

(b) $2621

(c) $3410

(d) $16,105

39   A manufacturer borrows $85,000 for machinery. The loan is for 10 years at 12% per year. What is the annual payment on the machinery?

(a) $4843

(b) $8500

(c) $13,834

(d) $15,045

40   How many years would you have to put $100 per year into an account that earns 15% annually to accumulate $6508?

(a) 17 years

(b) 21 years

(c) 30 years

(d) 65 years

41   A $10,000 face value municipal bond pays $1000 interest at the end of every year. If there are 12 more years of payments, at what price today would the bond yield 18% over the next 12 years?

(a) $1372

(b) $4793

(c) $6165

(d) $10,000

Gradient Cash Flows

42   Today $5000 is deposited in an account that earns 2.5% per quarter. Additional deposits are made at the end of every quarter for the next 20 years. The deposits start at $100 and increase by $50 each year thereafter. The amount that has accumulated in this account at the end of 20 years can be expressed as follows.

(a) $= 5000(P/F,2.5\%,20) + 100(F/A,2.5\%, 20\%) + 50(P/G,2.5\%,20 - 1)$

(b) $= 5000(F/P,10\%,80) + 150(P/G,10\%,80)(F/P, =10\%, 80)$

(c) $= 5000(P/F,10.38\%,20) + 100(P/A,10.38\%,20) + 50(P/G,10.38\%, 20)(P/F, 10.38\%, 1)$

(d) $= 5000(F/P,10.38\%,80) + 100(F/A,2.5\%,80) + 50(P/G,2.5\%,80)(F/P, 2.5\%,80)$

43   An investment returns the following end-of-year cash flows: Year 1, $0; Year 2, $1500; Year 3, $3000; Year 4, $4500; and Year 5, $6000. Given a 10% interest rate, what is the present worth?

(a) $5970

(b) $6597

(c) $9357

(d) $10,293

44   A project's annual revenues will be $50,000 the first year and will decrease by $1500 per year over its 20-year life. If the firm's interest rate is 12%, what is the project's present worth?

(a) $305,998

(b) $373,450

(c) $384,654

(d) $440,902

Nominal and Effective Interest Rates

45   A rate of 2% per quarter compounded quarterly is closest to what annual compounded interest rate?

(a) 2.00%

(b) 8.00%

(c) 8.24%

(d) 24.00%

46   A nominal interest rate of "12% per year compounded yearly" is closest to:

(a) 1% per month effective rate

(b) 3% per quarter effective rate

(c) 6% per 6 months effective rate

(d) 12% per year effective rate

47   An interest rate expressed as "1.5% per month" is exactly the same as:

(a) 4.5% per quarter effective interest

(b) 18% effective interest per year

(c) 18% per year compounded monthly

(d) None of the above

48   A deposit of $50,000 is made into an account that pays 10% compounded semiannually. How much would be in the account after 10 years?

(a) $81,445

(b) $129,685

(c) $132,665

(d) $336,375

49   A mining firm must deposit funds in a "reclamation" account each quarter. The account must have $25 million on deposit when a project reaches its horizon in 10 years. The account pays interest at a rate of 2% per quarter.

(a) $41,500

(b) $172,575

(c) $228,325

(d) $414,000

50   A deposit of $1000 compounds to $2500 in 5 years. The interest on this account compounds quarterly. What is the closest nominal annual rate of return?

(a) 2.50%

(b) 4.70%

(c) 18.75%

(d) 20.11%

Continuous Compounding Interest Rate

51   How long will it take money to triple in an account if it pays interest expressed as 8% nominal annual compounded continuously?

(a) 13.73 years

(b) 14.27 years

(c) Your money will never triple.

(d) None of the above

52   A deposit of $500 per year (beginning of year) is made for a period of 2 years in an account that earns 6% nominal interest compounded continuously. How much is in the account after 2 years?

(a) $917

(b) $1062

(c) $1092

(d) $1095

Chapter 5: *Present Worth Analysis*

53   A project is being considered that has a first cost of $12,500, creates $5000 in annual cost savings, requires $3000 in annual operating costs, and has a salvage value of $2000 after a project life of 3 years. If interest is 10% per year, which formula calculates the project's present worth?

(a) PW = $12,500(P/F, 10\%, 1) + (-5000 + 3000)$
$(P/A, 10\%, 3) - 2000(F/P, 10\%, 3)$

(b) PW = $-12,500 + (5000 - 3000)(P/A, 10\%, 3) -$
$2000(P/F, 10\%, 3)$

(c) PW = $12,500(F/P, 10\%, 3) + (5000 - 3000)$
$(F/A, 10\%, 3) + 2000$

(d) PW = $-12,500 + 5000(P/A, 10\%, 3) - 3000$
$(P/A, 10\%, 3) + 2000(P/F, 10\%, 3)$

54   A new packing machine will cost $57,000. The existing machine can be sold for $5000 now and the new machine for $7500 after its 10-year useful life. If the new machine reduces annual expenses by $5000, what is the present worth at 25% of this investment?

(a) −$18,388

(b) −$33,340

(c) −$34,145

(d) −$38,340

55 A vendor is offering an extended repair contract on a machine. The firm's experience is that this will cover repair costs over the next 4 years of $200, $200, $400, and $500. At 6%, what is the extended repair contract worth now?

(a) $1040

(b) $1089

(c) $1099

(d) $1300

56 Annual disbursements for maintenance of critical heavily used equipment will be $25,000 for the next 10 years, and then $35,000 into infinity. What is the present worth of the maintenance cost cash flow stream if interest is 15%?

(a) $166,667

(b) $183,147

(c) $192,367

(d) $233,334

57 Manufacturing costs from a scrapped poor-quality product are $6000 per year. An investment in an employee training program can reduce this cost. Program *A* reduces the cost by 75% and requires an investment of $12,000. Program *B* reduces the cost by 95% and will cost $20,000. Based on low turnover at the plant, either program should be effective for the next 5 years. If interest is 20%, the present worth of the two programs is nearest what values? (Consider cost reduction a positive cash flow.)

(a) *A:* −$25,460; *B:* −$37,049

(b) *A:* $1460; *B:* −$2951

(c) *A:* $5060; *B:* $1609

(d) *A:* $13,460; *B:* $17,049

Chapter 6: *Annual Worth Analysis*

58 New product tracking equipment costs $120,000 and will have a $10,000 salvage value when disposed of in 10 years. Annual repair costs begin at $5000 in the fifth year and increase by $500 per year thereafter until disposed of. If interest is 10%, what is the closest equivalent annual cost of ownership?

(a) $21,505

(b) $21,766

(c) $21,844

(d) $23,109

59 Your company is considering two alternatives:

	Alt. I	Alt. II
First cost	$42,500	$70,000
Annual maintenance	6,000	4,000
Annual savings if implemented	18,500	20,000
Salvage value	12,000	25,000
Useful life of alternative	3 years	6 years

What is the annual dollar advantage of Alt. II over Alt. I at an interest rate of 15%?

(a) Alt. II has no annual advantage over Alt I.

(b) $3020

(c) $3500

(d) $7436

60 Specialized bits (costing $50,000) used in the mining industry have a useful life of 5000 hours of operation and can be traded in when a new bit is purchased for 10% of first cost. The drilling machine that uses the bit is used 1000 hours per year. What is the equivalent uniform annual cost of these bits at 2.5%?

(a) $8559

(b) $9510

(c) $9828

(d) $10,920

61 A new chemical remediation tank is needed. Current technology tanks, which cost $150,000, must be drained and treated every 2 years at a cost of $30,000; the tanks will last 10 years, and each will have a salvage value of 5% of first cost. A tank with new technology has just come on the market. There are no periodic maintenance costs, and a tank will last 20 years. If the new tanks cost $325,000, what minimum salvage value, as a percentage of first cost, would be required for this technology to be a better option? Use a 12% interest rate.

(a) 10%

(b) 36%

(c) 57%

(d) 72%

62 A beautiful bridge is being built over the river that runs through a major city in your state. The cost of the bridge is estimated at $600 million. Annual costs of the bridge will be $200,000, and the bridge is estimated to last a very long time. If accountants in city hall use 3% as the interest rate for analysis, what is the annualized cost of the bridge project?

(a) $18 million

(b) $18.2 million

(c) $20,000 million

(d) $219,500 million

Chapter 7: *Rate of Return Analysis*

63   Which of the following equations can be used to find the internal rate of return (i) for a project that has initial investment of P, net annual cash flows of A, and salvage value of S after n years?

(a) $0 = -P + A(P/A, i\%, n) + S(P/A, i\%, n)$

(b) $(P - A)(P/A, i\%, n) = S(P/F, i\%, n)$

(c) $-A = -P(A/P, i\%, n) - S(A/F, i\%, n - 1)$

(d) $0 = -P(F/P, i\%, n) + A(F/A, i\%, n) + S$

64   The rate of return on an investment of $1500 that doubles in value over a 4-year period, and produces a $300 annual cash flow, is nearest to which value?

(a) 15%

(b) 20%

(c) 25%

(d) 30%

65   The interest rate that makes Alternatives A and B equivalent is in what range?

Year	Alt. A	Alt. B
0	–$1000	–$3000
1	100	500
2	100	550
3	100	600
4	200	650
5	200	700

(a) Less than 2%

(b) Between 2 and 5%

(c) Greater than 5%

(d) There is no interest rate that equates these two cash flow series.

66   A corporate bond with a face (par) value of $10,000 will mature 7 years from today (it was issued 3 years ago). The bond just after the 6th interest payment is being sold for $6950. The bond's interest rate is 4% nominal annual, payable semiannually. The yield of the bond if held to maturity is in what range?

(a) Less than 4%

(b) Between 4 and 6%

(c) Between 6 and 10%

(d) Greater than 10%

67   A firm borrowed $50,000 from a mortgage bank. The terms of the loan specify quarterly payments for a 10-year period. If payments to the bank are $3750 per quarter, what effective annual interest rate is the firm paying?

(a) Less than 1%

(b) 7%

(c) 28%

(d) 31%

Chapter 8
Incremental Cash Flows and Analysis

68   You are given the cash flow series for two projects, Alt. A and Alt. B.

Year	0	1	2	3	4	5	6
Alt. A ($)	–I1	X	X	X	X	X	X + S1
Alt. B ($)	–I2	Y	Y	Y	Y	Y	Y + S2

Assume I2 > I1 and X, Y, S1, and S2 are positive; the incremental rate of return (i) on the additional investment in Alt. B can be calculated with the following expression.

(a) $0 = -I2 + Y(P/A, i\%, 6) + S2(P/F, i\%, 6)$

(b) $0 = -(I2 - I1) + (Y - X)(P/A, i\%, 5) + (S2 - S1)(P/F, i\%, 6)$

(c) $0 = -(I2 - I1)(F/P, i\%, 6) + (Y - X)(F/A, i\%, 6) + (S2 - S1)$

(d) $0 = -(I2 - I1) + (Y - X) + [(Y + S2) - (X + S1)]$

69   A firm is considering two mutually exclusive alternatives ($i = 8\%$):

	Alt. Alpha	Alt. Omega
First cost	$10,000	$30,000
Annual maintenance	2,800	2,000
Annual savings if implemented	5,000	6,500
Salvage value	2,000	5,000
Useful life of alternative	4 years	8 years

If Alt. Alpha will be replaced with a "like alternative" at the end of 4 years, what is the present worth of the incremental cash flows associated with going from an investment in Alpha to an investment in Omega?

(a) –$6201

(b) –$5942

(c) –$5028

(d) $852

70   Project 1 requires an initial investment on $50,000 and has an internal rate of return (IRR) of 18%. A mutually exclusive alternative, Project 2, requires an investment of $70,000 and has an IRR of 23%. Which of the following statements is true concerning the rate of return on the incremental $20,000 investment?

(a) It is less than 18%.

(b) It is between 18 and 23%.

(c) It is greater than 23%.

(d) It cannot be determined from the data given.

Chapter 9

Future Worth Analysis

71   The future worth of a project with initial cost P, positive annual cash flows of A, salvage value S, and interest rate of i over a life of n years can be calculated using which statement?

(a) $\text{FW} = -P(F/P, i\%, n) + A(F/A, i\%, n) + S(F/P, i\%, n)$

(b) $\text{FW} = P(F/P, i\%, n) + A(F/A, i\%, n) + S$

(c) $\text{FW} = -P(P/F, i\%, n) + A(F/A, i\%, n) - A[(P/A, i\%, n)] + S$

(d) $\text{FW} = -P(F/P, i\%, n) + A(F/A, i\%, n) + S$

72   A firm has been investing retained earnings to establish a building fund. The firm has retained $1.2 million, $1.0 million, and $950,000, respectively, 3, 2, and 1 year ago. This year the firm has $1.8 million to invest. If the firm earns 18% on invested funds, what is the value of the project that can be undertaken using the funds as a 25% down payment?

(a) $6.28 million

(b) $7.42 million

(c) $25.1 million

(d) $29.7 million

73   A firm is considering the purchase of a software analysis package that costs $450,000. Annual licensing fees are $25,000 (payable at the beginning of each year, starting in Year 1). The firm is bidding on a large 4-year government project where the new software will be used. If the firm uses an interest rate of 20%, what value for software costs should be put in the bid?

(a) $514,725

(b) $527,650

(c) $1,067,540

(d) $1,094,346

Benefit–Cost Ratio Analysis

74   The annual benefits associated with construction of an outer belt highway are estimated at $10.5 million by a local planning commission. The initial construction costs will be $400 million, and the project's useful life is 50 years. Annual maintenance costs are $500,000 with periodic rebuilding costs of $10 million every 10 years. If interest is 2%, the benefit–cost ratio is closest to what value?

(a) 0.25

(b) 0.75

(c) 1.11

(d) 1.35

75   A city needs a new pedestrian bridge over a local stream. The city uses an interest rate of 5%, and the project life is 30 years. The following data (in millions of dollars) summarizes the bids that were received.

	Bid A	Bid B
Construction materials costs	$4.20	$6.20
Construction labor costs	0.60	0.70
Construction overhead costs	0.35	0.03
Initial administrative and legal costs	0.60	0.01
Annual operating costs	0.05	0.075
Annual revenue from operation	Unknown	0.40
Other annual benefits to the city	0.22	0.25

What would the annual revenue of Bid A have to be for the two projects to be equivalent? Choose the closest value.

(a) 0.10

(b) 0.20

(c) 0.30

(d) 0.40

Sensitivity and Breakeven Analysis

76   BVM manufactured and sold 25,000 small statues this past year. At that volume, the firm was exactly in a breakeven situation in terms of profitability. BVM's unit costs are expected to increase by 30% next year. What additional information is needed to determine how much the production volume/sales would have to increase next year to just break even in terms of profitability?

(a) Costs per unit

(b) Sales price per unit and costs per unit

(c) Total fixed costs, sales price per unit, and costs per unit (*d*) No data is needed, the volume increase is 25,000 + 25,000(.30) = 32,500 units.

77 Process *A* has fixed costs of $10,000 and unit costs of $4.50 each, and Process *B* has fixed costs of $25,000 and unit costs of $1.50 each. At what level of annual production would the two processes have the same cost?

(*a*) 50 units

(*b*) 500 units

(*c*) 5000 units

(*d*) 50,000 units

78 A seasonal bus tour firm has 5 buses with a capacity of 60 people each. Each customer pays $25 for a one-day tour. Records show $360,000 in fixed costs per season, incremental costs of $5 per customer, and an average daily occupancy of 80%. The number of days of operation necessary each season to break even is closest to which value?

(*a*) 50 days

(*b*) 75 days

(*c*) 100 days

(*d*) 120 days

Chapter 10: *Risk and Expected Value*

79 An interest rate of 15% is used to evaluate a new system that has a first cost of $212,400, annual operating and maintenance costs of $41,200, annual savings of $94,600, a life of 6 years, and a salvage value of $32,500. After initial evaluation, the firm receives word from the vendor that the first cost is 5% higher than originally quoted. The percentage error in the system's present worth from this is closest to what value?

(*a*) 5%

(*b*) 15%

(*c*) 100%

(*d*) 300%

80 A machine has a first cost of $10,000 and annual costs of $3500. There is no salvage value, and interest is 10%. If the project's useful life is described by the following data, what is the annual worth of costs?

Useful Life (years)

	4	5	6	7
Prob. of life (%)	5	22	41	32

(*a*) $3500

(*b*) $5127

(*c*) $5554

(*d*) $5796

81 Three estimators have estimated a project with a 10-year life.

	Estimate 1	Estimate 2	Estimate 3
First cost	$10,000	$17,500	$15,000
Net annual cash flows	7,500	8,000	6,000
Salvage value	3,500	0	10,000

Use an interest rate of 20% to determine the project's expected present worth. The value is closest to which of the following?

(*a*) $16,066

(*b*) $16,612

(*c*) $31,660

(*d*) $31,607

82 The first cost (FC), life (*n*) and annual benefits (A) for a prospective project are uncertain. Optimistic (OP), most likely (ML), and pessimistic (PS) estimates are given. If the interest rate is 25%, what is the present worth difference between a total worst-case scenario and a total best-case scenario?

	Estimate		
Parameter	**Pessimistic**	**Most Likely**	**Optimistic**
First cost	$150,000	$100,000	$80,000
Annual cash flows	25,000	45,000	50,000
Project life, in years	5	7	10

(*a*) $15.8

(*b*) $42.2

(*c*) $181.3

(*d*) $282.5

Chapter 11: *Depreciation of Capital Assets*

83 The correct percentage to use to calculate the depreciation allowance for a MACRS 3-year property for Year 2 is which of the following?

(*a*) 14.4%

(*b*) 32.0%

(*c*) 33.3%

(*d*) 44.5%

84   With reference to the straight-line depreciation method, which statement is false?

(a) An equal amount of depreciation is allocated in each year.

(b) The book value of the asset decrements by a fixed amount each year.

(c) The depreciation life (n) is set based on the MACRS property classes.

(d) The asset is depreciated down to a book value equal to the salvage value.

85   A 7-year MACRS property has a cost basis for depreciation of $50,000. The estimated salvage (market) value is $10,000 after its 10-year useful life. The depreciation charge for the 4th year and book value of the asset after the 4th year of depreciation are closest to what values?

(a) $5760; $44,240

(b) $6245; $12,496

(c) $6245; $15,620

(d) $6245; $43,755

86   A $100,000 asset has a $20,000 salvage value after its 10-year useful life. The depreciation allowance using straight-line depreciation is closest to what value?

(a) $2000

(b) $8000

(c) $10,000

(d) $12,000

87   The book value of an asset that is listed as a 10-year MACRS property is $49,500 after the first year. If the asset's estimated salvage (market) value is $5000 after its 15-year useful life, what was the asset's original cost basis?

(a) $50,000

(b) $52,105

(c) $55,000

(d) $61,875

Chapter 12: *Calculating Income Taxes*

88   Which of the following is true?

(a) Tax rates are based on two flat-rate schedules, one for individuals and one for businesses.

(b) When businesses subtract expenses, they always include capital costs.

(c) For businesses, taxable income is total income less depreciation and ordinary expenses.

(d) When quantifying depreciation allowance, one must always divide first cost by MACRS 3-year life.

89   If a corporation has taxable income of $60,000, which of the following expressions is used to calculate the federal tax owed?

(a) Flat 15% of taxable income

(b) Flat 25% of taxable income

(c) $7,500 + 25% over $50,000

(d) $13,750 + 34% over $75,000

90   This past year CLL Industries had income from operations of $8.2 million, and expenses of $1.8 million. Allowances for depreciated capital expenses were $400,000. What is CLL's taxable income and federal taxes owed for operations last year? Choose the closest values.

(a) $6.0 million; $1.93 million

(b) $6.0 million; $2.04 million

(c) $6.4 million; $1.93 million

(d) $6.4 million; $2.04 million

91   Annual data for a firm for this tax year is:

Revenues	$45 million
Operating and maintenance costs	7 million
Labor/Salary costs	15 million
Overhead and administrative costs	3 million
Depreciation allowance	8 million

Next year the firm can increase revenue by 20% and costs will increase by 2%. If the depreciation allowance stays the same, what will be the change in firm's after-tax net profit? Choose the closest answer.

(a) $5.2 million

(b) $5.4 million

(c) $7.9 million

(d) $13.3 million

92   Widget Industries recently erected a facility costing $1.56 million on land bought for $1 million. The firm used straight-line depreciation over a 39-year period; it installed $2.5 million worth of plant and office equipment (all classified as MACRS 7-year property). Gross income from the first year of operations (excluding capital expenditures) was $8.2 million, and $5.8 million was spent on labor and materials. How much did Widget pay in federal income taxes for the first year of operation?

(a) $680,935

(b) $1,002,750

(c) $1,321,815

(d) $2,788,000

Chapter 13: *Replacement Analysis*

93   Which of the following is a valid reason to consider replacing an existing asset?

    (*a*) It has become obsolete and does not perform its intended function.

    (*b*) There is a newer asset available that is technologically superior.

    (*c*) It has become very costly to keep the asset in working order.

    (*d*) All of the above.

94   The replacement of a typical existing asset (defender) that is beyond its minimum economic life with a new asset (challenger) should be done when?

    (*a*) It should never be done because the existing asset is working.

    (*b*) The defender's marginal cost is greater than the challenger's equivalent annual cost.

    (*c*) The challenger's average future cost becomes less than the existing asset.

    (*d*) The defender's equivalent annual cost equals the challenger's equivalent annual cost.

95   A factory asset has a first cost of $100,000, annual costs of $15,000 the first year and increasing by 7.5% per year, and a salvage (market) value that decreases by 20% per year over its 5-year life. The minimum cost economic life of the asset is what value? Interest is 10%.

    (*a*) 2 years

    (*b*) 3 years

    (*c*) 4 years

    (*d*) 5 years

96   The minimum cost life of a new replacement machine is 6 years with a minimum equivalent annual cost of $6000. Given the existing machine's marginal cost data for the next 4 years, when should the existing machine be replaced?

Year	Total Marginal Cost
1	$5400
2	5800
3	6200
4	8000

    (*a*) After Year 1

    (*b*) After Year 2

    (*c*) After Year 3

    (*d*) After 6 years

97   A material handling system was purchased 3 years ago for $120,000. Two years ago it required substantial upgrading at a cost of $15,000. It once again is requiring an upgrading cost at $25,000. Alternately, a new system can be purchased today at a cost of $200,000. The existing machine could be sold today for $50,000. In an economic analysis, what first cost should be assigned to the existing system?

    (*a*) $50,000

    (*b*) $65,000

    (*c*) $75,000

    (*d*) $80,000

Chapter 14: *Inflation Effects*

98   If the real growth of money interest rate for the past year has been 4% and the general inflation has been 2.5%, the combined (market) interest rate is closest to what value?

    (*a*) 1.5%

    (*b*) 6.5%

    (*c*) 6.6%

    (*d*) 10.0%

99   To convert actual (inflated) dollars to constant purchasing power dollars (where n = difference in time between today and purchasing base) that occur at the same point in time, one must:

    (*a*) Multiply by $(P/F$, inflation rate %, $n)$.

    (*b*) Multiply by $(F/P$, inflation rate %, $n)$.

    (*c*) Multiply by $(P/F$, combined rate %, $n)$.

    (*d*) Divide by $(P/F$, real interest rate %, $n)$.

100   The cost of a material was $2.00 per ounce 5 years ago. If prices have increased (inflated) at an average rate of 4% per year, what is the cost per ounce now?

    (*a*) $2.04

    (*b*) $2.08

    (*c*) $2.43

    (*d*) $8.00

101   A deposit of $1000 is made into an account that promises a minimum of 2% per year increase in purchasing power. If general price inflation is 3, 1, and 5%, respectively, over the next 3 years, the minimum value that will be in the account at Year 3 is closest to what amount?

    (*a*) $1061

    (*b*) $1092

    (*c*) $1157

    (*d*) $1177

Chapter 15: *Capital Budgeting*

102   Which of these statements can be said of projects considered to be mutually exclusive?

   (*a*) The projects are equivalent or mutual in terms of their cash flows.

   (*b*) Neither alternative should be chosen.

   (*c*) All projects can be chosen as long as they meet minimum economic criterion.

   (*d*) The selection of one in the set eliminates selection of others in that same set.

   Use the following data for Problems 103–105
   A firm has identified four projects for possible funding in the next budget cycle.

Project	First Cost ($M)	PW ($M)
A	12	225
B	18	250
C	24	320
D	30	400

103   The projects are independent, what is the total number of possible investment strategies (combinations) that the firm can use?

   (*a*) 4

   (*b*) 6

   (*c*) 10

   (*d*) 15

104   If the projects have the following contingencies, what is the total number of possible investment strategies (combinations) that the firm can use? Project *A* can be invested in only by itself; Project *D* is chosen only if Project *B* is chosen.

   (*a*) 0

   (*b*) 2

   (*c*) 4

   (*d*) 6

105   If the projects are independent, and the project budget (capital limit) is $32 million, what investment combination maximizes PW?

   (*a*) *A* and *C*

   (*b*) *D* alone

   (*c*) Invest in all projects to maximize PW.

   (*d*) *A* and *B*

106   A firm is considering the "make vs. buy" question for a subcomponent. If the part is made in-house, the production data would be: first cost = $350,000; annual costs for operation = $45,000; salvage value = $15,000; project life = 5 years; interest = 10%; and material cost per unit = $8.50. If annual production is 10,000 units, the maximum amount that the firm should be willing to pay to an outside vendor for the subcomponent is nearest?

   (*a*) $10 per unit

   (*b*) $16 per unit

   (*c*) $22 per unit

   (*d*) $28 per unit

107   A firm is considering whether to buy specialized equipment that would cost $200,000 and have annual costs of $15,000. After 5 years of operation, the equipment would have no salvage value. The same equipment can be leased for $50,000 per year (annual costs included in the lease), payable at the beginning of each year. If the firm uses an interest rate of 5% per year, the annual cost advantage of leasing over purchasing is nearest what value?

   (*a*) $2494

   (*b*) $8694

   (*c*) $11,200

   (*d*) $12,758

Chapter 17: *Accounting*

108   Which of the following summarizes a firm's revenues and expenses over a month, quarter, or year?

   (*a*) Balance sheet

   (*b*) Statement of assets and liabilities

   (*c*) Income statement

   (*d*) None of the above

109   Which of the following financial ratios provides insight into a firm's solvency over the short term by indicating its ability to cover current liabilities?

   (*a*) Acid-test ratio

   (*b*) Quick ratio

   (*c*) Current ratio

   (*d*) Net profit ratio

110   Which statement is most accurate related to an activity-based costing (ABC) system?

   (*a*) ABC spreads the firm's indirect costs based on volume-based activities.

   (*b*) ABC seeks to assign costs to the activities that drive those costs.

   (*c*) ABC gives an inaccurate view of a firm's costs and should never be used.

   (*d*) ABC is called ABC because it is an easy method to use.

111 A firm's balance sheet has the following data:

Cash on hand	$450,000
Market securities	25,000
Net accounts and notes receivable	125,000
Retailers' inventories	560,000
Prepaid expenses	48,500
Accounts and notes payable (short term)	700,000
Accumulated liabilities	120,000

The firm's current ratio and acid-test ratio are closest to what values?

(a) 1.42; 0.73

(b) 1.42; 0.79

(c) 1.47; 0.73

(d) 1.47; 0.79

112 A firm's income statement has the following data (in $10,000):

Total operating revenues	$1,200
Total nonoperating revenues	500
Total operating expenses	925
Total nonoperating expenses	125

If the firm's incomes taxes were $60,000 what was the net profit (loss)?

(a) All necessary data is not given, one cannot calculate net profit (loss).

(b) −$53,500

(c) $1,150,000

(d) $6,440,000

113 Annual manufacturing cost data (1000s) for four product lines are as follows:

Data	Line 1	Line 2	Line 3	Line 4
Annual production	4000	3500	9800	675
Cost of direct materials	$800	$650	$1200	$2500
Cost of direct labor	$3500	$3750	$600	$320

Rank the product lines from lowest to highest in terms of manufacturing cost per unit. Total indirect costs of $10.8 million are allocated based on total direct cost.

(a) 1-2-3-4

(b) 3-1-2-4

(c) 3-2-1-4

(d) 3-4-1-2

References

AASHTO. *A Manual on User Benefit Analysis of Highway and Bus-Transit Improvements.* Washington, DC: American Association of State Highway and Transportation Officials, 1978.

American Telephone and Telegraph Co. *Engineering Economy, 3rd.* McGraw-Hill, 1977.

Baasel, W. D. *Preliminary Chemical Engineering Plant Design, 2nd.* Van Nostrand Reinhold, 1990.

Benford, H. A. *Naval Architect's Introduction to Engineering Economics.* Ann Arbor, MI: Department of Naval Architecture and Marine Engineering Report 282, 1983.

Bernhard, R. H. "A Comprehensive Comparison and Critique of Discounting Indices Proposed for Capital Investment Evaluation," *The Engineering Economist.* Vol. 16, No. 3, 1971, pp. 157–186.

Block, S. "Are There Differences in Capital Budgeting Procedures between Industries? An Empirical Study," *The Engineering Economist.* Vol. 50, No. 1, 2005, pp. 55–67.

Elizandro, D. W., and J. O. Matson. "Taking a Moment to Teach Engineering Economy," *The Engineering Economist.* Vol. 52, No. 2, 2007, pp. 97–116.

The Engineering Economist. A quarterly journal of the Engineering Economy Divisions of ASEE and IIE.

Eschenbach, T. G., J. Hartman, and L. E. Bussey. *The Economic Analysis of Industrial Projects, 3rd.* Oxford University Press, 2009.

Eschenbach, T. *Engineering Economy: Applying Theory to Practice, 2nd.* Oxford University Press, 2003.

Eschenbach, T. G., and J. P. Lavelle. "Technical Note: MACRS Depreciation with a Spreadsheet Function: A Teaching and Practice Note," *The Engineering Economist.* Vol. 46, No. 2, 2001, pp. 153–161.

Eschenbach, T. G., and J. P. Lavelle. "How Risk and Uncertainty Are/Could/Should Be Presented in Engineering Economy," *Proceedings of the 11th Industrial Engineering Research Conference.* IIE, Orlando, May 2002, CD.

Fish, J. C. L. *Engineering Economics.* McGraw-Hill, 1915.

Jones, B. W. *Inflation in Engineering Economic Analysis.* John Wiley & Sons, 1982.

Lavelle, J. P., H. R. Liggett, and H. R. Parsaei, editors. *Economic Evaluation of Advanced Technologies: Techniques and Case Studies.* Taylor & Francis, 2002.

Liggett, H. R., J. Trevino, and J. P. Lavelle. "Activity-Based Cost Management Systems in an Advanced Manufacturing Environment," *Economic and Financial Justification of Advanced Manufacturing Technologies.* H. R. Parsaei, W. G. Sullivan, and T. R. Hanley, editors. Elsevier Science, 1992.

Lorie, J. H., and L. J. Savage. "Three Problems in Rationing Capital," *The Journal of Business.* Vol. 28, No. 4, 1955, pp. 229–239.

Newnan, D. G. "Determining Rate of Return by Means of Payback Period and Useful Life," *The Engineering Economist.* Vol. 15, No. 1, 1969, pp. 29–39.

Perry, H. P., D. W. Green, and J. O. Maloney. *Perry's Chemical Engineers' Handbook, 7th*. McGraw-Hill, 1997.

Peters, M. S., K. D. Timmerhaus, M. Peters, and R. E. West. *Plant Design and Economics for Chemical Engineers, 5th*. McGraw-Hill, 2002.

Sundaram, M., editor. *Engineering Economy Exam File*. Oxford University Press, 2003.

Steiner, H. M. *Public and Private Investments: Socioeconomic Analysis*. John Wiley & Sons, 1980.

Terborgh, G. *Business Investment Management*. Machinery and Allied Products Institute, 1967.

Tippet, D. D., and P. Hoekstra. "Activity-Based Costing: A Manufacturing Management Decision-Making Aid," *Engineering Management Journal*. Vol. 5, No. 2, June 1993, American Society for Engineering Management, pp. 37–42.

Wellington, A. M. *The Economic Theory of Railway Location*. John Wiley & Sons, 1887.

Photo credits

Chapter 1, Motoring Picture Library/Alamy; Chapter 2, Ron Niebrugge/Alamy; Chapter 3, Olivia Baumgartner/CORBIS SYGMA; Box 1, Photo courtesy of Nicholas D. Cotton; Box 2, Photo courtesy of Kimberly Brock; Chapter 4, © Pete Pacifica/Getty Images; Chapter 5, © Boeing Management Company; Chapter 7, RFID, Photo courtesy of Oliver Hedgepeth; Chapter 8, © Michael Nelson/Getty Images; Chapter 9, David L. Lawrence Convention Center in Pittsburgh, PA, Photo supplied by Andre Jenny/Alamy; Chapter 10, © Terry Donnelly/Getty Images; Chapter 11, © Internal Revenue Service; Chapter 12, © Getty Images; Chapter 13, © Richard T. Nowitz/CORBIS; Chapter 14, © Photo courtesy of Suncor Energy Inc.; Chapter 15, © Shephard Sherbell/CORBIS SABA; Chapter 16, © William Truran, March 2007; Chapter 17, © Steve Cole/Getty Images.

Index